How to Solve It: Modern Heuristics

T0172283

How to Solve it Modern Heuristics

Springer
Berlin
Heidelberg
New York
Hong Kong
London
Milan
Paris
Tokyo

Zbigniew Michalewicz • David B. Fogel

How to Solve It: Modern Heuristics

Second, Revised and Extended Edition

With 174 Figures and 7 Tables

Springer

Dr. Zbigniew Michalewicz

Department of Computer Science
University of North Carolina
9201 University City Boulevard
Charlotte, NC 28223-0001, USA
zbyszek@uncc.edu

Dr. David B. Fogel

Natural Selection, Inc.
Suite 200
3333 N. Torrey Pines Ct.
La Jolla, CA 92037, USA
dfogel@natural-selection.com

Artwork included in cover design by Ewa J. Michalewicz

ACM Computing Classification (1998): I.2.8, F.2.2, K.3.2

ISBN 978-3-642-06134-9

This work is subject to copyright. All rights are reserved, whether the whole or part of the material is concerned, specifically the rights of translation, reprinting, reuse of illustrations, recitation, broadcasting, reproduction on microfilm or in any other way, and storage in data banks. Duplication of this publication or parts thereof is permitted only under the provisions of the German Copyright Law of September 9, 1965, in its current version, and permission for use must always be obtained from Springer-Verlag. Violations are liable for prosecution under the German Copyright Law.

Springer is a part of Springer Science+Business Media

springeronline.com

© Springer-Verlag Berlin Heidelberg 2010
Printed in Germany

The use of general descriptive names, trademarks, etc. in this publication does not imply, even in the absence of a specific statement, that such names are exempt from the relevant protective laws and regulations and therefore free for general use.

Cover Design: Künkel + Lopka, Werbeagentur, Heidelberg

Printed on acid-free paper   45/3142SR – 5 4 3 2 1

To Ewa with love: the nicest 'puzzle'
I've been 'solving' for 25 years!

Z.M.

For my high school math teachers:
Mr. Flattum, Mr. McPhee,
Mr. Jackson, and Mr. Copperthwaite.

D.B.F.

Preface to the Second Edition

> No pleasure lasts long
> unless there is variety in it.
>
> Publilius Syrus, *Moral Sayings*

We've been very fortunate to receive fantastic feedback from our readers during the last four years, since the first edition of *How to Solve It: Modern Heuristics* was published in 1999. It's heartening to know that so many people appreciated the book and, even more importantly, were using the book to help them solve their problems. One professor, who published a review of the book, said that his students had given the best course reviews he'd seen in 15 years when using our text. There can be hardly any better praise, except to add that one of the book reviews published in a SIAM journal received the best review award as well. We greatly appreciate your kind words and personal comments that you sent, including the few cases where you found some typographical or other errors. Thank you all for this wonderful support.

One of the telltale signs that a book is having a significant impact is when it gets translated into another language. And we owe special thanks to Hongqing Cao, who prepared the Chinese edition of this book (and found a few remaining errors as well). An even more telling sign that a book is doing well is when it's translated into two languages, and we've learned recently that a Polish translation of the book is forthcoming. Still another good sign is when the publisher wants to publish a new edition of the book. Since the first printing, over 80 pages of the original edition have been replaced gradually by edited and corrected pages, so we hope this second edition will be error free!

You'll find that this second edition contains two new chapters in response to our readers' requests: one on coevolutionary systems and the other on multicriteria decision-making. Of course, in the spirit of our first edition, we've added some new puzzles for you as well. We hope you find these puzzles to be at the same challenging level as those from the first edition. You'll find a few other chapters have some changes, and a new section in Chap. 11.

We'd like to take this opportunity to thank everyone who took the time to share their thoughts on the text with us — they were most helpful. These thanks go to hundreds of readers who sent us e-mails commenting on different sections of the text, examples, references, and puzzles. We also express our gratitude to Rodney Johnson, Michael Melich, and Dave Schaffer for useful comments on parts of new chapters, Min Sun for her help in building some new figures included in the text, Brian Mayoh and Antoni Mazurkiewicz, who pointed us to the "pirates puzzle" and the "bicycle puzzle," respectively, both of which are placed in Chap. XIV. Special thanks are due to Carlos A. Coello Coello, who provided us with invaluable assistance and experimental results of a multiobjective optimization method on selected test cases. These are included in Sect. 15.2.2. We also thank Neville Hankins and Ronan Nugent for their assistance with writing style and Ingeborg Mayer (all from Springer-Verlag) for her help throughout the project.

As with the first edition, our purpose will be fulfilled if you find this book challenging, entertaining, and provocative.

Charlotte, NC   Zbigniew Michalewicz
La Jolla, CA   David B. Fogel
April 2004

Preface to the First Edition

'I will tell you'
the hermit said to Lancelot
'the right of the matter.'

Anonymous, *The Quest of the Holy Grail*

Gyorgy Polya's *How to Solve It* [357] stands as one of the most important contributions to the problem-solving literature in the twentieth century. Even now, as we move into the new millennium, the book continues to be a favorite among teachers and students for its instructive heuristics. The first edition of the book appeared in 1945, near the end of the Second World War and a few years before the invention of the transistor. The book was a quick success, and a second edition came out in 1957.

How to Solve It is a compendium of approaches for tackling problems as we find them in mathematics. That is, the book provides not only examples of techniques and procedures, but also instruction on how to make analogies, use auxiliary devices, work backwards from the goal to the given, and so forth. Essentially, the book is an encyclopedia of problem-solving methods to be carried out by hand, but more than that, it is a treatise on how to think about framing and attacking problems.

Current texts in heuristic problem solving provide detailed descriptions of the algorithmic procedures for each of a number of classic algorithms. Sadly, however, they often fail to provide the proper guidance on when to apply these algorithms, and more importantly, when not to apply them. Instead, they rely on a cookbook of recipes and leave the responsibility of determining whether or not a particular method is appropriate for the task at hand to the reader. Most often, the reader is completely unprepared to make this determination, having never been taught the issues involved, nor even that there are indeed issues to consider!

This situation is undoubtedly a consequence of the computer revolution. If there were any single event that could necessitate an updated approach to problem solving nearly 50 years after Polya's book, it would be the availability of powerful desktop computers at affordable prices. Solutions to real-world challenges are rarely computed with pencil and paper anymore. Instead, we use computer algorithms to numerically approximate and extend the range of questions for which we can generate useful answers. Because these computers are so effective, the problem solver often attempts to "hack" a solution, or at least what masquerades as a solution, without giving due consideration to the assumptions that underlie the implemented method.

As a result, despite the enormous progress that has been made toward the optimization of performance in medicine, defense, industry, finance, and so forth, we have achieved little of our potential. For example, the amount of money saved by using the common methods of linear programming as opposed to hand calculation and guesswork must rise into literally billions of dollars annually, and yet linear programming is almost always used inappropriately in real-world conditions. Individuals and corporations look incessantly, almost blindly, for the quick fix in commercial-off-the-shelf (COTS) solutions to problems that don't exist. Imagine how much money could be saved (or earned!) if truly appropriate techniques were applied to problems that go beyond the simple heuristics of linear programming.

As the world moves toward more open and free markets, competition is the driving force that will mandate more effective methods for solving problems. The analogy between Darwinian variation-and-selection and free-market dynamics is apt. Those entities that fail to acquire the necessary resources will go bankrupt, the economic equivalent of death and survival of the fittest. All that's needed is a slight edge over a competitor to drive that competitor out of business. Only the enlightened individuals, corporations, and agencies that make the effort to adopt modern heuristics to solve their problems will endure.

What is needed now is a two-fold update to Polya's contribution. First, the reader must learn about the specific techniques that are available, mainly through the application of computer algorithms. Second, the reader must come to understand when each of these methods should and should not be used, and how to frame their own problems so that they can best apply the heuristics that Polya's *How to Solve It* could not have anticipated.

This book is a first attempt toward a comprehensive view of problem solving for the twenty-first century. The main points are communicated by direct instruction, analogy, example, and through a series of problems and puzzles that are integrated throughout the discourse and inbetween chapters. The text is intended to serve as the main text for a class on modern heuristics. We believe that such a course is a *must* for students in science, business, or engineering. As such, it assumes some basic knowledge of discrete mathematics and a familiarity with computer programming. Problem solvers who do not have these basic skills should invest the time to acquire them, for it will be well worth the effort. Those who want more in-depth mathematical treatment of algorithms will find the material in the text to be a useful stepping stone to more advanced courses.

The book is organized into 15 chapters which begin after we present a general discussion of solving problems in the Introduction. Chapter 1 indicates the main sources of difficulty in problem solving and a short chapter 2 provides basic notation. Chapters 3 and 4 survey classical optimization algorithms. Together with chapter 5, which covers two modern search algorithms, these chapters constitute the first part of the book. We then move to an evolutionary approach to problem solving. Chapters 6 and 7 introduce intuitions and details about these new evolutionary algorithms. Chapters 8–10 then address some challenging issues concerning problems that require searching for a particular permutation of items, handling constraints, and tuning the algorithms to the problem. These chapters provide a detailed review of many efforts in these areas. Chapter 11 addresses the issue of time-varying environments and noise. Two additional chapters (12 and 13) provide tutorials for neural networks and fuzzy systems. Chapter 14 provides a short and general discussion on hybrid systems and extensions to evolutionary algorithms. Chapter 15 concludes the text by summarizing the material and providing some hints for practical problem solving. You'll note that each puzzle section (I, II, etc.) illustrates a point that is made in the subsequent chapter (1, 2, etc.) — problem solving should be fun and we hope this approach will be enjoyable and engaging.

In addition to the material in the main part of the text, two appendices offer supplemental information. Appendix A provides an overview of fundamental concepts in probability and statistics. It starts with the axioms of probability and proceeds to statistical hypothesis testing and linear regression. This appendix is not intended as a substitute for a thorough preparation in random processes, but can serve to refresh recollections and illuminate important issues in applying probability and statistics to problem solving. Appendix B offers a list of suggested problems and projects to tackle when the text is used as part of a university course. Even if you are reading the book without the benefit of a primary teacher, we strongly encourage you to take an active role in problem solving and implement, test, and apply the concepts that you learn here. The material in Appendix B can serve as a guide for such applications.

We thank everyone who took the time to share their thoughts on the text with us — they were most helpful. In particular, we express our gratitude to Dan Ashlock, Ole Caprani, Tom English, Larry Fogel, Larry Hall, Jim Keller, Kenneth Kreutz-Delgado, Martin Schmidt, and Thomas Stidsen. We also would like to express special appreciation to Kumar Chellapilla, who not only reviewed sections of the book but also provided invaluable assistance with the graphics for some figures. We are pleased to acknowledge the assistance of several co-authors who worked with us during the last two years; many results of this collaboration were included in this text. We thank Thomas Bäck, Hongqing Cao, Ole Caprani, Dipankar Dasgupta, Kalyan Deb, Gusz Eiben, Susana Esquivel, Raul Gallard, Özdemir Göl, Robert Hinterding, Sławomir Kozieł, Lishan Kang, Moutaz Khouja, Witold Kosiński, Thiemo Krink, Guillermo Leguizamón, Brian Mayoh, Maciej Michalewicz, Guo Tao, Krzysztof Trojanowski, Marc Schoenauer, Martin Schmidt, Thomas Stidsen, Roman Śmierzchalski, Martyna Weigl, Janek Wieczorek, Jing Xiao, and Lixin Zhang.

We thank the executive editor of Springer-Verlag, Hans Wössner, for his help throughout the project, Professor Leonard Bolc, for his encouragement, and Professor Antoni Mazurkiewicz for interesting discussions on many puzzles presented in this text. Special thanks are due to Joylene Vette, the English copy editor, Springer-Verlag, for her precious comments on the first draft of the text. The first author would also like to acknowledge the excellent working environments at Aarhus University where he spent his sabbatical leave (August 1998 — July 1999) and grants from the National Science Foundation (IRI-9322400 and IRI-9725424) and from the ESPRIT Project 20288 Cooperation Research in Information Technology (CRIT-2): "Evolutionary Real-time Optimization System for Ecological Power Control," which helped in preparing a few chapters of this text. Also, he would like to thank all of the graduate students from UNC-Charlotte (USA), Universidad Nacional de San Luis (Argentina), and Aarhus University (Denmark) who took part in various courses offered from 1997 to 1999 and went through a painful process of problem solving. The second author would like to thank the undergraduate students of a course he taught on machine learning and pattern recognition at UC San Diego in the winter quarter of 1999, as well as his fellow employees at Natural Selection, Bill Porto, Pete Angeline, and Gary Fogel, for their support and encouragement. Special thanks go to Jacquelyn Moore for giving up so many weekends and evenings so that this book could be completed.

Our purpose will be fulfilled if you find this book challenging, entertaining, and provocative. We hope that you enjoy the book at least as much as we enjoyed assembling it, and that you profit from it as well.

Charlotte, NC   Zbigniew Michalewicz
La Jolla, CA   David B. Fogel
September 1999

Table of Contents

Introduction

> The reasonable man adapts
> himself to the world;
> the unreasonable one persists
> in trying to adapt the world to himself.
> Therefore all progress depends
> on the unreasonable man.
>
> George Bernard Shaw, *Maxims for Revolutionists*

This is not a book about algorithms. Certainly, it is full of algorithms, but that's not what this book is about. This book is about possibilities. Its purpose is to present you not only with the prerequisite mandatory knowledge of the available problem-solving techniques, but more importantly to expand your ability to frame new problems and to think creatively — in essence, to solve the problem of how to solve problems, a talent that has become a lost art. Instead of devoting the necessary time and critical thinking required to frame a problem, to adjust our representation of the pieces of the puzzle, we have become complacent and simply reach for the most convenient subroutine, a magic pill to cure our ills. The trouble with magic is that, empirically, it has a very low success rate, and often relies on external devices such as mirrors and smoke. As with magic, most of the seemingly successful applications of problem solving in the real world are illusory, mere specters of what could have been achieved.

The importance of effective problem solving has never been greater. Technology has enabled us with the ability to affect the environment to such an extent that the decisions we make today may have irrevocable future consequences. This same technology continues to expand the number of people we interact with and who affect our lives. By consequence, problem solving grows increasingly more difficult because there are more factors to consider — and the potentially worst course of action is to ignore these factors, hoping that a solution found for a "simpler problem" will be effective in spite of these other concerns. Whether you are determining the proper operation of your business, the proper use of environmental resources for your country, or merely the best route to get to work in the morning, you simply cannot afford to assume away your interaction with the rest of world. As these interactions become ever more frequent and complex, procedures for handling real-world problems become imperative. There is a great deal to be gained from solving problems, and a great deal to be lost in solving them poorly.

We all face problems. A problem exists when there is a recognized disparity between the present and desired state. Solutions, in turn, are ways of allocating the available resources so as to reduce the disparity between the present and desired state. To understand the situation, we must first recognize that problems are the possessions of purpose-driven decision makers. If you don't have a purpose then you don't have a problem. The trouble is, not only does each of us have a purpose, but often times those purposes seem to be at odds, or at least not complementary. In determining how best to solve your problem, you must consider it not just in isolation but rather in the context of how others' actions affect your possible solution. It's likely that if they interact with you, they will change their allocation of resources — sometimes to assist you, other times to stand in your way. Implementing a solution without considering "what comes next" is much like playing poker as if it were a one-player game: you and your money would soon be parted.

A prerequisite to handling the real world as a multiplayer game is the ability to manipulate an arsenal of problem-solving techniques, namely, algorithms that have been developed for a variety of conditions. Unfortunately, it's almost always the case that the real world presents us with circumstances that are slightly or considerably different than are required by these methods. For example, one particular technique is available to calculate the minimum-cost allocation of resources for a problem where both the cost function and constraints (equality and inequality) are linear. The method is fast and reliable. And it is almost always applied *inappropriately* in real-world settings where the cost function and constraints are almost always nonlinear. In essence, anyone using this method runs the risk of generating the right answer to a problem that does not exist. The loss incurred using this solution is multifold: not only could a better solution be obtained by properly handling the available constraints in light of the real cost function, but a competitor who treats the situation appropriately may in fact find this better solution! It only takes a slight advantage to one side to bankrupt the other. (Las Vegas has operated on this principle for many years.) On the other hand, using a fast but approximate solution may in fact be the right thing to do: If a competitor is using a slow procedure to find the exact solution, you may be able to discover some inexact but useful solution before they do, and sometimes having any solution at all is better than having to wait for one that is superior. The proper manner in which to proceed requires judgment.

Regrettably, judgment seems to be waning, or at least is out of vogue. Perhaps this is because it requires hard work. In order to make proper judgments in problem solving you have to know what each particular problem-solving method assumes, its computational requirements, its reliability, and so forth. Gaining this information requires time and effort. It's easier to simply peruse the Internet or the commercially available software and obtain something that seems like it might (sort of) fit the problems at hand. Indeed, some businesses and agencies mandate the purchase of such commercial-off-the-shelf (COTS) software. Certainly, this often costs less at the initial outlay, but it's frequently penny wise and pound foolish. Years later, when the software has failed to address the business's real needs and when other competitors have brought their solutions forward in less time, the purchased software sits in an archive, unused, while the workers go back to solving the original problem, sometimes by hand!

Rather than lament this situation, it should be recognized as an opportunity. A great reward awaits those who can effectively solve problems, and you can develop this talent. It begins with a proper understanding of the purpose to be achieved. All too often, in an attempt to utilize a familiar algorithm, we change the purpose to be achieved simply to fit the algorithm. It's as if, armed with only a hammer, we needed to screw a screw into a board: we pound the screw into the board even though we know we'll never be able to unscrew it again. Armed with just a hammer, everything looks like a nail. Similarly, armed with only a small subset of classic problem-solving techniques, all cost functions are magically transformed into smooth, linear, convex mappings, and every model is optimized in light of the mean squared error. This results in "perfect" answers that have questionable relevance. Understanding when the decision maker's purpose should or shouldn't be approximated by linearization and other simplifying techniques is of paramount importance. Formalizing subjective purpose in objective measurable terms so that you can identify important differences between approximations and reality is a challenging task that requires practice. There's no substitute for this experience.

Having made a judgment as to the degree of fidelity required in modeling the decision maker's purpose, the next step is to choose a tactic that will find an optimal allocation of the available resources. For this task, you have lots of options. There are many books devoted to algorithms for problem solving and each presents an array of individual techniques. It's tempting to simply look up an algorithm as you would look up a recipe for a cake. But this natural tendency foregoes creativity; it doesn't afford you the possibility of being ingenious. Instead, it forces you to fit the problem into the constraints of a particular algorithm. In the rush to present problem-solving techniques, each is typically offered independently, thereby overlooking potentially important synergies between different procedures. It's as if a complex jigsaw puzzle could be completed without considering how the pieces fit together. Better solutions to real-world problems can often be obtained by usefully hybridizing different approaches. Effective problem solving requires more than a knowledge of algorithms; it requires a devotion to determining the best combination of approaches that addresses the purpose to be achieved within the available time and in light of the expected reactions of others who are playing the same game. Recognizing the complexity of real-world problems is prerequisite to their effective solution.

Unfortunately, our troubles begin early on in life, even as early as elementary school. We're taught to decompose problems and treat the smaller simpler problems individually. This is wonderful when it works. But complex real-world problems don't often decompose easily or meaningfully. What might even be worse, we are spoon fed the solutions to problems, chapter by chapter, never once being forced to think about whether or not the problems we are facing should be solved with the technique just described in our textbook. Of course they must! *Why else would this problem be in this chapter?!* This is the case not just with texts for elementary mathematics for primary and secondary schools, but also with most textbooks for university courses, and even many monographs that explain particular techniques and illustrate their application with various examples. The problem and its solution in these books are never far apart.

Let's consider a mathematics textbook for junior high school. There might be a chapter there on solving quadratic equations. At the end of the chapter there are some problems to solve. One of them reads: A farmer has a rectangular ranch with a perimeter of 110 meters and an area of 700 square meters. What are the dimensions of his ranch? The student *knows* that to solve the problem, he should use something that was discussed in the same chapter. So he tries to construct a quadratic equation. Sooner or later he comes up with two equations

$$\begin{cases} 2x + 2y = 110 \\ xy = 700, \end{cases}$$

that lead to a quadratic equation:

$$x(55 - x) = 700.$$

Then, following some given formulas, he easily calculates x, and solves the problem.

Some other chapter is on geometry and discusses properties of triangles. The Pythagorean theorem is explained, and the chapter concludes with yet another series of problems to solve. The student has no doubt that he should apply this theorem to solve these problems.

It seems, however, that this is not the right way to teach. The relationship between a problem and a method should be discussed from the perspective of a problem, and not the method. Doing otherwise actually causes more harm than good in the long run, because like all "hand outs," it leaves the students incapable of fending (i.e., thinking) for themselves.

Let's make this point by presenting two problems to solve. The first is as follows. There is a triangle ABC, and D is an arbitrary interior point of this triangle (see figure 0.1). All you have to do is prove that

$$AD + DB < AC + CB,$$

where AD denotes the distance between points A and D, and the other terms indicate the analogous distances between their representative vertices.

This problem is *so* easy, it seems that there is nothing to prove. It's so obvious that the sum of the two segments inside the triangle *must* be shorter than the sum of its two sides! But this problem now is removed from the context of its "chapter" and outside of this context, the student has no idea of whether to apply the Pythagorean theorem, build a quadratic equation, or do something else!

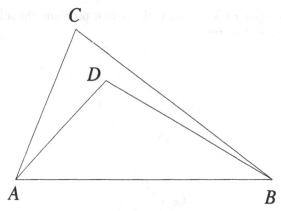

Fig. 0.1. Triangle ABC with an interior point D

The issue is more serious than it first appears. We have given this very problem to many people, including undergraduate and graduate students, and even full professors in mathematics, engineering, or computer science. Fewer than five percent of them solved this problem within an hour, many of them required several hours, and we witnessed some failures as well. What's interesting here is the fact that this problem was selected from a math text for fifth graders in the United States. It's easy to solve when placed at the end of the appropriate chapter! But outside of this context, it can be quite difficult.

At this stage we'd like to encourage you to put the book aside and try to solve this problem. To reduce your temptation to simply "keep reading" without *thinking*, we placed the solution to this problem in the final section of this text. If you can't solve it within an hour, this book is probably for you![1]

Let's consider another puzzle. There is a well, open at the top, with a diameter of three meters. We throw two sticks with lengths of four and five meters, respectively, into the well. Let's assume that these sticks are positioned as shown in figure 0.2, i.e., they stay in the same plane and they "cross each other" as shown. The problem is really two-dimensional, as the well can be replaced by a rectangle that is open at the top.

Now, the challenge is to determine the distance h from the bottom of the well to the point where the two sticks intersect. The position and the lengths of the sticks guarantee a unique solution, and it seems that the problem is easy. Indeed, it is easy; it takes less than a minute to solve and you shouldn't be surprised to learn that the problem is again taken from a primary school curriculum. However, the associated "chapter contents" for this problem are once again missing, and once more we encourage you, the reader, to solve it. If you solve it within an hour (!), you'll belong to the elite one percent of the people we tested who managed to get the right answer within that time. What's more, everyone we tested had at least an undergraduate degree either in mathematics,

---

[1]Note, however, that we already gave you a valuable piece of information: the problem is elementary, so there's no need to use any "sophisticated" tools. Without this hint it would be more difficult to solve. And you'll probably benefit from this book even if you can solve this problem within an hour!

engineering, or computer science. As with the first problem, the solution is given in the final section of this text.

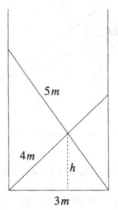

Fig. 0.2. Two sticks in a well

Our main point here is that we learn how to apply particular methods to particular problems, but only within the context of knowing that these exact methods ought to be perfect for these particular problems. We almost never learn how to think about solving problems in general — we're constrained to concentrate on each question in the back of each chapter using the information learned in that chapter. Hopefully, this will change. We look forward to the day when some problems in the math text for our first graders will be different. For example, when the problem

> It takes 48 hours for a rocket to travel from the Earth to the Moon.
> How long will this trip take if a new rocket is twice as fast?

is replaced by

> It takes 48 hours for a rocket to travel from the Earth to the Moon.
> How long will this trip take for two rockets?

which forces a child (or even an adult) to *think* (e.g., whether to multiply or divide 48 by 2, or whether it still take 48 hours). Even in higher grades there are some possibilities:

> A coin is flipped. It comes up heads. What is your guess for the next toss?

What is missing in most curriculums — on all levels, starting from primary schools through university education — is a development of problem-solving skills. Young people often have serious difficulties in solving any real problem because real problems aren't found in any chapter! This book is designed not only to present you with an array of algorithms and techniques for solving textbook problems, but it's also designed to get you to *think* about how to frame and solve problems, particularly those that you don't encounter at the end of some chapter in a book.

We humans like to tout our ability to solve problems. We have even gone so far as to define ourselves in terms of sapience: wisdom, enlightenment, genius. It would be an easy conclusion that such a wise and enlightened genius should be able to solve its problems. And yet, the world is full of people with problems. We hope this book helps you solve not only your own problems, but others' problems as well.

I. What Are the Ages of My Three Sons?

One of the reasons that problem solving is often difficult is that we don't know where to begin. Once we have some sort of course charted out in front of us, we can follow that path and hopefully arrive at a solution. But conjuring up that course to follow is a significant challenge. Often this is the most challenging part of problem solving because it is entirely creative: you have to literally *create* the plan for generating a solution. Without the plan, you're sunk. This is the time where you should revel in the opportunity to display your brilliance! One thing that may help in forging a path from the problem to the solution is to always consider all of the available data. This is good advice in any regard, but it's especially important when first considering how to find a solution. Failing to consider everything right at the start means that you might miss the one opportunity to begin the solution. Review all of the information that's provided, determine the implications of these data or facts, and then see if you can make a connection between the goal and what you are given.

Give it a try with the following problem. Two men meet on the street. They haven't seen each other for many years. They talk about various things, and then after some time one of them says: "Since you're a professor in mathematics, I'd like to give you a problem to solve. You know, today's a very special day for me: All three of my sons celebrate their birthday this very day! So, can you tell me how old each of them is?"

"Sure," answers the mathematician, "but you'll have to tell me something about them."

"OK, I'll give you some hints," replies the father of the three sons, "The product of the ages of my sons is 36."

"That's fine," says the mathematician, "but I'll need more than just this."

"The sum of their ages is equal to the number of windows in that building," says the father pointing at a structure next to them.
The mathematician thinks for some time and replies, "Still, I need an additional hint to solve your puzzle."

"My oldest son has blue eyes," says the other man.

"Oh, this is sufficient!" exclaims the mathematician, and he gives the father the correct answer: the ages of his three sons.

Your challenge now is to do the same: to follow the reasoning of the mathematician and solve the puzzle. Again, the puzzle is quite easy, yet most people have difficulties with it.

How did you do? If you didn't resist the temptation to simply flip the page and find the answer, do yourself a favor and turn back now before it's too late!

All right, to start let's examine all of the information supplied by the conversation carefully. What do we know on the basis of the first piece of information? If the product of the ages of the three sons is 36, there are only eight possibilities to consider; this reduces the search space to only eight cases:

Age of son 1	Age of son 2	Age of son 3
36	1	1
18	2	1
12	3	1
9	4	1
9	2	2
6	6	1
6	3	2
4	3	3

The second piece of information was that the sum of the sons' ages is the same as the number of windows in the building. We have to assume that the mathematician knew the number of windows, so he knew the total. What are the possibilities here? How can this be useful? Adding the numbers for all eight cases yields the following totals:

$$36 + 1 + 1 = 38$$
$$18 + 2 + 1 = 21$$
$$12 + 3 + 1 = 16$$
$$9 + 4 + 1 = 14$$
$$9 + 2 + 2 = 13$$
$$6 + 6 + 1 = 13$$
$$6 + 3 + 2 = 11$$
$$4 + 3 + 3 = 10$$

Suddenly, everything is clear. If the number of windows in the building had been 21 (or 38, 16, 14, 11, or 10), the mathematician would have given the answer immediately. Instead, he said that he still needed an additional piece of information. This indicates that the number of windows was 13, thereby leaving two and only two options:

(9, 2, 2) or (6, 6, 1).

As the second option does not have an *oldest* son ("the *oldest* son has blue eyes"), the ages of the three sons must be (9, 2, 2).

What makes this problem easier than it might otherwise be is that the order in which you should consider the information in the problem is the same as the order in which it's presented. It's almost as if the authors of the puzzle wanted you to find the right answer! Real-world problems aren't so neatly organized. Nevertheless, if you consider all of the available data then you'll give yourself every opportunity to find a useful starting point.

1. Why Are Some Problems Difficult to Solve?

Our problems are man-made;
therefore they may be solved by man.

John F. Kennedy, speech, June 10, 1963

At the risk of starting with a tautology, real-world problems are difficult to solve, and they are difficult for several reasons:

- The number of possible solutions in the *search space* is so large as to forbid an exhaustive search for the best answer.

- The problem is so complicated that just to facilitate any answer at all, we have to use such simplified models of the problem that any result is essentially useless.

- The *evaluation function* that describes the quality of any proposed solution is noisy or varies with time, thereby requiring not just a single solution but an entire series of solutions.

- The possible solutions are so heavily constrained that constructing even one feasible answer is difficult, let alone searching for an optimum solution.

- The person solving the problem is inadequately prepared or imagines some psychological barrier that prevents them from discovering a solution.

Naturally, this list could be extended to include many other possible obstacles. For example, we could include noise associated with our observations and measurements, uncertainly about given information, and the difficulties posed by problems that have multiple and possibly conflicting objectives (which may require a set of solutions rather than a single solution). The above list, however, is sufficient for now. Each of these are problems in their own right. To solve a problem, we have to understand the problem, so let's discuss each of these issues in turn and identify their inherent details.

1.1 The size of the search space

One of the elementary problems in logic is the Boolean satisfiability problem (SAT). The task is to make a compound statement of Boolean variables evaluate to TRUE. For example, consider the following problem of 100 variables given in conjunctive normal form:

$$F(\mathbf{x}) = (x_{17} \vee \overline{x}_{37} \vee x_{73}) \wedge (\overline{x}_{11} \vee \overline{x}_{56}) \wedge \ldots \wedge (x_2 \vee x_{43} \vee \overline{x}_{77} \vee \overline{x}_{89} \vee \overline{x}_{97}).$$

The challenge is to find the truth assignment for each variable x_i, for all $i = 1, \ldots, 100$ such that $F(\mathbf{x}) = $ TRUE. We can use 1 and 0 as synonyms for TRUE and FALSE, and note that $\overline{x}_i$ here is the negation of x_i (i.e., if x_i were TRUE or 1, then $\overline{x}_i$ would be FALSE or 0).

Regardless of the problem being posed, it's always useful to consider the space of possible solutions. Here, any binary string of length 100 constitutes a potential solution to the problem. We have two choices for each variable, and taken over 100 variables, this generates 2^{100} possibilities. Thus the size of the search space $\mathcal{S}$ is $|\mathcal{S}| = 2^{100} \approx 10^{30}$. This is a huge number! Trying out all of these alternatives is out of the question. If we had a computer that could test 1000 strings per second and could have started using this computer at the beginning of time itself, 15 billion years ago right at the Big Bang, we'd have examined fewer than one percent of all the possibilities by now!

What's more, the choice of which evaluation function to use isn't very clear. What we'd like is for the evaluation function to give us some guidance on the quality of the proposed solution. Solutions that are closer to the right answer should yield better evaluations than those that are farther away. But here, all we have to operate on is $F(\mathbf{x})$ which can either evaluate to TRUE or FALSE. If we try out a string $\mathbf{x}$ and $F(\mathbf{x})$ returns TRUE then we're done: that's the answer. But what if $F(\mathbf{x})$ returns FALSE? Then what? Furthermore, almost every possible string of 0s and 1s that we could try would likely evaluate to FALSE, so how could we distinguish between "better" and "worse" potential solutions? If we were using an enumerative search we wouldn't care because we'd simply proceed through each possibility until we found something interesting. But if we want the evaluation function to help us find the best solutions faster than enumeration, we need more than just "right" or "wrong." The way we could accomplish that for the SAT problem isn't clear immediately.

Some problems seem easier than SAT problems because they suggest a possible evaluation function naturally. Even so, the size of the search space can still be enormous. For example, consider a traveling salesman problem (TSP). Conceptually, it's very simple: the traveling salesman must visit every city in his territory exactly once and then return home covering the shortest distance. Some closely related problems require slightly different criteria, such as finding a tour of the cities that yields minimum traveling time, or minimum fuel cost, or a number of other possibilities, but the idea is the same. Given the cost of traveling between each pair of cities, how should the salesman plan his itinerary to achieve a minimum cost tour?

Figure 1.1 illustrates a simple symmetric 20-city TSP where the distance between each pair of cities i and j is the same in either direction. That is, $dist(i, j) = dist(j, i)$. The actual distances aren't marked on the figure, but we could assume this is the case. Alternatively, we could face an asymmetric TSP where $dist(i, j) \neq dist(j, i)$ for some i and j. These two classes of TSP present different obstacles for finding minimum cost paths.

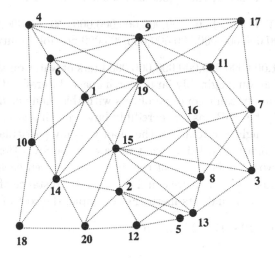

Fig. 1.1. A sample TSP. Textbook TSPs usually allow paths from every city to every other cities, but real-world problems don't always afford such opportunities.

So then, what is the search space for the TSP? One possibility would be to view it as the set of permutations of n cities. Any permutation of n cities yields an ordered list that defines the sequence of cities to be visited, starting at the salesman's home base, and continuing to the last location before returning home. The optimum solution is a permutation that yields the minimum cost tour. Note that tours such as:

$$2 - ... - 6 - 15 - 3 - 11 - 19 - 17,$$
$$15 - 3 - 11 - 19 - 17 - 2 - ... - 6,$$
$$3 - 11 - 19 - 17 \quad 2 - ... - 6 - 15, \text{etc.}$$

are identical because the circuit that each one generates is exactly the same regardless of the starting city, and there are n tours like that for any n-city TSP. It's easy to see then that every tour can be represented in $2n$ different ways (for a symmetrical TSP). And since there are $n!$ ways to permute n numbers, the size of the search space is then $|S| = n!/(2n) = (n-1)!/2$.

Again, this is a huge number! For any $n > 6$, the number of possible solutions to the TSP with n cities is larger than the number of possible solutions to the SAT problem with n variables. Furthermore, the difference between the sizes of these two search spaces increases very quickly with increasing n. For $n = 6$, there are $5!/2 = 60$ different solutions to the TSP and $2^6 = 64$ solutions to a SAT. But for $n = 7$ these numbers are 360 and 128, respectively.

To see the maddening rate of growth of $(n-1)!/2$, consider the following numbers:

- A 10-city TSP has about 181,000 possible solutions.

- A 20-city TSP has about 10,000,000,000,000,000 possible solutions.

- A 50-city TSP has about 100,000,000,000,000,000,000,000,000,000,000, 000,000,000,000,000,000,000,000,000,000,000 possible solutions.

There are only 1,000,000,000,000,000,000,000 liters of water on the planet, so a 50-city TSP has an unimaginably large search space. Literally, it's so large that as humans, we simply can't conceive of sets with this many elements.

Even though the TSP has an incredibly large search space, the evaluation function that we might use to assess the quality of any particular tour of cities is much more straightforward than what we saw for the SAT. Here, we can refer to a table that would indicate all of the distances between each pair of cities, and after n addition operations we could calculate the distance of any candidate tour and use this to evaluate its merit. For example, the cost of the tour

$$15 - 3 - 11 - 19 - 17 - 2 - \ldots - 6$$

is

$$cost = dist(15, 3) + dist(3, 11) + dist(11, 19) + \ldots + dist(6, 15).$$

We might hope that this, more natural, evaluation function would give us an edge in finding useful solutions to the TSP despite the size of the search space.

Let's consider a third example — a particular nonlinear programming problem (NLP). It's a difficult problem that has been studied in the scientific literature and no traditional optimization method has given a satisfactory result. The problem [254] is to maximize the function:[1]

$$G2(\mathbf{x}) = \left| \frac{\sum_{i=1}^{n} \cos^4(x_i) - 2\prod_{i=1}^{n} \cos^2(x_i)}{\sqrt{\sum_{i=1}^{n} ix_i^2}} \right|,$$

subject to

$\prod_{i=1}^{n} x_i \geq 0.75$, $\sum_{i=1}^{n} x_i \leq 7.5n$, and bounds $0 \leq x_i \leq 10$ for $1 \leq i \leq n$.

The function $G2$ is nonlinear and its global maximum is unknown, but it lies somewhere near the origin. The optimization problem poses one nonlinear constraint and one linear constraint (the latter one is inactive near the origin).

What's the size of the search space now? In a sense, it depends on the dimensionality of the problem — the number of variables. When treated as a purely mathematical problem, with n variables, each dimension can contain an infinity of possible values, so we have an infinitely large space — maybe even several degrees of infinity. But on a computer, everything is digital and finite, so if we were going to implement some sort of algorithm to find the optimum of $G2$, we'd have to consider the available computing precision. If our precision guaranteed six decimal places, each variable could then take on

---

[1]We're preserving the notation $G2$ used for this function in [319]. It's the second function in an 11-function testbed for nonlinear programming problems.

10,000,000 different values. Thus, the size of the search space would be $|\mathcal{S}| = 10,000,000^n = 10^{7n}$. That number is much much larger than the number of solutions for the TSP. Even for $n = 50$ there are 10^{350} solutions to the NLP with only six decimal places of precision. Most computers could give twice this precision.

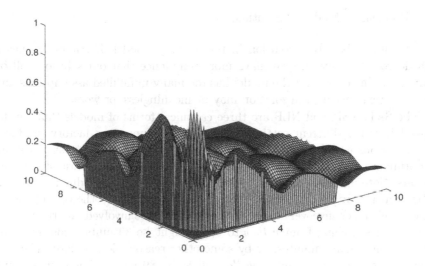

Fig. 1.2. The graph of function $G2$ for $n = 2$. Infeasible solutions were assigned a value of zero.

What about the evaluation function? How can we measure the quality of alternative solutions? One idea would be to use the function $G2$ itself as an evaluation function. Those solutions that yield higher values of $G2$ would be judged as being better than those that yield lower values. But there are some difficulties in this regard because, as illustrated in figure 1.2, there are infeasible regions, and all of the infeasible points were assigned a value of zero. The interesting boundary between feasible and infeasible regions is defined by the equation $\prod_{i=1}^{n} x_i = 0.75$, and the optimal solution lies on (or close to) this boundary. Searching boundaries of a feasible part of the search space isn't easy. It requires specialized operators that are tailored for just this purpose, on just this problem. This presents an additional level of difficulty that we didn't see in the SAT or TSP (although for a TSP that didn't allow transit between all possible pairs of cities, some permutations would also be infeasible). Even without this additional wrinkle, it's evident that problems that seem simple at first can offer significant challenges simply because of the number of alternative solutions. The means for devising ways to assess these solutions isn't always clear.

1.2 Modeling the problem

Every time we solve a problem we must realize that we are in reality only finding the solution to a *model* of the problem. All models are a simplification of the real world, otherwise they would be as complex and unwieldy as the natural setting itself. The process of problem solving consists of two separate general steps: (1) creating a model of the problem, and (2) using that model to generate a solution:

Problem $\Rightarrow$ Model $\Rightarrow$ Solution.

The "solution" is only a solution in terms of the model. If our model has a high degree of fidelity, we can have more confidence that our solution will be meaningful. In contrast, if the model has too many unfulfilled assumptions and rough approximations, the solution may be meaningless, or worse.

The SAT, TSP, and NLP are three canonical forms of models that can be applied in many different settings. For example, suppose a factory produces cars in various colors where there are n colors altogether. The task is to find an optimum production schedule that will minimize the total cost of painting the cars. Note, however, that each machine involved in the production line has to be switched from one color to another between jobs, and the cost of such a switch (called a changeover) depends on the two colors involved and their order. The cost of switching from *yellow* to *black* might be 30 units. This might be measured in dollars, minutes, or by some other reasonable standard. The cost of switching back from *black* to *yellow* might be 80 units.[2] The cost of going from *yellow* to *green* might be 35 units, and so forth. To minimize costs, we have to find a sequence of jobs that meets all of the production requirements for the number of cars of each color, in a timely fashion, while still keeping the costs of the operation as low as possible. This might be viewed in terms of a TSP, where each city is now a job that corresponds to painting a certain car with a particular color and the distance between cities corresponds to the cost of changing jobs. Here, the TSP would be asymmetric.

Consider the following scenario that illustrates how simplifications are inherent to modeling. Suppose a company has n warehouses that store paper supplies in reams. These supplies are to be delivered to k distribution centers. The warehouses and distribution centers can be viewed as *sources* and *destinations*, respectively. Every possible delivery route between a warehouse i and a distribution center j has a measurable transportation cost, which is determined by a function f_{ij}. The shape of this function depends on a variety of factors including the distance between the warehouse and the distribution center, the quality of the road, the traffic density, the number of required stops, the average speed limit, and so forth. For example, the transportation cost function between warehouse 2 and distribution center 3 might be defined as:

---

[2]Note the asymmetry here: the costs of switching between two colors need not be the same. Switching from *yellow* to *black* isn't usually as expensive as the obverse switch.

$$f_{23}(x) = \begin{cases} 0 & \text{if } x = 0 \\ 4 + 3.33x & \text{if } 0 < x \le 3 \\ 19.5 & \text{if } 3 < x \le 6 \\ 0.5 + 10\sqrt{x} & \text{if } 6 < x, \end{cases}$$

where x refers to the quantity of supplies transported from 2 to 3 (figure 1.3 displays the graph of the function f_{23}).[3]

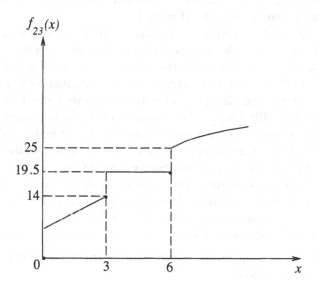

Fig. 1.3. An example transportation cost function for a given source and a given destination

Even though this function looks a bit unconventional, the justification for using it in the model of our transportation problem might be quite straightforward. If there's no delivery, the cost is zero, of course. If up to three reams of paper are transported we can use a special shipping container. This incurs an overhead cost of four units for the container and an additional cost of 3.33 units per ream. So the cost for this case increases linearly. However, if we transport more than three reams but not more than six, we can use a special wire mesh box. In this case, the cost is a flat 19.5 units, regardless of the number of reams being shipped. Finally, if we are shipping more than six reams we have to use a large reinforced crate, with a total transportation cost that depends on the number of reams being shipped and grows as a square root of that quantity plus a small overhead of 0.5 units.

Given these preliminaries, we can construct a model of the problem:

minimize $\sum_{i=1}^{n} \sum_{j=1}^{k} f_{ij}(x_{ij})$,

subject to

---

[3]Note that discontinuities are quite typical for most *real* transportation cost functions.

$$\sum_{j=1}^{k} x_{ij} \leq sour(i), \text{ for } i = 1, 2, \ldots, n,$$
$$\sum_{i=1}^{n} x_{ij} \geq dest(j), \text{ for } j = 1, 2, \ldots, k,$$
$$x_{ij} \geq 0, \text{ for } i = 1, 2, \ldots, n \text{ and } j = 1, 2, \ldots, k,$$

where *sour* is the source and *dest* is the destination. The constraints of the problem define a feasible solution: no transport from any warehouse exceeds the number of available reams in that warehouse, and the total transport to any distribution center must satisfy its demand (i.e., the total transport is at least equal to the number of ordered reams).

It might just be that this cost function describes the real-world situation faithfully (i.e., exactly), neglecting the costs for the other aspects we mentioned earlier such as the traffic density between the source and destination and so forth. And we might be able to construct similar exact functions that describe the costs of transporting the reams of paper from every warehouse to every distribution center. Still, such a precise model of the problem might be of only limited utility because these functions are too complex for many traditional optimization algorithms. For starters, they are discontinuous, and discontinuities present severe problems. The results that we would obtain after using some gradient-based methods on these functions would likely be quite poor [320]. Thus, we cannot derive a solution based on this model, so the model — as perfect as it is — is useless for deciding what to do!

What options do we have? There are at least two ways to proceed:

1. We can try to simplify the model so that traditional optimizers might return better answers.

2. We can keep the model as it is, and use a nontraditional approach to find a near-optimum solution.

The first idea is quite tempting. For example, we can *approximate* the function f_{23} as follows:

$$f'_{23}(x) = 2.66x + 8.25,$$

where x denotes the number of reams transported from 2 to 3 (figure 1.4 displays the graph of the approximate function f'_{23} together with the original f_{23}).

In this case, we simplified the transportation cost function f_{23}, and we can perform similar simplifications for the other functions. Note that if all of the f'_{ij}s were linear, we'd obtain a linear model of the problem that can be solved precisely by a linear programming method. But note that this exact solution would then be a solution for the simplified model and not for the real problem!

The second option is to leave the precise model as it is — with all of its discontinuities — and use a nontraditional method (e.g., simulated annealing or an evolutionary algorithm) to find a near-optimal solution. A large volume of experimental evidence shows that this latter approach can often be used to practical advantage.

Let's rephrase this discussion. There are two possible approaches. The first uses an approximate model, $Model_a$, of a problem, and then finds the precise solution $Solution_p(Model_a)$ for this approximate model:

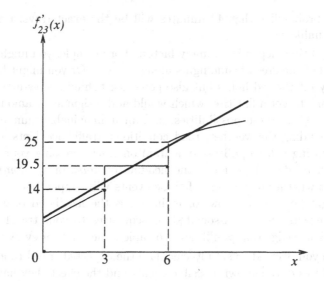

Fig. 1.4. An approximation of the transportation cost function (*bold line*) for a given source and a given destination

$$\text{Problem} \Rightarrow Model_a \Rightarrow Solution_p(Model_a).$$

The second approach uses a precise model, $Model_p$, of the problem, and then finds an approximate solution $Solution_a(Model_p)$ for this precise model:

$$\text{Problem} \Rightarrow Model_p \Rightarrow Solution_a(Model_p).$$

Of these two approaches, the latter one is often superior; i.e., $Solution_a(Model_p)$ is better than $Solution_p(Model_a)$ as a solution of the original problem.

But either way, this is the second source of difficulties we face in problem solving: it's difficult to obtain a precise solution to a problem because we either have to approximate a model or approximate the solution.

1.3 Change over time

As if the above concerns weren't enough, real-world problems often present another set of difficulties: they change. They change before you model them, they change while you are deriving a solution, and they change after you execute your solution. Let's look at some of the sources of trouble.

Think back to the traveling salesman problem from figure 1.1. Suppose you are the salesman and you are leaving your home at city 1 on your way to city 6. You've driven this route before and you know the distance between the two towns is, say, 20 miles. But do you know how long it will take you to travel between the cities? Not precisely. You might figure that typically you can average 30 miles per hour and so on average this trip will take 40 minutes. But what is the probability that 40 minutes will be the exact travel time today? That's very unlikely.

The travel time depends on many factors. For example, you might be lucky and make all of the green traffic lights along the way. Or you might be unlucky and not only hit the red lights but also get stuck behind a slow-moving truck. Worse, you might get a flat tire, which would add a significant amount of time to your trip. All of these possibilities, and an unimaginable number of other outcomes including the weather, road conditions, traffic accidents, emergency vehicles requiring you to pull aside, a train on tracks crossing your path, and so forth, can be described under the heading of *noise* or *randomness*. You can't predict whether or not any of these events will happen before you leave. All you might be able to know, or estimate, is the likelihood of each of the anticipated events and the associated consequences to your travel time. But you must acknowledge that you'll never be able to account for every possibility.

Suppose you decided to simply calculate the expected travel time based on the probabilities of the known possible events and the effect they have on your travel time. Let's simplify things a little. Say there are only two possibilities for your trip: (1) everything goes fine and you can travel to city 6 in 40 minutes, or (2) you get stuck behind a slow vehicle and it takes you 60 minutes. Furthermore, let's say that these two events are equally likely. So the expected time for the trip is 50 minutes. But note that neither of the two possibilities above takes 50 minutes. If you use 50 minutes as the approximate time you ensure using a value that *will never happen in reality*. You can be 100 percent certain that your value will be wrong. It might be easy to imagine that if you used a series of these incorrect approximations to determine your travel time to each of the cities in your route, compounding your errors between each city as you go, that your final estimated time might be very different from any of the possible times that it would really take. Any decision you make based on this *average* time fails to take into account the *variability* of the time, and this can be much more important in effecting good decisions.

Random chance isn't the only source of change in real-world problems. Sometimes there are purely deterministic troubles as well. You know, for example, that travel at rush hour will be more time consuming in each city than travel at midnight. You might not know exactly how much more time you'll spend — that's a random event — but you know there is a bias that rush hour will stall your progress in traffic. That bias is a regular, predictable pattern, and you need to consider it or else your model might not correspond sufficiently to reality, and your solutions with the model won't be useful. In the worst case, a particular route between two cities might be available only during certain times of day, and not available otherwise (this often happens in cities where city planners limit your options for turning into side streets at rush hour in order to increase traffic flow). Acting as if the unavailable route were still an option would lead to an infeasible solution, and that means no solution at all.

It's also important to be sure that the model reflects current knowledge about the problem. It might be that road improvements or a new freeway system between two of the cities on your list now allow for more rapid transit. If you fail to update your model to account for this change, you'll be deriving a solution to a problem that no longer exists.

The situation, however, is even more complex than this. The above vagaries might occur as a function of environmental changes or uncontrollable events, but none of them were conspiring against you. Unfortunately, in the real world, it's often the case that other people are trying to defeat your solution, and this requires you to continually update your model and anticipate other people's actions.

For example, suppose you are the owner of a major supermarket chain. You need to decide where to place a new store, so you calculate the cost of construction in each possible location, the demographics of the neighboring areas, the existing competition, etc., formulate an evaluation function, which would likely result in a nonlinear programming problem, and set out to decide where the best place for the store is. But there's more to the problem: as you are deciding where to put your new store, your competition is anticipating your choice of location, and they have their own new store to construct. They are actively trying to figure out how best to place their new store so as to minimize your success. If you only solve the problem of the best position for your store given the current conditions, you are treating the situation as if it were a one-player game. The real-world problem often changes while you are deriving a solution, and sometimes it changes in ways that are designed to make your life difficult.

1.4 Constraints

Unfortunately, things are often even worse than we've made it out so far because real-world problems don't offer you all of the possibilities that you might like to have. Almost all practical problems pose constraints, and if you violate the constraints you can't implement your solution. Think back to the NLP from section 1.1. There, we had a case where it wasn't enough to find the maximum of the function $G2$, we had to ensure that the solution we proposed was in the feasible region bounded by the product and summation constraints (see figure 1.2). Now at first you might think constraining problems like this would make life easier — after all, we have a smaller search space to worry about and therefore fewer possibilities to consider. That's true, but remember that to *search* for improved solutions we have to be able to move from one solution to the next. We need operators that will act on feasible solutions and hopefully in turn generate new feasible solutions that are an improvement over what we've already found. It's here where the geometry of the search space gets tricky.

For example, suppose we're faced with the problem of making a timetable for all of the classes at a college in one semester. Think about what this entails. First, we have to make a list of all the courses that will be offered. Next, we need a list of all the students assigned to each class, and let's not forget the professor assigned to each class too. Third, we need a list of available classrooms, noting the size and other facilities that each offers (e.g., a white board, a video projector, laboratory equipment, and so forth). So then, what are we trying to accomplish? There are three hard constraints:

- Each class must be assigned to an available room that has enough seats for every assigned student and has the requisite facilities for the type of instruction (e.g., a chemistry lab must have beakers, Bunsen burners, the appropriate chemicals, safeguards, etc.).

- Students who are enrolled in more than one class can't have their classes held at the same time on the same day.

- Professors can't be assigned to teach courses that overlap in time.

We said those are the *hard constraints*. By that, we mean these are the things that absolutely must be satisfied in order to have a feasible solution. Moreover, with what we've presented so far, any assignment that meets the constraints would solve our problem. So this means the task is quite similar to the SAT problem: we have to find an assignment of classes (as compared with Boolean variables) such that an overall evaluation function returns a value of TRUE. Anything that violates the constraints means our evaluation function returns a value of FALSE. But this alone doesn't give us sufficient information to guide the search for a feasible solution.

We might be able to employ some strategy that could provide this additional information. For example, we might judge the quality of the solution not just by whether or not it satisfies the constraints, but for those assignments that fail to meet the constraints, we could tally the number of times that the constraints are violated (e.g., each time a student is assigned to two classes that meet at the same time we increase the tally). This would give us a quantitative measure of how poor our infeasible solutions were, and it might be useful in guiding us toward successively better solutions, minimizing the number of constraint violations. We could apply different operators for reassigning courses to classrooms, professors to courses, and so forth, and over time we'd hope to generate a solution that met the available constraints.

But then there are the *soft constraints*, the things we hope to accomplish but aren't mandatory. These include:

- Courses that meet twice a week should preferably be assigned to Mondays and Wednesdays or Tuesdays and Thursdays. Having these courses meet on consecutive days or with two or more days inbetween is not desired.

- Courses that meet three times per week should preferably be assigned to Mondays, Wednesdays, and Fridays. Other assignments are not desired.

- Course times should be assigned so that students don't have to take final exams for multiple courses without any breaks in between (final exam times are typically based on the time for the course).

- If undergraduate prerequisite courses are scheduled for the same day as their counterpart graduate courses, they should preferably be given earlier than the graduate course (this facilitates learning foundational material prior to advanced material in the same day).

- If more than one room satisfies the requirements for a course and is available at the designated time, the course should be assigned to the room with the capacity that is closest to the class size (this means that large auditoriums aren't used for small classes, thus enhancing student participation).

Certainly we could imagine many more such soft constraints. Any assignment that meets the hard constraints is feasible, but not necessarily optimal in light of the soft constraints. Here is where the problem gets sticky. First, we have to quantify each of the soft constraints into mathematical terms so that we can evaluate any two candidate assignments and decide that one is better than the other. Next, we have to be able to modify one feasible solution and, hopefully, generate another feasible solution that better meets the soft constraints.

Let's take the first issue: each soft constraint has to be quantified. Considering the first soft constraint, we could say that for each case where a solution is feasible, we could count up the number of times twice-a-week courses become separated by two or more days, or are placed on consecutive days, and use this as a penalty term. The lower the term, the better the solution. In fact, we could employ a similar approach to each of the soft constraints. But what would we do when we're through? We'd still need an overall method for considering the degree of violation of each of these constraints. That is, we'd have to determine the answers to questions such as: which is worse, scheduling five students to have back-to-back final exams, or scheduling back-to-back classrooms at opposite ends of the campus? Each of these possible trade-offs would have to be considered and quantified in some evaluation function, which poses quite a challenge!

Of course it's worse than that because even after all of these soft constraints have been quantified, we are still left with the problem of searching for the best assignment: the solution that is both feasible and minimizes our evaluation function for the soft constraints. Suppose we have found a feasible solution, but it doesn't do very well with regard to the soft constraints. Say we apply some variation operators to this solution and we significantly improve the situation with respect to the soft constraints, but in so doing, we generate a solution that violates one hard constraint. Now what? We might choose to discard the solution since it's infeasible, or we might see if we can repair it to generate a feasible solution that still handles the soft constraints well. Either way, this is typically a difficult chore. It would be even better to devise variation operators that never corrupt a feasible solution into an infeasible solution while still searching vigorously over the space of feasible solutions to find those that best handle the soft constraints. That's a nice aspiration, but it's often not much more than wishful thinking. Effectively handling real-world constrained problems is one of the most challenging tasks we face.

1.5 The problem of proving things

Although it seems a bit strange, in our quest to solve problems, we sometimes make matters more difficult than they have to be. Having worked with students at many universities and in many countries, we've experienced the phenomenon that if you ask someone to *find* some solution to a problem, they'll typically find this much easier than if you had asked them to *prove* something about the solution, even when the two tasks are exactly the same mathematically.

As an example, think about any mathematical problem with just a moderate degree of difficulty. The task in the problem should be to find a particular value, whether it's the height of a building, the speed of a car, or the time to complete your homework. Regardless, the task should be to find the value of x. For example, you'll take four hours to fill a pool using a large pipe. You'll take six hours if you use a small pipe. How long would it take if you used both pipes? If the problem is formulated in terms of "find the value x" — such as, find the amount of time required to fill the pool using both pipes — this is a fairly easy task. But, for some reason, if we change the task and instead ask: prove that the amount of time required to fill the pool using both pipes is less than a, fewer students will manage this problem despite the fact that it's not the slightest bit more difficult. If you can find the time required to fill the pool, and if it's smaller than a constant a, then the proof is completed. To verify this for yourself, take the values given above for the times required to fill the pool using either pipe alone and test it out on your friends. Ask some to find the time required when using both pipes, and ask others to prove that the time required is less than 2 hours and 25 minutes.[4]

We believe that the reason for this aversion to proving things is that most people simply aren't experienced in proving things and don't know how to begin. Generalizing on this observation, many problems are apparently difficult simply because of the difficulty encountered when facing the question: "How should I start?"

Here's an example to help you exercise your ability to frame problems and get started on their solution:

> Prove that any polyhedron must have at least two faces with the same number of edges.

No doubt, the first inclination when reading this is to think "Do I have to?" Yes, you do. So let's get started.

First, it's helpful to remind ourselves of some basic ideas about proving things. One way to prove something is simply to show that the conclusion

---

[4] Actually, the required time is 2 hours and 24 minutes.

follows directly from the given knowledge. For example, suppose that $a > b$ and $b > c$, then prove that $a > c$. You simply cite the transitive law and you're done. Another way to prove something is to consider the contrapositive. Recall that "if p then q" is logically equivalent to "if not q then not p." Sometimes it might be easier to prove this relationship. Still another way to prove something is to assume that the condition you are trying to prove is false and then show that this is impossible. This is called *proof by contradiction*. Let's try that here.

Restating our problem, then, we need a proof that shows it's impossible to construct a polyhedron such that all of the faces have a different number of edges. The problem seems difficult because there aren't that many general theorems about polyhedrons. There is a famous one — Euler's theorem — which states that the following formula holds for every polyhedron:

$$v + f = e + 2,$$

where v, e, and f represent the number of vertices, edges, and faces of the polyhedron, respectively. But just how we might actually use this theorem in proving that all of the faces must have a different number of edges isn't very clear.

The initial step is the hardest part of the puzzle. Indeed, here the problem doesn't provide us with any convenient starting point. There's no number in the problem, nothing for us to factorize, divide by two, multiply by six, or anything else. In situations like this, it's often advantageous to introduce such a number ourselves: Let's consider a polyhedron with f faces. Now we have something, a variable, to work with.

We have to prove something about the edges of faces for this polyhedron. Each face has a number of edges. To say something about these numbers, it would be nice at least to know the range of values that we might expect. So, let's ask the question: what is the minimum number of edges that a face may have? The answer is three, and in that case the face is a triangle. This minimum number is independent of the total number f of faces of the polyhedron.

And what is the maximum number of edges a face may have? Well, each edge belongs to precisely two faces, so if we have a face with six edges, we know that the face is part of a polyhedron with at least seven faces (the current face plus six new faces, one for each edge; see figure 1.5). Thus, any face of a polyhedron that has f faces in total can't have more than $f - 1$ edges, since a new face originates from each edge.

This concludes the proof.

If this fact surprises you then take note that you have lost sight of what you are trying to prove. Go back and remind yourself of the problem. The reason the proof is complete is because if there are f faces in total, and if each face must have a number of edges between 3 and $f - 1$, then some repetitions in the number of edges must occur across the f faces. There are more faces than possibilities for the number of edges, so you'll have to use some of these numbers more than once. Therefore, at least two faces will have the same number of edges.

The keys to solving this problem were to invent a starting point to pursue and not lose sight of the goal.

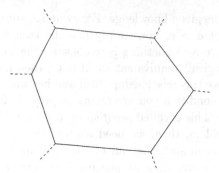

Fig. 1.5. A single face of a polyhedron with six edges

1.6 Your chance for glory

Now that you're primed for solving a problem, here's your chance to *prove* you can do it. This problem was offered to us by Peter Ross of the University of Edinburgh.

> Mr. Smith and his wife invited four other couples for a party. When everyone arrived, some of the people in the room shook hands with some of the others. Of course, nobody shook hands with their spouse and nobody shook hands with the same person twice.
>
> After that, Mr. Smith asked everyone how many times they shook someone's hand. He received different answers from everybody.
>
> How many times did Mrs. Smith shake someone's hand?

We encourage you to put the book aside and try to frame this problem for yourself. Try to think of a starting point, maybe a graphical image that describes the problem, and see if you can follow it through to a successful conclusion.

The reason that this problem is a challenge is that once again there's no obvious starting point, but it's really quite simple to develop a graphical model for the problem (see figure 1.6).

This is a good model because we can clearly see Mr. Smith (shaded circle) and the remaining nine people (including his wife). Now, what information do we have? (Remember, think about all the given information.) The only information provided in this problem is that Mr. Smith asked every person in the room how many times they shook someone's hand, and that all of the answers he received were different.

What answers did he get then? Well, the minimum number of handshakes is zero: it's possible that some antisocial person didn't shake anyone's hand at all. But what is the maximum number of handshakes for a single person? It is more difficult to think of *asking* this question than it is to answer it. The maximum number of handshakes for a person is eight, as there are ten people

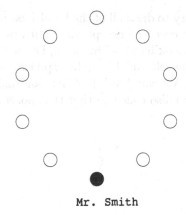

Mr. Smith

Fig. 1.6. Mr. Smith and the other people

in the room, and a person can't shake hands with himself or herself nor with his or her spouse.

Let's collect the facts we have now. Mr. Smith asked the question to nine people in the room and all of the answers were different. Furthermore, each answer was a number between zero and eight. Therefore, the answers he received were zero, one, two, three, four, five, six, seven, and eight (not necessarily in that order, of course!). So, we can update our model accordingly (figure 1.7).

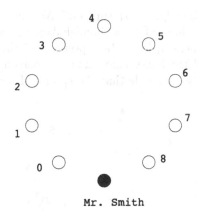

Mr. Smith

Fig. 1.7. Mr. Smith and the other people. The number of handshakes is indicated for each person (except Mr. Smith).

What's the next step, then? What can we infer? It seems, not much. We've already taken advantage of the fact that all of the answers were different, and we now know all of these answers. But how many times did Mrs. Smith shake hands? And which of the people in figure 1.7 is Mrs. Smith, anyway? If we only knew who has shaken hands with whom, then everything would be easier.

But, can't we? Let's try to draw all of the handshakes that were exchanged. The person 8 (we'll name everyone except Mr. Smith by the number of handshakes they exchanged) shook hands eight times, i.e., with everyone else in the room except himself or herself and his or her spouse. We can do two things based on this observation. We can draw all of the handshakes made by person 8 (see figure 1.8) and we can also conclude that the spouse of person 8 is person 0.

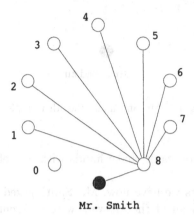

Fig. 1.8. Mr. Smith and the other people. Each handshake from person 8 is indicated.

Did the proverbial light bulb just turn on? We can turn our attention to person 7. This person exchanged seven handshakes, i.e., with everyone in the room except himself or herself, his or her spouse, and the spouse of person 8. Again, we can add all of the handshakes made by person 7 to our model (see figure 1.9) and we can also conclude that the spouse of person 7 is person 1.

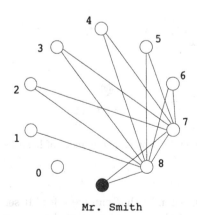

Fig. 1.9. Mr. Smith and the other people. All of the handshakes from persons 8 and 7 are indicated.

We can repeat this reasoning for persons 6 and 5. After adding the lines that correspond to the handshakes they exchanged we find an interesting graph shown in figure 1.10. The immediate conclusions are:

- The spouse of person 6 is person 2.

- The spouse of person 5 is person 3.

Using these conclusions, we know that the spouse of Mr. Smith is person 4. Therefore, Mrs. Smith exchanged four handshakes.

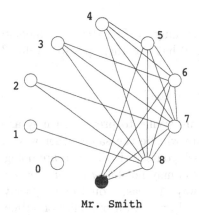

Fig. 1.10. Mr. Smith and the other people. All of the handshakes are indicated.

How did you do?

1.7 Summary

Problem solving is difficult for several reasons:

- Complex problems often pose an enormous number of possible solutions.

- To get any sort of solution at all, we often have to introduce simplifications that make the problem tractable. As a result, the solutions that we generate may not be very valuable.

- The conditions of the problem change over time and might even involve other people who want you to fail.

- Real-world problems often have constraints that require special operations to generate feasible solutions.

Furthermore, problem solving is often more difficult than it needs to be because we are threatened by the idea of *proving* things.

- Don't let the word *prove* intimidate you.

- Think about different ways to prove the solution. Sometimes it's useful to assume that what you are trying to prove is false, and then show that the false condition is impossible.

- Be ready to take a first step even if there doesn't seem to be anything in the problem for you to step on. Invent something, a variable, that describes a facet of the problem and see if that helps. Don't worry if you have to start over a few times before you finally find a path that takes you to the answer.

One way to facilitate taking the first step is to understand the search space: what are the variables? What are their possible values? What are the constraints? Most of all:

- Don't lose sight of the goal!

If you forget what you are trying to prove, you may as well watch television because your success rate will be the same either way: zero! Stay focused on the end result. Always ask yourself if what you are doing facilitates getting to where you want to go. You may not know the answer with certainty, but keep asking yourself this anyway. At least, you'll more quickly identify those paths that don't lead to the goal and you'll minimize the time spent on dead ends. At most, you'll readily identify the right path to choose and be on your way!

II. How Important Is a Model?

Whenever you solve a real-world problem, you have to create a model of that problem first. It is critical to make the distinction that the model that you work with isn't the same as the problem. Every model leaves something out. It has to — otherwise it would be as complicated and unwieldy as the real-world itself. We always work with simplifications of how things really are. We have to accept that. Every solution we create is, to be precise, a solution *only to the model* that we postulate as being a useful representation of some real-world setting that we want to capture.

The trouble with models is that every one of them has an associated set of assumptions. The most common mistake when dealing with a model is to forget the assumptions. For example, suppose you throw a ball into the air and plot its trajectory as it falls to the earth. What is the shape of that trajectory? The most likely answer is that it's parabolic, but that's incorrect. "Oh, yes, wind resistance, we forgot about that," you might say. True, we did omit that here, but that's not why the answer is wrong. Even with no wind resistance, the trajectory isn't parabolic. The correct answer is that the trajectory is elliptical. Why? Because it's only parabolic under the assumption that *the earth is flat.* Once we assume that the earth is curved, the trajectory is no longer parabolic.

Certainly, within your local neighborhood (i.e., the distance you could possibly throw a ball) the earth seems pretty flat. Thus there is very little difference in the model developed under the parabolic trajectory and the one described using an elliptical path. To illustrate the consequence of unstated assumptions more dramatically, we'll use a simple example that was offered to us by Jan Plaza of State University of New York at Plattsburgh.

There is a circle R with a radius of 1 on the ground in front of you. A straight line L of infinite length (in both directions) is thrown on the ground such that it intersects the circle. What is the probability p that the length l of the chord created by the intersection of the circle and the line is greater than or equal to $\sqrt{3}$?

To calculate this probability, we have to envision a space of possible outcomes and then assign an appropriate likelihood to the event that contains all of the outcomes such that the length of the intersecting chord is greater than or equal to $\sqrt{3}$.

Before we consider a few possibilities for constructing the space of outcomes, let's remind ourselves of some elementary geometrical facts: (1) the length of a side of an equilateral triangle T inscribed in the circle R is $\sqrt{3}$, and (2) the radius of a circle r inscribed into the triangle T is $1/2$ (see figure II.1).

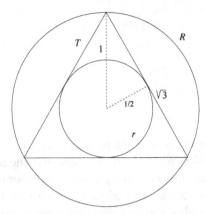

Fig. II.1. Circle R, equilateral triangle T, and inscribed circle r

Possibility 1.

We know that the length l of the chord may vary from 0 to 2, since 2 is the diameter of R. Let's consider the midpoint m of the chord. If m is in the interior of r then $l \geq \sqrt{3}$, otherwise m is in the interior of R but not in the interior of r and $l < \sqrt{3}$ (see figure II.2a). Since the midpoint could turn up anywhere inside R, the probability that it turns up inside r, implying $l \geq \sqrt{3}$, is determined as the ratio of the area of r to the area of R. Thus

$$p = \frac{\pi(\frac{1}{2})^2}{\pi(1)^2} = \frac{1}{4}$$

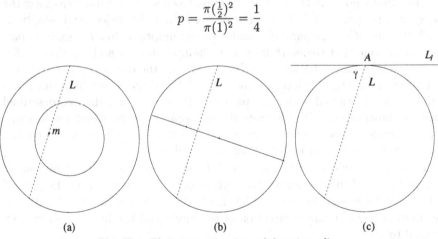

Fig. II.2. Three interpretations of throwing a line

Possibility 2.

Let's draw an arbitrary diameter of R. When a line L is thrown, we can rotate R until the diameter is perpendicular to the line, as shown in (figure II.2b). Now we only have to look at half the circle because the length of the chord must be less than or equal to the diameter of R. The midpoint of the chord will either be less than a $1/2$ unit away from the center of R or not. If it is, the length of the chord is greater than or equal to $\sqrt{3}$. Otherwise, it's less than $\sqrt{3}$. Thus, the probability p can be determined as a ratio between the diameters of r and R:

$$p = \frac{\frac{1}{2}}{1} = \frac{1}{2}$$

Possibility 3.

Let's fix an arbitrary point A on the boundary of R. When a line L is thrown, we can rotate R until A belongs to L. Let's also draw a tangent line L_1 to R at A (see figure II.2c). From the figure we can see that the angle γ between lines L and L_1 determines the outcome of the trial. If γ is between $\pi/3$ and $2\pi/3$ then the midpoint of L will be inside r and therefore l will be at least $\sqrt{3}$. On the other hand, if $0 \leq \gamma < \pi/3$ or $2\pi/3 < \gamma \leq \pi$ then l will be smaller than $\sqrt{3}$. Since there are π possible radians for γ and a range of $\frac{1}{3}\pi$ generate lines of length $l \geq \sqrt{3}$, the probability is:

$$p - \frac{\frac{1}{3}\pi}{\pi} - \frac{1}{3}$$

Which of these three possibilities is correct? It seems that all three cases are reasonable and all three values found for p make sense! Yet they're all different! How can this be?

Still, all of these possible solutions are correct. It just depends on how you model the possibilities for how you'll generate a random line. The reason is that each model makes implicit assumptions about the way in which a line is "thrown" at the circle. Can you make them explicit in each case? We'll give you the answer for the first possibility: the assumption is that the midpoint of the line L can occur at any interior point of R, and that no points or regions are favored over any others. If you can make the assumptions for the second and third possibilities explicit, you'll see why the answers to the question differ in each case. If you can't, you should take away a healthy respect for models, because even in this very simple situation, an implicit assumption can lead to drastically different results. Just imagine what sort of trouble you can get into when things are more complicated!

A puzzle from the early sixteenth century (this problem has staying power!) also illustrates the importance of having the right model.

In front of you is a small 3×3 chess board (figure II.3). Two black and two white knights are placed in the corners. Your task is to make the two white knights exchange places with the black knights in as few moves as possible. Of course, the knights have to move the same way they do in the game of chess.

Fig. II.3. Two black (B) and two white (W) knights on a 3 × 3 chess board

This is a great puzzle to start out with a "trial and error" approach — the problem is that you'll find mostly "errors." If you have the right model in mind, however, everything turns out to be straightforward.

If we redraw the board marking all of the squares from 1 to 9 (figure II.4a), we can indicate every possible move for a knight between these squares. For a knight to cycle all the way back to its original position it requires a particular path of eight steps (moves): starting from the top-left square, move to the middle-right square, then (to avoid moving back) continue to bottom-left square, top-middle, etc. (figure II.4b). If we "untie" this path to make a simple loop (figure II.4c), it's easy to see that every knight must follow a particular direction and no knight can pass any other along the way. Since it takes 4 moves for a knight to arrive in its required position, the total number of moves that's necessary and sufficient is 16. The order of the moves is also apparent: we move each knight one step at a time and repeat this sequence of four moves, four times.

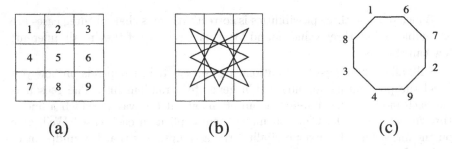

(a)   (b)   (c)

Fig. II.4. (a) The board and numbers of the squares. **(b)** The path (loop) of possible moves. **(c)** The untied loop

There's also another version of this problem where any number of successive moves by the same piece is considered as one move. In that case seven moves are required. Can you see how?

2. Basic Concepts

If we can really understand the problem,
the answer will come out of it,
because the answer is not separate
from the problem.

Krishnamurti, *The Penguin Krishnamurti Reader*

Three basic concepts are common to every algorithmic approach to problem solving. Regardless of the technique that you employ, you'll need to specify: (1) the representation, (2) the objective, and (3) the evaluation function. The representation encodes alternative candidate solutions for manipulation, the objective describes the purpose to be fulfilled, and the evaluation function returns a specific value that indicates the quality of any particular solution given the representation (or minimally, it allows for a comparison of the quality of two alternative solutions). Let's consider each of these aspects of problem solving in turn.

2.1 Representation

The three problems that we examined in the previous chapter provide a good basis for examining alternative representations. Starting with the satisfiability problem (SAT) where we have n variables that are logical bits, one obvious way to represent a candidate solution is a binary string of length n. Each element in the string corresponds to a variable in the problem. The size of the resulting search space under this representation is 2^n, as there are 2^n different binary strings of length n. Every point in this search space is a feasible solution.

For the n-city traveling salesman problem (TSP), we already considered one possible representation: a permutation of the natural numbers $1, ..., n$ where each number corresponds to a city to be visited in sequence. Under this representation, the search space S consists of all possible permutations, and there are $n!$ such orderings. However, if we're concerned with a symmetric TSP, where the cost of traveling from city i to j is the same in either direction, then we don't care if we proceed down the list of cities from right to left, or left to right. Either way, the tour is the same. This means that we can shrink the size of the search space by one-half. Furthermore, the circuit would be the same regardless of which city we started with, so this reduces the size of the search space by factor of n. Consequently, the real size of the search space reduces to $(n-1)!/2$, but we've seen that this is a huge number for even moderately large TSPs.

For a nonlinear programming problem (NLP), the search space consists of all real numbers in n dimensions. We normally rely on floating-point representations to approximate the real numbers on a computer, using either single or double precision. We saw in the previous chapter that for six digits of precision on variables in a range from 0 to 10 there are 10^{7n} different possible values.

For each problem, the representation of a potential solution and its corresponding interpretation implies the search space and its size. This is an important point to recognize: The size of the search space is not determined by the problem; it's determined by your representation and the manner in which you handle this encoding.

Choosing the right search space is of paramount importance. If you don't start by selecting the correct domain to begin your search, you can either add numerous unviable or duplicate solutions to the list of possibilities, or you might actually preclude yourself from ever being able to find the right answer at all.

Here's a good example. There are six matches on a table and the task is to construct four equilateral triangles where the length of each side is equal to the length of a match. It's easy to construct two such triangles using five matches (see figure 2.1), but it's difficult to extend it further into four triangles, especially since only one match remains. Here, the formulation of the problem is misleading because it suggests a two-dimensional search space (i.e., the matches were placed on a table). But to find a solution requires moving into three dimensions (figure 2.2). If you start with the wrong search space, you'll never find the right answer.[1]

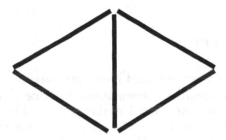

Fig. 2.1. Two triangles, five matches used, one match left

2.2 The objective

Once you've defined the search space, you have to decide what you're looking for. What's the objective of your problem? This is a mathematical statement of the task to be achieved. It's not a function, but rather an expression. For

---

[1]If you give this problem to another person, you can experiment by giving them some additional "hints" like: you can break the matches into smaller pieces. Such "hints" are *harmful* in the sense that (1) they don't contribute anything towards obtaining the solution, and (2) they expand the search space in a significant way! But don't try this on your friends.

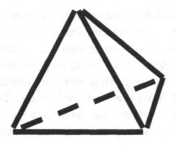

Fig. 2.2. Four triangles, six matches

example, consider the TSP. The objective is typically to minimize the total distance traversed by the salesman subject to the constraint of visiting each city once and only once and returning to the starting city. In mathematical terms, the objective is:

$$\min \sum dist(x, y).$$

For the SAT, the objective is to find the vector of bits such that the compound Boolean statement is satisfied (i.e., made TRUE). For the NLP, the objective is typically to minimize or maximize some nonlinear function of candidate solutions.

2.3 The evaluation function

The objective, however, is not the same thing as the evaluation function. The latter is most typically a mapping from the space of possible candidate solutions under the chosen representation to a set of numbers (e.g., the reals), where each element from the space of possible solutions is assigned a numeric value that indicates its quality. The evaluation function allows you to compare the worth of alternative solutions to your problem as you've modeled it. Some evaluation functions allow for discerning a ranking of all possible solutions. This is called an *ordinal* evaluation function. Alternatively, other evaluation functions are *numeric* and tell you not only the order of the solutions in terms of their perceived quality, but also the degree of that quality.

For example, suppose we wanted to discover a good solution to a TSP. The objective is to minimize the sum of the distances between each of the cities along the route while satisfying the problem's constraints. One evaluation function might map each tour to its corresponding total distance. We could then compare alternative routes and not only say that one is better than another, but exactly how much better. On the other hand, it's sometimes computationally expensive to calculate the exact quality of a particular solution, and it might only be necessary to know approximately how good or bad a solution is, or simply whether or not it compares favorably with some other choice. In such a case, the evaluation function might operate on two candidate solutions and merely return an indication as to which solution was favored.

In every real-world problem, you choose the evaluation function; it isn't given to you with the problem. How should you go about making this choice? One obvious criterion to satisfy is that when a solution meets the objective completely it also should have the best evaluation. The evaluation function shouldn't indicate that a solution that fails to meet the objective is better than one that meets the objective. But this is only a rudimentary beginning to designing the evaluation function.

Often, the objective suggests a particular evaluation function. For example, just above, we considered using the distance of a candidate solution to a TSP as the evaluation function. This corresponds to the objective of minimizing the total distance covered: the evaluation function is suggested directly from the objective. This often occurs in an NLP as well. For the problem of maximizing the function depicted in figure 1.2, the function itself can serve as an evaluation function with better solutions yielding larger values. But you can't always derive useful evaluation functions from the objective. For the SAT problem, where the objective is to satisfy the given compound Boolean statement (i.e., make it TRUE), every approximate solution is FALSE, and this doesn't give you any useful information on how to improve one candidate solution into another, or how to search for any better alternative. In these cases, we have to be more clever and adopt some surrogate evaluation function that will be appropriate for the task at hand, the representation we choose, and the operators that we employ to go from one solution to the next.

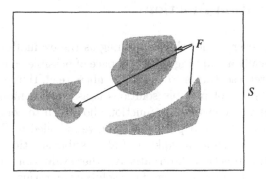

Fig. 2.3. A search space S and its feasible part $\mathcal{F}$. Note that the feasible regions may be disjoint.

When you design the evaluation function, you also have to consider that for many problems, the only solutions of interest will be in a subset of the search space S. You're only interested in *feasible* solutions, i.e., solutions that satisfy problem-specific constraints. The NLP of figure 1.2 constitutes a perfect example: we're only interested in floating-point vectors such that the product of their component values isn't smaller than 0.75 and their total doesn't exceed the product of 7.5 and the number of variables. In other words, the search space S for the NLP can be defined as the set of all floating-point vectors such that their coordinates fall between 0 and 10. Only a subset of these vectors satisfies the additional constraints (concerning the product and the sum of components). This subset constitutes the feasible part of the search space $\mathcal{F}$. In the remainder of the book, we will always distinguish between the search space S and the feasible search space $\mathcal{F} \subseteq S$ (see figure 2.3). Note, however, that for the SAT and TSP, $S = \mathcal{F}$; all of the points from the search space are feasible (presuming for the TSP that there is a path from every city to every other city).

2.4 Defining a search problem

We can now define a search problem.[2] Given a search space S together with its feasible part $\mathcal{F} \subseteq S$, find $x \in \mathcal{F}$ such that

$$eval(x) \leq eval(y),$$

for all $y \in \mathcal{F}$. Note that here we're using an evaluation function for which solutions that return smaller values are considered better (i.e., the problem is one of minimization), but for *any* problem, we could just as easily use an evaluation function for which larger values are favored, turning the search problem into one of maximization. There is nothing inherent in the original problem or an approximate model that demands that the evaluation function be set up for minimization or maximization. Indeed, the objective itself doesn't appear anywhere in the search problem defined above. This problem statement could just as easily be used to describe a TSP, a SAT, or an NLP. The search itself doesn't know what problem you are solving! All it "knows" is the information that you provide in the evaluation function, the representation that you use, and the manner in which you sample possible solutions. If your evaluation function doesn't correspond with the objective, you'll be searching for the right answer to the wrong problem!

The point x that satisfies the above condition is called a *global* solution. Finding such a global solution to a problem might be very difficult. In fact, that's true for all three of the example problems discussed so far (i.e., the SAT, TSP, and NLP), but sometimes finding the best solution is easier when we can concentrate on a relatively small subset of the total (feasible or perhaps also infeasible) search space. Fewer alternatives can sometimes make your life easier. This is a fundamental observation that underlies many search techniques.

---

[2]The terms "search problem" and "optimization problem" are considered synonymous. The search for the best feasible solution is the optimization problem.

2.5 Neighborhoods and local optima

If we concentrate on a region of the search space that's "near" some particular point in that space, we can describe this as looking at the *neighborhood* of that point. Graphically, consider some abstract search space S (figure 2.4) together with a single point $x \in S$. The intuition is that a neighborhood $N(x)$ of x is a set of all points of the search space S that are close in some measurable sense to the given point x.

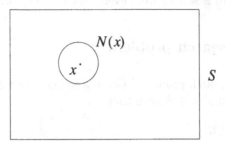

Fig. 2.4. A search space S, a potential solution x, and its neighborhood $N(x)$ denoted by the interior of the circle around x

We can define the nearness between points in the search space in many different ways. Two important possibilities include:

- We can define a distance function *dist* on the search space S:

$$dist : S \times S \longrightarrow \mathbb{R},$$

and then define the neighborhood $N(x)$ as

$$N(x) = \{y \in S \; : \; dist(x, y) \leq \epsilon\},$$

for some $\epsilon \geq 0$. Another way to say this is that if y satisfies the condition for $N(x)$ above then it is in the ϵ-neighborhood of x. Note that when we're dealing with search spaces defined over continuous variables (e.g., the NLP) a natural choice for defining a neighborhood is the Euclidean distance:

$$dist(\mathbf{x}, \mathbf{y}) = \sqrt{\sum_{i=1}^{n}(x_i - y_i)^2}.$$

This is also the case for the SAT problem where the distance between two binary strings can be defined as a Hamming distance, i.e., the number of bit positions with different truth assignments. Be wary, though, that just because something is a "natural" choice doesn't necessarily mean it's the best choice in light of other considerations. Natural choices do, however, often provide a good starting position.

- We can define a mapping m on the search space $\mathcal{S}$ as

$$m : \mathcal{S} \longrightarrow 2^{\mathcal{S}},$$

and this mapping defines a neighborhood for any point $x \in \mathcal{S}$. For example, we can define a *2-swap* mapping for the TSP that generates a new set of potential solutions from any potential solution x. Every solution that can be generated by swapping two cities in a chosen tour can be said to be in the neighborhood of that tour under the 2-swap operator. In particular, a solution x (a permutation of $n = 20$ cities):

$$15 - 3 - 11 - 19 - 17 - 2 - \ldots - 6,$$

has $\frac{n(n-1)}{2}$ neighbors. These include:

$15 - 17 - 11 - 19 - 3 - 2 - \ldots - 6$ (swapping cities at the second and fifth locations),

$2 - 3 - 11 - 19 - 17 - 15 - \ldots - 6$ (swapping cities at the first and sixth locations),

$15 - 3 - 6 - 19 - 17 - 2 - \ldots - 11$ (swapping cities at the third and last locations),

etc.

In contrast, for the SAT problem we could define a *1-flip* mapping that generates a set of potential solutions from any other potential solution x. These are all the solutions that can be reached by flipping a single bit in a particular binary vector, and these would be in the neighborhood of x under the 1-flip operator. For example, a solution x (a binary string of, say, $n = 20$ variables):

01101010001000011111

has n neighbors. These include:

11101010001000011111 (flipping the first bit),
00101010001000011111 (flipping the second bit),
01001010001000011111 (flipping the third bit),
etc.

In this case we could also define the neighborhood in terms of a distance function: the neighborhood contains all of the strings with a Hamming distance from x that is less than or equal to one.

With the notion of a "neighborhood" defined, we can now discuss the concept of a *local* optimum. A potential solution $x \in \mathcal{F}$ is a local optimum with respect to the neighborhood N, if and only if

$$eval(x) \leq eval(y),$$

for all $y \in N(x)$ (again assuming a minimization criterion). It's often relatively easy to check whether or not a given solution is a local optimum when the size of the neighborhood is very small, or when we know something about the derivative of the evaluation function (if it exists).

Many search methods are based on the statistics of the neighborhood around a given point; that is, the sequence of points that these techniques generate while searching for the best possible solution relies on *local* information at each step along the way. These techniques are designed to locate solutions within a neighborhood of the current point that have better corresponding evaluations. Appropriately, they are known as "neighborhood" or "local" search strategies [476]. For example, suppose you're facing the problem of maximizing the function $f(x) = -x^2$ and you've chosen to use $f(x)$ itself as an evaluation function. Better solutions generate higher values of $f(x)$. Say your current best solution x is 2.0, which has a worth of -4.0. You might define an ϵ-neighborhood with an interval of 0.1 on either side of 2.0 and then search within that range. If you sampled a new point in [1.9, 2.1] and found that it had a better evaluation than 2.0, you'd replace 2.0 with this better solution and continue on from there; otherwise, you'd discard the new solution and take another sample from [1.9, 2.1].

Naturally, most evaluation functions in real-world problems don't look like a quadratic bowl such as $f(x) = -x^2$. The situation is almost always more difficult than this. The evaluation function defines a response surface that is much like a topography of hills and valleys (e.g., as shown in figure 1.2), and the problem of finding the best solution is similar to searching for a peak on a mountain range while walking in a dense fog. You can only sample new points in your immediate vicinity and you can only make local decisions about where to walk next. If you always walk uphill, you'll eventually reach a peak, but this might not be the highest peak in the mountain range. It might be just a "local optimum." You might have to walk downhill for some period of time in order to find a position such that a series of local decisions within successive neighborhoods leads to the global peak.

Local search methods (chapter 3) present an interesting trade-off between the size of the neighborhood $N(x)$ and the efficiency of the search. If the size of the neighborhood is relatively small then the algorithm may be able to search the entire neighborhood quickly. Only a few potential solutions may have to be evaluated before a decision is made on which new solution should be considered next. However, such a small range of visibility increases the chances of becoming trapped in a local optimum! This suggests using large neighborhoods: a larger range of visibility could help in making better decisions. In particular, if the visibility were unrestricted (i.e., the size of the neighborhood were the same as the size of the whole search space), then eventually we'd find the best series of steps to take. The number of evaluations, however, might become enormous, so we might not be able to complete the required calculations within the lifetime of the universe (as noted in chapter 1). When using local search methods, the appropriate size of the neighborhood can't be determined arbitrarily. It has to fit the task at hand.

2.6 Hill-climbing methods

Let's examine a basic *hill-climbing* procedure and its connection with the concept of a neighborhood. Hill-climbing methods, just like all local search methods (chapter 3), use an iterative improvement technique.[3] The technique is applied to a single point — the current point — in the search space. During each iteration, a new point is selected from the neighborhood of the current point. If that new point provides a better value in light of the evaluation function, the new point becomes the current point. Otherwise, some other neighbor is selected and tested against the current point. The method terminates if no further improvement is possible, or we run out of time or patience.

It's clear that such hill-climbing methods can only provide locally optimum values, and these values depend on the selection of the starting point. Moreover, there's no general procedure for bounding the relative error with respect to the global optimum because it remains unknown. Given the problem of converging on only locally optimal solutions, we often have to start hill-climbing methods from a large variety of different starting points. The hope is that at least some of these initial locations will have a path that leads to the global optimum. We might choose the initial points at random, or on some grid or regular pattern, or even in the light of other information that's available, perhaps as a result of some prior search (e.g., based on some effort someone else made to solve the same problem).

There are a few versions of hill-climbing algorithms. They differ mainly in the way a new solution is selected for comparison with the current solution. One version of a simple iterated hill-climbing algorithm is given in figure 2.5 (steepest ascent hill-climbing). Initially, all possible neighbors of the current solution are considered, and the one $\mathbf{v}_n$ that returns the best value $eval(\mathbf{v}_n)$ is selected to compete with the current string $\mathbf{v}_c$. If $eval(\mathbf{v}_c)$ is worse than $eval(\mathbf{v}_n)$, then the new string $\mathbf{v}_n$ becomes the current string. Otherwise, no local improvement is possible: the algorithm has reached a local or global optimum ($local$ = TRUE). In such a case, the next iteration ($t \leftarrow t + 1$) of the algorithm is executed with a new current string selected at random.

The success or failure of a single iteration (i.e., one complete climb) of the above hill-climbing algorithm is determined completely by the initial point. For problems with many local optima, particularly those where these optima have large *basins of attraction*, it's often very difficult to locate a globally optimal solution.

---

[3]The term *hill-climbing* implies a maximization problem, but the equivalent *descent* method is easily envisioned for minimization problems. For convenience, the term hill-climbing will be used here to describe both methods without any implied loss of generality.

procedure iterated hill-climber
begin
   $t \leftarrow 0$
   initialize *best*
   repeat
      local $\leftarrow$ FALSE
      select a current point $\mathbf{v}_c$ at random
      evaluate $\mathbf{v}_c$
      repeat
         select all new points in the neighborhood of $\mathbf{v}_c$
         select the point $\mathbf{v}_n$ from the set of new points
            with the best value of evaluation function *eval*
         if $eval(\mathbf{v}_n)$ is better than $eval(\mathbf{v}_c)$
            then $\mathbf{v}_c \leftarrow \mathbf{v}_n$
            else *local* $\leftarrow$ TRUE
      until *local*
      $t \leftarrow t + 1$
      if $\mathbf{v}_c$ is better than *best*
         then *best* $\leftarrow \mathbf{v}_c$
   until $t = MAX$
end

Fig. 2.5. A simple iterated hill-climber

Hill-climbing algorithms have several weaknesses:

- They usually terminate at solutions that are only locally optimal.

- There is no information as to the amount by which the discovered local optimum deviates from the global optimum, or perhaps even other local optima.

- The optimum that's obtained depends on the initial configuration.

- In general, it is *not* possible to provide an upper bound for the computation time.

On the other hand, there is one alluring advantage for hill-climbing techniques: they're very easy to apply! All that's needed is the representation, the evaluation function, and a measure that defines the neighborhood around a given solution.

    Effective search techniques provide a mechanism for balancing two apparently conflicting objectives: *exploiting* the best solutions found so far and at the same time *exploring* the search space.[4] Hill-climbing techniques exploit the best available solution for possible improvement but neglect exploring a large

---

[4]This balance between exploration and exploitation was noted as early as the 1950s by the famous statistician G.E.P. Box [54].

portion of the search space S. In contrast, a random search — where points are sampled from S with equal probabilities — explores the search space thoroughly but foregoes exploiting promising regions of the space. Each search space is different and even identical spaces can appear very different under different representations and evaluation functions. So there's no way to choose a single search method that can serve well in every case. In fact, this can be proved mathematically [499, 156].

Getting stuck in local optima is a serious problem. It's one of the main deficiencies that plague industrial applications of numerical optimization. Almost every solution to real-world problems in factory scheduling, demand forecasting, land management, and so forth, is at best only locally optimal.

What can we do? How can we design a search algorithm that has a chance to escape local optima, to balance exploration and exploitation, and to make the search independent from the initial configuration? There are a few possibilities, and we'll discuss some of them in chapter 5, but keep in mind that the proper choices are always problem dependent. One option, as we discussed earlier, is to execute the chosen search algorithm for a large number of initial configurations. Moreover, it's often possible to use the results of previous trials to improve the choice of the initial configuration for the next trial. It might also be worthwhile to introduce a more complex means for generating new solutions, or enlarge the neighborhood size. It's also possible to modify the criteria for accepting transitions to new points that correspond with a *negative* change in the evaluation function. That is, we might want to accept a worse solution from the local neighborhood in the hope that it will eventually lead to something better.

But before we move to more sophisticated search methods, we need to review the basic, classic problem-solving techniques, such as dynamic programming, the A^* algorithm, and a few others. Before we do that, you get another chance to display your prowess by solving an interesting problem in geometry.

2.7 Can you sink this trick shot?

Now that you've seen a number of different search strategies and have been introduced to the issues of choosing a representation, the objective, and an evaluation function, let's see if you can solve this challenging puzzle. Remember, you might have to rely on some information that we presented in an earlier section — or maybe not! We're not telling!

> A single ball rests calmly on a square-shaped billiard table. The ball is hit and it moves on the table for an infinite amount of time, obeying an elementary law of physics: The angles α and β (figure 2.6) are always equal. Under what circumstances will the movement of the ball be cyclic?

Please put the book aside and think about it.

One way we might approach the problem is to restrict our attention to the surface of the billiard table and try to reason "within" this square. It's very

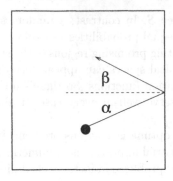

Fig. 2.6. A square-shaped billiard table and the movement of the ball

difficult to see any pattern of the ball's movement there, however, as the number of reflections of the ball might be very high (infinitely high in the limit), and the ball's path may consist of many segments.

Let's "expand" the space for reasoning about the problem. Instead of imagining the ball to reflect off of each side of the billiard table, let's imagine the ball as if it were traveling in a straight line and consider a *reflection of the table!* Figure 2.7 illustrates the concept.

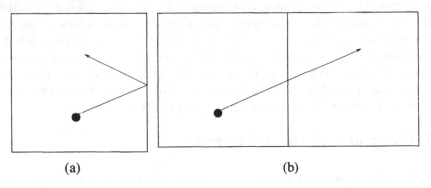

(a)   (b)

Fig. 2.7. (a) Reflection of a ball on the billiard table, and (b) the corresponding reflection of the billiard table

This is a significant departure from the way we solve most problems. We usually think about manipulating objects within a framework. However, it's possible to change the orientation and think about manipulating the framework that surrounds the objects. If you made this leap without cheating and reading ahead, then congratulations, your future is bright!

Each time the ball strikes a wall of the table, we can draw a reflection of the table against that wall. Note that now the ball can only strike the top or right wall, and it will cycle if it returns to its initial position (see figure 2.8).

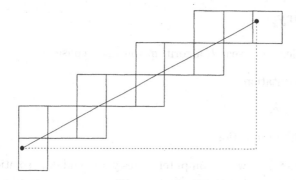

Fig. 2.8. Getting a cycle. The ball returns to its initial position on a subsequent reflection of the billiard table.

Suppose that the ball does cycle. This means that it "traveled" some number of tables "up" (say, p tables) and some number of tables "right" (say, q tables). For angle $\alpha = 0$, $p = 0$. For $\alpha = \pi/2$, $q = 0$. For $0 < \alpha < \pi/2$, $p > 0$ and $q > 0$.

With this in mind we can answer the question. The ball will cycle on the billiard table if and only if either $\alpha = \pi/2$ or the tangent of the angle α is a rational number (i.e., $\tan(\alpha) = p/q$ for $0 \le \alpha < \pi/2$).

In the first case this is clear: The ball will just move up and down forever. In the second case, if it's not clear, then remember that in a right triangle, the tangent of an acute angle is equal to the length of the opposite side over the adjacent side. Take a look at figure 2.8 and imagine making the path of the ball corresponding to the hypotenuse of a right triangle. The base is eight tables long and the height is four tables. To reach the identical point on some future reflected table, the ball will have had to travel some whole, even number of tables up and right (the numbers must be even to preserve the ball's original angle of motion). Thus the tangent of the angle α must be a rational number. And in the case that α is equal to zero, then the ball simply moves horizontally *ad infinitum*.

Remembering "SOHCAHTOA" from high school trigonometry is really the easy part of this exercise.[5] The hard part is to avoid getting trapped in the description of the problem which lures you into using a framework that doesn't help you find the answer. It's all the more challenging to escape this trap because in real life, billiard tables don't move and billiard balls do! But in solving problems, you have to learn to "think outside the box" — or outside the billiard table.

---

[5]SOHCAHTOA is a common mnemonic device for remembering that, in a right triangle, the Sine of an acute angle is the length of the Opposite side over the Hypotenuse, the Cosine is the length of the Adjacent side over the Hypotenuse, and the Tangent is the length of the Opposite side over the Adjacent side.

2.8 Summary

In order to implement a search algorithm you must consider

- The representation.

- The objective.

- The evaluation function.

The representation is how the computer stores your candidate solutions and the objects that it manipulates in searching for new ones. The objective is what you are trying to accomplish. The evaluation function is a means for measuring the quality of any potential solution to the problem. Taken together, these three elements form a strategic troika. You have to give careful consideration to each facet.

Sometimes it's useful to look at a search space in geometric terms, thinking about the neighborhood that surrounds a particular solution and whether or not we can find better solutions within that neighborhood. If we can't, then we've discovered a local optimum. It might be that this solution is really a global optimum; that is, no matter what size neighborhood we'd pick, the solution would still be the best. It's often too computationally expensive to expand the search beyond small neighborhoods. As a result, we face a danger of getting trapped in local optima, particularly when using a hill-climbing method.

Finally, when framing a problem, don't be constrained by the physics of the real world. Allow yourself the freedom to ask questions like: what if it was the billiard table that was moving relative to the ball? What if I went outside the plane of the table to construct a pyramid with these match sticks?

III. What Are the Prices in 7–11?

There are many ways to search for the answer to a problem. To be effective, particularly when the domain of possible solutions is very large, you have to organize the search in a systematic way. What's more, the structure of your search has to in some way "match" the structure of the problem. If it doesn't, then your search may perform even worse than random guesswork.

Some of the heuristics for searching the spaces of solutions involve pruning out those cases that can't possibly be correct. In this way, the size of the total space is reduced and with luck, you'll be able to more quickly zoom in on feasible solutions. So think about how you might go about eliminating the impossibilities in this perplexing case of prices in the so-called "7–11" problem, which goes like this.

There's a chain of stores in the United States called 7–11. They're probably called this because they used to be open from 7 a.m. until 11 p.m., but now they're usually open all the time. One day a customer arrived in one of these 7–11 shops and selected four items. He then approached the counter to pay for these items. The salesman took his calculator, pressed a few buttons and said, "The total price is $7.11."
The customer tried a joke, "Why? Do I have to pay $7.11 because the name of your shop is 7–11?"
The salesman didn't get the joke and answered, "Of course not! I have multiplied the prices of these four items and I have just given you the result!"
The customer was very surprised. "Why did you multiply these numbers? You should have added them to get the total price!"
The salesman said, "Oh, yes, I'm sorry, I have a terrible headache and I pressed the wrong button!"
Then the salesman repeated all the calculations, i.e., he added prices of these four items, but to his and customer's great surprise, the total was still $7.11.
Now, the task is to find the prices of these four items!

This is a typical search problem. We have four variables, x, y, z, and t, that represent the prices of four items. Each variable takes a value from its domain, and in this case every domain is the same: from 0.01 up to 7.08. The smallest price might be 0.01 and since the total price for all four items is 7.11, the largest possible price of any single item is 7.08. Thus, the domain of each variable is the set

$$\{0.01, 0.02, 0.03, \ldots, 7.06, 7.07, 7.08\}.$$

To avoid dealing with fractional numbers, let's change its formulation into integers. We'll talk about prices in cents rather than in dollars and cents. In that way the domain of each variable is the set of integer numbers

$$P = \{i \in N \ : \ 1 \leq i \leq 708\}.$$

The problem is then to find x, y, z, and t, all from the set P, such that

$$x + y + z + t = 711, \text{ and}$$
$$xyzt = 711000000.$$

Note that during the conversion from dollars and cents into just cents, each variable was multiplied by 100. Thus their product 7.11 increased 10^8 times, resulting in $7.11 \cdot 10^8 = 711000000$.

As $711000000 = 2^6 \cdot 3^2 \cdot 5^6 \cdot 79$, there are many cases to consider. Exhaustively searching every possible combination would give us a solution, but it would also take some time — a great deal of time. Let's investigate some shortcuts that are similar to what's called a *branch and bound* approach.

One thing to think about here, particularly because we have two equations where the second one is a product of terms, is the idea of factoring. We can factor 711000000 into its prime factors. This might help because the variables in question would have to be multiples of those factors. In this case, 79 is the largest prime factor of 711000000. It's clear then that the price (in cents, of course) of one of the items (say, x), is divisible by 79.

Let's consider a few cases that revolve around 79. We choose 79 because, as the largest prime factor of 711000000, it means we have to consider the minimum number of cases.

• If $x = 79$, the problem for the remaining three variables is

$$y + z + t = 632, \text{ and}$$
$$yzt = 2^6 \cdot 3^2 \cdot 5^6.$$

Here, it's impossible that all of variables y, z, and t are divisible by 5 (as their total is not), so the product of two of the variables (say, y and z) is divisible by $5^6 = 15625$. Now we have several cases:

Case (a): $y = 15625y'$
Case (b): $y = 3125y'$ and $z = 5z'$
Case (c): $y = 625y'$ and $z = 25z'$
Case (d): $y = 125y'$ and $z = 125z'$

for some $y' \geq 1$ and $z' \geq 1$. For cases (a)–(c), $y + z > 632$ (which is impossible), so the only case we really have to consider is case (d). In this case, the problem becomes:

$$125(y' + z') + t = 632, \text{ and}$$
$$y' \cdot z' \cdot t = 2^6 \cdot 3^2.$$

The first equation implies that $y' + z' \leq 5$, so all we have to do now is consider six additional possibilities because we can assume that $y' \leq z'$:

Case (d1): $y' = 1$ and $z' = 1$
Case (d2): $y' = 1$ and $z' = 2$
Case (d3): $y' = 1$ and $z' = 3$
Case (d4): $y' = 1$ and $z' = 4$
Case (d5): $y' = 2$ and $z' = 2$
Case (d6): $y' = 2$ and $z' = 3$.

None of these six cases gives a solution. For example, for case (d1) we have

$x = 79$,
$y = 125 \cdot y' = 125$, and
$z = 125 \cdot z' = 125$,

which requires $t = 382$ (so four variables would sum to 711), but then their product wouldn't be equal to 711000000.

• If $x = 2 \cdot 79 = 158$, the problem is

$y + z + t = 553$, and
$yzt = 2^5 \cdot 3^2 \cdot 5^6$.

In this case we can repeat the same reasoning as we did with the previous case: y must be $125y'$ and $z = 125z'$, and

$$125(y' + z') + t = 553,$$

thus $y' + z' \leq 4$. None of the possibilities

$y' = 1$ and $z' = 1$
$y' = 1$ and $z' = 2$
$y' = 1$ and $z' = 3$
$y' = 2$ and $z' = 2$

leads to a solution either though.

It's an easy exercise to see that if

• $x = 3 \cdot 79 = 237$
• $x = 5 \cdot 79 = 395$
• $x = 6 \cdot 79 = 474$
• $x = 7 \cdot 79 = 553$
• $x = 8 \cdot 79 = 632$,

the problem has no solutions. In each of these cases, the number of subcases to consider shrinks. For example, if $x = 553$, then for $y = 125y'$ and $z = 125z'$ we have $125(y' + z') + t = 158$, which gives an immediate contradiction because $y' + z' \geq 2$).

If the problem actually has a solution, then $x = 316$. In this (final) case, the problem becomes

$$y + z + t = 395, \text{ and}$$
$$yzt = 2^4 \cdot 3^2 \cdot 5^6.$$

Again, we have to consider a few cases. Since the total of the three remaining variables is divisible by 5, either one or all three variables are divisible by 5. If only one variable (say, y) is divisible by 5, then $y = 5^6 y'$ and we get a contradiction, $y > 395$. Thus, we know that all of the variables y, z, and t are divisible by 5:

$$y = 5y', \ z = 5z', \text{ and } t = 5t'.$$

Now the problem becomes

$$y' + z' + t' = 79, \text{ and}$$
$$y'z't' = 2^4 \cdot 3^2 \cdot 5^3.$$

As $y' + z' + t' = 79$, one of these primed variables is not divisible by 5, so some other variable must be divisible by 25, say, $y' = 25y''$. In that case we have

$$25y'' + z' + t' = 79, \text{ and}$$
$$y''z't' = 2^4 \cdot 3^2 \cdot 5 = 720.$$

This means that y'' is either 1, or 2, or 3 (otherwise $25y'' > 79$). The second two choices don't lead to solution (e.g., if $y'' = 2$, $z' + t' = 29$ and $z't' = 360$, which doesn't have solution in the real numbers, and things are "complex" enough, aren't they?). Thus $y'' = 1$ (i.e., $y = 125$). In that case we get

$$z' + t' = 54, \text{ and}$$
$$z't' = 720,$$

which gives us $z' = 24$ and $t' = 30$, or the other way around. Thus $z = 5 \cdot 24 = 120$ and $t = 5 \cdot 30 = 150$.

That solves the problem. Indeed, it has a unique solution (we've converted the prices back into dollars and cents):

$$x = \$3.16, \ y = \$1.25, \ z = \$1.20, \text{ and } t = \$1.50.$$

It's important to point out the systematic approach to solving this problem. Step by step, we eliminated the impossible solutions and even more importantly, we started by pruning the space of the solution dramatically by examining the prime factors of the product of the prices.

Of course, it's possible to use a different approach. It's also possible to do some guesswork. Usually some guesswork is desirable as it may provide insight into a problem. We'll see this more clearly in the next chapter. It's also worthwhile to observe that simply guessing may lead to some amazing "discoveries," like getting a set of prices

($3.16, $1.25, $1.25, $1.44) or ($2.37, $2.00, $2.00, $0.75).

For both sets, the product is $7.11, however, the former set results in a sum of $7.10, while for the latter, the sum is $7.12. The problem with guessing is that it doesn't often lead you to the next solution, which is hopefully better than the last. The concept of generating improvements in the quality of a solution is intrinsic to a great many classic and modern methods of optimization. And hey, it's comforting to know that things are getting better!

3. Traditional Methods — Part 1

> ... are all alike in their promises.
> It is only in their deeds that they differ.
>
> Molière, *The Miser*

There are many classic algorithms that are designed to search spaces for an optimum solution. In fact, there are so many algorithms that it's natural to wonder why there's such a plethora of choices. The sad answer is that none of these traditional methods is robust. Every time the problem changes you have to change the algorithm. This is one of the primary shortcomings of the well-established optimization techniques. There's a method for every problem — the problem is that most people only know one method, or maybe a few. So they often get stuck using the wrong tool to attack their problems and consequently generate poor results.

The classic methods of optimization can be very effective when appropriate to the task at hand. It pays for you to know when and when not to use each one. Broadly, they fall into two disjoint classes:

- Algorithms that only evaluate complete solutions.

- Algorithms that require the evaluation of partially constructed or approximate solutions.

This is an important difference. Whenever an algorithm treats complete solutions, you can stop it at any time and you'll always have at least one potential answer that you can try. In contrast, if you interrupt an algorithm that works with partial solutions, you might not be able to use any of its results at all.

Complete solutions mean just that: all of the decision variables are specified. For example, binary strings of length n constitute complete solutions for an n-variable SAT. Permutations of n cities constitute complete solutions for a TSP. Vectors of n floating-point numbers constitute complete solutions for an NLP. We can easily compare two complete solutions using an evaluation function. Many algorithms rely on such comparisons, manipulating one single complete solution at a time. When a new solution is found that has a better evaluation than the previous best solution, it replaces that prior solution. Examples include exhaustive search, local search, hill climbing, and gradient-based numerical optimization methods. Some modern heuristic methods such as simulated annealing, tabu search, and evolutionary algorithms also fall into this class as well (see chapters 5 and 6).

Partial solutions, on the other hand, come in two forms: (1) an incomplete solution to the problem originally posed, and (2) a complete solution to a reduced (i.e., simpler) problem.

Incomplete solutions reside in a subset of the original problem's search space. For example, in an SAT, we might consider all of the binary strings where the first two variables were assigned the value 1 (i.e., TRUE). Or in the TSP, we might consider every permutation of cities that contains the sequence $7 - 11 - 2 - 16$. Similarly, for an NLP, we might consider all of the solutions that have $x_3 = 12.0129972$. In each case, we fix our attention on a subset of the search space that has a particular property. Hopefully, that property is also shared by the real solution!

Alternatively, we can often decompose the original problem into a set of smaller and simpler problems. The hope is that in solving each of these easier problems, we can eventually combine the partial solutions to get an answer for the original problem. In a TSP, we might only consider k out of n cities and try to establish the shortest path from city i to city j that passes through all k of these cities (see section 4.3). Or, in the NLP, we might limit the domains of some variables x_i, thereby reducing the size of the search space significantly, and then search for a complete solution within that restricted domain. Such partial solutions can sometimes serve as building blocks for the solution to the original problem.

Making a complex problem simpler by dividing it up into little pieces that are easy to manage sounds enticing. But algorithms that work on partial solutions pose additional difficulties. You have to: (1) devise a way to organize the subspaces so that they can be searched efficiently, and (2) create a new evaluation function that can assess the quality of partial solutions. This latter task is particularly challenging in real-world problems: what is the value of a remote control channel changer? The answer depends on whether or not there is a television to go along with it! The utility of the remote control device depends heavily on the presence or absence of other items (e.g., a television, batteries, perhaps a cable connection). But when you evaluate a partial solution you intrinsically cannot know what the other external conditions will be like. At best, you can try to estimate them, but you could always be wrong.

The former task of organizing the search space into subsets that can be searched efficiently is also formidable, but many useful ideas have been offered. You might organize the search space into a tree, where complete solutions reside at the leaves of the tree. You could then progress down each branch in the tree, constructing the complete solution in a series of steps. You might be able to eliminate many branches where you know the final solutions will be poor. And there are other ways to accomplish this sort of "branch and bound" procedure. Each depends on how you arrange the solutions in the search space, and that in turn depends on your representation.

Remember, there's never only one way to represent different solutions. We've seen three common representations for the SAT, TSP, and NLP: (1) binary strings, (2) permutations, and (3) floating-point vectors, respectively. But we don't have to limit ourselves to these choices. We could, for example, represent solutions to an NLP in binary strings. Each solution would then be of length kn, where n still represents the number of variables in the problem, and k indicates the number of bits used to describe a value for each variable. It's straightforward to map a binary string $\langle b_{k-1} \ldots b_0 \rangle$ into a floating-point value x_i from some range $[l_i, u_i]$ in two steps:

- Convert the binary string $\langle b_{k-1} \ldots b_0 \rangle$ from base 2 to base 10:

$$(\langle b_{k-1} \ldots b_0 \rangle)_2 = (\textstyle\sum_{i=0}^{k-1} b_i \cdot 2^i)_{10} = x'.$$

- Find a corresponding floating-point value x within the required range:

$$x = l_i + x' \cdot \frac{u_i - l_i}{2^k - 1}.$$

For example, if the range were $[-2, 3]$ and we chose to use binary strings of length 10, then the string $\langle 1001010001 \rangle$ would map to 0.89833822 as:
$$0.89833822 = -2 + 593 \cdot \frac{5}{1023}.$$
Note that as we increase the value of k, we generate more precision in approximating a particular floating-point value. Here, with 10 bits, the successor to $\langle 1001010001 \rangle$ is $\langle 1001010010 \rangle$, which maps to 0.90322581 — the gap between floating-point numbers is almost 0.005. If we needed more precision, we'd have to increase the value of k. This would also increase the computation time because we'd have to decode longer binary strings each time we wanted to determine the value of a particular variable in the NLP. But there's nothing that forces us to solve NLPs with floating-point values directly, even though that's the domain over which they're defined.

Actually, this freedom to choose the representation allows you to become really creative. If we're solving a TSP, we might use the straightforward permutation representation we've discussed earlier. But even then, we might consider interpreting that permutation in a little different manner. For instance, suppose the i-th element of the permutation is a number in the range from 1 to $n - i + 1$ and indicates a city from the remaining part of some reference list of cities C for a given TSP. Say $n = 9$ and the reference list C is

$$C = (1 \ 2 \ 3 \ 4 \ 5 \ 6 \ 7 \ 8 \ 9).$$

Then the following tour,

$$5 - 1 - 3 - 2 - 8 - 4 - 9 - 6 - 7,$$

is represented as a vector,

$$(5 \ 1 \ 2 \ 1 \ 4 \ 1 \ 3 \ 1 \ 1).$$

This is interpreted as follows: (1) take the fifth city from the current reference list C and remove it (this would be city 5); (2) take the first city from the current reference list C and remove it (this is city 1); (3) take the second city from the current reference list C and remove it (this is now city 3); and so on.

This representation seems to be only remotely connected to the problem itself, but you can indeed use it to identify every possible tour, and perhaps it could offer some interesting possibilities when it comes to organizing the search space that wouldn't be found in the typical permutation representation (see chapter 8 for a discussion on other possible representations for the TSP).

There are lots of possible representations even for the SAT. We might represent a solution as a vector of floating-point values from the range $[-1, 1]$, and then assign the condition TRUE to each Boolean variable if its corresponding floating-point variable is non-negative and FALSE otherwise (see, for example, section 9.1 of this book).

Whatever representation you choose for a particular problem, it's possible to consider the implied search space as a set of complete solutions and search through them in a particular manner. You can also divide the space into subsets and work with partial or incomplete solutions. Since there are so many different algorithms to examine, we've divided them up. We'll look at those that work with complete solutions in this chapter and then turn our attention to those that work with partial or incomplete solutions in chapter 4.

3.1 Exhaustive search

As the name implies, exhaustive search checks each and every solution in the search space until the best global solution has been found. That means if you don't know the value that corresponds to the evaluated worth of the best solution, there's no way to be sure that you've found that best solution using exhaustive search unless you examine everything. No wonder it's called *exhaustive*. You'll be tired and old before you get an answer to problems of even modest size! Remember, the size of the search space of real-world problems is usually enormous. It might require centuries or more of computational time to check every possible solution. From section 1.1, we saw that a 50-city TSP generates 10^{62} different possible tours. You'd be waiting a very long time for an exhaustive search of that space!

But exhaustive algorithms (also called *enumerative*) are interesting in some respects. At the very least, they are simple. The only requirement is for you to generate every possible solution to the problem systematically. In addition, there are ways to reduce the amount of work you have to do. One such method is called *backtracking* and we'll take a closer look at that later. What's more, some classic optimization algorithms that construct a complete solution from partial solutions (e.g., branch and bound or the A^* algorithm) are based on an exhaustive search, so let's at least give a little bit more time to this approach.

The basic question to answer when applying an exhaustive search is: How can we generate a sequence of every possible solution to the problem? The order in which the solutions are generated and evaluated is irrelevant because the plan is to evaluate every one of them. The answer for the question depends on the selected representation.

3.1.1 Enumerating the SAT

To enumerate the search space for the SAT with n variables, we have to generate every possible binary string of length n. There are 2^n such strings, from $\langle 0...000 \rangle$ to $\langle 1...111 \rangle$. All of these binary strings correspond to whole numbers in a one-to-one mapping (see table 3.1).

Table 3.1. The correspondence between a binary string of length $n = 4$ and its decimal equivalent

0000	0	0100	4	1000	8	1100	12
0001	1	0101	5	1001	9	1101	13
0010	2	0110	6	1010	10	1110	14
0011	3	0111	7	1011	11	1111	15

All we do is generate the non-negative integers from 0 to $2^n - 1$. We then convert each integer into the matching binary string of length n and the bits of that string are taken as the truth assignments of the decision variables.

Another possibility is to operate on the binary strings directly and add a binary one $\langle 0...001 \rangle$ to get each string in succession. Once a new string is generated, it's evaluated and the string with the best score is kept as the best solution found so far. This still leaves us with the problem of choosing an evaluation function, which could simply be

1, if the string satisfies the compound Boolean statement, and
0, otherwise.

In this case, we could actually halt the search whenever the string at hand had a corresponding evaluation of 1.

Even in this simple case we can take advantage of an opportunity to organize the search and prune the number of alternative strings that we really *need* to examine. Suppose we partition the search space of the SAT into two disjoint subspaces: the first contains all of the strings such that $x_1 = F$ (FALSE), whereas the second contains all of the vectors where $x_1 = T$ (TRUE). We could then organize the entire space as a tree (see figure 3.1).

On the next level down in the tree we again divide the space into two disjoint subspaces based on the assignment to x_2, and continue like this for all n variables. The resulting binary tree has n levels and the i-th variable is assigned on the i-th level.

Having organized the space into a tree like this, we can use an exhaustive search technique that is designed especially for trees: depth-first search. This simple recursive procedure (figure 3.2) traverses a directed graph (a tree, in particular) starting from some node and then searches adjacent nodes recursively.

The advantage of organizing the search in this way is that sometimes we can deduce that the remaining branch below a particular node can lead to nothing

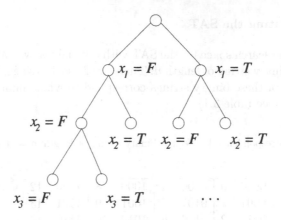

Fig. 3.1. A binary search tree for the SAT

procedure depth-first(v)
begin
    visit v
    for each child w of v **do**
        dept-first(w)
end

Fig. 3.2. Depth-first search

but a dead end. We can then "backtrack" to the parent of this node at the next higher level and proceed further with the search in another branch. For example, it might happen that after assigning truth values to a few variables, the whole SAT formula evaluates to FALSE. In that case, there's no need to go further down that branch. Thus, if the SAT formula contains a clause

$$\ldots \wedge (x_1 \vee \overline{x}_3 \vee x_4) \wedge \ldots,$$

and the assignments of the first four variables are

$x_1 = $ FALSE; $x_2 = $ TRUE; $x_3 = $ TRUE; $x_4 = $ FALSE, or

$x_1 = $ FALSE; $x_2 = $ FALSE; $x_3 = $ TRUE; $x_4 = $ FALSE,

then no matter what we assign to x_5 or any other variable, the whole SAT formula will evaluate to FALSE. It's time to pick another branch.

Depth-first search eliminates parts of the search space by checking whether or not a given node (subspace) can contain a solution. If it can, then the search proceeds further, but if it can't, the search returns to the parent of the given node. This backtracking is a common practice in many tree-based algorithms. We'll see a similar idea later in chapter 4 when discussing branch and bound methods (section 4.4).

3.1.2 Enumerating the TSP

Applying an exhaustive algorithm to the TSP isn't as straightforward. Here the question is again first and foremost: How can we generate all of the possible permutations? The primary difficulty is that unless the TSP is "fully connected," some permutations might not be feasible. City i might not be connected to city j, at least not directly, so every permutation where these two cities are adjacent would be illegal. We might get around this obstacle by allowing these infeasible connections anyway, but then giving any such infeasible solution a very high cost, much higher than the length of any legal tour. In that case, the best solution among all of the permutations would also be the best legal solution, provided that one exists.

The second problem is the simple nuts and bolts of generating all possible permutations of n numbers. Let's consider a few possibilities. We could use an approach that was similar to what we used for the SAT, where there was a mapping between the whole numbers and binary strings. Since there are exactly $n!$ different permutations of n numbers, we can define a function that ranks all permutations from 0 to $n! - 1$ as follows. Note that any integer k between 0 and $n! - 1$ can be uniquely represented by an array of integers:

$c[n], \ldots, c[2]$, where $0 \le c[i] \le i - 1$ (for $i = 2, \ldots, n$), as $k = \sum_{i=2}^{n} c[i] \cdot (i - 1)!$.

To generate a permutation corresponding to k (or an array $c[i]$), we can use the following procedure [426]:

> **for** $i = 1$ **to** n **do**
> $\quad P[i] \leftarrow i$
> **for** $i = n$ **step** -1 **to** 2 **do**
> $\quad$ swap $(P[i], P[c[i] + 1])$

The above method can be used in various ways. In particular, instead of generating permutations for consecutive numbers (from 0 to $n! - 1$), we could generate these integers randomly; however, if we do that then we also have to keep track of the numbers that we've already generated (e.g., by maintaining a binary vector of length $n!$).

Another possibility [425] is to use the following recursive procedure (figure 3.3). The procedure gen1_permutation, when called with k initialized to -1, the parameter i set to 0, and all of the entries of the array P initialized to 0, prints every permutation of $(1, \ldots, n)$. It does this by fixing 1 in the first position and generating the remaining $(n - 1)!$ permutations of numbers from 2 to n. The next $(n - 1)!$ permutations are then listed with 1 in the second position, and so forth. For $n = 3$, the procedure produces the following output: (1 2 3), (1 3 2), (2 1 3), (3 1 2), (2 3 1), (3 2 1). Of course, in real applications, we don't usually print the current permutation but instead do something useful with it (e.g., evaluate the sum of all distances between all of the cities in the permutation).

The problem with the previous methods is that they aren't very practical. If you try any of them for $n > 20$ you'll see why! A more fruitful approach

```
procedure gen1_permutation(i)
begin
    k ← k + 1
    P[i] ← k
    if k = n then
        for q = 1 to n do
            print P[q]
    for q = 1 to n do
        if P[q] = 0 then gen1_permutation(q)
    k ← k − 1
    P[i] ← 0
end
```

Fig. 3.3. A procedure to generate all permutations of integer numbers from 1 to n

relies on transforming one permutation into another new one (this is the easy part) and doesn't repeat any permutation until all of them have been generated (that's the tricky part). The advantage of this approach is that there's no need to keep track of which permutations have already been generated. The current permutation always implies the next one! Many algorithms can perform this task, and you'll find one that prints one permutation at a time in figure 3.4 [80].

Another possibility that is specific to the TSP is to enumerate all possible permutations by successively exchanging adjacent elements. This saves a significant amount of time because all that you have to do in order to calculate the cost of each new permutation is determine the incremental effect of the exchange. The rest of the permutation remains the same so you don't have to recalculate its cost. You can find more information on other algorithms to generate permutations in [426], as well as some hints on how to implement them efficiently.

3.1.3 Enumerating the NLP

Strictly speaking, it isn't possible to search exhaustively in continuous domains because there is an infinite number of alternatives, but it is possible to divide the domain of each continuous variable into a finite number of intervals and consider the Cartesian product of these intervals. For example, when finding the maximum of a function $f(x_1, x_2)$ where $x_1 \in [-1, 3]$ and $x_2 \in [0, 12]$, we can divide the range of x_1 into 400 intervals, each of length 0.01, and divide the range of x_2 into 600 intervals, each of length 0.02. This creates $400 \times 600 = 240,000$ cells that we can search enumeratively. The merit of each cell can be defined by applying the evaluation function to a particular corner of the cell, or perhaps the midpoint. Once the cell with the best merit is selected we have the choice of going back and doing the same thing all over again within that cell.

```
procedure gen2_permutation(I, P, n)
    begin
        if not I then
        begin
            for i ← 0 step 1 until n do
                x[i] ← 0
                x[n] ← −1
                goto E
        end
        for i ← n step −1 until 0 do
        begin
            if x[i] ≠ i then goto A
            x[i] ← 0
        end
        P[0] ← −1
        goto E
    A:  x[i] ← x[i] + 1
        P[0] = 0
        for i ← 1 step 1 until n do
        begin
            P[i] ← P[i − x[i]]
            P[i − x[i]] ← i
        end
    E:  end
```

Fig. 3.4. This procedure produces all permutations of integers from 0 to n. Upon entry with I = FALSE, the procedure initializes itself producing no permutation. Upon each successive entry into the procedure with I = TRUE, a new permutation is stored in $P[0]$ through $P[n]$. A sentinel is set, $P[0] = -1$, when the process has been exhausted.

This is a form of exhaustive search in the sense that we're trying to "cover" all possible solutions. But it isn't really exhaustive since only one solution of the many solutions in each cell is considered and evaluated. The size of the cells is important here: the smaller the cell, the better our justification for calling this method "exhaustive." But the disadvantages of the approach are also important:

- Using a fine granularity of the cells increases the total number of cells significantly. If the domain of the first variable were divided into 4,000 intervals, each of length 0.001, and the domain of the second variable were divided into 6,000 ranges, each of length 0.002, the total number of cells would grow by two orders of magnitude from 240,000 to 24,000,000.

- A low granularity of the cells increases the probability that the best solution will not be discovered, i.e., the cell that contains the best solution wouldn't have the best merit because we'd be looking at a level that was too coarse.

- When facing a large number of dimensions (variables), the method becomes impractical because the number of cells can be very large. An optimization problem with 50 variables, each divided into only 100 intervals, yields 10^{100} cells.

When you face a small problem and have the time to enumerate the search space, you can be guaranteed to find the best solution. Don't consider this approach for larger problems, however, because you'll simply never complete the enumeration.

3.2 Local search

Instead of exhaustively searching the entire space of possible solutions, we might focus our attention within a local neighborhood of some particular solution. This procedure can be explicated in four steps:

1. Pick a solution from the search space and evaluate its merit. Define this as the *current* solution.

2. Apply a transformation to the current solution to generate a new solution and evaluate its merit.

3. If the new solution is better than the current solution then exchange it with the current solution; otherwise discard the new solution.

4. Repeat steps 2 and 3 until no transformation in the given set improves the current solution.

The key to understanding how this *local search* algorithm works lies in the type of transformation applied to the current solution. At one extreme, the transformation could be defined to return a potential solution from the search space selected uniformly at random. In this case, the current solution has no effect on the probabilities of selecting any new solution, and in fact the search becomes essentially enumerative. Actually, it's possible for this search to be even worse than enumeration because you might resample points that you've already tried. At the other extreme lies the transformation that always returns the current solution — and this gets you nowhere!

From a practical standpoint, the right thing to do lies somewhere in between these extremes. Searching within some local neighborhood of the current solution is a useful compromise. In that way, the current solution imposes a bias on where we can search next, and when we find something better we can update the current point to this new solution and retain what we've learned. If the size of the neighborhood is small then we might be able to search that neighborhood very quickly, but we also might get trapped at a local optimum. In contrast, if the size of the neighborhood is very large then we'll have less chance of getting stuck, but the efficiency of the search may suffer. The type of transformation that we apply to the current solution determines the size of the neighborhood, so we have to choose this transformation wisely in light of what we know about the evaluation function and our representation.

3.2.1 Local search and the SAT

Local search algorithms are surprisingly good at finding satisfying assignments for certain classes of SAT formulas [428]. One of the best-known (randomized) local search algorithms for the SAT (given in conjunctive normal form) is the GSAT procedure shown in figure 3.5.

```
procedure GSAT
begin
    for i ← 1 step 1 until MAX-TRIES do
    begin
        T ← a randomly generated truth assignment
        for j ← 1 step 1 until MAX-FLIPS do
            if T satisfies the formula then return(T)
            else make a flip
    end
    return('no satisfying assignment found')
end
```

Fig. 3.5. The GSAT procedure. The statement "make a flip" flips the variable in T that results in the largest decrease in the number of unsatisfied clauses.

GSAT has two parameters: MAX-TRIES, which determines the number of new search sequences (trials), and MAX-FLIPS, which determines the maximum number of moves per try.

GSAT begins with a randomly generated truth assignment. If we happen to be lucky and this assignment satisfies the problem, the algorithm terminates. Rarely are we so lucky. At the next step, it flips each of the variables from TRUE to FALSE or FALSE to TRUE in succession and records the decrease in the number of unsatisfied clauses. After trying all of these possible flips, it updates the current solution to the new solution that had the largest decrease in unsatisfied clauses. Again, if this new solution satisfies the problem, then we're done; otherwise, the algorithm starts flipping again.

There's an interesting feature of this algorithm. The best available flip might actually *increase* the number of unsatisfied clauses. The selection is only made from the neighborhood of the current solution. If every neighbor (defined as being one flip away) is worse than the current solution, in that each neighbor satisfies fewer clauses, then GSAT will take the one that is the least bad. This procedure might remind you of a presidential election! But as we'll see in chapter 5, this can actually be very beneficial: the algorithm has a chance to escape from a local optimum. It's only a chance, however, not a guarantee, because the algorithm might oscillate between points and never escape from some plateaus. In fact, experiments with GSAT indicate that plateau moves dominate the search [428]. Many variations of the procedure have been offered to avoid this problem [182, 429]. For example, we can assign a weight to each clause [429] and then, at the end of a try, we can increase the weights for all of the clauses that remain unsatisfied. A clause that remains unsatisfied over many tries will have a higher weight than other clauses, therefore, any new solution that satisfies such a clause will be judged to have an accordingly higher score.

3.2.2 Local search and the TSP

There's a huge variety of local search algorithms for the TSP. The simplest is called *2-opt*. It starts with a random permutation of cities (call this tour T) and tries to improve it. The neighborhood of T is defined as the set of all tours that can be reached by changing two nonadjacent edges in T. This move is called a *2-interchange* and is illustrated in figure 3.6.

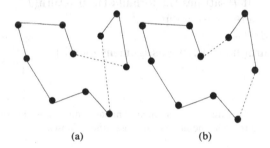

(a)   (b)

Fig. 3.6. A tour (a) before and (b) after the *2-interchange* move is applied. Broken lines indicate affected edges.

A new tour T' replaces T if it is better.[1] If none of the tours in the neighborhood of T is better than T, then the tour T is called *2-optimal* and the algorithm terminates. As with GSAT, the algorithm should be restarted from several random permutations before we really give up.

The *2-opt* algorithm can be generalized easily to a *k-opt* procedure, where either k, or up to k, edges are selected for removal and are replaced by lower-cost connections. The trade-off between the size of the neighborhood and the efficiency of the search is directly relevant here. If k is small (i.e., the size of the neighborhood is small), the entire neighborhood can be searched quickly, but this small range increases the likelihood of yielding only a suboptimal answer. On the other hand, for larger values of k the number of solutions in the neighborhood become enormous — the number of subsets grows exponentially with k. For this reason, the *k-opt* algorithm is seldomly used for $k > 3$.

---

[1]Note that we replace the tour *every* time we find an improvement. In other words, we terminate the search in the neighborhood of T when the "first improvement" is found.

The best-known local search procedure for the TSP is the Lin-Kernighan algorithm [284]. In a sense, this procedure refines the *k-opt* strategy by allowing k to vary from one iteration to another. Moreover, instead of taking the "first improvement" as is done with *k-opt*, it favors the largest improvement. We don't immediately replace the current best tour when we find something better.

There are several versions of the basic Lin-Kernighan method that differ in their details. Here we describe one of these versions briefly (also see [78]) in which the representation isn't based on a Hamiltonian path but rather a so-called δ-path. You'll see how this works below.

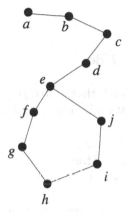

Fig. 3.7. A δ-path

A δ-path is defined as a path where the number of nodes is one greater than the number of edges, and all of the nodes appear only once in the path except for the last node which appears earlier in the path. A picture makes the explanation much easier so look at figure 3.7, which illustrates a δ-path

$$a - b - c - d - e - f - g - h - i - j - e.$$

This is a δ-path because there are 10 edges, 11 nodes, and the nodes are all distinct except for the last node, e.

Any δ-path can be repaired into a legal tour by replacing a single edge: the last one. For example, replacing the edge $(j\ e)$ by $(j\ a)$ generates a legal tour. Also, by replacing the last edge $(j\ e)$ by any other edge that originates in j we can also obtain a new δ-path, e.g., replacing the edge $(j\ e)$ by $(j\ f)$. Let's call such a replacement of one δ-path by another a *switch*(e, f), where the labels e and f indicate the switched nodes. The order of the nodes is important: The first node (e) is removed and replaced by the second (f). If the labels weren't essential, we'd just refer to this move as a *switch*.

The cost of a new δ-path increases by

$$cost(j, f) - cost(j, e),$$

and this "increase" might be negative. It's important to note that a δ-path has two "last edges." In figure 3.7, we could replace the edge $(j\ e)$ or the edge $(f\ e)$ when building a legal tour or generating a new δ-path.

Now we're ready to describe the basic steps of Lin-Kernighan. It starts with a random tour T and generates a sequence of δ-paths. The first is constructed from T and the next one is always constructed from the previous one by using a *switch* move.

procedure Lin-Kernighan
begin
    generate a tour T
    Best_cost $= cost(T)$
    for each node k of T **do**
        for each edge $(i\ k)$ of T **do**
C: **begin**
        if there is $j \neq k$ such that $cost(i, j) \leq cost(i, k)$
        then create a δ-path p by removing $(i\ k)$ and adding $(i\ j)$
        else goto B
A: **construct** a tour T from p
        if $cost(T) <$ Best_cost **then**
            Best_cost $\leftarrow cost(T)$
            store T
        if there is a *switch* of p resulting in a
            δ-path with cost not greater than $cost(T)$
        then
            make the *switch* getting a new δ-path p
            goto A
B: **if** $cost(T) <$ Best_cost **then**
            Best_cost $\leftarrow cost(T)$
            store T
        if there remain untested node/edge combinations
        then goto C
    end
end

Fig. 3.8. The structure of Lin-Kernighan algorithm

Figure 3.8 provides the basic outline of the Lin-Kernighan algorithm without the details. A few lines of code,

    if there is cost improving *switch* of p
    then
        make the *switch* getting a new δ-path p

display the general idea, but they also hide important details. The process of generating a new δ-path is based on adding and removing edges. For a given node k and an edge $(i\ k)$ (i.e., for any single iteration of the **for** loop), an edge can either be added or removed but not both. This requires maintaining two lists with *added* edges and *deleted* edges, respectively. If an edge is already present in one of these lists, then the switch moves are restricted. If an edge is in the *added* list, it can only be removed in a switch move. Thus, this part of the algorithm is limited to n switches, where n is the number of nodes. Note also that each generated δ-path has a cost that is at most $cost(T)$. Another important detail is connected with the selection of a node in both **if there is** statements for converting a tour into a δ-path and for generating a new δ-path from the current one (figure 3.8). Since it might be too expensive to check all possible nodes, it's reasonable to limit this part of the search, and this can be done in a variety of ways [242].

The Lin-Kernighan procedure can run very fast and generate near-optimal solutions (e.g., within two percent of the optimum path [243]) for TSPs with up to a million cities. It can be done in under an hour on a modern workstation [239].

3.2.3 Local search and the NLP

The majority of numerical optimization algorithms for the NLP are based on some sort of local search principle. However, there's quite a diversity of these methods. Classifying them neatly into separate categories is difficult because there are many different options. Some incorporate heuristics for generating successive points to evaluate, others use derivatives of the evaluation function, and still others are strictly local, being confined to a bounded region of the search space. But they all work with complete solutions and they all search the space of complete solutions.

Recall that the NLP concerns finding the vector $\mathbf{x}$ so as to

$$\text{optimize } f(\mathbf{x}), \ \mathbf{x} = (x_1, \ldots, x_n) \in \mathbf{R}^n,$$

where $\mathbf{x} \in \mathcal{F} \subseteq \mathcal{S}$. The *evaluation function* f is defined on the *search space* $\mathcal{S} \subseteq \mathbf{R}^n$ and the set $\mathcal{F} \subseteq \mathcal{S}$ defines the *feasible region*. Usually, the search space $\mathcal{S}$ is defined as an n-dimensional rectangle in $\mathbf{R}^n$, where the domains of the variables are defined by their lower and upper bounds,

$$l(i) \leq x_i \leq u(i), \ \ 1 \leq i \leq n,$$

whereas the feasible region $\mathcal{F}$ is defined by a set of m additional constraints $(m \geq 0)$,

$$g_j(\mathbf{x}) \leq 0, \text{ for } j = 1, \ldots, q, \text{ and } h_j(\mathbf{x}) = 0, \text{ for } j = q+1, \ldots, m.$$

At any point $\mathbf{x} \in \mathcal{F}$, the constraints g_k that satisfy $g_k(\mathbf{x}) = 0$ are called the *active* constraints at $\mathbf{x}$. By extension, equality constraints h_j are also described as active at all points $\mathcal{S}$.

One reason that there are so many different approaches to the NLP is that no single method is superior to all others. In general, it's impossible to develop a deterministic method for finding the best global solution in terms of f that would be better than exhaustive search. Gregory [204] stated:

> It's unrealistic to expect to find one general NLP code that's going to work for every kind of nonlinear model. Instead, you should try to select a code that fits the problem you are solving. If your problem doesn't fit in any category except 'general', or if you insist on a globally optimal solution (except when there is no chance of encountering multiple local optima), you should be prepared to have to use a method that boils down to exhaustive search, i.e., you have an intractable problem.

Nevertheless, it's possible to classify the proposed methods from two perspectives: (1) the type of problem being addressed, and (2) the type of technique being used [219]. The first perspective provides a few orthogonal categories of problems: constrained[2] vs. unconstrained, discrete vs. continuous, and special cases of evaluation functions that are convex, quadratic, linearly separable, etc. The second perspective provides a classification of the methods, such as derivative-based[3] vs. derivative-free, analytical vs. numerical, deterministic vs. random, and so forth.

There are several fundamental problems within the broader category of NLPs. One problem involves finding the extremum of a functional, as described above. Another concerns finding the zeros of a function. Although this seems like a completely different problem, and is often treated as if it were different, we could imagine posing a surrogate evaluation function that would judge the merit of any possible root of a function. We could then perform an optimization over that surrogate function and if we chose the evaluation function wisely then we'd get the same answer either way. Let's take a look at some local methods for addressing tasks that arise within the framework of the NLP.

Bracketing methods. Consider the function in figure 3.9 for which $f(x^*) = 0$. Our task is to find x^*. Suppose that we know how to bracket x^* with two other numbers a and b. Then one very simple way to find x^* is the method of bisecting the range between a and b, finding this midpoint, m, determining the value of $f(m)$ and then resetting the left or right limit of the range to m depending on whether or not $f(m)$ is positive. If we continue to iterate this procedure, we eventually converge arbitrarily close to a value x^* such that $f(x^*) = 0$. (This is a perfect example of the divide and conquer approach discussed later in section 4.2.)

---

[2]Note that constrained problems can be further divided into those with equality constraints, inequality constraints, or both, as well as the type of constraints (e.g., linear).

[3]This can be further divided into first-order and second-order derivatives.

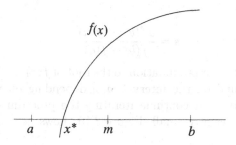

Fig. 3.9. The bisection method finds the zero of a function by first bracketing that zero and then iteratively dividing the region of interest in half. Here, a and b are the left and right limits of the bracketing interval, m is the midpoint that serves as the next point to try, and x^* is the actual answer. The method terminates when the width of the interval shrinks below a given tolerance.

Note that this *bisection* is a local search method. It relies on two bracketing points, a and b, but it still updates only one point at a time and has one current best solution at any iteration.

The assumption is that $f(a)f(b) < 0$. That is, we must be certain that $f(a)$ and $f(b)$ lie on opposite sides of zero. The bisection algorithm is simple:

1. Choose a and b such that $f(a)f(b) < 0$ and $x^* \in [a, b]$.

2. Generate $m \leftarrow (a + b)/2$.

3. If $f(m) \neq 0$ then either $f(a)f((a + b)/2) < 0$ or $f((a + b)/2)f(b) < 0$. If the former case is true, then set $b \leftarrow m$, otherwise set $a \leftarrow m$.

4. If $(b - a) < \epsilon$ then stop; otherwise, proceed to step 2.

The value of ϵ is chosen as a small number that represents the degree of error we're willing to tolerate.

We are interested in how quickly the error bound on a zero of $f(x)$ shrinks because this will determine how many iterations we'll have to make with the algorithm. If we start with an interval of length $L_0 = b - a$, the length will be $L_1 = (b - a)/2$ after the first step. After the next step, it will be $L_2 = (b-a)/2^2$, and so forth. Thus, the length of the interval decreases geometrically as a function of the number of iterations: $L_n = (b - a)/2^n$. As we'll see shortly, although this seems attractive, there are many other methods that will converge much faster.

One such method is another bracketing technique called *Regula Falsi*. This method also starts with an interval $[a, b]$ that bounds a zero of $f(x)$ where $f(a)f(b) < 0$. Instead of bisecting the interval, however, the idea is to construct the secant line between $(a, f(a))$ and $(b, f(b))$ and find the point s where this line intersects the x-axis. The secant line is found by

$$\frac{y - f(b)}{x - b} = \frac{f(a) - f(b)}{a - b}.$$

Therefore,

$$s = \frac{af(b) - bf(a)}{f(b) - f(a)}.$$

The value of s is the first approximation to the root of $f(x)$. As before, we replace the left or right bound on the interval $[a, b]$ depending on whether $f(a)f(s)$ is positive or negative and continue iterating the procedure until the length of the interval is sufficiently small. Figure 3.10 shows how the method works graphically.

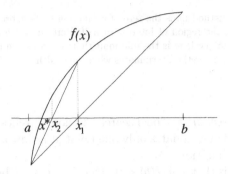

Fig. 3.10. The method of *Regula Falsi* relies on a bracket around the zero of the function. Instead of generating the midpoint between a and b, it connects the line between a and b and takes the intersection of that line with the x-axis as the next point to try. The interval is reduced over successive iterations and the method terminates when the interval is below a given tolerance.

The length of the interval L_n at the n-th iteration under *Regula Falsi* may not converge to zero as $n \to \infty$. It depends on the function $f(x)$. But the method may converge faster than bisection. The underlying reason is that if $|f(a)| > |f(b)|$ then we would expect that the zero of $f(x)$ is closer to b than to a, and vice versa. Using the secant line to determine the next bounding point will take advantage of this condition when it is true. It can be shown that if the second derivative $f''(x)$ is continuous then *Regula Falsi* converges at a rate of $|s_n - x^*| \leq \lambda^{n-1}\epsilon$, where $0 < \lambda < 1$ and ϵ is a fixed constant.

Fixed-point methods. Although the bisection and *Regula Falsi* methods will converge for continuous functions, they are slow in comparison to other methods such as the following so-called *Newton's method*. This procedure, however, may not always converge even on continuous functions so we must be more careful in applying it. The concept is shown in figure 3.11. Given a function $f(x)$ and an initial starting guess at a value for which $f(x) = 0$, say s, find the line tangent at $f(s)$ and locate where this tangent line intersects the x-axis. Then use this crossing point as the next guess for the root, and so forth, until the distance between guesses over two successive iterations is sufficiently small.

Mathematically, the sequence of guesses at the zero of $f(x)$ proceeds as

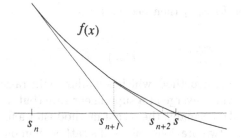

Fig. 3.11. An illustration of Newton's method for finding the zero of a function. Instead of bracketing the zero, as is done in bisection or *Regula Falsi*, the method uses the slope of the function at a trial point and calculates the intersection of the line tangent at that point with the x-axis. That intersecting point is the next point to try. The method halts when the difference between successive points falls below a threshold.

$$s_{n+1} = s_n - \frac{f(s_n)}{f'(s_n)},$$

for $n = 0, 1, \ldots$. We can view this as a special case of a sequence $s_{n+1} = g(s_n)$, which is a fixed-point problem: we want to find a function $g(x)$ such that if $f(s) = 0$, then $s = g(s)$. Often, there are many choices for this auxiliary function $g(x)$. For example, if $f(x) = x^2 - 4x + 4$ then $g(x)$ could be $g(x) = \frac{-4}{x-4}$, $g(x) = x^2 - 3x + 4$, or $g(x) = \sqrt{4x - 4}$. In all three cases, when $x = 2$, which yields $f(x) = 0$, then $g(2) = 2$.

To find a suitable auxiliary function $g(x)$, it's helpful to impose some constraints. The first is that for any interval $L = [a, b]$, for every $x \in L$, $g(x) \in L$. In words, when x is in the interval $[a, b]$, so is $g(x)$. If we also require $g(x)$ to be continuous then it's guaranteed to have a fixed point in L.

Newton's method, then, consists of finding the auxiliary function $g(x)$ subject to the above constraints and an interval $L = [a, b]$ that bounds the root of $f(x)$, and then iterating on the update equation, $s_{n+1} = g(s_n)$. If $|g'(x)| \leq B < 1$, there will be only one fixed point in L, and if x^* is the root of $f(x)$, the error: $e_n = s_n - x^*$ satisfies $|e_n| \leq \frac{B^n}{1-B} \cdot |s_1 - s_0|$. The method can be stopped when $|s_{n+1} - s_n| < \epsilon$. If we can make a good guess at the location of x^*, Newton's method can converge to it very quickly. Under the condition where $f''(x)$ is continuous and $f'(x) \neq 0$ in an interval L that contains x^*, there will be some ϵ such that Newton's method converges *quadratically* whenever the starting position s_0 is such that $|s_0 - x^*| < \epsilon$.

Newton's method, although it converges quickly under the above conditions, has two main drawbacks: (1) it may require a close initial guess for x^*, and (2) the update equation for s_{n+1} requires computing $f'(x)$, which may be computationally expensive. Instead of computing this derivative, we can substitute an approximation for it:

$$\frac{f(s_n) - f(s_{n-1})}{s_n - s_{n-1}}.$$

The update equation for s_{n+1} then becomes

$$s_{n+1} = s_n - \frac{f(s_n)(s_n - s_{n-1})}{f(s_n) - f(s_{n-1})}.$$

This is the so-called *secant* method, which is similar to the rationale of the *Regula Falsi* technique, but will converge at a superlinear rate (but not a quadratic rate like Newton's method). Formally, the secant method isn't a fixed-point iteration because it depends on two previous values of s rather than only the most recent value. Note too that for the method to be valid, $f(s_n) - f(s_{n-1})$ can never equal zero.

Gradient methods of minimization. Returning to the typical problem of finding the extremum of some functional $f(x)$, as we've seen with Newton's method, one thing that we might try to take advantage of is the gradient of the evaluation function at a particular candidate solution. We could use this information to direct our search toward improved solutions, at least in a local neighborhood. There are so many methods that rely on the gradient in some manner that it would require a complete textbook to review them all. Instead, we'll focus here on the main underlying principles, and refer you to [35] for more detailed information on particular variants of gradient methods.

The basic idea is that we need to find a *directional derivative* so that we can proceed in the direction of the steepest ascent or descent, depending on whether we are maximizing or minimizing. We will only consider minimization here, but we'll do that without loss of generality. If the evaluation function is sufficiently smooth at the current candidate solution, the directional derivative exists. Our challenge is to find the angle around the current solution for which the magnitude of the derivative of the evaluation function with respect to some step size s is maximized. By applying calculus, we can derive that this maximum occurs in the direction of the negative gradient $-\nabla f(\mathbf{x})$ [245]. Recall that

$$\nabla f(\mathbf{x}) = \left[\frac{\delta f}{\delta x_1}, \frac{\delta f}{\delta x_2}, \ldots, \frac{\delta f}{\delta x_n} \right]_{\mathbf{x}}.$$

The method of steepest descent, then, is essentially this: Start with a candidate solution $\mathbf{x}_k$, where $k = 1$. Then, generate a new solution using the update rule:

$$\mathbf{x}_{k+1} = \mathbf{x}_k - \alpha_k \nabla f(\mathbf{x}_k),$$

where $k \geq 1$, $\nabla f(\mathbf{x}_k)$ is the gradient at $\mathbf{x}_k$, and α_k is the step size. The bigger α_k is, the less *local* the search will be. The method is designed to ensure reductions in the evaluation function for small enough steps, so the idea is to find the right step size to guarantee the best rate of reduction in the evaluations of candidate solutions over several iterations.

If we incorporate second-order derivative information into the update, we get an update equation that is essentially Newton's method:

$$\mathbf{x}_{k+1} = \mathbf{x}_k - (H(f(\mathbf{x}_k)))^{-1}\nabla f(\mathbf{x}_k),$$

where $H(f(\mathbf{x}_k))$ is the *Hessian* matrix of f:[4]

$$H(f(\mathbf{x}_k)) = \begin{bmatrix} \frac{\delta^2 f}{\delta x_1^2} & \frac{\delta^2 f}{\delta x_1 \delta x_2} & \cdots & \frac{\delta^2 f}{\delta x_1 \delta x_n} \\ \frac{\delta^2 f}{\delta x_2 \delta x_1} & \frac{\delta^2 f}{\delta x_2^2} & \cdots & \frac{\delta^2 f}{\delta x_2 \delta x_n} \\ \cdot & \cdot & \cdot & \cdot \\ \cdot & \cdot & \cdot & \cdot \\ \cdot & \cdot & \cdot & \cdot \\ \frac{\delta^2 f}{\delta x_n \delta x_1} & \frac{\delta^2 f}{\delta x_n \delta x_2} & \cdots & \frac{\delta^2 f}{\delta x_n^2} \end{bmatrix}_{\mathbf{x}_k}$$

This method is designed to find the minimum of a quadratic bowl in one step. That's fast convergence! Let's try it out.

Suppose $f(x_1, x_2) = x_1^2 + x_2^2$. Let's take our starting guess to be $\mathbf{x}_1 = [5\ 1]^T$. To apply the method we have to compute the gradient vector

$$\nabla f(\mathbf{x}_1) = [2x_1\ 2x_2]^T = [2(5)\ 2(1)]^T = [10\ 2]^T$$

and the Hessian matrix

$$H(f(\mathbf{x}_1)) = \begin{bmatrix} 2 & 0 \\ 0 & 2 \end{bmatrix}.$$

Then, we have to take the inverse of the Hessian, yielding

$$(H(f(\mathbf{x}_1)))^{-1} = \begin{bmatrix} \frac{1}{2} & 0 \\ 0 & \frac{1}{2} \end{bmatrix}.$$

To complete the update equation and generate the next point,

$$\mathbf{x}_{k+1} = [5\ 1]^T - \begin{bmatrix} \frac{1}{2} & 0 \\ 0 & \frac{1}{2} \end{bmatrix} [10\ 2]^T$$

and that's $[0\ 0]^T$, which is the minimum all right! It's easy to see that for any $\mathbf{x}_k$, the update equation will generate the next solution $\mathbf{x}_{k+1}$ at the origin if the evaluation function is a quadratic bowl. For functions that aren't quadratic bowls, however, things aren't as nice. The primary consideration is that since the method relies on gradient information, we could get stuck at a local optimum. We also have to choose an error tolerance ϵ and halt the algorithm when the difference between $\mathbf{x}_{k+1}$ and $\mathbf{x}_k$ is less than ϵ.

Admittedly, calculating the inverse of the Hessian matrix is rarely as easy as in this example. Many so-called *quasi-Newton* methods use different mechanisms for estimating the Hessian. Again, there is a cornucopia of different techniques here, so please see [35] for more details on their derivation. If Gaussian elimination is used to generate the inverse Hessian then the method is called Newton-Gauss. In the remainder of the book we'll use Newton's method and Newton-Gauss synonymously unless otherwise noted.

---

[4]The Hessian matrix $H(f(\mathbf{x}_k))$ is often denoted by $\nabla^2 f(\mathbf{x}_k)$.

Newton-Gauss, and all of the quasi-Newton methods are essentially local search procedures. They update a single complete solution with another complete solution. They use information that is local to the current best point to generate a new point. And they also work to the boundary of a neighborhood, typically defined by the inverse Hessian.

Another somewhat related technique is *Brent's method*, which is a bracketing technique designed to find the minimum of a quadratic function. It works with three points — instead of just two as with bisection and *Regula falsi* — and fits a parabola to those points. It then uses the minimum of that parabola as the next guess for the minimum of the overall function. The method iterates by narrowing the bracket based on each new point tested. Figure 3.12 illustrates the procedure and you'll find the algorithm coded in [363].

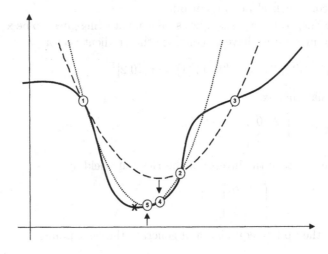

Fig. 3.12. Brent's method is a bracketing technique that relies on three points. In order to find the minimum of a function, it calculates a parabolic auxiliary function through the three points and locates the minimum of that parabola. That minimum is then used as one of the three points to define the next parabola, and so forth. The method terminates when the difference between guesses at the minimum of the function fall below a threshold.

3.3 Linear programming: The simplex method

At last we come to the final traditional technique for processesing complete solutions that we'll discuss here: the simplex method for linear programming (LP) problems. Problems in LP require finding the extremum (let's say the maximum) of a linear combination of variables,

$$eval = a_{01}x_1 + \ldots + a_{0n}x_n,$$

subject to the primary constraints,

$$x_1 \geq 0, \ldots, x_n \geq 0,$$

and also subject to $M = m_1 + m_2 + m_3$ additional constraints; m_1 of the form:

$$a_{i1}x_1 + \ldots + a_{in}x_n \leq b_i, (b_i \geq 0), i = 1, \ldots, m_1,$$

m_2 of the form

$$a_{j1}x_1 + \ldots + a_{jn}x_n \geq b_j, (b_j \geq 0), j = m_1 + 1, \ldots, m_1 + m_2,$$

and m_3 of the form

$$a_{k1}x_1 + \ldots + a_{kn}x_n = b_k, (b_k \geq 0), k = m_1 + m_2 + 1, \ldots, M.$$

Any vector $\mathbf{x}$ that satisfies all of the constraints is said to be *feasible* and our goal is to find the feasible vector that yields the best result in terms of the evaluation function.

The simplex method constructs the best vector $\mathbf{x}$ in a series of operations that take advantage of the linear characteristics of the problem. Figure 3.13 depicts the situation. Potentially, there are equality and inequality constraints imposed over the evaluation function, but since the evaluation function is linear, the optimum solution lies at one of the vertexes of the simplex (the feasible region), or in a degenerate case, anywhere along a boundary. The simplex procedure is guaranteed to find the best vertex of the simplex.

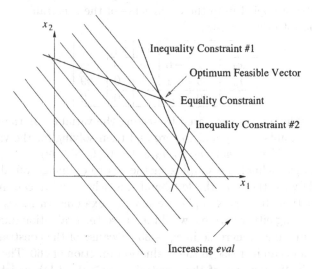

Fig. 3.13. A graphical interpretation of the linear programming problem. The contour lines indicate the values of the function *eval* which increase into the first quadrant. The bold lines indicate constraints (either equality or inequality). The optimum vector lies on the equality constraint and at a vertex of the simplex.

The method is best illustrated on a simple problem. Consider the following scenario. A factory produces two types of items: small fancy chairs and large simple tables. The profit per chair is $20, whereas the profit per table is $30. Each chair requires a single unit of wood and three hours of labor. Each table, on the other hand, requires six units of wood and one hour of labor. The production process has some restrictions: There are 288 units of wood available and 99 hours of labor available during a particular period of time. The task is to maximize the factory's profit.

We can formulate this problem as a linear programming problem:

$$\text{maximize } eval = 20x_1 + 30x_2,$$

such that

$$x_1 + 6x_2 \leq 288 \text{ (limitation for units of wood)},$$
$$3x_1 + x_2 \leq 99 \text{ (limitation for hours of labor)}.$$

Additionally, $x_1 \geq 0$ and $x_2 \geq 0$.
Note that we can reformulate the above problem in the following way:

$$\text{maximize } 20x_1 + 30x_2,$$

such that

$$x_3 = -x_1 - 6x_2 + 288,$$
$$x_4 = -3x_1 - x_2 + 99, \text{ and} \qquad (3.1)$$
$$x_1 \geq 0, \ x_2 \geq 0, \ x_3 \geq 0, \ x_4 \geq 0.$$

We proceed by making a *tableau*, which is just a fancy name for a table, where each entry is copied from the coefficients of the constraints and the evaluation function into the last row.

	x_1	x_2	x_3	x_4	
x_3	−1	−6	0	0	288
x_4	−3	−1	0	0	99
eval	20	30	0	0	0

Our initial feasible point is generated by setting the variables on the right-hand side of the constraints (3.1) equal to zero and then solving for the variables on the left-hand side. Here we get $(x_1, x_2, x_3, x_4) = (0, 0, 288, 99)$.

The main idea behind the simplex method is to exchange variables on the right-hand side with those on the left-hand side wherever we can improve the evaluation function. In our example, we have to examine variables x_1 and x_2 and see if increasing either variable would increase the evaluation function *eval*. Note that the maximum increase in x_1 is 33 (because of the constraint $x_4 \geq 0$, (3.1)), which yields an increase in the evaluation function of 660. The maximum increase in x_2 is 48 (because of the constraint $x_3 \geq 0$, (3.1)), which yields an increase in the evaluation function of 1440. Thus we set

$$x_2 \leftarrow x_2 + 48.$$

Note that at this stage x_3 becomes zero and we have to rewrite (3.1) to keep all of the "zero" variables (e.g., x_1 and x_3) on the right-hand side. Thus we have to rewrite the original problem expressing variable x_2 as a function of variables x_1 and x_3. This leads us to a new problem:

maximize $15x_1 - 5x_3 + 1440$,

such that

$$x_2 = -\tfrac{1}{6}x_1 - \tfrac{1}{6}x_3 + 48,$$
$$x_4 = -\tfrac{17}{6}x_1 + \tfrac{1}{6}x_3 + 51, \text{ and} \qquad (3.2)$$
$$x_1 \geq 0, \ x_2 \geq 0, \ x_3 \geq 0, \ x_4 \geq 0.$$

Now our tableau is:

	x_1	x_2	x_3	x_4	
x_2	$-1/6$	0	$-1/6$	0	48
x_4	$-17/6$	0	$1/6$	0	51
eval	15	0	-5	0	1440

This time we have to examine variables x_1 and x_3 and see if increasing either variable would increase the evaluation function *eval*. This is only true for x_1; note that the maximum increase in x_1 is 18 because of the constraint $x_4 \geq 0$, (3.2). Thus we set

$$x_1 \leftarrow x_1 + 18$$

and this leads us to a new problem:

maximize $-\tfrac{100}{17}x_3 - \tfrac{90}{17}x_4 + 1710$,

such that

$$x_1 = 18 + \tfrac{1}{17}x_3 - \tfrac{6}{17}x_4,$$
$$x_2 = 45 - \tfrac{3}{17}x_3 + \tfrac{1}{17}x_4, \text{ and}$$
$$x_1 \geq 0, \ x_2 \geq 0, \ x_3 \geq 0, \ x_4 \geq 0.$$

Now our tableau is:

	x_1	x_2	x_3	x_4	
x_1	0	0	$-1/17$	$-6/17$	18
x_2	0	0	$-3/17$	$1/17$	45
eval	0	0	$-100/17$	$-90/17$	1710

In this case, we're stuck. Any increase for x_3 or x_4 would *decrease* the value of *eval* — and that means we've found the right answer, $x_1 = 18$, $x_2 = 45$, for which *eval* = 1710 (representing the maximum profit).

Note that in general, we focused on all the "right-hand variables" where we could increase *eval*. This process is illustrated in figure 3.14.

There are many tricks for converting LP problems into the form required for this procedure, and there are variations of the procedure for different cases. The basic idea, however, lies somewhere between working with complete solutions and working with partial solutions. Through analytic means, we decide which variable to increase and over a series of swaps with other variables, we alternately examine other dimensions of the problem to see if we can obtain

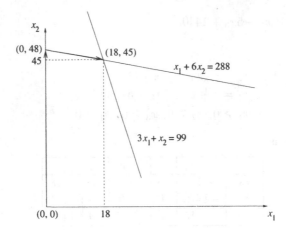

Fig. 3.14. Moving from one vertex to another with the simplex method. The *boldface arrows* indicate the steps of the procedure. The feasible area lies below both lines $x_1 + 6x_2 = 288$ and $3x_1 + x_2 = 99$.

improvements. The linear nature of the problem is what makes this a feasible routine. Once the evaluation function is nonlinear you have to compensate with other tricks that are beyond the discourse here. In the limit, when the evaluation function is potentially multimodal or even discontinuous, this class of method breaks down. It leaves you with the option of trying to approximate the problem as if it were linear or perhaps quadratic or restricted to integer values, but if you do that then you will generate the right answer to the wrong problem, and there's no rule for just how bad that answer might be.

3.4 Summary

There are many classic algorithms that operate on complete solutions. You can interrupt these procedures at any iteration and they will provide some potential solution. The quality of that solution, however, depends largely on the degree to which the procedure matches the problem. For example, if your task is to maximize an evaluation function that's a linear combination of variables that are subject to linear equality and inequality constraints, then linear programming methods are appropriate. If the evaluation function is smooth (i.e., differentiable), unimodal, and you want to find its minimum, then methods based on gradients and higher-order functions can be useful. If you face a combinatorial problem, like a TSP, there are many different local operators you can employ to take advantage of the characteristics that are inherent to circuits in the Euclidean plane. These will iteratively improve complete tours and may quickly find one that is nearly optimal.

The drawback that you face is that if you really don't know much about your problem, or if it doesn't fit well within the class of problems that each of these algorithms can tackle, you are left with enumerating all possible solutions. For any real-world problems of concern, this approach is almost always impractical because the number of alternative solutions is prohibitively large.

We mentioned that there are two classes of classic algorithms: (1) those that operate on complete solutions, and (2) those that evaluate partially constructed or approximate solutions. When the methods of this chapter break down, it is sometimes reasonable to think about the latter class of techniques and attempt to decompose problems into simpler subproblems that can be reassembled, or perhaps to solve complex problems in a series of simpler stages. These are treated in detail in the subsequent chapter.

IV. What Are the Numbers?

Let's consider a puzzle which illustrates that the process of finding all the solutions to some problems isn't all that simple. The problem was offered to us by Peter Ross from the University of Edinburgh. It's a seemingly simple problem: Find the natural numbers a, b, c, and d, such that

$$ab = 2(c + d), \text{ and}$$
$$cd = 2(a + b).$$

For example, one of the solutions to this problem is $(a, b) = (1, 54)$ and $(c, d) = (5, 22)$. Clearly, some solutions are far from trivial.

It's important here to analyze the problem before we attempt to solve it. It might be a good idea to get four random integers which we can use to check if the above equations hold. Certainly, the chances for that occurrence are very slim! But we can then change one number at the time to get "a feel" for the problem. After a short time we'd discover that it's unlikely that all of these numbers are large. If a, b, c, and d are sufficiently large, then the product of a and b is much greater than twice their sum. The same is true for c and d. For example, if $a \geq 5$ and $b \geq 5$, then

$$ab = (5 + a')(5 + b') = 25 + 5a' + 5b' + a'b' = (5 + 3a' + 3b' + a'b')$$
$$+(2a' + 10) + (2b' + 10) > (2a' + 10) + (2b' + 10) = 2a + 2b,$$

for any $a' \geq 0$ and $b' \geq 0$ (we set $a = 5 + a'$ and $b = 5 + b'$). This simple observation leads to the following conclusion: If all of the numbers a, b, c, and d are 5 or more, then

$$ab > 2(a + b) = cd > 2(c + d) = ab,$$

resulting in a contradiction!

Assuming, then, that $a = \min\{a, b, c, d\}$, it's sufficient to consider only four cases, namely $1 \leq a \leq 4$. So, if $a = 1$, the problem reduces to

$$b = 2(c + d), \text{ and}$$
$$cd = 2(1 + b) = 2 + 2b.$$

Clearly, b must be even, so $b = 2b'$ and we have

$$b' = c + d, \text{ and}$$
$$cd = 2 + 4b'.$$

Replacing b' in the second equation, we get

$$cd = 2 + 4c + 4d,$$

which is easy to solve:

$$c = \frac{4d + 2}{d - 4} = 4 + \frac{18}{d - 4}.$$

As c is a natural number, d must be 5, 6, 7, 10, 13, or 22. For these we get three pairs of (symmetrical) solutions:

$$(c, d) = (5, 22); \ (c, d) = (6, 13); \ \text{or} \ (c, d) = (7, 10).$$

These three solutions imply $b = 54$, $b = 38$, and $b = 34$, respectively, and in that way we have found the first three solutions of the problem:

$(1, 54), (5, 22),$
$(1, 38), (6, 13),$ and
$(1, 34), (7, 10).$

The second case is for $a = 2$. In that case the problem reduces to:

$2b = 2(c + d),$ and
$cd = 2(2 + b),$

which simplifies to

$b = c + d,$ and
$cd = 4 + 2b.$

Again, replacing b in the second equation we get

$$cd = 4 + 2c + 2d,$$

which gives us

$$c = \frac{2d + 4}{d - 2} = 2 + \frac{8}{d - 2}.$$

As before, d must be 3, 4, 6, or 10. For these we get two pairs of symmetrical solutions

$$(c, d) = (3, 10) \ \text{and} \ (c, d) = (4, 6).$$

These two solutions imply $b = 13$ and $b = 10$, respectively, and in that way we have found two additional solutions of the problem:

$(2, 13), (3, 10),$ and
$(2, 10), (4, 6).$

The third case to consider is for $a = 3$. In that case b must be even ($b = 2b'$ as the product of a and b is even) and we have

$3b' = c + d,$ and
$cd = 6 + 4b'.$

In that case (replacing b' in the second equation) we get

$$c = \frac{4d + 18}{3d - 4} = 1 + \frac{d + 22}{3d - 4}.$$

We can see that $d > 6$ implies $c < 3$ (as $\frac{d+22}{3d-4} < 2$) which gives a contradiction (as $a = 3 < c$). Thus all we have to do is consider cases for d equal to 3, 4, 5, and 6. The first and last cases give the same (i.e., symmetrical) solution,

$$(c, d) = (3, 6),$$

which also implies $b = 6$. Thus, we found the sixth solution of the problem:

$$(3, 6), (3, 6).$$

The final case is for $a = 4$. Here

$$4b = 2(c + d), \text{ and}$$
$$cd = 2(4 + b),$$

which leads to

$$cd - 8 + c + d.$$

Thus,

$$c = \frac{d + 8}{d - 1} = 1 + \frac{9}{d - 1}.$$

Since $d \geq a = 4$, d is either 4 or 10. The latter value results in $c = 2$, which is a contradiction, whereas $d = 4$ implies $c = 4$ and $d = 4$. This case yields the final solution of the problem,

$$(4, 4), (4, 4),$$

and the problem has seven solutions in total.

It's important to keep in mind the way the search space was reduced here. "Playing" with the problem may give us important insights into its nature and provides the opportunity to discover ways for considering only a fraction of the possible solutions. Again, the key point is don't give up! And don't settle for some solution that's less than perfect if you don't have to.

You can practice these skills right away. Another similar puzzle was offered to us by Antoni Mazurkiewicz from the Polish Academy of Sciences.

The problem is to find the natural numbers a, b, and c, such that

$$a + b + 1 \text{ divides } c, \ a + c + 1 \text{ divides } b, \text{ and } b + c + 1 \text{ divides } a.$$

The problem has 12 solutions. For example, one of them is

$$(21, 14, 6),$$

as

21 + 14 + 1 = 36, which divides 6,
21 + 6 + 1 = 28, which divides 14, and
14 + 6 + 1 = 21, which, of course, divides 21.

Try to find all 12 solutions!

If you prefer something simple, you may consider the case where two "wrongs" make a "right"! The puzzle

$$
\begin{array}{ccccc}
W & R & O & N & G \\
W & R & O & N & G \\
\hline
R & I & G & H & T \\
\end{array}
$$

has a few solutions (addition is assumed). One of them is

$$
\begin{array}{ccccc}
2 & 5 & 7 & 3 & 4 \\
2 & 5 & 7 & 3 & 4 \\
\hline
5 & 1 & 4 & 6 & 8 \\
\end{array}
$$

Keep this in mind for some ripe opportunity!

4. Traditional Methods — Part 2

Two and two make four
Four and four make eight
Eight and eight make sixteen
Repeat! says the teacher.

Jacques Prévert, *Page d'écriture*

We've seen a slew of techniques that manipulate complete solutions. Let's now consider some of the algorithms that work with partial or incomplete solutions and solve problems by constructing solutions one piece at a time. We begin with the best-known category: greedy algorithms.

4.1 Greedy algorithms

Greedy algorithms attack a problem by constructing the complete solution in a series of steps. The reason for their popularity is obvious: simplicity! The general idea behind the greedy approach is amazingly simple: assign the values for all of the decision variables one by one and at every step make the best available decision. Of course, this approach assumes a heuristic for decision making that provides the best possible move at each step, the best "profit," thus the name *greedy*. But it's clear that the approach is also shortsighted since taking the optimum decisions at each separate step doesn't always return the optimum solution overall.

4.1.1 Greedy algorithms and the SAT

Let's consider a few examples by starting with the SAT problem. We'll assign a truth value (TRUE or FALSE) to each variable, but we need some heuristic to guide our decision-making process. Consider the following approach:

- For each variable from 1 to n, in any order, assign the truth value that results in satisfying the greatest number of currently unsatisfied clauses. If there's a tie, choose one of the best options at random.

This heuristic is indeed greedy: at every step it tries to satisfy the largest number of currently unsatisfied clauses! Unfortunately, we can easily show that the performance of such unrestrained greed is quite poor, even for some simple problems. For example, consider the following test case:

$$\overline{x}_1 \wedge (x_1 \vee x_2) \wedge (x_1 \vee x_3) \wedge (x_1 \vee x_4).$$

If we consider the variable x_1 first, then we choose $x_1 =$ TRUE because this assignment satisfies three clauses. But the first clause is not and cannot be satisfied when x_1 is TRUE, so all of the remaining effort of the algorithm is wasted. The formula will never be satisfied for any assignment of x_2, x_3, and x_4, as long as x_1 remains TRUE.

We can easily identify the problem with the above greedy approach. It's too greedy in the sense that we haven't given enough care as to what order to consider the variables. We'd be better off using some additional heuristic for choosing the order. In particular, we might start with those variables that appear in only a few clauses (like x_2, x_3, or x_4 in the formula above), leaving more commonly occurring variables (like x_1) for later. So, the next, upgraded greedy algorithm for the SAT might be:

- Sort all the variables on the basis of their frequency, from the smallest to the largest, for all clauses.

- For each variable, taken in the above order, assign a value that would satisfy the greatest number of currently unsatisfied clauses. In the event of a draw, make an arbitrary decision.

After some additional thought, we might discover some weaknesses of this new algorithm. For example, consider the following case:

$$(x_1 \vee x_2) \wedge (x_1 \vee x_3) \wedge (\overline{x}_1 \vee x_4) \wedge (\overline{x}_2 \vee \overline{x}_4) \wedge (x_2 \vee x_5) \wedge (x_2 \vee x_6) \wedge F,$$

where we assume that the formula F does not contain variables x_1 and x_2, but contains many occurrences of the remaining variables. Thus, x_1 is present in only three clauses, and x_2 is in four clauses, with all the other variables having a higher frequency. The above greedy algorithm would assign x_1 the value of TRUE because this assignment satisfies two clauses. Similarly, it would assign x_2 the value of TRUE because it satisfies two additional clauses. However, it wouldn't be possible to satisfy both the third and fourth clauses,

$$(\overline{x}_1 \vee x_4) \wedge (\overline{x}_2 \vee \overline{x}_4),$$

and the algorithm would fail to find the required truth assignment.

Of course, not to be defeated just yet, there's plenty of room for further "upgrades." We could introduce an extra rule that would forbid a particular assignment if it makes any clause FALSE. Or we might consider the frequency of variables in the *remaining* (i.e., still unsatisfied) clauses. We might also try to be more sophisticated. Instead of ordering the variables solely on the basis of their frequencies, we might take into account the length of a clause. The intuition is that an occurrence of a variable in a short clause has a different significance than an occurrence of the variable in a long clause! Unfortunately, none of these approaches leads to an algorithm that will work for all SAT problems. *There is no such greedy algorithm for the SAT!*

4.1.2 Greedy algorithms and the TSP

The most intuitive greedy algorithm for the TSP is based on the nearest-neighbor heuristic: starting from a random city, proceed to the nearest unvisited city and continue until every city has been visited, at which point return to the first city. Such a tour can be far from perfect. There's often a high price to pay for making greedy choices at the beginning of the process. A simple example to illustrate the case is given in figure 4.1, where starting from city A, this greedy heuristic constructs a tour A – B – C – D – A, with the total cost of $2+3+23+5 = 33$, but the tour A – C – B – D – A costs only $4+3+7+5 = 19$.

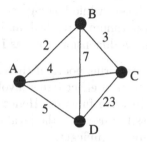

Fig. 4.1. A sample TSP with four cities. Note that the costs for each edge do not correspond to the distance between the cities as drawn.

Another greedy possibility is to construct a tour by choosing the shortest available connections (i.e., edges) between cities, avoiding a situation where a city is involved in more than two connections. In other words, we start by finding the cheapest among all available connections between every pair of cities, then we select the next cheapest available connection, and so on. Of course, once we select the connections between, say, cities A – B and cities B – E, then we can't select any more connections that involve city B. Note that this greedy approach would produce the same answer as the previous greedy approach on the four-city TSP in figure 4.1.

Yet another greedy possibility [73] is based on the following concept. A single city is selected and the initial "tour"[1] is constructed by starting from the selected city, visiting another city, returning to the selected city, going to some other unvisited city, returning to the selected city, and so on. In other words, the initial tour consists of a path that is a union of uniform segments leading from the selected city to some other city and back. Then — this is where greedy choices are made — we consider pairs of cities (excluding the selected city). A pair of cities (say, i and j) is selected for which the largest savings can be made by going directly from i to j instead of returning inbetween to the selected city. The algorithm considers pairs of unselected cities in a non-increasing order of

---

[1]This initial tour is not a tour in the sense of TSP, since the selected city is visited many times.

savings, modifying the tour if possible (e.g., avoiding premature cycles). Again, this approach will fail on the simple example given in figure 4.1. For every common sense heuristic you can invent, you can find a pathological case that will make it look very silly.

4.1.3 Greedy algorithms and the NLP

Let's also consider the final case of the NLP. There really are no efficient greedy algorithms for the NLP, but we can design an algorithm that displays some greedy characteristics. For example, to optimize a function of, say, two variables, x_1 and x_2, we could set one of the variables, say, x_1, at a constant and vary x_2 until we reach an optimum. Then, while holding the new value of x_2 constant, we could vary x_1 until a new optimum is reached, and so on. Naturally, this *line search* can be generalized into n dimensions, but as Himmelblau stated [219]:

> This process, however, performs poorly if there are interactions be-
> tween x_1 and x_2, that is, if in effect terms involving products of x_1
> and x_2 occur in the objective function. Hence the method cannot
> be recommended unless the user has the [evaluation] function such
> that the interactions are insignificant.

Strictly speaking, line searches aren't really greedy algorithms because they only evaluate complete solutions. Nevertheless, the intention of choosing the best available opportunity with respect to a single dimension at a time does follow the general notion of greedy algorithms.

Greedy methods, whether applied to the SAT, TSP, NLP, or other domains, are conceptually simple, but they normally pay for that simplicity by failing to provide good solutions to complex problems with interacting parameters. And those are the problems that we face routinely in the real world.

4.2 Divide and conquer

Sometimes it's a good idea to solve a seemingly complicated problem by break-ing it up into smaller simpler problems. You might be able to then solve each of those easier problems and find a way to assemble an overall answer out of each part. This "divide and conquer" approach is really only cost effective if the time and effort that's required to complete the decomposition, solve all of the decomposed problems, and then reassemble an answer is less than the cost of solving the problem as it originally stands with all its inherent complex-ity. Also, you have to take care that when you assemble the solution from the decomposed pieces, you actually do get the answer you were looking for. Some-times the chance for assembling an overall solution disintegrates as you break the problem apart.

The outline of the divide and conquer (D&C) approach is given in figure 4.2.

procedure D&C (P)
begin
    split the problem P into subproblems $P_1, \ldots, P_k$
    for $i = 1$ **to** k **do**
      if size(P_i) $< \rho$ **then** solve P_i (getting s_i)
                 else $s_i \leftarrow$ D&C (P_i)
    combine s_i's into the final solution.
end

Fig. 4.2. Divide and conquer recursive procedure for solving problem P

The original problem P is replaced by a collection of subproblems, each of which is further decomposed into sub-subproblems, and so forth, often in a recursive manner. The process continues until the problems are reduced to being trivial (i.e., smaller than a conceptual constant ρ in the outline given in figure 4.2) so that they can be solved "by hand." The algorithm then spirals "upward," combining the solutions to construct a solution to larger subproblems.

The divide and conquer principle is intrinsic to many sorting algorithms, such as quicksort and mergesort. Polynomial and matrix multiplications can also be accomplished using this approach. To give you an explicit example here, consider the following problem that illustrates the idea of the divide and conquer approach nicely. There is a chess board with a dimension of 2^m (i.e., it consists of 2^{2m} squares) with a hole, i.e., one arbitrary square is removed. We have a number of L shaped tiles (see figure 4.3) and the task is to cover the board by these tiles. (The orientation of a tile isn't important.)

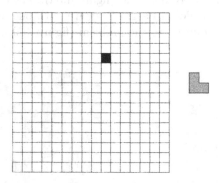

Fig. 4.3. A chess board $2^m \times 2^m$ ($m = 4$) and a tile

In this case the search space consists of many possible arrangements of tiles on the board and we have to find the right one. Many methods can be considered, however, the divide and conquer technique is ideal here. We can simply divide the board into four equal areas: the hole is located in one of them. Thus, we can place the first tile in a such way that it would cover three squares: one from each area that doesn't contain the original hole (figure 4.4a). Then, instead of one square (with a hole), there are four smaller squares. Each of them has a hole, whether it's the original one or one created by the placement of a tile. Thus we can continue this process, dividing each of the smaller squares into four even smaller squares and place new tiles such that each of the new smaller squares has precisely one "hole" (figure 4.4b).

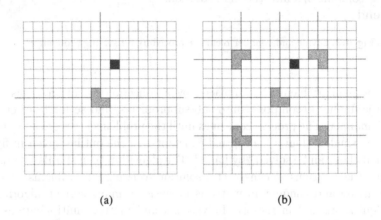

(a)   (b)

Fig. 4.4. The placement of the first tile (a) and the next four tiles (b)

Note that every square, which is now just 4×4, has just one missing piece, i.e., one hole. Continuing our procedure, we can divide each of them into four 2×2 squares and by an appropriate placement of tiles, each of these new small squares would have precisely one hole (figure 4.5a). Then we face the easiest possible task: to place a tile into a 2×2 square with one missing piece! Thus, we easily get the final solution to the problem (figure 4.5b).

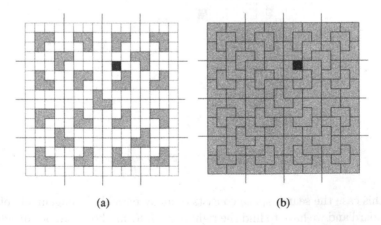

(a)   (b)

Fig. 4.5. (a) The placement of the next 16 tiles and (b) the final solution

4.3 Dynamic programming

Dynamic programming works on the principle of finding an overall solution by operating on an intermediate point that lies between where you are now and where you want to go.[2] The procedure is recursive, in that each next intermediate point is a function of the points already visited. A prototypical problem that is suitable for dynamic programming has the following properties:

- The problem can be decomposed into a sequence of decisions made at various stages.

- Each stage has a number of possible states.

- A decision takes you from a state at one stage to some state at the next stage.

- The best sequence of decisions (also known as a *policy*) at any stage is independent of the decisions made at prior stages.

- There is a well-defined cost for traversing from state to state across stages. Moreover, there is a recursive relationship for choosing the best decisions to make.

The method can be applied by starting at the goal and working backward to the current state. That is, we can first determine the best decision to make at the last stage. From there, we determine the best decision at the next to last stage, presuming we will make the best decision at the last stage, and so forth.

The simplest example is the *stagecoach problem* developed by Harvey Wagner [217]. The problem has a lot of window dressing involving a salesman taking his stagecoach through territory with hostile Native American Indians, but we'll dispense with that to cut straight to the core of the issue.

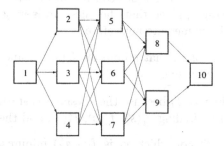

Fig. 4.6. The flow diagram indicating the choices for the salesman as he takes his stagecoach through hostile territory.

---

[2]The term *dynamic* was used to identify the approach as being useful for problems "in which times plays a significant role, and in which the order of operations may be crucial" [40].

The salesman must start from his current position and arrive at a fixed destination. He effectively has three stages at which he can make alternative choices about how to proceed (see figure 4.6). At the first stage, he has three alternative paths. Likewise, at the second stage, there are three alternatives. Finally, at the third stage there are two alternatives. The fourth stage does not offer a choice. The costs for traversing from each state at each stage to each next possible state are shown in figure 4.7. The problem is to find the least-cost path from the first state (1) to the last state (10).

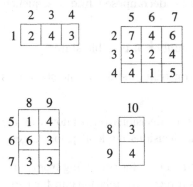

Fig. 4.7. The cost for making each possible choice at each decision stage. The row indicates the current state and the column indicates the next state. The four matrices correspond to the four stages. Note that only the first three stages offer a choice of which state to choose.

We need to introduce some notation here before proceeding to work out the answer:

- There are n stages and the decision of which state to choose at the n-th stage is x_n. Here, $n = 4$.

- $f_n(s, x_n)$ is the cost of the best sequence of decisions (policy) for all of the remaining stages, given that the salesman is in state s, stage n, and chooses x_n as the immediate next state.

- $x_n^*(s)$ is the value of x_n that minimizes $f_n(s, x_n)$ and $f_n^*(s)$ is the corresponding minimum value.

- The goal is to find $f_1^*(1)$ because the salesman is starting in state 1. This is accomplished by finding $f_4^*(s)$, $f_3^*(s)$, $f_2^*(s)$, and then finally $f_1^*(1)$.

What then is $f_4^*(s)$? For which x_4 is $f_4(s, x_4)$ minimized? This is almost a trick question because there's only one state to choose while in the fourth stage: $x_4 = 10$. Thus we easily compute the values shown below, and we have completed the first phase of the dynamic programming procedure.

s	$f_4^*(s)$	$x_4^*(s)$
8	3	10
9	4	10

The above matrix displays the cost of going from states 8 or 9 to a state at the final stage (in this case, there is only one such state).

The next phase asks for the value of $f_3^*(s)$. Recall that since the decisions are independent, $f_3(s, x_3) = c_{sx_3} + f_4^*(x_3)$, where c_{sx_3} is the cost of traveling from s to x_3. The matrix shown below indicates the relevant values of $f_3(s, x_3)$ for every possible choice of destination presuming s is 5, 6, or 7.

s	$f_3(s, 8)$	$f_3(s, 9)$	$f_3^*(s)$	$x_3^*(s)$
5	4	8	4	8
6	9	7	7	9
7	6	7	6	8

We can see, for example, that if we're in state 5, then state 8 minimizes the total remaining cost. State 9 minimizes the cost if $s = 6$, and state 8 minimizes the cost if $s = 7$.

Working backward again, we must find the value of $f_2^*(s)$. In an analogous manner to what appears directly above, the matrix shown below provides the relevant values of $f_2(s, x_2)$. Note that $f_2(s, x_2) = c_{sx_2} + f_3^*(x_2)$.

s	$f_2(s, 5)$	$f_2(s, 6)$	$f_2(s, 7)$	$f_2^*(s)$	$x_2^*(s)$
2	11	11	12	11	5 or 6
3	7	9	10	7	5
4	8	8	11	8	5 or 6

Finally, with regard to $f_1^*(s)$, the last matrix offers the costs of proceeding to states 2, 3, or 4:

s	$f_1(s, 2)$	$f_1(s, 3)$	$f_1(s, 4)$	$f_1^*(s)$	$x_1^*(s)$
1	13	11	11	11	3 or 4

This is the final table for the stagecoach problem. Now the dynamic programming method has been applied all the way back to the first stage. The values in the above matrix indicate the cost of traveling from state 1 to 2, 3, or 4.

At this point we can identify the best answer to the overall problem. In fact, here there are three best answers (policies), each with an equal cost of 11 units:

- $1 \to 3 \to 5 \to 8 \to 10$
- $1 \to 4 \to 5 \to 8 \to 10$
- $1 \to 4 \to 6 \to 9 \to 10$.

Note that the obverse procedure of choosing the least-cost path moving forward through the stages does not give the optimum solution. That greedy solution has a cost of 13 units (you can verify this). Thus, here is a case where the greedy approach fails, but dynamic programming generates the right answer. One drawback of the approach, however, is that it can be computationally intensive. If N is both the number of stages and states per stage, then the required number of operations scales as N^3. The method can, however, be extended to handle a variety of optimization problems including the order of multiplying matrices and finding the best organization for a tree to minimize the cost of searching [425].

Dynamic programming algorithms tend to be somewhat complicated to understand. This is because, in practice, the construction of a dynamic program depends on the problem. It is a sort of "artistic intellectual activity depending in part on the specific structure of the sequential decision problem" [432]. This is why we'll illustrate this technique by two additional examples: matrix multiplication and the TSP.

Suppose that the dimensions of the matrices A_1, A_2, A_3, and A_4 are 20×2, 2×15, 15×40, and 40×4, respectively, and that we want to know the optimum way to compute $A_1 \times A_2 \times A_3 \times A_4$, i.e., we would like to compute this product with a minimum number of multiplications. Here we assume that to multiply two matrices, P, which is $n \times k$, by Q, which is $k \times m$, it takes nkm multiplications. The resulting matrix R is $n \times m$ and

$$r_{ij} = \sum_{v=1}^{k} p_{iv} q_{vj},$$

for all $1 \leq i \leq n$ and $1 \leq j \leq m$.
Note that different orders of multiplications have different costs. For example,

- $A(B(CD))$ requires $15 \cdot 40 \cdot 4 + 2 \cdot 15 \cdot 4 + 20 \cdot 2 \cdot 4 = 2680$ multiplications,

- $(AB)(CD)$ requires $20 \cdot 2 \cdot 15 + 15 \cdot 40 \cdot 4 + 20 \cdot 15 \cdot 4 = 4200$ multiplications, whereas

- $((AB)C)D$ requires $20 \cdot 2 \cdot 15 + 20 \cdot 15 \cdot 40 + 20 \cdot 40 \cdot 4 = 15800$ multiplications!

In a dynamic programming approach we create a structure $M(i, j)$ where we maintain a record of the minimum number of multiplications required to multiply matrices from A_i to A_j $(i \leq j)$. Clearly, $M(1, 1) = M(2, 2) = M(3, 3) = M(4, 4) = 0$, as no multiplications are required in these cases. Note that the problem involves finding $M(1, 4)$.

The connection between solutions of the smaller problems and a bigger one is

$$M(i, j) = \min_{i \leq k < j}\{M(i, k) + M(k + 1, j) + cost_{ij}^{k}\},$$

where $cost_{ij}^{k}$ is the number of multiplications required for multiplying the product $A_i \ldots A_k$ by $A_{k+1} \ldots A_j$. The point is that to multiply a sequence from A_i to A_j in the optimal way, we must find the optimal breaking point k such that the total number of multiplications required for calculating the product $A_i \ldots A_k$, which is $M(i, k)$, the product $A_{k+1} \ldots A_j$, which is $M(k + 1, j)$, and these two products together (which is $cost_{ij}^{k}$) is minimal.

Keeping this in mind, we can proceed as follows. It's a basic exercise to get:

$M(1, 2) = 600,$
$M(2, 3) = 1200,$
$M(3, 4) = 2400,$

because there's no room for a breaking point in multiplying two matrices. Then,

$M(1,3) = 2800$
$M(2,4) = 1520.$

Note that $M(1,3)$ is the smaller of two possibilities

$M(1,1) + M(2,3) + cost_{13}^1 = 0 + 1200 + 1600 = 2800,$ and
$M(1,2) + M(3,3) + cost_{13}^2 = 600 + 0 + 12000 = 12600.$

Similarly, $M(2,4)$ is the smaller of two possibilities

$M(2,2) + M(3,4) + cost_{24}^2 = 0 + 2400 + 120 = 2520,$ and
$M(2,3) + M(4,4) + cost_{24}^3 = 1200 + 0 + 320 = 1520.$

Finally, we find

$M(1,4) = 1680,$

as the smallest value from three possibilities

$M(1,1) + M(2,4) + cost_{14}^1 = 0 + 1520 + 160 = 1680,$
$M(1,2) + M(3,4) + cost_{14}^2 = 600 + 2400 + 1200 = 4200,$ and
$M(1,3) + M(4,4) + cost_{14}^3 = 2800 + 0 + 3200 = 6000.$

Thus, the minimal cost in terms of the number of multiplications is 1680, but we still need to find the corresponding order of matrix multiplications. To find this best order we need an additional data structure, $O(i,j)$, where we keep the index of the best breaking point. In other words, $O(i,j) = k$ if and only if $M(i,j)$ attains the minimum value for $M(i,k)+M(k+1,j)+cost_{ij}^k$. The indices kept in the matrix O reveal the order we were searching for: $A_1((A_2A_3)A_4)$.

Let's conclude this section by considering an example of a dynamic program for the TSP. It's a lengthy example, so be forewarned. To make things meaningful, let's tackle a five-city TSP. The matrix L of distances between cities is given below.

$$L = \begin{bmatrix} 0 & 7 & 12 & 8 & 11 \\ 3 & 0 & 10 & 7 & 13 \\ 4 & 8 & 0 & 9 & 12 \\ 6 & 6 & 9 & 0 & 10 \\ 7 & 7 & 11 & 10 & 0 \end{bmatrix}$$

The cost of traveling from city i to city j is given in the i-th row and j-th column of the matrix L. For example, the distance between cities 1 and 3 is $L(1,3) = 12$. This distance is different than the distance between cities 3 and 1, which is $L(3,1) = 4$; therefore, we're facing an asymmetric TSP. Note that $L(i,i) = 0$ for all $1 \leq i \leq n$. Figure 4.8 illustrates the case.

The tour for the TSP is a cycle that visits *every* city once and only once, so the starting city is not of any real significance. Suppose we start from city 1. Now, we're ready to split the problem into smaller problems. Let $g(i, S)$ denote the length of the shortest path from city i to city 1 that passes through *each* city listed in the set S exactly once. Thus, in particular,

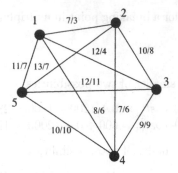

Fig. 4.8. A TSP with five cities. For each pair of cities, two numbers are given separated by a slash. The first number provides the distance from the city with the lower index to the city with the larger index, whereas the second number represents the distance in the opposite direction.

$$g(4, \{5, 2, 3\})$$

represents the shortest path that leads from city 4 through cities 5, 2, and 3 (in some unspecified order) and then returns to city 1. To find the length of the shortest complete tour for the TSP we need to find

$$g(1, V - \{1\}),$$

where V represents the set of *all* the cities for the TSP. In other words, we're trying to find the shortest path that starts from city 1, visits every city in the set $V - \{1\}$ exactly once, and returns to city 1.

The dynamic programming formulation provides a connection between solutions of the smaller problems and a bigger one. In general, we claim that

$$g(i, S) = \min_{j \in S}\{L(i, j) + g(j, S - \{j\})\}.$$

In other words, to select the shortest path that originates at city i and visits every city in S before returning to city 1, we have to find a city j from the set S such that the sum of $L(i, j)$ and its continuation, $g(j, S - \{j\})$, is minimized. If we know the solutions for the smaller problems (i.e., the problems where the cardinality of S is k), we can obtain a solution for a bigger problem (i.e., where the cardinality of S is $k + 1$).

Let's return to the example. The problem is to find

$$g(1, \{2, 3, 4, 5\}).$$

Note also that

$$g(2, \emptyset) = L(2, 1) = 3,$$
$$g(3, \emptyset) = L(3, 1) = 4,$$
$$g(4, \emptyset) = L(4, 1) = 6, \text{ and}$$
$$g(5, \emptyset) = L(5, 1) = 7,$$

since the set S is empty and we proceed directly from the city i ($i = 2, 3, 4, 5$) to city 1. The next iteration is also straightforward: we need to find the solutions to all the problems where the cardinality of S is one. This would involve solving 12 subproblems, since we'll start from cities 2, 3, 4, and 5, and for each of these cities, we'll consider three possible singleton sets. For example, for city 2 we have

$$g(2, \{3\}) = L(2, 3) + g(3, \emptyset) = 10 + 4 = 14,$$
$$g(2, \{4\}) = L(2, 4) + g(4, \emptyset) = 7 + 6 = 13, \text{ and}$$
$$g(2, \{5\}) = L(2, 5) + g(5, \emptyset) = 13 + 7 = 20.$$

Note that the result $g(2, \{3\}) = 14$ means that the length of the shortest path from city 2 to 1, which before arriving at 1 goes through city 3, is 14. This result is obvious since S consists of only a single element and there is no room for any choices. Similarly, for city 3 we have

$$g(3, \{2\}) = L(3, 2) + g(2, \emptyset) = 8 + 3 = 11,$$
$$g(3, \{4\}) = L(3, 4) + g(4, \emptyset) = 9 + 6 = 15,$$
$$g(3, \{5\}) = L(3, 5) + g(5, \emptyset) = 12 + 7 = 19,$$

for city 4,

$$g(4, \{2\}) = L(4, 2) + g(2, \emptyset) = 6 + 3 = 9,$$
$$g(4, \{3\}) = L(4, 3) + g(3, \emptyset) = 9 + 4 = 13,$$
$$g(4, \{5\}) = L(4, 5) + g(5, \emptyset) = 10 + 7 = 17,$$

and for city 5,

$$g(5, \{2\}) = L(5, 2) + g(2, \emptyset) = 7 + 3 = 10,$$
$$g(5, \{3\}) = L(5, 3) + g(3, \emptyset) = 11 + 4 = 15,$$
$$g(5, \{4\}) = L(5, 4) + g(4, \emptyset) = 10 + 6 = 16.$$

Now we're ready for the next iteration of the dynamic program where the cardinality of S is two. In this iteration we have to solve 12 subproblems as well, since each of four cities can be considered along with any two other cities (out of three). Thus, for city 2 we have

$$g(2, \{3, 4\}) = \min\{L(2, 3) + g(3, \{4\}), L(2, 4) + g(4, \{3\})\}$$
$$= \min\{10 + 15, 7 + 13\} = \min\{25, 20\} = 20,$$
$$g(2, \{3, 5\}) = \min\{L(2, 3) + g(3, \{5\}), L(2, 5) + g(5, \{3\})\}$$
$$= \min\{10 + 19, 13 + 15\} = \min\{29, 28\} = 28,$$
$$g(2, \{4, 5\}) = \min\{L(2, 4) + g(4, \{5\}), L(2, 5) + g(5, \{4\})\}$$
$$= \min\{7 + 17, 13 + 16\} = \min\{24, 29\} = 24.$$

The result $g(2, \{3, 4\}) = 20$ means that the length of the shortest path from city 2 to 1, which before arriving at 1 goes through cities 3 and 4 (in some order), is 20. This time we have to consider two possibilities (going first to city 3 or going first to city 4), and select the better option. Similarly, for city 3,

$$g(3, \{2,5\}) = \min\{L(3,2) + g(2, \{5\}), L(3,5) + g(5, \{2\})\}$$
$$= \min\{8 + 20, 12 + 10\} = \min\{28, 22\} = 22,$$
$$g(3, \{2,4\}) = \min\{L(3,2) + g(2, \{4\}), L(3,4) + g(4, \{2\})\}$$
$$= \min\{8 + 13, 9 + 9\} = \min\{21, 18\} = 18,$$
$$g(3, \{4,5\}) = \min\{L(3,4) + g(4, \{5\}), L(3,5) + g(5, \{4\})\}$$
$$= \min\{9 + 17, 12 + 16\} = \min\{26, 28\} = 26,$$

for city 4,

$$g(4, \{2,3\}) = \min\{L(4,2) + g(2, \{3\}), L(4,3) + g(3, \{4\})\}$$
$$= \min\{6 + 14, 9 + 15\} = \min\{20, 24\} = 20,$$
$$g(4, \{2,5\}) = \min\{L(4,2) + g(2, \{5\}), L(4,5) + g(5, \{2\})\}$$
$$= \min\{6 + 20, 10 + 10\} = \min\{26, 20\} = 20,$$
$$g(4, \{3,5\}) = \min\{L(4,3) + g(3, \{5\}), L(4,5) + g(5, \{3\})\}$$
$$= \min\{9 + 19, 10 + 15\} = \min\{28, 25\} = 25.$$

and for city 5,

$$g(5, \{2,3\}) = \min\{L(5,2) + g(2, \{3\}), L(5,3) + g(3, \{2\})\}$$
$$= \min\{7 + 14, 11 + 11\} = \min\{21, 22\} = 21,$$
$$g(5, \{2,4\}) = \min\{L(5,2) + g(2, \{4\}), L(5,4) + g(4, \{2\})\}$$
$$= \min\{7 + 13, 10 + 19\} = \min\{20, 29\} = 20,$$
$$g(5, \{3,4\}) = \min\{L(5,3) + g(3, \{4\}), L(5,4) + g(4, \{3\})\}$$
$$= \min\{11 + 15, 10 + 13\} = \min\{26, 23\} = 23.$$

Now we're ready for the next step, where the cardinality of the set S is three. Here we face only four subproblems: one for each city. Thus, for city 2 we have to select the shortest path among three paths:

$$g(2, \{3,4,5\})$$
$$= \min\{L(2,3) + g(3, \{4,5\}), L(2,4) + g(4, \{3,5\}), L(2,5) + g(5, \{3,4\})\}$$
$$= \min\{10 + 26, 7 + 25, 13 + 23\} = \min\{36, 32, 34\} = 32.$$

The result $g(2, \{3,4,5\}) = 32$ means that the length of the shortest path from city 2 to 1, which before arriving at 1 goes through cities 3, 4, and 5 (in some order), is 32. Now we have to consider three possibilities, going first to city 3, 4, or 5, and select the best option. Similarly, for cities 3, 4, and 5 we have

$$g(3, \{2,4,5\})$$
$$= \min\{L(3,2) + g(2, \{4,5\}), L(3,4) + g(4, \{2,5\}), L(3,5) + g(5, \{2,4\})\}$$
$$= \min\{8 + 24, 9 + 20, 12 + 20\} = \min\{32, 29, 32\} = 29,$$
$$g(4, \{2,3,5\})$$
$$= \min\{L(4,2) + g(2, \{3,5\}), L(4,3) + g(3, \{2,5\}), L(4,5) + g(5, \{2,3\})\}$$
$$= \min\{6 + 28, 9 + 22, 10 + 21\} = \min\{34, 31, 31\} = 31, \text{ and}$$
$$g(5, \{2,3,4\})$$
$$= \min\{L(5,2) + g(2, \{3,4\}), L(5,3) + g(3, \{2,4\}), L(5,4) + g(4, \{2,3\})\}$$
$$= \min\{7 + 20, 11 + 18, 10 + 20\} = \min\{27, 29, 30\} = 27.$$

Now, it's time for the final iteration where we start from city 1 and close the cycle. Thus we return to the original problem:

$$g(1, \{2, 3, 4, 5\}) = ?$$

Starting from city 1 we have four options: to go first either to city 2, city 3, city 4, or city 5. The other subproblems (i.e., the shortest lengths for each continuation of each possible beginning) are already known, so it's easy to make the final decision:

$$g(1, \{2, 3, 4, 5\}) = \min\{L(1, 2) + g(2, \{3, 4, 5\}), L(1, 3) + g(3, \{2, 4, 5\}),$$
$$L(1, 4) + g(4, \{2, 3, 5\}), L(1, 5) + g(5, \{2, 3, 4\})\} =$$
$$\min\{7 + 32, 12 + 29, 8 + 31, 11 + 27\} = \min\{39, 41, 39, 38\} = 38.$$

Thus the shortest tour has a length of 38.

Which tour is that? As with the matrix multiplication problem, we've found the optimum value of the evaluation function but we still have to find out which tour has given us this result. Again, this is easy if we keep track of our calculations. We need an additional data structure, W, that provides information on the next city that should be chosen to minimize a path. Thus, for example,

$$W(5, \{2, 3, 4\}) = 2,$$

since the shortest path from city 5 to city 1, which has to pass through cities 2, 3, and 4 before arriving at 1, must go to city 2 first. Similarly,

$$W(1, \{2, 3, 4, 5\}) = 5.$$

Thus, the global solution for this TSP is given by the tour

$$1 - 5 - 2 - 4 - 3 - 1,$$

and the length of this tour is, of course,

$$11 + 7 + 7 + 9 + 4 = 38.$$

Does the thought of using dynamic programming to solve a 50-city TSP strike fear in your heart?

4.4 Branch and bound

We've seen that the size of real-world problems can grow very large as the number of variables in the problem increases. Recall that there are $(n-1)!/2$ different solutions for the symmetric TSP. Exhaustive search is out of the question for $n > 20$, so it would be helpful to have some heuristic that eliminates parts of the search space where we know that we won't find an optimum solution.

Branch and bound is one such heuristic that works on the idea of successive partitioning of the search space. We first need some means for obtaining a lower bound on the cost for any particular solution (or an upper bound depending on

whether we're minimizing or maximizing). The idea is then that if we have a
solution with a cost of say, c units, we know that the next solution to try has
a lower bound that is greater than c, and we are minimizing, we don't have to
compute just how bad it actually is. We can forget about that one and move
on to the next possibility.

It helps to think about the search space as being organized like a tree. The
heuristic of branch and bound prunes away branches that aren't of interest. For
example, let's consider a five-city symmetric TSP. The entire search space S
can be partitioned on the basis of, say, whether or not the edge (1 2) belongs to
a tour. Similarly, the space can be partitioned further with respect to whether
or not the edge (2 3) is included, and so forth. Figure 4.9 shows a tree that
represents this principle. The leaves of the tree represent $(5-1)!/2 = 12$ possible
tours.

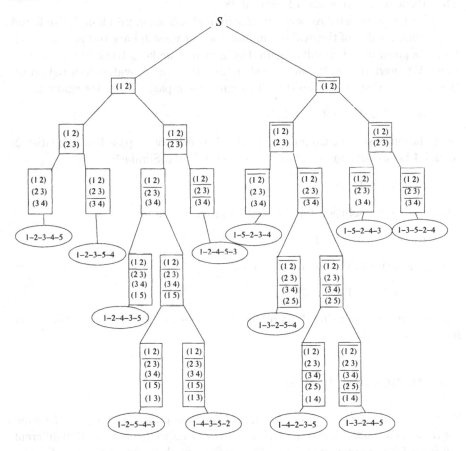

Fig. 4.9. A partition of the search space S of the five-city TSP according to whether a given
edge $(i\ j)$ belongs to the tour, or not. In the latter case, a horizontal bar is indicated over the
edge. The edges were selected in random order. The implied tours are given as well.

Suppose that the cost of traveling between cities was described by the following cost matrix:

$$\begin{bmatrix} 0 & 7 & 12 & 8 & 11 \\ 7 & 0 & 10 & 7 & 13 \\ 12 & 10 & 0 & 9 & 12 \\ 8 & 7 & 9 & 0 & 10 \\ 11 & 13 & 12 & 10 & 0 \end{bmatrix}$$

where each entry is the cost of traveling from a city in the i-th row to one in the j-th column. The zeros down the main diagonal indicate that we can't travel from a city to itself.

Given the organization of the search space as a tree, we need a heuristic for estimating a lower bound on the cost for any final solution, or even any node. If that lower bound is higher than the cost of the best solution we've found so far, we can discard the new solution and keep looking without having to actually compute its exact cost.

Here's a simple but perhaps not very effectual way to compute a lower bound for a tour. Consider an instance of a complete solution to the TSP. Every tour comprises two adjacent edges for every city: one edge gets you to the city, the other edge takes you on to the next city. Thus, if we select the two shortest edges that are connected to each city and take the sum of the lengths of these edges divided by two, we'll obtain a lower bound. Indeed, we couldn't possibly do any better because this rule selects the very best edges for every city. With respect to the above matrix, the lower bound over all possible tours is:

$$[(7+8) + (7+7) + (9+10) + (7+8) + (10+11)]/2 = 84/2 = 42.$$

Note that 7 and 8 in the first parentheses correspond to the lengths of the two shortest edges connected to city 1, whereas the entries 7 and 7 in the second parentheses correspond to the two shortest edges connected to city 2, and so forth.

Once some edges are specified for a tour we can incorporate that information and calculate a lower bound for that partial solution. For instance, if we knew that edge (2 3) was included but edge (1 2) was not, then the lower bound for this partial solution would be

$$[(8+11) + (7+10) + (9+10) + (7+8) + (10+11)]/2 = 91/2 = 45.5.$$

We can improve the lower bound by including the implied edges or excluding those that cannot occur. If edges (1 2) and (2 4) were included in a tour, from what's above we'd determine the lower bound to be 42. Upon closer examination, however, we can exclude the edge (1 4) here, so a better lower bound is

$$[(7+11) + (7+7) + (9+10) + (7+9) + (10+11)]/2 = 88/2 = 44.$$

The better the lower bound, the faster the algorithm for searching through the tree will be because it will eliminate more solutions. On the other hand, there is a cost for computing these lower bounds. For the method to be useful, the cost for computing the bounds has to be made up for by the time saved in pruning the tree. So we really want the lower bounds to be as "tight" as possible.

There are many possibilities for improving the lower bound in the TSP but they're beyond our scope here (see [344]). Instead, let's turn our attention back to numerical optimization and consider how branch and bound can be applied.

Assume that the task is to

minimize $f(\mathbf{x})$,

where $\mathbf{x} \in D \subseteq \mathbb{R}^n$. In other words, we search for $\mathbf{x}^* = \{\mathbf{x} \in D \,|\, f(\mathbf{x}) = f^*\}$, where $f^* = \inf_{\mathbf{x} \in D} f(\mathbf{x})$. Using the notation from [296],

- D_i is an n-dimensional box in $\mathbb{R}^n$, i.e., D_i is defined by $\mathbf{l}^i$ and $\mathbf{u}^i$, and $\mathbf{x} \in D_i$ iff $l_k^i \leq x_k \leq u_k^i$ for all $1 \leq k \leq n$.

- C, termed the candidate set, is a set of boxes D_i, $i = 1, \ldots, p$. For each box D_i, we obtain a lower bound of the function values of f in the box, i.e., a lower bound for f on D_i.

- f_{bound} represents the current upper limit of f^* found by the algorithm. This value is obtained, for example, as the lowest function value calculated so far.

The idea is that initially $C = \{D\}$ and the branch and bound method will reduce the set C and make it converge to the set of global minimizers $\mathbf{x}^*$. Note, however, that the issue of convergence is not straightforward here. The process depends on the selection mechanism of boxes. If the box with the smallest lower bound is always selected, the algorithm may never converge. On the other hand, if the largest box is always selected — according to some size measure such as the total length of all box's dimensions — the algorithm will converge, but the running time might be prohibitive. Also, the term "convergence" is topology dependent, as a sequence of boxes may only approximate a curve.

Figure 4.10 displays the skeleton of the branch and bound algorithm. At any iteration, C maintains a set of active boxes D_i's; at any time the set of global minima $\mathbf{x}^*$ is included in the union of these boxes:

$$\mathbf{x}^* \in \bigcup D_i \subseteq D.$$

A single box D_i is selected and removed from the candidate set C at each iteration. There are many methods for choosing which box to remove, but one common method is to select a box with the smallest lower bound. We then try to reduce the size of D_i or to eliminate it altogether. This is done by testing the evaluation function f for monotonicity in D_i (i.e., its partial derivatives are always positive or negative). If this is the case for some dimension i then the box is reduced to a single point along that dimension on the boundary of the

procedure branch and bound
begin
    initialize C
    initialize f_{bound}
    while (not termination-condition) **do**
        remove-best-box $C \rightarrow D_i$
        reduce-or-subdivide $D_i \rightarrow D_i'$
        update f_{bound}
        $C \leftarrow C \cup D_i'$
        for all $D_i \in C$ **do**
            if (lower bound of $f(D_i)$) $> f_{bound}$ **then** remove D_i from C
end

Fig. 4.10. Branch and bound algorithm

box.[3] If it isn't possible to reduce D_i, then it's replaced by two or more boxes. In any case, the candidate set C is extended by D_i', which represents either the reduced box D_i or a subdivision of D_i. Of course, new lower bounds are obtained before D_i' is inserted into C. Note also that if a lower bound of a box is greater than f_{bound}, the box can be eliminated from the candidate set.

There are many variants of this generic branch and bound algorithm. It's possible to use an interval arithmetic version of this approach where all of the calculations are performed on intervals rather than on real numbers, or a stochastic version where the values of f are calculated at a number of random points. For more details, see [296]. For general information on interval methods for global optimization, see [378].

4.5 *A** algorithm

We've seen that taking the best available move at any given time can lead to trouble. Greedy algorithms don't always perform very well. The problem is that what's better now might not be what's needed later, but if we had an evaluation function that was sufficiently "informative" to avoid these traps, we could use such greedy methods to a better advantage. This simple idea leads to a concept called *best-first* search, and its extension, the *A** algorithm.

When we can organize the search space as a tree, which is a special case of a directed graph and we'll return to this point shortly, we have the option of how to search that tree. We can use depth-first search and proceed down the tree to some prescribed level before making any evaluations and then backtrack our way through the tree. Or we might instead try to order the available nodes

---

[3]Additional tests can be done as well, e.g., Newton-reduction can be applied [296]. This can result in a decision that D_i does not contain stationary points — all iterative sequences of a search for stationary points go out of D_i — or that there is one stationary point, e.g., all iterative sequences converge to the same point.

according to some heuristic that corresponds with our expectations of what we'll find when we get to the deepest level of our search. We'd want to search first those nodes that offer the best chance of finding something good. A skeleton of the best-first search algorithm is offered in figure 4.11.

procedure best-first(v)
begin
    visit v
    for each *available* w **do**
        assign a heuristic value for w
    $q \leftarrow$ the best *available* node
    best-first(q)
end

Fig. 4.11. Best-first search

The key issue is hidden in the concept of *available* nodes. The algorithm maintains two lists of nodes: *open* and *closed*. The former indicates all of the available nodes, whereas the latter indicates those nodes that have already been processed. When the best-first procedure starts from a particular node, which is placed on the *closed* list, all its children are moved to the *open* list (i.e., they become available). They are evaluated by some heuristic and the best is selected for further processing.

If the search space is a general directed graph, one of the children of a node might already be present in the open or closed lists. If that happens, you have to re-evaluate that node, and you might need to move it from the closed list to the open list. We'll ignore this possibility and focus solely on search spaces that are organized as a tree. For example, figure 4.12 provides the tree for the TSP in figure 4.8.

The main differences between depth-first search, its sister breadth-first search, and best-first search are

- Best-first search explores the most promising next node, whereas depth-first search goes as deep as possible in an arbitrary pattern and breadth-first search explores all the nodes on one level before moving to the next.

- Best-first search uses a heuristic that provides a merit value for each node, whereas depth-first search and breadth-first search do not.

The efficiency of the best-first algorithm is closely connected with the quality of the heuristic, so properly designing such heuristics demands more attention.

In the process of evaluating a partially constructed solution, q, we should take into account its two components:

- The merit of the decisions already made, $c(q)$.

- The potential inherent to the remaining decisions, $h(q)$.

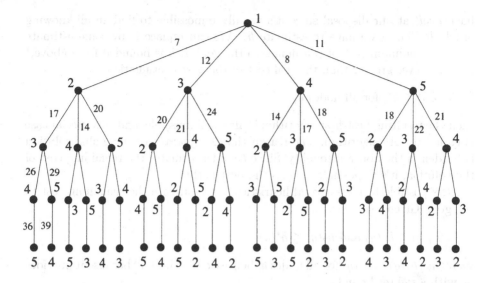

Fig. 4.12. A search space for the TSP with five cities. A number given next to some edges represents the cost of this partial solution including the node at the end of this edge, defined later as a value of c for this node. Note that the cost of the final solution requires the addition of the final edge to close the cycle.

Thus, the evaluation function *eval* for a partial solution q is

$$eval(q) = c(q) + h(q).$$

Sometimes the problem statement provides an exact measure for c. It's relatively easy to evaluate a merit of the decisions already made in a TSP. A partial tour

$$q = 1 - 3 - 5$$

can be evaluated as

$$c(q) = dist(1,3) + dist(3,5).$$

It's much harder to estimate the potential quality of the remaining decisions. This is why we need a good heuristic h. But how can we judge the quality of a particular heuristic?

One criterion for judging a heuristic h relies on what's termed *admissibility*. An algorithm is *admissible* if it always terminates at the optimum solution. We would like to choose h so that we guarantee admissibility. Suppose we had some *oracle* that could provide the real cost h^* of continuing on to the goal. The *ideal* evaluation function *eval** would be

$$eval^*(q) = c(q) + h^*(q),$$

which returns the real cost of the optimum solution having reached node q. It's easy to show that the best-first algorithm using *eval** is admissible, but we don't have *eval** at our disposal since it's usually impossible to find an all-knowing oracle h^*. Thus, we have to estimate h^*. We can replace it by some estimate h of the minimum cost of continuing to the goal. If h is bounded from above,[4] i.e., it's never greater than the real cost of the best continuation,

$$h(q) \leq h^*, \text{ for all nodes } q,$$

then the resulting best-first algorithm is always admissible, and this special case is called the A^* algorithm. The reason that A^* guarantees the global solution is hidden in the above inequality. Since h underestimates the remaining cost of the solution, it's impossible to miss the best path!

It's possible to have several heuristics h_i to estimate the remaining cost to the goal. Given two heuristics,

$$h_1(q) \leq h^*(q) \text{ and } h_2(q) \leq h^*(q),$$

such that $h_1(q) \leq h_2(q)$ for all nodes q, heuristic h_2 is better[5] because it provides us with a tighter bound.

It's interesting to indicate some connections among various search techniques discussed earlier. For example, a best-first search with the evaluation function

$$eval(q) = c(q) + h(q)$$

for a partial solution q can be viewed as a breadth-first search, if

$h(q) = 0$ for all q, and
$c(q') = c(q) + 1$, where q' is a successor of q,

whereas it can be viewed as a depth-first search, if

$h(q) = 0$ for all q, and
$c(q') = c(q) - 1$, where q' is a successor of q.

Note also that when

$h(q) = 0$ for all q, and $c(q) =$ the real cost of the path from the start to node q,

we have a branch and bound method!

We limited the discussion to the case when the search space is organized as a tree. But you can generalize the above algorithms to handle arbitrary graphs [354]. Doing so requires considering the cases of revisiting a node, possibly updating its evaluation value if the cost of a newly discovered path is better then the cost of the best-known path.

---

[4]For minimization problems.
[5]In artificial intelligence terminology, h_2 is *more informed*.

4.6 Summary

If you're observant, you might have noticed that the topic of "traditional meth-ods" spans 50 pages of this book over two chapters. And yet, we argued forcefully in the Preface and Introduction that the traditional methods of problem solv-ing leave much to be desired. The length of chapters 3 and 4 does not imply value, rather in this case it implies the weakness of each of the methods. Each is designed for a particular type of problem and treats that problem, and only that problem, efficiently. Once you change the problem, the traditional methods start to falter, and sometimes they fail altogether.

If you face a problem that poses a quadratic evaluation function you should use Newton-Gauss. If your situation concerns a linear evaluation function with linear equality and inequality constraints then the simplex method will yield the best solution. You need to know when each of the methods in chapters 3 and 4 will yield the global optimum solution to your problem and when they will fall short.

More than learning a set of individual methods, however, you should no-tice that, leaving the details aside, there are two main approaches to problem solving. You can either work with complete solutions or partial solutions. The former offers the advantage that you always have some answer ready to put into practice, even when you run out of time or if your algorithm converges to some-thing less than best. The latter offers the potential payoff of taking advantage of the structure that's inherent in some problems. You might be able to obtain a tremendous speed up in generating an answer to a seemingly complicated prob-lem by decomposing it into subproblems that are trivial. You might be able to organize your search space in the form of a tree and then use something like A^* to help you find the best solution in the least amount of time.

But again, you have to know that if you decompose a problem you may find that generating answers to each of the smaller subproblems doesn't yield any nice way of assembling those answers into an overall global solution. You can disassemble a 100-city TSP into twenty 5-city TSPs, each of which is trivial to solve. But what will you do with these 20 answers?! Most real-world problems do not yield to the traditional methods. This statement is almost tautological: If they could be solved by classic procedures they wouldn't be problems anymore. Thus, we usually face the case where the problem seems intractable, where there are lots of locally optimal solutions and most of these are unacceptable. For these problems, we need tools that go beyond the traditional methods.

V. What's the Color of the Bear?

Once you've found a solution to a problem, it's very tempting to quit. After all, the problem's solved, right? Well, not quite. It could be that there are multiple solutions, and it could be that some are more correct than others. Just like solutions that are only locally optimal, everything can seem fine with the solution that you have, but if you could only see a way to generate some other solution you might escape your current trap and leap into something better. Sometimes it helps to stand a problem on its head to find those other solutions. With that in mind, let's try the following puzzle about the color of a bear and a famous explorer, Mr. Brown.

One morning Mr. Brown left his tent and walked one mile south. Then he turned east and continued for the second mile. At this point he turned north, walked for another mile and arrived right back at his tent. While he was out walking, Mr. Brown saw a bear. What color was it?

As with some other problems, it might be hard to find a starting point. But you can quickly imagine that the key must be hidden in the special characteristics of Mr. Brown's sojourn. If we could find out where in the world Mr. Brown was, maybe that would give us the crucial information that we need.

After a bit of consideration about geometry on the globe, you'll likely come up with the answer that Mr. Brown had his tent on the North Pole. This seems like the only way to complete a loop going one mile south, east, and north (see figure V.1).

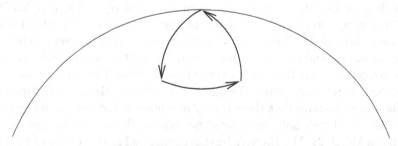

Fig. V.1. Mr. Brown's sojourn

Clearly, the bear must have been white: there are no other bears around the North Pole! If you were a betting person, you'd be thinking: "Well, they wouldn't make the color of the bear the same as the guy's name now, would they?"

Ah, but is this a unique solution? At first inspection, it seems to be unique. As we depart south, there's no other way of completing a loop by going one mile south, east, and north.

Not so fast! After some additional thought we can indeed identify another solution!

Consider a parallel on the Southern Hemisphere that has a circumference of one mile (parallel A in figure V.2). We can select an arbitrary point on this parallel and move 1 mile north to another parallel (parallel B in figure V.2). This is a possible location of Mr. Brown's tent. After leaving this place he goes south for one mile (thus arriving on parallel A), turns east and continues for one mile (thus completing a full circle), and returns north directly to his tent. Mr. Brown's tent can be located, of course, at any point on parallel B. It's not that there's just one more solution, there's an *infinitude* of solutions!

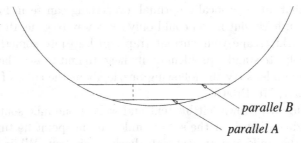

parallel B

parallel A

Fig. V.2. Possible locations of Mr. Brown's tent

So, is that it then? Have we found the complete infinity of solutions?

Nope. After some additional pondering it's clear that there are still more solutions because parallel A may have a circumference of $\frac{1}{2}$ mile, $\frac{1}{3}$ mile, etc. And in every one of these cases Mr. Brown would complete a whole number of circles always returning to the original point! Since there aren't any bears near the South Pole, however, these infinities of additional solutions don't change the answer to the real question, which we recall was to find the color of the bear.

By the way, there's another famous (and very flawed) version of the above puzzle. It goes like this: Mr. Brown, a famous explorer, celebrated his 70th birthday. During the party, a lady asked him what he did exactly 10 years ago. After some thought Mr. Brown replied, "I remember that day very well. I left my tent at sunrise and walked one mile south. Then I turned east and continued for the second mile. At this point I turned north, walked for another mile and arrived back at my tent." Now, the question is: what is Mr. Brown's birthday?

The puzzle assumes that there is only one possible location of the tent: the North Pole. As there's only one sunrise per year at this special location, which happens on March 21, Mr. Brown's birthday must be March 21. But as we know now, there are many more possible locations for his tent!

The point here is don't give up once you find a solution. Keep thinking! See if you have really exhausted all the possibilities. You might find something that you hadn't considered originally, and it might be even better than your first solution. You'll never know if you give up right away. Also, it helps to get a mental picture of the situation. Have in mind a picture of Mr. Brown out with a walking stick and a telescope exploring the wilderness. You'll more quickly find that wilderness turning into ice floes and glaciers. If you don't have a great imagination, then draw a picture on a piece of paper. Just having something to look at can help you frame the problem and think of ways to solve it.

Let's try this out and start with a picture while introducing an additional puzzle.

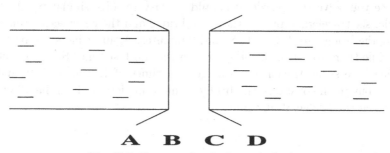

Fig. V.3. Four travelers in front of a bridge

Four travelers (let's call them Mr. A, B, C, and D) have to cross a bridge over a deep ravine (see Figure V.3). It's a very dark night and the travelers have but a single source of light: an old fashioned oil lamp. The light is essential for successfully crossing the ravine because the bridge is very old and has plenty of holes and loose boards. What's worse, its construction is really quite weak, and in its dilapidated condition, it can only at best support two of the men at any time. The question is how should the men arrange themselves to cross the bridge? The oil lamp has a limited supply of fuel. Time is running out.

Well, since only two travelers can be on the bridge at any one time, we know that they have to go across either individually or in pairs, but since they also must have some light to ensure that they don't fall through a hole in the bridge, we also know that it would be impossible for any one individual to carry the lamp without having a partner go with them. So no matter what the solution, it has to involve a sequence of pairs of travelers going across the bridge. But which pairs?

Here comes the really interesting part of the puzzle. It turns out that each traveler needs a different amount of time to cross the bridge. Mr. A is young and healthy and needs but a minute to quickly traverse the bridge. Mr. D, on the other hand, is an old man who recently had a hip replacement and will need 10 minutes to get across the bridge. Mr. B and Mr. C need two minutes and five minutes, respectively. And since each traveler needs the light to cross, whenever a pair of travelers go together, it is the slower man who determines the total time required to make the crossing.

So, now that you have all the information (think to yourself: do I really have all the information? Is there something else I might need to know?), what is the best way to schedule the travelers to get to the other side of the ravine in the shortest possible time?

Your first thought might be to send the quickest traveler, Mr. A, across the bridge with each man in turn. Mr. A could carry the lamp. So Mr. A and Mr. B could go across together – this would take two minutes – then Mr. A would return with the lamp which would take an additional minute. Then Mr. A could go across with Mr. C and then come back to do the same thing again with Mr. D. In all, this would require 19 minutes. But actually, there is a better way for them to accomplish their task. Can you find the solution?

Once you solve this puzzle, it should be easy to solve similar puzzles. For example, six travelers approach the same bridge and their respective times for crossing the bridge are 1, 3, 4, 6, 8, and 9 minutes. Again, what's the best way to schedule them to get to the other side in the shortest time? Suppose instead that there are seven travellers, with crossing times of 1, 2, 6, 7, 8, 9, and 10 minutes, but in this instance the bridge is more modern and can handle three travelers on the bridge at any time....

5. Escaping Local Optima

O! for a horse with wings!

Shakespeare, *Cymbeline*

We've discussed a few traditional problem-solving strategies. Some of them guarantee finding the global solution, others don't, but they all share a common pattern. Either they guarantee discovering the global solution, but are too expensive (i.e., too time consuming) for solving typical real-world problems, or else they have a tendency of "getting stuck" in local optima. Since there is almost no chance to speed up algorithms that guarantee finding the global solution, i.e., there is almost no chance of finding polynomial-time algorithms for most real problems (as they tend to be NP-hard), the other remaining option aims at designing algorithms that are capable of escaping local optima.

How can we do this? How can we design "a horse with wings"? We have already seen one possibility in chapter 2, where we discussed a procedure called an "iterated hill-climber." After reaching a local optimum, a new starting point is generated and the search is commenced all over again. Clearly, we can apply such a strategy with any other algorithm, but let's discuss some possibilities of escaping local optima within a single run of an algorithm.

Two main approaches that we'll discuss in this chapter are based on (1) an additional parameter (called *temperature*) that changes the probability of moving from one point of the search space to another, or (2) a memory, which forces the algorithm to explore new areas of the search space. These two approaches are called *simulated annealing* and *tabu search*, respectively. We discuss them in detail in the following two sections, but first let's provide some intuition connected with these methods and compare them to a simple local search routine.

We discussed the iterated hill-climber procedure in chapter 2. Let's now rewrite this procedure by omitting several lines of programming details and by listing only the most significant concepts. Also, let's restrict the algorithm to a single sequence of improvement iterations (i.e., $MAX = 1$, see figure 2.5). In such a case, a single run of a simple *local search* procedure can be described as follows:

```
procedure local search
begin
    x = some initial starting point in S
    while improve(x) ≠ 'no' do
        x = improve(x)
    return(x)
end
```

The subprocedure improve(x) returns a new point y from the neighborhood of x, i.e., $y \in N(x)$, if y is better than x, otherwise it returns a string "no," and in that case, x is a local optimum in S.

In contrast, the procedure *simulated annealing* (discussed fully in section 5.1) can be described as follows:

```
procedure simulated annealing
begin
    x = some initial starting point in S
    while not termination-condition do
        x = improve?(x, T)
        update(T)
    return(x)
end
```

There are three important differences between simulated annealing and local search. First, there is a difference in how the procedures halt. Simulated annealing is executed until some external "termination condition" is satisfied as opposed to the requirement of local search to find an improvement. Second, the function "improve?(x, T)" doesn't have to return a better point from the neighborhood of x.[1] It just returns an *accepted* solution y from the neighborhood of x, where the acceptance is based on the current temperature T. Third, in simulated annealing, the parameter T is updated periodically, and the value of this parameter influences the outcome of the procedure "improve?" This feature doesn't appear in local search. Note also that the above sketch of simulated annealing is quite simplified so as to match the simplifications we made for local search. For example, we omitted the initialization process and the frequency of changing the temperature parameter T. Again, the main point here is to underline similarities and differences.

Tabu search (discussed fully in section 5.2) is almost identical to simulated annealing with respect to the structure of the algorithm. As in simulated annealing, the function "improve?(x, H)" returns an *accepted* solution y from the neighborhood of x, which need not be better than x, but the acceptance is based on the history of the search H. Everything else is the same (at least, from a high-level perspective). The procedure can be described as follows:

---

[1] This is why we added a question mark in the name of this procedure: "improve?".

```
procedure tabu search
begin
    x = some initial starting point in S
    while not termination-condition do
        x = improve?(x, H)
        update(H)
    return(x)
end
```

With these intuitions in mind and an understanding of the basic concepts, we're
ready to discuss the details of these two interesting search methods.

5.1 Simulated annealing

Let's recall the detailed structure of the iterated hill-climber (chapter 2) because
simulated annealing doesn't differ from it very much.

```
procedure iterated hill-climber
begin
    t ← 0
    initialize best
    repeat
        local ← FALSE
        select a current point v_c at random
        evaluate v_c
        repeat
            select all new points in the neighborhood of v_c
            select the point v_n from the set of new points
                with the best value of evaluation function eval
            if eval(v_n) is better than eval(v_c)
                then v_c ← v_n
                else local ← TRUE
        until local
        t ← t + 1
        if v_c is better than best
            then best ← v_c
    until t = MAX
end
```

Fig. 5.1. Structure of an iterated hill-climber

Note again that the inner loop of the procedure (figure 5.1) *always* returns
a local optimum. This procedure only "escapes" local optima by starting a new
search (outer loop) from a new (random) location. There are MAX attempts

altogether. The best solution overall (*best*) is returned as the final outcome of the algorithm.

Let's modify this procedure in the following way:

- Instead of checking *all* of the strings in the neighborhood of a current point $\mathbf{v}_c$ and selecting the best one, select only one point, $\mathbf{v}_n$, from this neighborhood.

- Accept this new point, i.e., $\mathbf{v}_c \leftarrow \mathbf{v}_n$, with some probability that depends on the relative merit of these two points, i.e., the difference between the values returned by the evaluation function for these two points.

By making such a simple modification, we obtain a new algorithm, the so-called *stochastic hill-climber* (figure 5.2):

procedure stochastic hill-climber
begin
    $t \leftarrow 0$
    select a current string $\mathbf{v}_c$ at random
    evaluate $\mathbf{v}_c$
    repeat
        select the string $\mathbf{v}_n$ from the neighborhood of $\mathbf{v}_c$
        select $\mathbf{v}_n$ with probability $\dfrac{1}{1+e^{\frac{eval(\mathbf{v}_c)-eval(\mathbf{v}_n)}{T}}}$
        $t \leftarrow t+1$
    until $t = MAX$
end

Fig. 5.2. Structure of a stochastic hill-climber

Let's discuss some features of this algorithm (note that the probabilistic formula for accepting a new solution is based on maximizing the evaluation function). First, the stochastic hill-climber has only one loop. We don't have to repeat its iterations starting from different random points. Second, the newly selected point is "accepted" with some probability p. This means that the rule of moving from the current point $\mathbf{v}_c$ to the new neighbor, $\mathbf{v}_n$, is probabilistic. It's possible for the new accepted point to be *worse* than the current point. Note, however, that the probability of acceptance depends on the difference in merit between these two competitors, i.e., $eval(\mathbf{v}_c) - eval(\mathbf{v}_n)$, and on the value of an additional parameter T. Also, T remains constant during the execution of the algorithm. Looking at the probability of acceptance:

$$p = \frac{1}{1+e^{\frac{eval(\mathbf{v}_c)-eval(\mathbf{v}_n)}{T}}},$$

what is the role of the parameter T? Assume, for example, that the evaluation for the current and next points are $eval(\mathbf{v}_c) = 107$ and $eval(\mathbf{v}_n) = 120$ (remember, the formula is for maximization problems only; if you are minimizing, you have to reverse the minuend and subtrahend of the subtraction operation). The difference in our example is $eval(\mathbf{v}_c) - eval(\mathbf{v}_n)$ or -13, meaning that the new point $\mathbf{v}_n$ is better than $\mathbf{v}_c$. What is the probability of accepting this new point based on different values of T? Table 5.1 details some individual cases.

Table 5.1. Probability p of acceptance as a function of T for the case where the new trial $\mathbf{v}_n$ is 13 points better than the current solution $\mathbf{v}_c$

T	$e^{\frac{-13}{T}}$	p
1	0.000002	1.00
5	0.0743	0.93
10	0.2725	0.78
20	0.52	0.66
50	0.77	0.56
10^{10}	0.9999...	0.5...

The conclusion is clear: the greater the value of T, the smaller the importance of the relative merit of the competing points $\mathbf{v}_c$ and $\mathbf{v}_n$! In particular, if T is huge (e.g., $T = 10^{10}$), the probability of acceptance approaches 0.5. The search becomes random. On the other hand, if T is very small (e.g., $T = 1$), the stochastic hill-climber reverts into an ordinary hill-climber! Thus, we have to find an appropriate value of the parameter T for a particular problem: not too low and not too high.

To gain some additional perspective, suppose that $T = 10$ for a given run. Let's also say that the current solution $\mathbf{v}_c$ evaluated to 107, i.e., $eval(\mathbf{v}_c) = 107$. Then, the probability of acceptance depends only on the value of the new string $\mathbf{v}_n$ as shown in table 5.2.

Table 5.2. Probability of acceptance as a function of $eval(\mathbf{v}_n)$ for $T = 10$ and $eval(\mathbf{v}_c) = 107$

$eval(\mathbf{v}_n)$	$eval(\mathbf{v}_c) - eval(\mathbf{v}_n)$	$e^{\frac{eval(\mathbf{v}_c)-eval(\mathbf{v}_n)}{10}}$	p
80	27	14.88	0.06
100	7	2.01	0.33
107	0	1.00	0.50
120	-13	0.27	0.78
150	-43	0.01	0.99

Now the picture is complete: if the new point has the same merit as the current point, i.e., $eval(\mathbf{v}_c) = eval(\mathbf{v}_n)$, the probability of acceptance is 0.5. That's reasonable. It doesn't matter which you choose because each are of equal quality. Furthermore, if the new point is better, the probability of acceptance is greater than 0.5. Moreover, the probability of acceptance grows together with the (negative) difference between these evaluations. In particular, if the new solution is *much* better than the current one (say, $eval(\mathbf{v}_n) = 150$), the probability of acceptance is close to 1.

The main difference between the stochastic hill-climber and simulated annealing is that the latter changes the parameter T during the run. It starts with high values of T making the procedure more similar to a purely random search, and then gradually decreases the value of T. Towards the end of the run, the values of T are quite small so the final stages of simulated annealing merely resemble an ordinary hill-climber. In addition, we always accept new points if they are better than the current point. The procedure *simulated annealing* is given in figure 5.3 (again, we've assumed a maximization problem).

> **procedure** simulated annealing
> **begin**
> $t \leftarrow 0$
>   initialize T
>   select a current point $\mathbf{v}_c$ at random
>   evaluate $\mathbf{v}_c$
>   **repeat**
>   **repeat**
>   select a new point $\mathbf{v}_n$
>   in the neighborhood of $\mathbf{v}_c$
>   **if** $eval(\mathbf{v}_c) < eval(\mathbf{v}_n)$
>   **then** $\mathbf{v}_c \leftarrow \mathbf{v}_n$
>   **else if** $random[0,1) < e^{\frac{eval(\mathbf{v}_n)-eval(\mathbf{v}_c)}{T}}$
>   **then** $\mathbf{v}_c \leftarrow \mathbf{v}_n$
>   **until** (termination-condition)
> $T \leftarrow g(T,t)$
> $t \leftarrow t+1$
>   **until** (halting-criterion)
> **end**

Fig. 5.3. Structure of simulated annealing

Simulated annealing — also known as Monte Carlo annealing, statistical cooling, probabilistic hill-climbing, stochastic relaxation, and probabilistic exchange algorithm — is based on an analogy taken from thermodynamics. To grow a crystal, you start by heating a row of materials to a molten state. You then reduce the temperature of this *crystal melt* until the crystal structure is *frozen in*. Bad things happen if the cooling is done too quickly. In particular, some irregularities are locked into the crystal structure and the trapped energy level is much higher than in a perfectly structured crystal.[2] The analogy between the physical system and an optimization problem should be evident. The basic "equivalent" concepts are listed in table 5.3.

---

[2] A similar problem occurs in metallurgy when heating and cooling metals.

Table 5.3. Analogies between a physical system and an optimization problem

Physical System	Optimization Problem
state	feasible solution
energy	evaluation function
ground state	optimal solution
rapid quenching	local search
temperature	control parameter T
careful annealing	simulated annealing

As with any search algorithm, simulated annealing requires the answers for the following problem-specific questions (see chapter 2):

- What is a solution?

- What are the neighbors of a solution?

- What is the cost of a solution?

- How do we determine the initial solution?

These answers yield the structure of the search space together with the definition of a neighborhood, the evaluation function, and the initial starting point. Note, however, that simulated annealing also requires answers for additional questions:

- How do we determine the initial "temperature" T?

- How do we determine the cooling ratio $g(T, t)$?

- How do we determine the termination condition?

- How do we determine the halting criterion?

The temperature T must be initialized before executing the procedure. Should we start with $T = 100$, $T = 1000$, or something else? How should we choose the termination condition after which the temperature is decreased and the annealing procedure reiterates? Should we execute some number of iterations or should we instead use some other criterion? Then, how much or by what factor should the temperature be decreased? By one percent or less? And finally, when should the algorithm halt, i.e., what is the "frozen" temperature?

Most implementations of simulated annealing follow a simple sequence of steps:

STEP 1: $T \leftarrow T_{max}$
        select $\mathbf{v}_c$ at random

STEP 2: pick a point $\mathbf{v}_n$ from the neighborhood of $\mathbf{v}_c$
        if $eval(\mathbf{v}_n)$ is better than $eval(\mathbf{v}_c)$
            then select it $(\mathbf{v}_c \leftarrow \mathbf{v}_n)$
            else select it with probability $e^{-\frac{\Delta eval}{T}}$
        repeat this step k_T times

STEP 3: set $T \leftarrow rT$
        if $T \geq T_{min}$
            then goto STEP 2
            else goto STEP 1

Here we have to set the values of the parameters T_{max}, k_T, r, and T_{min}, which correspond to the initial temperature, the number of iterations, the cooling ratio, and the frozen temperature, respectively. Let's examine some possibilities by applying the simulated annealing technique to the SAT, TSP, and NLP, respectively.

Spears [444] applied simulated annealing (SA) on hard satisfiability problems. The procedure SA-SAT described in [444] is displayed in figure 5.4. Its control structure is similar to that of the GSAT (chapter 3). The outermost loop variable called "tries" of SA-SAT corresponds to the variable i of GSAT (see figure 3.5 from chapter 3). These variables keep track of the number of independent attempts to solve the problem. T is set to T_{max} at the beginning of each attempt in SA-SAT (by setting the variable j to zero) and a new random truth assignment is made. The inner repeat loop tries different assignments by probabilistically flipping each of the Boolean variables. The probability of a flip depends on the improvement δ of the flip and the current temperature. As usual, if the improvement is negative, the flip is quite unlikely to be accepted and vice versa: positive improvements make accepting flips likely.

There's a major difference between GSAT and SA-SAT, and this is the essential difference between any local search technique and simulated annealing: GSAT can make a backward move (i.e., a decrease in the number of satisfied clauses) if other moves are not available. At the same time, GSAT cannot make two backward moves in a row, as one backward move implies the existence of the next improvement move! On the other hand, SA-SAT can make an arbitrary sequence of backward moves, thus it can escape local optima!

The remaining parameters of SA-SAT have the same meaning as in other implementations of simulated annealing. The parameter r represents a decay rate for the temperature, i.e., the rate at which the temperature drops from T_{max} to T_{min}. The drop is caused by incrementing j, as

$$T = T_{max} \cdot e^{-j \cdot r}.$$

Spears [444] used

procedure SA-SAT
begin
    tries ← 0
    repeat
        v ← random truth assignment
        $j \leftarrow 0$
        repeat
            if v satisfies the clauses **then** return **v**
            $T = T_{max} \cdot e^{-j \cdot r}$
            for $k = 1$ **to** the number of variables **do**
            begin
                compute the increase (decrease) δ in the
                    number of clauses made true if v_k was flipped
                flip variable v_k with probability $(1 + e^{-\frac{\delta}{T}})^{-1}$
                v ← new assignment if the flip is made
            end
            $j \leftarrow j + 1$
        until $T < T_{min}$
        tries ← tries +1
    until tries = MAX-TRIES
end

Fig. 5.4. Structure of the SA-SAT algorithm

$T_{max} = 0.30$ and $T_{min} = 0.01$

for the maximum and minimum temperatures, respectively. The decay rate r depended on the number of variables in the problem and the number of the "try" (r was the inverse of the product of the number of variables and the number of the try). Spears commented on the choice of these parameters:

> Clearly these choices in parameters will entail certain tradeoffs. For a given setting of MAX-TRIES, reducing T_{min} and/or increasing T_{max} will allow more tries to be made per independent attempt, thus decreasing the number of times that 'tries' can be incremented before the MAX-TRIES cutoff is reached. A similar situation occurs if we decrease or increase the decay rate. Thus, by increasing the temperature range (or decreasing the decay rate) we reduce the number of independent attempts, but search more thoroughly during each attempt. The situation is reversed if one decreases the temperature range (or increases the decay rate). Unfortunately it is not at all clear whether it is generally better to make more independent attempts, or to search more thoroughly during each attempt.

The experimental comparison between GSAT and SA-SAT on hard satisfiability problems indicated that SA-SAT appeared to satisfy at least as many formulas as GSAT, with less work [444]. Spears also presented experimental evidence that the relative advantage of SA-SAT came from its backward moves, which helped escape local optima.

It's interesting to note that the TSP was one of the very first problems to which simulated annealing was applied! Moreover, many new variants of simulated annealing have also been tested on the TSP. A standard simulated annealing algorithm for the TSP is essentially the same as the one displayed in figure 5.3, where $\mathbf{v}_c$ is a tour and $eval(\mathbf{v}_c)$ is the length of the tour $\mathbf{v}_c$. The differences between implementations of simulated annealing are in (1) the methods of generating the initial solution, (2) the definition of a neighborhood of a given tour, (3) the selection of a neighbor, (4) the methods for decreasing temperature, (5) the termination condition, (6) the halting condition, and (7) the existence of a postprocessing phase.

All the above decisions are important and far from straightforward. They are also not independent. For example, the number of steps at each temperature should be proportional to the neighborhood size. For each of these facets there are many possibilities. We might start from a random solution, or instead take an output from a local search algorithm as the initial tour. We can define neighborhoods of various sizes and include a postprocessing phase where a local search algorithm climbs a local peak (as simulated annealing doesn't guarantee that this will happen for all schedules for reducing T over time).

For more details on various possibilities of applying simulated annealing to the TSP, with an excellent discussion on techniques for accelerating the algorithm and other improvements, see [242].

Simulated annealing can also easily be applied to the NLP. Since we are now concerned with continuous variables, the neighborhood is often defined on the basis of a Gaussian distribution (for each variable), where the mean is kept at the current point and the standard deviation is set to one-sixth of the length of the variable's domain (so that the length from the midpoint of the domain to each boundary equals three standard deviations). Thus, if the current point is

$$\mathbf{x} = (x_1, \ldots, x_n),$$

where $l_i \leq x_i \leq u_i$ for $i = 1, \ldots, n$, then the neighbor $\mathbf{x}'$ of $\mathbf{x}$ is selected as

$$x_i' \leftarrow x_i + N(0, \sigma_i),$$

where $\sigma_i = (u_i - l_i)/6$ and $N(0, \sigma_i)$ is an independent random Gaussian number with mean of zero and standard deviation σ_i.

At this stage the change in the function value can be calculated:

$$\Delta eval = eval(\mathbf{x}) - eval(\mathbf{x}').$$

If $\Delta eval > 0$ (assume we face a minimization problem), the new point $\mathbf{x}'$ is accepted as a new solution; otherwise, it is accepted with probability

$$e^{\Delta eval/T}.$$

The parameter T again indicates the current temperature.

In the area of numerical optimization, the issues of generating the initial solution, defining the neighborhood of a given point, and selecting particular neighbors are straightforward. The usual procedure employs a random start and Gaussian distributions for neighborhoods. But implementations differ in the methods for decreasing temperature, the termination condition, the halting condition, and the existence of a postprocessing phase (e.g., where we might include a gradient-based method that would locate the local optimum quickly). Note that continuous domains also provide for an additional flexibility: the size of the neighborhood can decrease together with the temperature. If parameters σ_i decrease over time, the search concentrates around the current point resulting in better fine tuning.

5.2 Tabu search

The main idea behind tabu search is very simple. A "memory" forces the search to explore new areas of the search space. We can memorize some solutions that have been examined recently and these become tabu (forbidden) points to be avoided in making decisions about selecting the next solution. Note that tabu search is basically deterministic (as opposed to simulated annealing), but it's possible to add some probabilistic elements to it [186].

The best way to explain the basic concepts of tabu search is by means of an example. We'll start with tabu search applied to the SAT. Later we'll provide a general discussion on some interesting aspects of this method and we'll conclude by providing an additional example of this technique regarding the TSP.

Suppose we're solving the SAT problem with $n = 8$ variables. Thus, for a given logical formula F we're searching for a truth assignment for all eight variables such that the whole formula F evaluates to TRUE. Assume further that we have generated an initial assignment for $\mathbf{x} = (x_1, \ldots, x_8)$; namely,

$$\mathbf{x} = (0, 1, 1, 1, 0, 0, 0, 1).$$

As usual, we need some evaluation function that provides feedback for the search. For example, we might calculate a weighted sum of a number of satisfied clauses, where the weights depend on the number of variables in the clause. In this case, the evaluation function should be maximized (i.e., we're trying to satisfy all of the clauses), and let's assume that the above random assignment provides the value of 27. Next, we have to examine the neighborhood of $\mathbf{x}$, which consists of eight other solutions, each of which can be obtained by flipping a single bit in the vector $\mathbf{x}$. We evaluate them and select the best one. At this stage of the search, this is the same as in a hill-climbing procedure.

Suppose that flipping the third variable generates the best evaluation (say, 31), so this new vector yields the current best solution. Now the time has come to introduce the new facet of tabu search: a memory. In order to keep a record of our actions (moves), we'll need some memory structures for bookkeeping.

We'll remember the index of a variable that was flipped, as well as the "time" when this flip was made, so that we can differentiate between older and more recent flips. In the case of the SAT problem, we need to keep a time stamp for each position of the solution vector: the value of the time stamp will provide information on the recency of the flip at this particular position. Thus, a vector M will serve as our memory. This vector is initialized to 0 and then at any stage of the search, the entry

$$M(i) = j \text{ (when } j \neq 0)$$

might be interpreted as "j is the most recent iteration when the i-th bit was flipped" (of course, $j = 0$ implies that the i-th bit has never been flipped). It might be useful to change this interpretation so we can model an additional aspect of memory: After some period of time (i.e., after some number of iterations), the information stored in memory is erased. Assuming that any piece of information can stay in a memory for at most, say, five iterations, a new interpretation of an entry

$$M(i) = j \text{ (when } j \neq 0)$$

could be: "the i-th bit was flipped $5-j$ iterations ago." Under this interpretation, the contents of the memory structure M after one iteration in our example is given in figure 5.5. Recall that the flip on the third position gave the best result. Note that the value "5" can be interpreted as "for the next five iterations, the third bit position is not available (i.e., tabu)."

0	0	5	0	0	0	0	0

Fig. 5.5. The contents of the memory after iteration 1

It might be interesting to point out that the main difference between these two equivalent interpretations is simply a matter of implementation. The latter approach interprets the values as the number of iterations for which a given position is not available for any flips. This interpretation requires that *all* nonzero entries of the memory are decreased by one at every iteration to facilitate the process of forgetting after five iterations. On the other hand, the former interpretation simply stores the iteration number of the most recent flip at a particular position, so it requires a current iteration counter t which is compared with the memory values. In particular, if $t - M(i) > 5$, i.e., the flip on the i-th position occurred earlier than five iterations ago (if at all), it should be forgotten. This interpretation, therefore, only requires updating a *single* entry in the memory per iteration, and increasing the iteration counter. Later on we'll assume the latter interpretation in an example, but most *implementations* of tabu search use the former interpretation for efficiency.

Let's say that after four additional iterations of selecting the best neighbor — which isn't necessarily better than the current point, and this is why we can escape from local optima — and making an appropriate flip, the memory has the contents as shown in figure 5.6.

3	0	1	5	0	4	2	0

Fig. 5.6. The contents of the memory after five iterations

The numbers present in the memory (figure 5.6) provide the following information:

> Bits 2, 5, and 8 are available to be flipped any time. Bit 1 is not available for the next three iterations, bit 3 isn't available but only for the next iteration, bit 4 (which was just flipped) is not available for the next five iterations, and so on.

In other words, the most recent flip (iteration 5) took place at position 4 (i.e., $M(4) = 5$: bit 4 was flipped $5 - 5 = 0$ iterations ago). Then, the other previously flipped bits are:

> bit 6 (at iteration 4), (as $M(6) - 4$),
> bit 1 (at iteration 3), (as $M(1) = 3$),
> bit 7 (at iteration 2), (as $M(7) = 2$) and, of course,
> bit 3 (at iteration 1).

By consequence, the current solution is $\mathbf{x} = (1, 1, 0, 0, 0, 1, 1, 1)$ and let's say that it evaluates out at 33. Now, let's examine the neighborhood of $\mathbf{x}$ carefully. It consists of eight solutions,

> $\mathbf{x}_1 = (0, 1, 0, 0, 0, 1, 1, 1)$,
> $\mathbf{x}_2 = (1, 0, 0, 0, 0, 1, 1, 1)$,
> $\mathbf{x}_3 = (1, 1, 1, 0, 0, 1, 1, 1)$,
> $\mathbf{x}_4 = (1, 1, 0, 1, 0, 1, 1, 1)$,
> $\mathbf{x}_5 = (1, 1, 0, 0, 1, 1, 1, 1)$,
> $\mathbf{x}_6 = (1, 1, 0, 0, 0, 0, 1, 1)$,
> $\mathbf{x}_7 = (1, 1, 0, 0, 0, 1, 0, 1)$, and
> $\mathbf{x}_8 = (1, 1, 0, 0, 0, 1, 1, 0)$,

which correspond to flips on bits 1 to 8, respectively. After evaluating each we know their respective merits, but tabu search utilizes the memory to force the search to explore new areas of the search space. The memorized flips that have been made recently are tabu (forbidden) for selecting the next solution. Thus at the next iteration (iteration 6), it's impossible to flip bits 1, 3, 4, 6, and 7, since all of these bits were tried "recently." These forbidden (tabu) solutions (i.e., the solutions that are obtained from the forbidden flips) are not considered, so the next solution is selected from a set of $\mathbf{x}_2$, $\mathbf{x}_5$, and $\mathbf{x}_8$.

Suppose that the best evaluation of these three possibilities is given by $\mathbf{x}_5$ which is 32. Note that this value represents a decrease in the evaluation between the current best solution and the new candidate. After this iteration the contents of the memory change as follows: All nonzero values are decreased by one to reflect the fact that all of the recorded flips took place one generation earlier. In particular, the value $M(3) = 1$ is changed to $M(3) = 0$; i.e., after five generations the fact that the third bit was flipped is removed from the memory. Also, since the selected new current solution resulted from flipping the fifth bit, the value of $M(5)$ is changed from zero to five (for the next five iterations, this position is tabu). Thus, after the sixth iteration the contents of memory now appear as shown in figure 5.7.

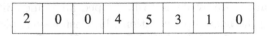

Fig. 5.7. The contents of the memory after six iterations

Further iterations are similarly executed. At any stage, there is a current solution being processed which implies a neighborhood, and from this neighborhood, tabu solutions are eliminated from possible exploration.

Upon reflection, such a policy might be too restrictive. It might happen that one of the tabu neighbors, say $\mathbf{x}_6$, provides an excellent evaluation score. One that's much better than the score of any solution considered previously. Perhaps we should make the search more flexible. If we find an outstanding solution, we might forget about the principles![3] In order to make the search more flexible, tabu search considers solutions from the *whole* neighborhood, evaluates them all, and under "normal" circumstances selects a non-tabu solution as the next current solution, whether or not this non-tabu solution has a better evaluation score than the current solution. But in circumstances that aren't "normal," i.e., an outstanding tabu solution is found in the neighborhood, such a superior solution is taken as the next point. This override of the tabu classification occurs when a so-called *aspiration criterion* is met.

Of course there are also other possibilities for increasing the flexibility of the search. For example, we could change the previous deterministic selection procedure into a probabilistic method where better solutions have an increased chance of being selected. In addition, we could change the memory horizon during the search: sometimes it might be worthwhile to remember "more," and at other times to remember "less" (e.g., when the algorithm hill-climbs a promising area of the search space). We might also connect this memory horizon to the size of the problem (e.g., remembering the last $\sqrt{n}$ moves, where n provides a size measure of the instance of the problem).

Another option is even more interesting! The memory structure discussed so far can be labeled as *recency-based* memory, as it only records some ac-

---

[3]This is quite typical in various situations in ordinary life. When seeing a great opportunity it's easy to forget about some principles!

tions of the last few iterations. This structure might be extended by a so-called *frequency-based* memory, which operates over a much longer horizon. For example (referring back to the SAT problem considered earlier), a vector H may serve as a long-term memory. This vector is initialized to 0 and at any stage of the search the entry

$$H(i) = j$$

is interpreted as "during the last h iterations of the algorithm, the i-th bit was flipped j times." Usually, the value of the horizon h is quite large, at least in comparison with the horizon of recency-based memory. Thus after, say 100 iterations with $h = 50$, the long-term memory H might have the values displayed in figure 5.8.

5	7	11	3	9	8	1	6

Fig. 5.8. The contents of the frequency-based memory after 100 iterations (horizon $h = 50$)

These frequency counts show the distribution of moves throughout the last 50 iterations. How can we use this information? The principles of tabu search indicate that this type of memory might be useful to *diversify* the search. For example, the frequency-based memory provides information concerning which flips have been under-represented (i.e., less frequent) or not represented at all, and we can diversify the search by exploring these possibilities.

The use of long-term memory in tabu search is usually restricted to some special circumstances. For example, we might encounter a situation where all non-tabu moves lead to a worse solution. This is a special situation indeed: all legal directions lead to inferior solutions! Thus, to make a meaningful decision about which direction to explore next, it might be worthwhile to refer to the contents of the long-term memory.

There are many possibilities here for incorporating this information into the decision-making process. The most typical approach makes the most frequent moves less attractive. Usually the value of the evaluation score is decreased by some penalty measure that depends on the frequency, and the final score implies the winner.

To illustrate this point by an example, assume that the value of the current solution $\mathbf{x}$ for our SAT problem is 35. All non-tabu flips, say on bits 2, 3, and 7, provide values of 30, 33, and 31, respectively, and none of the tabu moves provides a value greater than 37 (the highest value found so far), so we can't apply the aspiration criterion. In such circumstances we refer to the frequency-based memory (see figure 5.8). Let's assume that the evaluation formula for a new solution $\mathbf{x}'$ used in such circumstances is

$$eval(\mathbf{x}') - penalty(\mathbf{x}'),$$

where *eval* returns the value of the original evaluation function (i.e., 30, 33, and 31, respectively, for solutions created by flipping the 2nd, 3rd, and 7th bit), and

$$penalty(\mathbf{x}') = 0.7 \cdot H(i),$$

where 0.7 serves as a coefficient and $H(i)$ is the value taken from the long-term memory H:

   7 — for a solution created by flipping the 2nd bit,
   11 — for a solution created by flipping the 3rd bit, and
   1 — for a solution created by flipping the 7th bit.

The new scores for the three possible solutions are then

$$30 - 0.7 \cdot 7 = 25.1, \ 33 - 0.7 \cdot 11 = 25.3, \text{ and } 31 - 0.7 \cdot 1 = 30.3,$$

and so the third solution (i.e., flipping the 7th bit) is the one we select.

The above option of including frequency values in a penalty measure for evaluating solutions diversifies the search. Of course, many other options might be considered in connection with tabu search. For example, if we have to select a tabu move, we might use an additional rule (so-called *aspiration by default*) to select a move that is the "oldest" of all those considered. It might also be a good idea to memorize not only the set of recent moves, but also whether or not these moves generated any improvement. This information can be incorporated into search decisions (so-called *aspiration by search direction*). Further, it's worthwhile to introduce the concept of *influence*, which measures the degree of change of the new solution either in terms of the distance between the old and new solution, or perhaps the change in the solution's feasibility if we're dealing with a constrained problem. The intuition connected with influence is that a particular move has a larger influence if a corresponding "larger" step was made from the old solution to the new. This information can be also incorporated into the search (so-called *aspiration by influence*). For more details on the many available options in tabu search plus some practical hints for implementation see [186].

We conclude this section by providing an additional example of possible structures used in tabu search while approaching the TSP. For this problem, we can consider moves that swap two cities in a particular solution.[4] The following solution (for an eight-city TSP),

$$(2, 4, 7, 5, 1, 8, 3, 6),$$

has 28 neighbors, since there are 28 different pairs of cities, i.e., $\binom{8}{2} = \frac{7 \cdot 8}{2} = 28$, that we can swap. Thus, for a recency-based memory we can use the structure given in figure 5.9, where the swap of cities i and j is recorded in the i-th row and the j-th column (for $i < j$). Note that we interpret i and j as *cities*, and not as their positions in the solution vector, but this might be another possibility

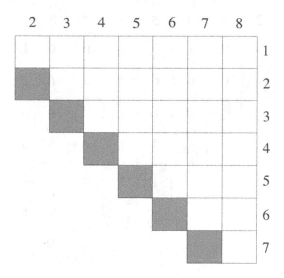

Fig. 5.9. The structure of the recency-based memory for the TSP

to consider. Note that the same structure can also be used for frequency-based memory.

For clarity, we'll maintain the number of remaining iterations for which a given swap stays on the tabu list (recency-based memory) as with the previous SAT problem, while the frequency-based memory will indicate the totals of all swaps that occurred within some horizon h. Assume both memories were initialized to zero and 500 iterations of the search have been completed. The current status of the search then might be as follows. The current solution is

$$(7, 3, 5, 6, 1, 2, 4, 8)$$

with the total length of the tour being 173. The best solution encountered during these 500 iterations yields a value of 171. The status of the recency-based and frequency-based memories are displayed in figures 5.10 and 5.11, respectively.

As with the SAT problem, it's easy to interpret the numbers in these memories. The value $M(2, 6) = 5$ indicates that the most recent swap was made for cities 2 and 6, i.e., the previous current solution was

$$(7, 3, 5, 2, 1, 6, 4, 8).$$

Therefore, swapping cities 2 and 6 is *tabu* for the next five iterations. Similarly, swaps of cities 1 and 4, 3 and 7, 4 and 5, and 5 and 8, are also on the tabu list. Out of these, the swap between cities 1 and 4 is the oldest (i.e., it happened five iterations ago) and this swap will be removed from the tabu list after the

---

[4]We do not make any claim that this decision is the best. For example, it's often more meaningful to swap edges of a tour.

	2	3	4	5	6	7	8	
	0	0	1	0	0	0	0	1
		0	0	5	0	0	0	2
			0	0	0	4	0	3
				3	0	0	0	4
					0	0	2	5
						0	0	6
							0	7

Fig. 5.10. The contents of the recency-based memory M for the TSP after 500 iterations. The horizon is five iterations.

	2	3	4	5	6	7	8	
	0	2	3	3	0	1	1	1
		2	1	3	1	1	0	2
			2	3	3	4	0	3
				1	1	2	1	4
					4	2	1	5
						3	1	6
							6	7

Fig. 5.11. The contents of the frequency-based memory F for the TSP after 500 iterations. The horizon is 50 iterations.

next iteration. Note that only 5 swaps (out of 28 possible swaps) are forbidden (tabu).

The frequency-based memory provides some additional statistics of the search. It seems that swapping cities 7 and 8 was the most frequent (it happened

6 times in the last 50 swaps), and there were pairs of cities (like 3 and 8) that weren't swapped within the last 50 iterations.

The neighborhood of a tour was defined by a swap operation between two cities in the tour. This neighborhood is not the best choice either for tabu search or for simulated annealing. Many researchers have selected larger neighborhoods. For example Knox [263] considered neighborhoods based on a *2-interchange* move (see figure 3.6) that's defined by deleting two nonadjacent edges from the current tour and adding two other edges to obtain a new feasible tour. The outline of a particular implementation of tabu search reported in [263] is given in figure 5.12.

```
procedure tabu search
begin
    tries ← 0
    repeat
        generate a tour
        count ← 0
        repeat
            identify a set T of 2-interchange moves
            select the best admissible move from T
            make appropriate 2-interchange
            update tabu list and other variables
            if the new tour is the best-so-far for a given 'tries'
            then update local best tour information
            count ← count +1
        until count = ITER
        tries ← tries +1
        if the current tour is the best-so-far (for all 'tries')
        then update global best tour information
    until tries = MAX-TRIES
end
```

Fig. 5.12. Particular implementation of tabu search for the TSP

Knox [263] made a tour tabu if *both* added edges of the interchange were on the tabu list. The tabu list was updated by placing the added edges on the list (deleted edges were ignored). What's more, the tabu list was of fixed size. Whenever it became full, the oldest element in the list was replaced by the new deleted edge. At initialization, the list was empty and all of the elements of the aspiration list were set to large values. Note that the algorithm examines *all* neighbors, that is, all of the *2-interchange* tours.

Knox [263] indicated that the best results were achieved when

- The length of the tabu list was $3n$ (where n is the number of cities of the problem).

- A candidate tour could override the tabu status if both edges passed an aspiration test, which compared the length of the tour with aspiration values for both added edges. If the length of the tour was better (i.e., smaller) than *both* aspiration values, the test was passed.

- The values present on the aspiration list were the tour costs prior to the interchange.

- The number of searches (MAX-TRIES) and the number of interchanges (ITER) depended on the size of the problem. For problems of 100 cities or less, MAX-TRIES was 4, and ITER was set to $0.0003 \cdot n^4$.

Of course, there are many other possibilities for implementing tabu lists, aspiration criteria, generating initial tours, etc. The above example illustrates only one possible implementation of tabu search for the TSP. You might generate initial tours at random, or by some other means to ensure diversity. The maximum number of iterations might be some function of the improvement made so far. No doubt you can imagine any number of other possibilities.

5.3 Summary

Simulated annealing and tabu search were both designed for the purpose of escaping local optima. However, they differ in the methods used for achieving this goal. Tabu search usually makes uphill moves only when it is stuck in local optima, whereas simulated annealing can make uphill moves at any time. Additionally, simulated annealing is a stochastic algorithm, whereas tabu search is deterministic.

Compared to the classic algorithms offered in chapter 4, we can see that both simulated annealing and tabu search work on complete solutions (as did the techniques in chapter 3). You can halt these algorithms at any iteration and you'll have a solution at hand, but you might notice a subtle difference between these methods and the classic techniques. Simulated annealing and tabu search have more parameters to worry about, such as temperature, rate of reduction, a memory, and so forth. Whereas the classic methods were designed just to "turn the crank" to get the answer, now you have to start really thinking, not just whether or not the algorithm makes sense for your problem, but how to choose the parameters of the algorithm so that it performs optimally. This is a pervasive issue that accompanies the vast majority of algorithms that can escape local optima. The more sophisticated the method, the more you have to use your judgment as to how it should be utilized.

VI. How Good Is Your Intuition?

It's human nature to try to guess at the solution to a complex problem. Some guesses seem more intuitive than others. Even some *people* seem more intuitive than others! But they aren't, are they? They just seem that way because of the small sample size that you consider. We seem to remember the lucky person who guessed the number on the roulette wheel three times in a row, and we forget the hundreds of others who didn't perform that marvelous feat. But the lucky person isn't clairvoyant. It's all just a matter of random sampling. If you could somehow take that lucky person back to the roulette and ask them to call out the next three numbers, the probability of their success would be exactly the same as before: $(\frac{1}{38})^3$, assuming a roulette wheel with both 0 and 00.

One famous trick that gets played on people who want to make money in the stock market goes like this. A fraudulent stock broker picks 1024 people as possible clients — "suckers" might be a better term. Each day, for the next 10 days, he mails each one of these people a prediction about whether or not the stock market will increase or decrease that next day. Of course, the trick is that in 10 days, there are 2^{10} different possible outcomes — and guess what — he mails each person one of these possible outcomes, played out over 10 days.

At the end of the 10 days, one unfortunate victim will have seen this stock broker predict the market's direction correctly an amazing 10 out of 10 times! The victim's intuition must be that this guy is a psychic! Even those who received 8 or 9 correct predictions out of 10 have to be impressed with the broker's prowess. But what about the others who get about as many right guesses as wrong ones, or even the lucky person who sees the broker get 10 incorrect predictions in a row? They forget about him, of course! So now the broker fixes in on the people he gave all the correct predictions to and tries to bilk them out of some money. Watch out for your intuition and hold on to your wallet!

Sometimes, you have to try really hard to reject your intuition. Even seemingly simple straightforward everyday events demand some serious scrutiny. Try these out for size.

This first puzzle is very simple, but it illustrates that the term "average" often isn't well understood. You drive a car at a constant speed of 40 km/h from Washington D.C. to New York City and return immediately, but at a higher speed of 60 km/h. What was your average speed for the whole trip?

It's likely that at least 90 percent of all people would say that the average was 50 km/h, as the intuition of "average" is very clear. The correct answer of 48 km/h would seem very counterintuitive! Note that the "average speed" is defined as a ratio between the distance and the time, thus

$$v_{avg} = \frac{D}{t},$$

where D and t represent the total distance and the time of the whole trip, respectively. In that case $D = 2d$ (where d is the distance between Washington D.C. and New York City) and $t = t_{WN} + t_{NW}$ (where t_{WN} and t_{NW} are the times to get from Washington D.C. to New York City and from New York City to Washington D.C., respectively). Note also that

$$t_{WN} = d/40 \text{ and } t_{NW} = d/60.$$

Thus,

$$v_{avg} = \frac{D}{t} = \frac{2d}{t_{WN} + t_{NW}} = \frac{2d}{d/40 + d/60} = \frac{2}{1/40 + 1/60} = 48.$$

Here's an additional upgrade for this puzzle. Suppose that you go from Washington D.C. to New York City at a constant speed of 40 km/h. What should your constant speed be for the return trip if you want to obtain an average speed of 80 km/h?

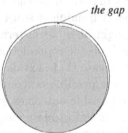

the gap

Fig. VI.1. The Earth and the ring

Here's another interesting counterintuitive result. Suppose that in the times of Ancient Greece, Zeus commissioned a blacksmith to make an iron ring that would go around the Earth, and the blacksmith was asked to make the diameter of the ring match the diameter of the Earth exactly.[5] The poor blacksmith, however, made a mistake. He made a ring that was just one meter longer in circumference than it was supposed to be. Nevertheless, Zeus placed the ring "around" the Earth, and it was made to touch the Earth at one point (see figure VI.1). The question is, how much did the ring "stick out" on the other side of the Earth? What kind of animal would be able to squeeze under the gap between the Earth and the ring? An ant? A mouse? A rat? A cat?

The counterintuitive answer is that that gap is almost 32 cm! That's over one foot! Even a small child could squeeze under the ring! This is so because the difference in circumference between the Earth and the ring is 100 cm:

---

[5] Assuming here that the Earth is a perfect sphere!

$$2\pi r_1 - 2\pi r_2 = 100,$$

so the difference in their diameters $(2r_1 - 2r_2)$ is $100/\pi \approx 31.83$ cm.

The third puzzle is even more tricky, and it illustrates the difficulty many people have confusing "average" with "probability." Which event is more likely: to throw 1 six when 6 dice are rolled, or to throw 2 sixes when 12 dice are rolled? Of course, it's possible to extend the question to "or throw 3 sixes when 18 dice are rolled," etc. Also, we can ask a similar question. Which event is more likely: to throw *at least* 1 six when 6 dice are rolled, or to throw *at least* 2 sixes when 12 dice are rolled?

Usually our intuition fails here. It doesn't seem like there's that much difference because the expected values are "1 six" and "2 sixes" in these two scenarios, but the probabilities of these events are in fact very different!

In the former case, the probability of throwing precisely 1 six out of 6 dice is

$$\binom{6}{1} \cdot \frac{1}{6} \cdot \frac{5}{6} \cdot \frac{5}{6} \cdot \frac{5}{6} \cdot \frac{5}{6} \cdot \frac{5}{6},$$

as

- $\binom{6}{1}$ is the number of combinations of the 6 dice that have exactly one six showing. It may appear on the first die, the second die, etc.

- $\frac{1}{6}$ is the probability of throwing a six on any die.

- $\frac{5}{6}$ is the probability of *not* throwing a six on a specific die; this must happen on 5 remaining dice.

So, the probability of throwing precisely 1 six out of 6 dice is 0.4019.

On the other hand, the probability of throwing precisely 2 sixes out of 12 dice is

$$\binom{12}{2} \cdot \left(\frac{1}{6}\right)^2 \cdot \left(\frac{5}{6}\right)^{10} = 0.2961,$$

which is quite a difference! By the way, the probability of throwing precisely k sixes with n dice is

$$\binom{n}{k} \cdot \left(\frac{1}{6}\right)^k \cdot \left(\frac{5}{6}\right)^{n-k}.$$

OK, here's your last chance! There are three large identical boxes. One of them contains an expensive item (e.g., a 1953 classic Corvette convertible, with a red interior and loaded with features), whereas the other two boxes contain prizes of much smaller value (e.g., a pencil and a few candies, respectively). Of course, you don't know which box contains which items.

You're asked to point to one of the boxes, which you do. At this time a person who runs the show and knows exactly which box contains the car, opens one of the other boxes. This one, much to your relief, doesn't contain the car. Now you're told to make your final pick: either you can stay with your original choice, or you can switch to the other unopened box. Whatever you do, this is final and you get what's in the box you select. What should you do? Should you stay with the original guess or should you switch?

At first it seems that it doesn't matter. After all, the car can be in either of the unopened boxes, so what would be the point of changing the initially selected box? A simple argument should convince you, however, that changing your selection will double your chances of winning the car! When you made your initial selection, the probability of pointing to the winning box was precisely 1/3. Thus, the probability that the car was in one of the other two boxes was 2/3. If one of the other boxes were opened, the chance that the car is in one of the other two boxes would still be 2/3, so the probability that the car is in the other unopened box is 2/3. In other words, if you stay with your original choice, your chances of winning stay at 1/3; however, if you switch to the other unopened box, the chances of winning jump to 2/3. Amazing? You can easily verify this by taking three playing cards, one king and two small cards, mixing them, and placing them face down on a table. Select a card and play it out. Do 20 such experiments where you stick with the original choice, and 20 experiments where you switch. Of course, you'll need a friend who knows the location of the winning card (the king) and turns the other card face up. Preferably, a friend with a 1953 classic Corvette!

6. An Evolutionary Approach

The works of nature must all be accounted good.

Cicero, *De Senectute*

In the previous three chapters we discussed various classic problem-solving methods, including dynamic programming, branch and bound, and local search algorithms, as well as some modern heuristic methods like simulated annealing and tabu search. Some of these techniques were seen to be deterministic. Essentially you "turn the crank" and out pops the answer. For these methods, given a search space and an evaluation function, some would always return the same solution (e.g., dynamic programming), while others could generate different solutions based on the initial configuration or starting point (e.g., a greedy algorithm or the hill-climbing technique). Still other methods were probabilistic, incorporating random variation into the search for optimal solutions. These methods (e.g., simulated annealing) could return different final solutions even when given the same initial configuration. No two trials with these algorithms could be expected to take exactly the same course. Each trial is much like a person's fingerprint: although there are broad similarities across fingerprints, no two are exactly alike.

There's an interesting observation to make with all of these techniques: each relies on a single solution as the basis for future exploration with each iteration. They either process complete solutions in their entirety, or they construct the final solution from smaller building blocks. Greedy algorithms, for example, successively build solutions according to maximizing locally available improvements. Dynamic programming solves many smaller subproblems before arriving at the final complete solution. Branch and bound methods organize the search space into several subspaces, then search and eliminate some of these in a systematic manner. In contrast, local search methods, simulated annealing, and tabu search process complete solutions, and you could obtain a potential answer (although quite likely a suboptimal one) by stopping any of these methods at any particular iteration. They always have a *single* "current best" solution stored that they try to improve in the next step. Despite these differences, every one of these algorithms works on or constructs a single solution at a time.

Speaking quite generally, the approach of keeping the best solution found so far and attempting to improve it is intuitively sound, remarkably simple, and often quite efficient. We can use (1) deterministic rules — e.g., hill-climbing uses the rule: if an examined neighbor is better, proceed to that neighbor and continue searching from there; otherwise, continue searching in the current neighborhood; (2) probabilistic rules — e.g., simulated annealing uses the rule: if an examined neighbor is better, accept this as the new current position; otherwise, either probabilistically accept this new weaker position anyway or continue to search in the current neighborhood; or even (3) the history of the search up to the current time — e.g., tabu search uses the rule: take the best available neighbor, which need not be better than the current solution, but which isn't listed in memory as a restricted or "tabu" move.

Now let's consider a revolutionary idea, one that doesn't appear in any of the methods that we discussed previously. Let's abandon the idea of processing only a single solution. What would happen if we instead maintained several solutions simultaneously? That is, what would happen if our search algorithm worked with a *population* of solutions?

At first blush, it might seem that this idea doesn't provide us with anything really new. Certainly, we can process several solutions in parallel if a parallel computer is available! If this were the case, we could implement a parallel search method, like a parallel simulated annealing or tabu search, where each processor would maintain a single solution and each processor would execute the same algorithm in parallel. It would seem that all there is to be gained here is just the speed of computation: instead of running an algorithm k times to increase the probability of arriving at the global optimum, a k-processor computer would complete this task in a single run, and therefore uses much less real time.

But there's an additional component that can make population-based algorithms essentially different from other problem-solving methods: the concept of competition between solutions in a population. That is, let's simulate the evolutionary process of competition and selection and let the candidate solutions in our population fight for room in future generations! Moreover, we can use random variation to search for new solutions in a manner similar to natural evolution.

In case you haven't heard this analogy before, let's digress with some poetic license to a somewhat whimsical example of rabbits and foxes. In any population of rabbits,[1] some are faster and smarter than others. These faster, smarter rabbits are less likely to be eaten by foxes, and therefore more of them survive to do what rabbits do best: make more rabbits! This surviving population of rabbits starts breeding. Breeding mixes the rabbits' genetic material, and the behaviors of the smarter and faster rabbits are heritable. Some slow rabbits breed with fast rabbits, some fast with fast, some smart ones with not-so-smart ones, and so on. On top of that, nature throws in a "wild hare" every once in a while because all of the genes in each rabbit have a chance of mutating along the way. The resulting baby rabbits aren't copies of their parents, but are instead random perturbations of these parents. Many of these babies will grow up to express the behaviors that help them compete even better with foxes, and other rabbits! Thus, over many generations, the rabbits may be expected to become smarter and faster than those that survived the initial generation. And don't forget that the foxes are evolving at the same time, putting added pressure on

---

[1]An individual rabbit represents a solution for a problem. It represents a single point of the search space.

the rabbits. The interaction of the two populations would seem to push both of them to their physical limits. Foxes are forced to get better at finding a meal and rabbits get better at avoiding being someone else's lunch!

Let's re-tell the above story in terms of search spaces, evaluation functions, and potential solutions. Suppose we start with a population of initial solutions, perhaps generated by sampling randomly from the search space S. Sometimes, to emphasize a link with genetics, these solutions are called *chromosomes*, but there is no need to think of biological terms.[2] The algorithms that we will explore are just that — algorithms — mathematical procedures based on analogies to natural evolutionary processes, and there is no need for our so-called chromosomes to behave in any way like nature's chromosomes. We may simply describe the candidate solutions as vectors, or whatever representation and data structure we select.

The evaluation function can be used to determine the relative merit of each of our initial solutions. This isn't always straightforward because multiple objectives need to be combined in a useful way. For example, which rabbit is better, the one that is fast and dumb, or slow but smart? Natural selection makes this decision by using a quality control method that might be called "trial by foxes." The proof of which rabbit is better able to meet the demands imposed by the environment rests in which ones can survive encounters with the foxes, or evade them all together. But whereas nature doesn't have a mathematical function to judge the quality of an individual's behavior, we often have to determine such a function. At least, the chosen evaluation function must be capable of differentiating between two individuals, i.e., it has to be able to rank one solution ahead of another. Those solutions that are better, as determined by the evaluation function, are favored to become parents for the next generation of offspring.

How should we generate these offspring? Well, actually we've seen the basic concept in most other algorithms: new solutions are generated probabilistically in the neighborhood of old solutions. When applying an evolutionary algorithm, however, we don't have to rely only on the neighborhoods of each individual solution. We can also examine the neighborhoods of pairs of solutions. That is, we can use more than one parent solution to generate a new candidate solution. One way we can do this is by taking parts of two parents and putting them together to form an offspring. For example, we might take the first half of one parent together with the second half of another. The first half of one parent might be viewed as an "idea," and similarly so with the second half of the second parent. Sometimes this recombination can be very useful: You put cookie dough together with ice cream and you get a delicious treat. But sometimes things don't work out so well: You put the first half of Shakespeare's *Hamlet* together with the second half of *King Lear* and you still get Shakespeare but it just doesn't make any sense any more. Another way of using two solutions to generate a new possibility occurs when facing continuous optimization problems. In these cases we can blend parameters of both parents, essentially performing

---

[2]In fact, the use of biological terms may offer the illusion that there is more of direct connection to natural evolution than can be justified in a particular example.

a weighted average component by component. Again, sometimes this can be advantageous, and others times it can be a waste of time. The appropriateness of every search operator depends on the problem at hand.

Since our evolutionary algorithms only exist in a computer, we don't have to rely on real biological constraints. For example, in the vast majority of sexual organisms (not including bacteria), mating occurs only between pairs of individuals, but our evolutionary searches can rely on "mating" or "blending" more than two parents, perhaps using *all* of the individuals in the population of each generation to determine the probabilities for selecting each new candidate solution.

With each generation, the individuals compete — either only among themselves or also against their parents — for inclusion in the next iteration. It's very often the case that after a series of generations, we observe a succession of improvements in the quality of the tested solution and a convergence toward the neighborhood of a nearly optimum solution.

Without any doubt, this is an appealing idea! Why should we labor to solve a problem by calculating difficult mathematical expressions or developing complicated computer programs to address approximate models of a problem if we can simulate the essential and basic process of evolution and discover nearly optimal solutions using models of much greater fidelity, particularly when the difficult part of the search essentially comes "for free." Things that are free are usually too good to be true, so it is appropriate to ask some probing questions. What kind of effort is expected from a user of an evolutionary algorithm? In other words, how much work does it take to implement these concepts in an algorithm? Is this cost effective? And can we *really* solve real-world problems using such an evolutionary approach? We'll answer these questions in the remainder of the book.

In the following sections we revisit the three problems introduced in chapter 1: the satisfiability problem (SAT), the traveling salesman problem (TSP), and a nonlinear programming problem (NLP), and we demonstrate some possibilities for constructing evolutionary algorithms to address each one.

6.1 Evolutionary approach for the SAT

Suppose that we wanted to use an evolutionary algorithm to generate solutions for an SAT of 100 variables. Let's say that the problem is essentially like that of chapter 1, section 1.1, where we have a compound Boolean statement in the form

$$F(\mathbf{x}) = (x_{17} \vee \overline{x}_{37} \vee x_{73}) \wedge (\overline{x}_{11} \vee \overline{x}_{56}) \wedge \ldots \wedge (x_2 \vee x_{43} \vee \overline{x}_{77} \vee \overline{x}_{89} \vee \overline{x}_{97}),$$

and we need to find a vector of truth assignments for all 100 variables x_i, $i = 1, \ldots, 100$, such that $F(\mathbf{x}) = 1$ (TRUE). Since we are dealing with variables that can only take on two states, TRUE or FALSE, a natural representation to choose is a binary string (vector) of length 100. Each component in the string signifies the truth assignment for the variable which corresponds to that location in the string. For example, the string $\langle 1010...10 \rangle$ would assign TRUE to all odd variables, $x_1, x_3, \ldots, x_{99}$, and FALSE to all even variables, $x_2, x_4, \ldots, x_{100}$, under the coding of 1 being TRUE and 0 being FALSE. Any binary string of 100 variables would constitute a possible solution to the problem. We could use $F(\mathbf{x})$ to check any such vector and determine whether it satisfied the necessary conditions to make $F(\mathbf{x}) = 1$.

We need an initial population to start this approach, so our next consideration is the population size. How many individuals do we want to have? Let's say that we want 30 parent vectors. One way to determine this collection would be to have a 50–50 chance of setting each bit in each vector to be either 1 or 0. This would, hopefully, give us a diverse set of parents. Typically, this diversity is desirable, although in some cases we might want to initialize all of the available parents to some best-known solution and proceed from there. For the sake of our example, however, let's say that we use this randomized procedure to determine the initial population. Note immediately, then, that every trial of our evolutionary algorithm will have a potentially different initial population based on the random sampling that we generate.

Suppose that we have the initial randomized population. What we need to do next is evaluate each vector to see if we have discovered the optimum solution and if not, we need to know the quality of the solutions we have at hand; that is, we need an evaluation function. This presents a difficulty. It is straightforward to use $F(\mathbf{x})$ to check if any of our initial vectors are perfect solutions, but let's suppose that none of these vectors turns out to be perfect; i.e., for each randomly selected initial vector, $\mathbf{x}$, $F(\mathbf{x}) = 0$. Now what? This doesn't provide any useful information for determining how close a particular vector might be to the perfect solution. If we use $F(\mathbf{x})$ as the evaluation function, and if the perfect solution were $\langle 11...11 \rangle$ and two solutions in the initial population were $\langle 00...00 \rangle$ and $\langle 11...10 \rangle$, these latter solutions would both receive an evaluation of 0, but the first vector has 100 incorrectly set variables while the second is only one bit away from perfection! Clearly $F(\mathbf{x})$ does not provide the information we need to guide a search for successively improved strings.

This is a case where the objective is not suggestive of a suitable evaluation function. The objective of making the compound Boolean statement $F(\mathbf{x}) = 1$ is insufficient. What *is* sufficient? At first, we might be tempted to simply compare the number of bits that differ between each candidate solution and the perfect solution and use this as an error function. The more bits that are incorrectly specified, the worse the error, and then we could seek to minimize the error. The trouble with this is that it presumes we already know the perfect vector in order to make the comparison, but if we knew the perfect vector, we wouldn't need to search for a solution! Instead, what we might do is consider each part of the compound Boolean expression $F(\mathbf{x})$ that is separated by an "and" symbol ($\wedge$), e.g., $(x_{17} \vee \overline{x}_{37} \vee x_{73})$, and count up how many of these separate subexpressions are TRUE. We could then use the number of TRUE subexpressions as a evaluation function, with the goal of maximizing the number of TRUE subexpressions, because, given $F(\mathbf{x})$ as it appears above, if all of the subexpressions are TRUE then it follows that $F(\mathbf{x}) = 1$, just as we desire. Do we know that this evaluation function will lead us to the perfect solution? Unfortunately, we don't. All we can say before some experimentation is that it appears to be a reasonable choice for the evaluation function.

Let's presume that we use this function to judge the quality of our 30 initial vectors, and let's further presume that there are 20 subexpressions that comprise $F(\mathbf{x})$; therefore, the best score we can get is 20 and the worst is 0. Suppose that the best vector in the population after this initial scoring has a score of 10, the worst has a score of 1, and the average has a score somewhere between 4 and 5. What we might do next is think about how we want to generate the individuals to survive as parents for the subsequent generation. That is, we need to apply a selection function.

One method for doing this assigns a probability of being selected to each individual in proportion to their relative fitness. That is, a solution with a score of 10 is 10 times more likely to be chosen than a solution with a score of 1. This *proportional selection* is also sometimes called *roulette wheel selection* because a common method for accomplishing this procedure is to think of a roulette wheel being spun once for each available slot in the next population, where each solution has a slice of the roulette wheel allocated in proportion to their fitness score (see figure 6.1). If we wanted to keep the population at a size of 30, we would allocate 30 slices to the wheel, one for each existing individual, and we would then spin the wheel 30 times, once for each available slot in the next population. Note, then, that we might obtain multiple copies of some solutions that happened to be chosen more than once in the 30 spins, and some solutions very likely wouldn't be selected at all. In fact, even the best solution in the current population might not be selected, just by random chance. The roulette wheel merely biases the selection toward better solutions, but just like spinning the roulette wheel in a casino, there are no guarantees!

Suppose that we have spun the wheel 30 times and have chosen our 30 candidates for the next generation. Perhaps the best solution is now represented three times, and the worst solution failed to be selected even once. It has now disappeared from consideration. Regardless of the resulting distribution of our solutions, we haven't generated anything new, we have only modified the distribution of our previous solutions. Next we need to use random variation to search for new solutions.

One possibility for doing this would be to randomly include the chance that some of our new 30 candidates will recombine. For each individual we might have a probability of, say, 0.9 that it will not pass unchanged into the next generation, but instead will be broken apart and reattached with another candidate solution chosen completely at random from the remaining 29 other solutions. This two-parent operator would essentially be a cut-and-splice procedure designed, hopefully, to ensure that successive components in candidate strings that tend to generate better solutions have a chance of staying together. Whether or not this would be effective partly depends on the degree of independence of the com-

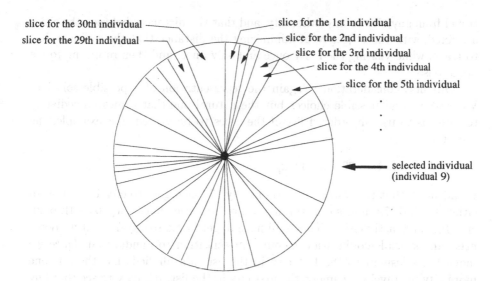

slice for the 30th individual
slice for the 29th individual

slice for the 1st individual
slice for the 2nd individual
slice for the 3rd individual
slice for the 4th individual
slice for the 5th individual

selected individual
(individual 9)

Fig. 6.1. A roulette wheel with 30 slices. The size of each slice corresponds to the fitness of the appropriate individual. From the graphic here, individual 9 had the highest fitness.

ponents. If the components of the candidate vectors aren't independent in light of the evaluation function, this recombination operator might end up metaphorically cutting and splicing *Hamlet* and *King Lear*. Another potential difficulty is that this recombination operator can only generate new solutions for evaluation if there is sufficient diversity in the current population. If we are unlucky and the roulette wheel happened to generate a homogeneous population, we'd be completely stuck, even if our current best solution wasn't a local optimum! As some insurance, we might also include another operator that flips a single bit in a chosen string at random, and then apply this with a low probability. In this way, we'd ensure that we always have the potential to introduce novel variations.

These five steps of specifying (1) the representation, (2) the initial population, (3) the evaluation function, (4) the selection procedure, and (5) the random variation operators define the essential aspects of one possible evolutionary approach to the SAT. By iterating over successive generations of evaluation, selection, and variation, our hope is that the population will discover increasingly appropriate vectors that satisfy more and more of the subexpressions of $F(\mathbf{x})$, and ultimately satisfy the condition that $F(\mathbf{x}) = 1$.

6.2 Evolutionary approach for the TSP

Suppose that we wanted to use an evolutionary algorithm to generate solutions for a TSP of 100 cities. To make things easier, let's say that it's possible to travel from any city to any other city, and that the distances between the cities are fixed, with the objective of minimizing the distance that the salesman has to travel while visiting each city once and only once, and then returning to his home base.

Our first consideration is again the representation of a possible solution. We have several possible choices but the natural one that comes immediately to mind is to use an ordered list of the cities to be visited. For example, the vector

$$[17, 1, 43, 4, 6, 79, ..., 100, 2]$$

would mean that the salesman starts at city 17, then goes to city 1, then on to cities 43, 4, 6, 79, and so on through all of the cities until they pass through city 100, and lastly city 2. They still have to get home to city 17, but we don't need any special modification to our representation to handle that. Once we move the salesman to the last city in the list, it's implied that there is one more city to travel to, namely, the first city in the list. It's easy to see that any permutation of this list of cities generates a new candidate solution that meets the constraints of visiting each city one time and returning home.

To begin, we have to initialize the population, so let's again figure that we have 30 candidate tours. We can generate 30 random permutations of the integers [1...100] to serve as the starting population. If we had some hints about good solutions, maybe from some salesmen who have traveled these routes before, we could include these, but let's say for the sake of example that we don't have any such prior information.

Once we have generated our 30 candidate tours, we have to evaluate their quality. In this case, the objective of minimizing the total distance of the tour suggests using the evaluation function that assigns a score to each tour that is equivalent to its total length. The longer the tour, the worse the score, and our goal then is to find the tour that minimizes the evaluation function. This part of the evolutionary approach appears much more straightforward for the TSP than for the SAT.

Let's say that we use the total distance to assess our 30 initial tours. We are now left with the tasks of applying selection and random variation, but we don't have to apply them in this order. As an alternative to the example for the SAT, suppose that we decide to generate new solutions first, evaluate them, and then apply selection.

Our first decision then concerns how to generate new solutions from the existing candidate tours. It's always nice to take advantage of knowledge about the problem we face [309], and this is one case where we can do this. Our evaluation function is based on distances in Euclidean space, so consider this question: Which is longer, the sum of the lengths of the diagonals of quadrilateral, or the sum of the lengths of either pair of opposite sides? Figure 6.2 illustrates the case. By Euclidean geometry, the sum of the diagonals (e and f) are always longer than the sum of the opposing sides (a and c or b and d), and by consequence this means that tours that cross their own path are longer than those that don't. So perhaps we can devise a variation operator that would remove instances where a parent tour crosses its own path.

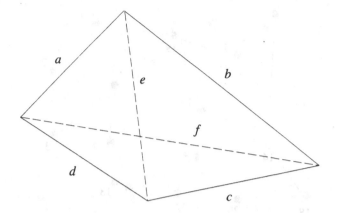

Fig. 6.2. A quadrilateral. Note that $a + c < e + f$ and $b + d < e + f$.

Indeed, this is very easily done for the representation we have chosen. For any ordered list of cities, if we pick two cities along the list and reverse all of the cities between the starting and the ending point, we have the ability to repair a tour that crosses its own path. Of course, if we apply this operator to solutions that don't cross their own paths, then we may introduce the same problems! But we can't know beforehand which solutions will be crossing over on themselves and which won't, so one reasonable approach is to simply pick two cities on the list at random and generate the new solution. If the variation isn't successful, we'll rely on selection to eliminate it from further consideration.

Like the example of the SAT before, we might think about using two or more tours to generate a new solution, but this would require some careful consideration. It's plain that we couldn't just take the first part of one solution and splice on the second part of another. In that case we would very likely generate a solution that had multiple instances of some cities and no instances of others (figure 6.3). It would violate the constraints of the problem that demand we visit every city once and only once. We could in turn simply eliminate these sorts of solutions, essentially giving them an error score of infinity, but they would be generated so frequently that we'd spend most of our time in vain, discovering infeasible solutions and then rejecting them. Under our chosen representation, we'd have to be fairly inventive to construct variation operators that would rely on two parents and generate feasible solutions.

For example, we might take two parents and begin at the first location in each tour by randomly selecting the first city in either the first parent or the second parent to be the first city in the offspring's tour list. Then we could proceed to the second position in both parents, again choose randomly from either parent, and continue on down the list. We'd have to ensure at each step that we didn't copy a city that was already copied in a previous step. To do

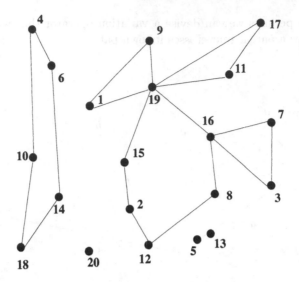

Fig. 6.3. An illegal solution for the TSP. The depicted routing is infeasible because it fails to meet the constraints of visiting each city once and only once while completing a circuit.

that, we could instead simply choose to copy from the other parent, or in the case where the cities in both parents at a particular location had already been placed in the list, we could simply choose a new city at random such that we didn't violate the constraints of visiting each city once and only once.

Suppose that instead of generating a number of offspring that is equal to the number of parents, we decide to generate three offspring per parent. We can choose any number we believe to be appropriate. For each of the 30 parents, three times in succession, we pick two cities at random in the parent, flip the list of cities between these points, and enter this new solution in the population. When we are through, we have 90 offspring tours.

We now have a choice: we can put the 90 offspring in competition with the 30 parents and select new parents from among all $30 + 90 = 120$ solutions, or we can discard the parents and just focus on the offspring. Although discarding the parents might not seem prudent, there are some supposed justifications that can be offered for this choice, but these will have to wait until chapter 7. For now, let's say that we will focus on the 90 offspring, and from these we will keep the best 30 to be parents for the next generation. This form of selection is described as (μ, λ) where there are μ parents and λ offspring, and then selection is applied to maintain the best μ solutions considering only the offspring. Here, we have a (30,90) procedure.

We could execute this process of random variation by using the two-city reversal to each parent, or the modified two-parent operator, generating three offspring per parent, and then applying selection to cull out the 60 worst new solutions. Over time, our expectation is that the population should converge toward nearly optimal tours.

6.3 Evolutionary approach for the NLP

Suppose we wanted to solve the NLP given in chapter 1 (figure 1.2) for the case of 10 variables. Given the above descriptions of possible ways to address the SAT and TSP, the NLP should be pretty straightforward. After all, the first thing to choose is a representation, and here the most intuitive choice is a vector of 10 floating-point values, each corresponding to a parameter in the problem. When we think about initializing the population, we have the constraints to worry about, but even then we could just choose values for the parameters uniformly at random from [0,10] and then check to be certain that we didn't violate the problem constraints. If we did, we could just discard that initial vector and generate a new one until we filled the initial population. The volume in $[0, 10]^{10}$ that violates the constraints is relatively small so this shouldn't waste too much computational time.

Next we'll need to generate offspring. For floating-point vectors, a common variation operator is to add a Gaussian[3] random variable to each parameter. You will recall that a Gaussian density function looks like a bell curve (figure 6.4). To define a Gaussian random variable we need a mean and a variance. For our example here, we choose the mean to be 0 and the variance to be 1 (i.e., a standard Gaussian). We can prove later that there are better choices, but this will suffice for now. To generate an offspring, we add a different Gaussian random variable to each component of a parent, making 10 changes, each one independent of the previous one. When we are done, we have an offspring that on average looks a lot like its parent, and the probability of it looking radically different in any parameter drops quite quickly because most of the probability density in a Gaussian lies within one standard deviation of the mean. This is often a desirable feature because we need to be able to maintain a link between parents and offspring. If we set the variance too high, the link can be broken, resulting in essentially a purely random search for new points. This is to be avoided because it fails to take into account what has already been learned over successive generations of evolution. The Gaussian variation also has the advantage of being able to generate any possible new solution, so the evolutionary process will never stall indefinitely at a suboptimum. Given a sufficiently long waiting time, all points in the space will be explored.

Once all of the offspring are generated, we have to evaluate them. Here again, as with the TSP, the objective suggests an evaluation function. We could just use the mathematical expression to be maximized as the evaluation function directly except that we also have constraints to deal with. And while the expression for the objective is defined over all the reals in each dimension, the feasible region is a small finite subset of the entire real space. One possibility in

---

[3]Also known as a *normal*.

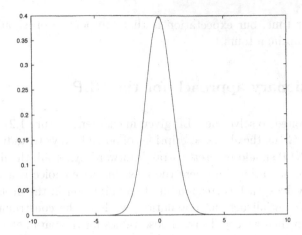

Fig. 6.4. A normal distribution $f(x) = \frac{1}{\sigma\sqrt{2\pi}}e^{-(x-m)^2/(2\sigma^2)}$ with standard deviation $\sigma = 1$ and mean $m = 0$

such cases is to apply a penalty function where the score an individual receives is the value generated by plugging it into the mathematical expression of the objective. This objective is also modified by a penalty for the degree to which it violates the applicable constraints. The further the individual moves into an infeasible region, the greater the penalty. Scaling the penalty function appropriately is a matter of judgment and experience, but for now we might just say that the evaluation function is the mathematical expression of the objective minus the distance from the individual to a boundary of the feasible region (if the individual is infeasible). The goal then is to maximize this sum.[4]

For selection, we could apply the forms we saw in the SAT or TSP, but we can introduce a third choice here: ranking selection using a tournament. Suppose that we started with 30 individuals and generated 30 more through our Gaussian variation (one offspring per parent). We'd then have a collection of 60 vectors. Suppose further that we'll keep the population at the start of each generation constant at 30. To choose these 30 new parents, one tournament method might be to pick pairs of solutions at random and determine the better one, which is placed in the next generation. After 30 such pairings we have a new population where those individuals that are better on average have a greater chance of surviving, but are not completely ensured of passing on into the next generation.

Over successive iterations of variation, evaluation (with penalties), and this form of tournament selection, we could expect to evolve an individual that meets the constraints and resides close to the global optimum. There are a number of ways that might accelerate the rate at which good solutions could be discovered

---

[4]Alternatively, for a minimization problem, the penalty would be added rather than subtracted.

by changing parameters of the evolutionary operators, but these can be deferred to later chapters.

6.4 Summary

The structure of an evolutionary algorithm is shown in figure 6.5.

> **procedure** evolutionary algorithm
> **begin**
> $t \leftarrow 0$
>   initialize $P(t)$
>   evaluate $P(t)$
>   **while** (**not** termination-condition) **do**
>   **begin**
> $t \leftarrow t + 1$
>   select $P(t)$ from $P(t-1)$
>   alter $P(t)$
>   evaluate $P(t)$
>   **end**
> **end**

Fig. 6.5. The structure of an evolutionary algorithm

The evolutionary algorithm maintains a population of individuals, $P(t) = \{x_1^t, \ldots, x_n^t\}$ for iteration t. Each individual represents a potential solution to the problem at hand and, in any evolutionary algorithm, is implemented as some data structure S. Each solution x_i^t is evaluated to give some measure of its "fitness." A new population at iteration $t+1$ is then formed by selecting the more fit individuals (the "select" step). Some members of the new population undergo transformations (the "alter" step) by means of variation operators to form new solutions. There are unary transformations u_i that create new individuals by a change in a single individual ($m_i : S \rightarrow S$), and higher-order transformations c_j that create new individuals by combining parts from several (two or more) individuals ($c_j : S \times \ldots \times S \rightarrow S$). The algorithm is executed until a predefined halting criterion is reached, which might include finding a suitable solution, generating a specified number of candidate solutions, or simply running out of available time.

The history of applying these sorts of algorithms dates back into the early 1950s and was actually invented as many as ten times by different scientists, completely independent of each other [149]. Each procedure was slightly different from the others. Some of the researchers in the 1960s gave their algorithms different names like *genetic algorithms, evolution strategies,* or *evolutionary programming,* but not everyone chose to christen their inventions. Over time, more specialized techniques developed that were associated with particular data structures, such as *classifier systems* and *genetic programming.* In the

1990s, however, there has been considerable theoretical and empirical evidence that none of these canonical approaches can offer anything that another approach cannot achieve. What's more, as interest in evolutionary algorithms has rapidly increased, ideas have been borrowed, exchanged, and modified across all of these approaches. As a result, there is no longer any scientific basis for discriminating between evolutionary approaches. In light of the current state-of-the-art, we'll adopt the term *evolutionary algorithm* to describe any of the algorithms that use population-based random variation and selection throughout the remainder of the text, and we encourage you to think in these broad terms as well.

Let's summarize some of the ideas discussed in earlier sections of this chapter and point out a few additional possibilities. To begin, we want to emphasize that evolutionary algorithms may incorporate any representation that appears suitable for the task at hand, whether it's a binary string as suggested in the SAT problem, a permutation of integers as offered for the TSP, or a vector of floating-point numbers as we saw in the NLP. Of course, we're not limited to these representations. We could use matrices, graphs, finite state machines, and other more complex data structures to represent potential solutions. In particular, individuals may represent a set of rules (having a variable-length structure), a blueprint of some complex design, or even a computer program. Indeed, the possibilities are limitless!

Note, however, that the choice of variation operators, which are responsible for changing parents into offspring, depends strongly on the chosen representation. We have already seen (section 6.1) a flip-mutation, which flips the value of a bit, and the so-called "1-point crossover," two operators that have commonly been applied to binary representations. We have also looked (section 6.2) at a swap or reversal operator for integer representations together with special crossover that builds a permutation of integer numbers out of two distinct permutations. We also described (section 6.3) a Gaussian mutation for floating-point representations.

Despite the many differences in approach to these problems, the implementation of an evolutionary algorithm was seen to be fairly easy. Whether or not the problem was an SAT, TSP, or NLP, the recipe was essentially the same: (1) create a population of individuals that represent potential solutions to the problem at hand, (2) evaluate them, (3) introduce some selective pressure to promote better individuals or eliminate those of lesser quality, and (4) apply some variation operators that generate new solutions to be tested. Then repeat the evolutionary loop of evaluation, selection, and variation (steps 2–4) several times.[5]

There are a few other interesting observations to make. We won't provide a detailed analysis here, but rather a few remarks. First, the broadly encompassing approach of the evolutionary technique should be apparent. For example, we can argue that simulated annealing is a special case of an evolutionary algo-

---

[5]Alternatively, the loop can proceed with variation, selection, and evaluation, depending on the particular implementations.

rithm, where the population size is limited to a single individual, the variation operator relies on a single parent, and the selection procedure is based on an extrinsic parameter called *temperature*. In chapter 10 we'll discuss evolutionary algorithms that change the scope or rates of their mutation operators, and this connection between simulated annealing and evolutionary algorithms will be made more directly.

In contrast to annealing, tabu search is based on very different concepts, but even there, it's possible to extend an evolutionary algorithm using the concept of memory and then tabu search becomes a special case. It's relatively easy to imagine adding either a memory unit to each individual (e.g., in terms of adding additional parameters to the represented structure and some interpretation rule that would operate on these additional parameters to guide the search; see also chapter 11), or a global memory (like a library of past ideas) for the whole population (e.g., cultural algorithms [388]). Evolutionary algorithms can in fact incorporate the main concepts underlying tabu search quite naturally.

In chapter 3, we argued that one particular classification of optimization algorithms can be based on how an algorithm "perceives" the search space. One class of algorithms (e.g., greedy algorithms, dynamic programming, branch and bound, A^*) evaluates subspaces, rather than individual solutions. Algorithms in the other class (e.g., hill climbers, simulated annealing, tabu search) consider the whole search space as a uniform set of potential solutions. Only single complete solutions are evaluated and there is no concept of a subspace.[6] On the other hand, evolutionary algorithms may combine these two categories by allowing individuals to describe subspaces as well as particular solutions.

There are also various additional possibilities to explore. Many of these are discussed in length later in this text, but we will mention some of them right here to give you a better perspective on this exciting field!

Most algorithms require a setup of some parameters (e.g., to apply simulated annealing, you have to fix the starting temperature, cooling rate, the maximum number of iterations per given temperature, etc.). Similarly, evolutionary algorithms require a few parameters as well. These parameters include the population size, probabilities (rates) of various operators, a parameter to regulate the selective pressure of the system, penalty coefficients (if the penalty function approach is used for solving a constrained optimization problem), and others such as the mutation step size, the number of crossover points, etc. The values of such parameters may determine whether or not the algorithm will find a near-optimum solution and, what's also important, whether or not it will find a reasonable solution efficiently. Choosing effective parameter values, however, is a time-consuming task and considerable effort has gone into developing good heuristics for these choices across many problems.

Recall, though, that evolutionary algorithms implement the idea of evolution, and that evolution itself must have evolved to reach its current state of sophistication. It is thus natural to expect adaptation to be used not only for finding solutions to a problem, but also for tuning the algorithm to the par-

---

[6]Except of course a neighborhood, which constitutes a subspace.

ticular problem. Technically speaking, this amounts to modifying the values of parameters during the execution of the algorithm taking the actual search process into account. As discussed later in the text, the issue of controlling the values of various parameters of an evolutionary algorithm has the potential of tuning the algorithm to the problem while solving the problem!

Further, many real-world problems require multiple solutions. This is common when there are multiple objectives to meet and it is difficult to aggregate them with a normalizing function. It would be great if we could have an algorithm that returned more than one good solution! Having a few alternatives to choose from is often very useful!

Evolutionary algorithms can be adapted easily to meet this new challenge. It's possible to utilize feedback on the diversity of the population, and in particular, on the presence of other individuals in a neighborhood of a chosen individual. Such feedback, duly incorporated into the evaluation function, can be used to control the number of returned solutions.

In chapter 1, we mentioned that many problems require time-adaptive solutions. We often get a new piece of information that contributes to the evaluation function while solving the problem. This might occur, for example, while traversing the best tour for the TSP. Perhaps we discover that, due to unforeseen circumstances, a particular segment of the tour is no longer available. Or while scheduling a factory, we learn that one of the machines has broken down. These unexpected events occur all the time in the real world, and they demand quick adaptation of current solutions to meet the new challenges.

A vast majority of classic optimization algorithms assume a fixed evaluation function. Any change in the evaluation function requires restarting the algorithm from scratch! In contrast, evolutionary algorithms are inherently adaptive; individuals in the population *adapt* to the current environment, and the quality of that adaptation is measured by the evaluation function. Consequently, an unexpected change in the model, or even in the evaluation function itself, can cause a short-lived disturbance of the evolutionary process, but after some period of time, which varies by problem, individuals in the population can adjust to the requirements of the new setting. Evolutionary techniques don't require restarting every time something changes. Instead, they continue their adaptation in light of similarities between previous experience and current demands. This illustrates a great potential of evolutionary algorithms.

Another advantage of the evolutionary approach to problem solving is that it's often easy to hybridize evolutionary algorithms with other methods. Special variation and selection operators that do the work of hill climbing or other methods can be incorporated into the process directly.

Here's another possible advantage. Recall our discussion from chapter 1, section 1.3 concerning the owner of a supermarket chain who needs to decide where to place a new store, while at the same time, his competitors are making their decisions as well. Of course, they are trying to maximize their profits, in turn quite likely minimizing the profits of the owner of the supermarket chain. Thus, the proper decision making process is very complex (he thinks that they think that he thinks that they think...), but we can model this situation by running two opposing evolutionary processes. The first evolutionary algorithm optimizes the strategy for the supermarket chain's owner, while the other evolutionary algorithm handles the opponent's strategies. Note that the evaluation of each side's strategy is on the basis of the current (or perhaps projected) strategy of the opponent. This type of evolutionary competition is called a *coevolutionary model*. These sorts of models can often be used to discover solutions to problems that appear too unwieldy to describe in precise mathematics due to the joint relationship of the players.

Another item should be mentioned while discussing advantages of evolutionary algorithms: parallelism. Parallelism promises to put within our reach solutions to problems that were heretofore intractable. Evolutionary algorithms are very suitable for parallel implementations as they are explicitly parallel "in nature"! It's relatively straightforward to assign a processor to each individual in the population, or to split a population into several subpopulations (the latter can be enhanced further by adding the concept of migration of individuals from one subpopulation to another).

While listing potential advantages of evolutionary algorithms, it's worthwhile to mention the final "argument," which is related more to the psychology of programming than the efficiency of the approach. When running an artificial evolutionary process, you are playing the role of *creator*, and you determine the rules of the simulation, much like determining the physics of the real world. You may choose to pay close attention to specific mechanisms in natural evolution, or perhaps consider things not found in biological systems (e.g., mating more than two parents). You can even violate the laws of nature completely and introduce *Lamarckian evolution*, where individuals acquire characteristics during their "lifetime" and then pass those characteristics along to their offspring. The barrier between the *phenotype* (an individual's expressed behavior) and the *genotype* (an individual's genetic composition) doesn't have to exist for you!

You might introduce the concept of gender among the individuals, assign them an age (some number of generations), and then use this to determine how long they might survive. You might include a memory in an individual, or explore the concept of families of individuals and social learning. You could even put entire populations in competition with each other and perhaps migrate solutions between populations. You might explore including the *Baldwin effect* in which individuals learn during their lives and in doing so can affect their fitness and alter the course of evolution. All of these avenues lead to other choices. For example, which solutions should migrate: the best, the average, the worst? Indeed, the possibilities open to you as a programmer and problem solver are virtually unlimited!

VII. One of These Things Is Not Like the Others

A tall man walks into a jewelry store and places a bag full of old mint coins on the counter. The jeweler picks up the bag and empties the coins out in front of them.

"How much will you give me for these coins?" asks the man.

"Ah, 1907 Indian head pennies. And they all look to be in very good condition," says the jeweler as he examines them with an eye piece, "I'd say each was worth $100. You've got 12 coins here, right?"

"Well, yes, but there's a problem you should know about. You see, when I bought these coins, the guy who sold them to me said that one of them was counterfeit," admits the man.

"How about that? An honest thief?" exclaims the jeweler.

A moment of silence passes as the tall man tries to discern if the jeweler was talking about him or the guy who sold him the coins.

"Right, well, how can we go about finding the counterfeit?" asks the tall man, now fidgeting slightly.

"Hmmm. I've got an idea. Most counterfeit coins are made with cheaper materials and therefore don't weigh the same amount as the real thing. I've got a balancing scale right here. Now all we have to do is figure out how to weigh the coins so that we can identify the fake."

The tall man smiles "Oh, this is easy. We can do it in just three weighings. Not only that, but we can tell whether or not the fake coin is lighter or heavier than the rest too."

."Really? How do you know that?" asks the jeweler, surprised at the apparent talent of this man.

"Yes, well, I've recently read a very good book on heuristics for problem solving," replies the man, "The only trouble is that I can't remember how they did it."

The two men slump down over the scale and start fiddling with the coins. Can you figure out to find the fake?

Put the book aside for a minute and see if you can get a headstart on this problem by yourself.

Here we are faced with a problem of locating a single item with a certain characteristic from an assortment of other items that don't have that same characteristic. The tool we have in front of us is a balancing scale (figure VII.1). This device can only tell us that one side is heavier than the other, or that both sides are equal.

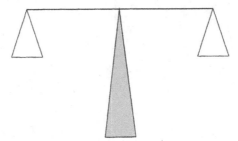

Fig. VII.1. A balancing scale

The natural thing to do in the problem is to start dividing the 12 coins into groups where one group could be weighed against another. The difficulty, however, is that if you divide the coins into two groups of six and compare those, then no matter what result you get, you'll have no idea how to partition those groups to obtain the answer with only two more weighings. Remember, the goal here is to get the answer with just three uses of the scale.

Let's consider an alternative approach of grouping the coins in three sets of four. Suppose that the items were arranged into three groups:

$$G_1 = (1,2,3,4), \ G_2 = (5,6,7,8), \ \text{and} \ G_3 = (9,10,11,12).$$

In the first weighing we compare groups G_1 and G_2. Either they balance or one group is heavier. Let's consider these two cases in turn.

If the total weights of the coins in G_1 and G_2 are the same, then the fake coin must be in G_3, and straight away we know that all of the coins in G_1 and G_2 are genuine. Thus, in our second weighing, we can compare three arbitrary genuine coins, say, 1, 2, and 3, with three coins from group G_3:

$$(1,2,3) \ \text{versus} \ (9,10,11).$$

Again, there are two possible outcomes:

- The coins balance. This means that the fake coin is 12 because it's the only item from G_3 that's not involved in the second weighing. The third weighing (say, 1 against 12) would determine whether the fake coin is heavier or lighter than the rest.

- The coins don't balance. This means that the fake coin is either coin 9, 10, or 11. At this stage we also know if the fake is lighter or heavier. If $(1,2,3)$ is heavier than $(9,10,11)$ then the fake coin is lighter, and vice versa. The third weighing (say, 9 against 10) would determine which coin is counterfeit. If coin 9 balances with 10 then the fake is coin 11. If they don't balance then from the previous information on whether or not the fake was lighter or heavier, we could choose appropriately.

So, if we're lucky and strike a balance in the first weighing of G_1 against G_2, everything is simple.

What should we do if there's an imbalance in the first weighing?

In this case we know two things: (1) the counterfeit coin is in G_1 or G_2, and (2) coins 9, 10, 11, and 12 are genuine.

This is probably the most difficult step in getting the solution. It's important to arrange the second weighing in a meaningful way. We might consider splitting the coins from set G_1 onto the left and right sides of the scale for the second weighing (say, leaving coins 1 and 2 on the left side and moving coins 3 and 4 to the right side), but this wouldn't be sufficient. If they balance, we'd know that the counterfeit is among those in G_2; we'd also know if the counterfeit is lighter or heavier. We wouldn't, however, be able to find the fake in the third and final weighing because we have four candidates (5, 6, 7, and 8) at this stage, so we have to eliminate one of these potential candidates.

We can do this by moving one of the coins from G_2 (say, item 5) to the left side of the balance, and by adding a genuine coin (say, item 12) to the right side. Thus, the second weighing is

$(1, 2, 5)$ versus $(3, 4, 12)$.

Assume that in the first weighing, coins $(1, 2, 3, 4)$ were heavier than coins $(5, 6, 7, 8)$. There are then three possible outcomes of the second weighting:

- Coins $(1, 2, 5)$ are heavier. This means that coins 3, 4, and 5 are genuine because we changed their locations on the scale but the outcome of the weighing remained the same (i.e., the left-hand side is heavier). Since coin 12 is also genuine, the fake is either coin 1 or 2. We also know here that the fake coin is heavier. A third weighing (coin 1 versus 2) determines the fake.

- Coins $(3, 4, 12)$ are heavier. Since the outcomes of the first and second weighings changed (i.e., the left-hand side was heavier first, but now the right-hand side is heavier), the fake coin had to have been moved from one side to the other. Thus, either coin 3 or 4 is fake and heavier, or coin 5 is fake and lighter. A third weighing (coin 3 versus 4) determines the counterfeit. If there's a balance, the fake is coin 5.

- Coins $(1, 2, 5)$ and $(3, 4, 12)$ balance. In that case the fake wasn't included in the second weighing, so it must be one of coins 6, 7, or 8. We also know that the fake is lighter (from the first weighing), thus a third weighing (say, coin 6 versus 7) will determine the counterfeit.

Thus, in three weighings the man and the jeweler can determine the counterfeit coin and also know if it's lighter or heavier than the rest.

The approach to solving the problem relies on partitioning the space of solutions so that a series of comparisons can eliminate many of the possibilities and leave only those few that demand further attention. This strategy is often quite effective. Who wants to spend time searching for the answer to a problem when the answer lies somewhere else? Just be wary of the natural tendency to divide things in half. There are times when other partitions are more effective.

"Terrific!" exclaims the jeweler, "You remembered after all!"

"A bit fortunate, but all's well that ends well," smiles the tall man as he leaves with $1,100 and one counterfeit coin.

As he walks out the door, he bumps into an old woman coming in the other direction.

"Oh goodness, excuse me!" he apologizes, as the woman approaches the jeweler.

"Can I help you?"

"Oh, I certainly hope so. You see I have 120 mint Liberty head nickels from the early 1900s in my bag here," says the woman, patting the bag that jingles back in response.

"And you want to sell them?"

"Yes, but you see, one of them is a fake..."

See if you can figure out which one is the counterfeit with only five weighings. Oh, yes, and also figure out whether it's lighter or heavier!

7. Designing Evolutionary Algorithms

A thousand things advance;
nine hundred and ninety-nine retreat:
that is progress.

Henri Fréderic Amiel, *Journal*

The essential idea of evolutionary problem solving is quite simple. A population of candidate solutions to the task at hand is evolved over successive iterations of random variation and selection. Random variation provides the mechanism for discovering new solutions. Selection determines which solutions to maintain as a basis for further exploration. Metaphorically, the search is conducted on a landscape of hills and valleys (see figure 7.1), which is also called a "response surface" in that it indicates the response of the evaluation function to every possible trial solution. The goal of the exploration is most often to locate a solution, or set of solutions, that possesses sufficient quality as measured by the evaluation function. It's not enough to simply find these solutions, however, we need to find them quickly. After all, enumeration will find such solutions too, but for real problems we'd grow old waiting for the answers. The speed with which suitable solutions can be discovered is in part dependent on the choices we make in determining the representation of trial solutions, the evaluation function, the specific variation and selection operators, and the size and initialization of the population, among other facets. The key to designing successful evolutionary algorithms lies in making the appropriate choices for these concerns in light of the problem at hand.

Wouldn't it be nice if there were one best choice for each of these parameters that would give us the optimum performance regardless of the problem we are trying to solve? Once we found this superior set up, we could just implement it directly and not have to be concerned with issues of how to represent solutions, how to construct variation operators, or how to decide which solutions to maintain as a basis for further exploration. Unfortunately, it's possible to prove (within some limited assumptions) that, in fact, such an optimum set of choices doesn't exist [499]. Essentially, all search procedures that don't resample points in the state space of possible solutions perform, on average, exactly the same across all problems. Although it might be counterintuitive at first, a hill-climbing algorithm that never resamples points will do exactly as well on average as blind random search without resampling when trying to find the maximum of all possible functions. This holds true for all evolutionary algorithms as well. If you don't know something about the problem you are facing, you have no justification for choosing any particular search algorithm.

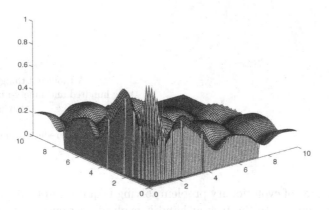

Fig. 7.1. The evaluation function can present a landscape of hills and valleys (see Figure 1.2).

At first, this mathematical result might seem a bit depressing. It implies that if you don't utilize some specific knowledge in solving your problem, it doesn't matter what representation you use, which variation operators, how big your population might be, what form of selection you implement, and so forth. Actually, in a way, the situation is even worse than that. It's possible to show that no bijective[1] representation offers any capability not found in another representation [156]. Anything you can do in binary strings, you can also do in octal, or hex, or base-13 for that matter. What's more, no matter which search operators you pick, their effects on searching the space won't be unique. There is always another variation operator that will mimic the one you choose, regardless of whether it's based on a single parent, or two or more parents. If you choose to use, say, one-point crossover on binary strings, someone else (if they're clever) can come along and construct an exactly equivalent algorithm using, say, base-17 encoding and a different two-parent variation operator that's not based on crossing over. Thus, not only is there no best choice for the design parameters of a search algorithm on average across all problems, there isn't a single best way to search for a solution in any individual problem either. Metaphorically, there are many ways to skin a cat, and as long as the cat gets skinned, it doesn't matter how you do it!

Rather than view these results as pouring cold water over our enthusiasm for finding good search algorithms, let's look at things more optimistically. First, we know that we must use knowledge about our problem in order to design an algorithm that will be more effective than, say, blind random sampling. Thus, it is incumbent on us to take responsibility for knowing as much about the problem as we can and try to incorporate that knowledge into useful procedures. Second, we know that there won't be one best way to solve the problem. Indeed,

---

[1] A bijective mapping between two sets is one-to-one and onto.

there may be many best ways and even more exceedingly good ways! Whatever problem we might face, we needn't be concerned with forcing a particular representation or search operator onto that problem. Our quest is to design the right tool that matches up with the right problem.

Evolutionary algorithms are incredibly versatile. By varying the representation, the variation operators, the population size, the selection mechanism, the initialization, the evaluation function, and other aspects, we have access to a diverse range of search procedures. Evolutionary algorithms are much like a Swiss Army knife [125]: a handy set of tools that can be used to address a variety of tasks. If you've ever used such a device, you know that this versatility comes at a cost. For each application there is usually a better tool devised specifically for that task. If you have to pry a nail out of a board, the claw of a hammer will probably do a better job than anything you'll find in a Swiss Army knife. Similarly, if you have to remove a splinter from your finger, a carefully crafted set of tweezers will usually do a better job than the makeshift version you'll find in the Swiss Army knife. Yet, if you don't know exactly what task you'll be faced with, the flexible nature of the Swiss Army knife comes in very handy. Imagine trying to remove a nail with tweezers, or to extract a splinter with the claw of a hammer! Having the Swiss Army knife provides you with the ability to address a wide variety of problems quickly and effectively, even though there might be a better tool for the job, and you don't have to carry a hammer, tweezers, a screwdriver, a bottle opener, an awl, a file, a toothpick, and so forth.

This analogy carries directly over to the application of evolutionary algorithms. Consider the problem of finding the minimum of an n-dimensional quadratic bowl:

$$f(\mathbf{x}) = \sum_{i=1}^{n} x_i^2.$$

Suppose you were to use a simple evolutionary algorithm where you had one parent, $\mathbf{x}$, and you generated an offspring, $\mathbf{x}'$, as follows:

$$x_i' = x_i + N(0, \sigma),$$

for each dimension $i = 1, \ldots, n$, where $N(0, \sigma)$ is a Gaussian random variable with mean zero and standard deviation σ. For this algorithm, Rechenberg proved over 25 years ago that the maximum expected rate of convergence can be obtained when σ is proportional to the value $\sqrt{f(\mathbf{x})}$. Specifically [383],

$$\sigma^* = 1.224\sqrt{f(\mathbf{x})}/n.$$

Moreover, the expected rate of convergence under these conditions is geometric in the number of iterations [18].

It's interesting to compare this rate of convergence with what can be obtained by an algorithm designed especially for the quadratic bowl. The perfect algorithm for this function is the Newton–Gauss procedure (chapter 3). It's designed to transition from any point on the quadratic bowl to the minimum of the bowl in just one step. That's fast convergence! Indeed, speaking of the rate of convergence for such an algorithm isn't very meaningful. You just can't do better than getting the perfect solution in only a single iteration! Thus, we know that the Newton–Gauss procedure is the one to choose if we're searching a quadratic bowl. Evolutionary algorithms, even when other tricks are employed to speed up local convergence (see [46]) don't match this level of performance on this particular function.

But consider what happens when we switch to the function shown in figure 7.1 and apply the Newton–Gauss procedure. This procedure assumes that every function being searched has a quadratic surface. From any starting point, it operates as if the entire function were a quadratic, calculates the supposed global optimum, and then continues iterating until it converges to a local optimum. With a bumpy function like that of figure 7.1, however, such a local optimum may not be sufficient. In contrast, an evolutionary algorithm can overcome these local optima, escape entrapment, and converge to a global optimum, and it can do this across a wide variety of functions.

Essentially, the Newton–Gauss procedure has a narrow assumption: everything in the world looks like a bowl. Evolutionary algorithms take a broader view, often making in essence no assumptions before casting the search problem into a set of data structures with variation and selection operators. As a consequence, when the function being searched really looks like a bowl, they still work effectively, but not as efficiently as we might like. Obversely, when the function at hand doesn't look anything like a bowl, Newton–Gauss may not work effectively at all. It becomes a pair of tweezers trying to pull a nail out of a board. Once you've picked the structures, search operators, and selection procedures for your algorithm, you've made an inherent assumption about what types of problems you are hoping to face. If your assumptions are overly narrow, you may find yourself with a very efficient search that is completely ineffective.

It would be nice if we could place all the real-world problems in a framework of search on a quadratic bowl! Then we could solve all our problems in just one step! Unfortunately, all the problems like that were solved long ago. Those were the easy ones. Certainly, it's not impossible to transform difficult problems into trivial ones, but it's rarely practical. Consider the function from figure 7.1 again. There's a transformation that could be applied to every point in the feasible region that will result in all of these points conforming to a quadratic surface, but this is a point-to-point mapping that essentially requires knowing the right answer before you start! In that case, you wouldn't need to do any searching at all, even for just one iteration. We shouldn't preclude the possibility of choosing evaluation functions that make the problems we face a bit easier, but we can't hope to solve the world's problems using Newton–Gauss.

Likewise, we can't hope to solve the world's problems with *any* single procedure or operator. For example, just as Newton–Gauss works under a particular assumption, so does one-point crossover, two-point crossover, blending, Gaussian variation, and so forth. Each operator carries a set of assumptions with it. The key issue to grasp is how broad or narrow the implicit assumptions of each operator are, and in turn, which combination of evaluation functions and representations will pose suitable or unsuitable problems. The trick to effective problem solving is to fit as much about the problem into the method of solution as you can without overfitting. If you don't fit enough information into your solution, you sacrifice efficiency. If you overfit, you may preclude obtaining an effective solution.

Let's take a closer look at each of the issues in designing evolutionary algorithms. Before beginning this effort, let's re-emphasize that none of the choices you make for any part of the algorithm should be made in isolation from the choices you make for the other parts. Everything goes hand in hand. For example, the appropriateness of an evaluation function cannot be judged except in the light of the representation and variation operators that are being adopted. Yet, the search operators can only be judged when we are given the representation and evaluation function. Moreover, it would seem difficult to prescribe an evaluation function and search operators without having a representation in mind. Examining the effects of each facet of an evolutionary algorithm separately can only provide limited information. Even worse, it may be misleading.

7.1 Representation

A representation is a mapping from the state space of possible solutions to a state space of encoded solutions within a particular data structure. For example, consider the TSP. The state space of possible solutions contains the tours that the salesman may traverse. The choice of how to represent these tours in the computer is not obvious. One possibility we've seen is to make an ordered list of the cities to visit, with a return to the home city being implied. Thus a ten city tour might be represented as

[1 3 5 10 4 2 8 9 7 6].

Here, the representation is a permutation of the first ten natural numbers, and any such permutation would provide a feasible tour, but this isn't the only possible representation.

There are 181,440 different tours. We could assign each one a number from 1 to 181,440. This would be a decimal representation. In fact, there are 181,440! ways we could do this. Thus for representations in base-10, for the ten city TSP, there are 181,440! different mappings we could consider. For that matter, we could encode every decimal value as a binary number (or any other base), and again have 181,440! different such mappings for that cardinality.[2] Yet, we have to admit that these sorts of representations appear poorly suited for this problem. We can't easily envision a search operator that could move from one or more tours to another and maintain improvements in the quality of solution over successive iterations. Immediately we are faced with the problem of integrating

---

[2]Again, it's clear that the cardinality of alphabet alone is insufficient to base the choice of representation.

the representation with the search operators in light of some possible evaluation function!

A "good" representation allows for the application of search operators that maintain a functional link between parents and their offspring. There has to be a useful relationship between parents and offspring that is exploited by the search operator(s). A mismatch between search operators and the representation leads to an algorithm that may be equivalent to a blind random search, essentially sampling new trials from the state space of possible solutions without regard for previous samples. In fact, it's possible that such a procedure could perform worse than a purely random search. Depending on the choice of representation, the utility might be in the form of (1) having a graded range of variation operators that lead to a corresponding graded range of differing behaviors in the offspring (i.e., it's possible to tune the effective "step-size" of the search operator), (2) combining useful independent parts of different solutions, (3) providing for operators that always generate feasible solutions, and so forth.

Recall that Fogel and Ghozeil [156] proved that, within the class of representations that are bijections, no choice for representation offers a unique advantage over another. Thus the best recommendation for selecting a representation is to pick a structure that appears intuitive from the statement of the problem. If you're concerned with optimizing continuous parameters, such as temperature or pressure, floating-point vectors provide an intuitive data structure. If you instead need to determine the optimum subset of features to use in a pattern classifier, it may be more intuitive to use a binary string where each bit indicates that the feature either is or is not used, or perhaps you might think of a variable-length list of the features to be included (e.g., [1, 3, 5] or [1, 4, 7, 9]). In each case, as soon as you have the representation in mind, you also will have possible variation operators in mind, for if you don't, then you have certainly chosen your representation poorly.

Below are a variety of representations that are commonly employed in evolutionary algorithms. They are grouped, with illustrations taken from problems that you might face. In all, the list contains most of the common representations found in the literature.[3]

7.1.1 Fixed-length vectors of symbols

Let's consider a few possibilities:

Possibility (i). Suppose you are given a subset selection problem where you must determine which of a number of items you wish to choose. This occurs in statistical modeling (i.e., which independent variables to include), transportation optimization (i.e., which items should be transported by each particular vehicle), and so forth. A typical representation for these problems is a list of bits

---

[3]There are more complicated representations involving self-adaptive parameters that determine the search operators used by the algorithm, but this will be delayed until chapter 10.

$\{0, 1\}$ in a fixed-length vector (i.e., a binary string). The length of the vector corresponds to the number of items or parameters at hand (e.g., 1 indicates that the indexed item is included, 0 indicates that it's omitted). This choice is also intuitive for problems where digital code in machine language is evolved, particularly for implementation in hardware.

Possibility (ii). Say you're faced with the problem of optimizing a set of rotational angles of a chemical compound. In this case, the possible angles range over $[0, 2\pi)$ radians, so a natural representation is a list of floating-point values constrained to $[0, 2\pi)$. Alternatively, given a list of independent variables to include in a model with the challenge of finding the continuous parameters for these variables such that some error function of the model output and the dependent variable is minimized, you might again choose a floating-point representation, but here the values would be unconstrained, ranging over $\pm\infty$.

Possibility (iii). Combining ideas from (i) and (ii) above, suppose you need to find a linear predictive model that operates on some set of past values of a variable, e.g., you want to predict the price of IBM's stock based on its history. You could postulate a model of the form

$$y[t] = a_1 y[t-1] + \ldots + a_k y[t-k], \tag{7.1}$$

where $y[i]$ is the sequence of stock prices, t is the time, k is the number of price values prior to the current value that will be included in the model, and $a_1, \ldots, a_k$ are real-valued coefficients. The problem entails choosing the optimal value of k such that the subsequent optimal values of $a_1, \ldots, a_k$ yield the minimum error in predicting $y[t]$ over all possible values of k. Here, you might choose a compound representation of

$$[k, a_1, \ldots, a_{max}],$$

where max is the maximum number of previous values you will consider. The first component of your vector, k, indicates how many of the successive components will be included in the model, according to (7.1). You might note that the manner in which you will implement your search operators for k will differ from those operating on $a_1, \ldots, a_{max}$.

7.1.2 Permutations

Suppose you are faced with a scheduling problem. You must order a list of j jobs to be processed in a factory. Each job must eventually be processed, and once processed it is removed from further consideration. Under these circumstances, a permutation of the jobs,

$$[1, 2, \ldots, j],$$

is a convenient representation. The order in the permutation corresponds to the order of application of the jobs. Note that it may be necessary to have multiple instances of certain jobs; each could be represented by a different number or simply by including numbers multiple times. Permutations are also useful in variations of the TSP.

7.1.3 Finite state machines

Say you're faced with a problem of forecasting a sequence of symbols in time and you must devise a procedure for making useful predictions. One such procedure can be a finite state machine (e.g., a Mealy machine), which has a finite number of "states," with one being defined as the start state, and for every possible input symbol in every state there is an associated output symbol and next-state transition. Given input and output alphabets of symbols $A = \{a_1, \ldots, a_m\}$ and $B = \{b_1, \ldots, b_n\}$, respectively, a useful representation for such a machine is

$$(O_{ij}, S_{ij}, s),$$

where O_{ij} and S_{ij} are the matrices of output symbols and next-state transitions, respectively, for each input symbol i coupled with being in state j, and s is the starting state, taking a value from 1 to the maximum number of states. Note that the representation could be extended to:

$$(O_{ij}, S_{ij}, s, k),$$

where there are k states to the machine; therefore, k defines the size of the matrices O_{ij} and S_{ij}, as well as the maximum value of s. In this case, the representation may take on a variable length.

7.1.4 Symbolic expressions

Suppose you're faced with a problem of constructing a function to perform a particular mapping task, e.g., implement a control sequence to balance a broom that's mounted upside down on a cart (figure 7.2) solely by pushing on the cart. The function must operate on the current state of the system x, $\dot{x}$, θ, $\dot{\theta}$, these variables representing the position and velocity of the cart and pole, respectively, and output the corresponding force to apply to the cart. In this case it may be appropriate to use a symbolic expression in the form of a parse tree. The tree is a specific structure connecting functions and terminals. Here, the functions might be $+$, $-$, $*$, $/$, sin, cos, and so forth, and the terminals might be x, $\dot{x}$, θ, $\dot{\theta}$, and r, a real number that can take on any specified value. Functions operate on other functions and terminals. Thus, one simple example is

$$(* (+ \sin(\dot{x})\ 2.13)\ (\theta)),$$

which would translate into the mathematical expression

$(\sin(\dot{x}) + 2.13) * \theta.$

As with finite state machines, it's easy to imagine these structures as variable-length representations of trees in different configurations.

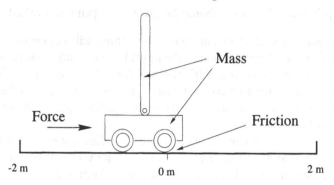

Fig. 7.2. A cart with a pole hinged to its top. The object is to push or pull on the cart so as to return it to the center of the track without having the pole fall over or the cart hit either end of the track.

Certainly, the above listing is not exhaustive and it's easy to imagine combining representational forms. For example, suppose you have a number of targets, T, to shoot at with different weapons, W. Each weapon can only fire at a subset of the targets at each time step. If a weapon fires, then it must be reloaded, preventing the weapon from being used for some period of time. Thus a representation for a fire control strategy could be a matrix of $|W|$ rows, one for each weapon, by T columns, one for each time step, with elements in the matrix being permutation vectors indicating an order of priority for shooting at available targets or choosing not to fire at all. Note that the permutation vectors in the matrix would be of different lengths, corresponding to the number of targets available for the particular weapon in question. (This representation was used in [147].) Essentially, all common representations are fundamentally fixed- or variable-length vectors or matrices comprising symbols. The interpretation of these structures can lead to implementations of neural networks, fuzzy controllers, cellular automata, or virtually any other computational structure.

7.2 Evaluation function

The evaluation function is your sole means of judging the quality of the evolved solutions.[4] The interaction between the evaluation of the solutions and the variation operators to a large part determines the effectiveness of the overall search.

---

[4]Some evolutionary algorithms, however, do not use an extrinsic evaluation function. Instead, they evaluate only the relative worth of solutions in pairwise competition (e.g., when playing a game like checkers or backgammon). In this *co-evolutionary* model, the evaluation function is intrinsic to the evolutionary process itself.

Carefully designing suitable evaluation functions requires considerable insight into how the evolutionary algorithm is likely to proceed, but there is at least one common sense guideline to follow:

The optimum solution(s) should be given the optimum evaluation(s).

Typical implementations of evolutionary algorithms will converge to a global optimum of the evaluation function in the limit. It seems only reasonable to have these global optima correspond with the solutions that you are really looking for. Moreover, you'd like to have a strong correlation between your subjective quality of any proposed solution and its corresponding quality determined by the evaluation function.

To illustrate, consider the simple problem of finding a vector of ten symbols from $\{0, 1\}$ such that the sum of the symbols in the vector is maximized. The evident best solution is [1111111111]. And the evaluation function you would probably think of straight away is

$$f(\mathbf{x}) = \sum_{i=1}^{n} x_i,$$

so that the more 1s there are in the vector, the "closer" it is to the perfect solution and the higher its evaluated quality. In contrast, consider the evaluation function

$$f(\mathbf{x}) = \prod_{i=1}^{n} x_i.$$

Here, the global optimum solution still provides the highest possible evaluated quality, but all solutions other than [1111111111] receive a score of zero. Thus, there is no correlation between the perceived quality of less than perfect solutions and their evaluated "fitness." Using the first evaluation function allows for a guided search toward successively better solutions, while using the second is much like searching for a needle in a haystack: you only solve it if you stumble across the right answer.

In practice, you'll rarely be able to find perfect solutions to real-world problems. You simply don't have the time to let the evolutionary algorithm run long enough. You have to admit the goal of obtaining high-quality solutions in a short enough period of time for them to be useful, rather than hold out for perfection. Thus it's imperative that your evaluation function provides information that can guide the search over solutions that are less than perfect. An all-or-nothing evaluation function isn't practical.

Designing evaluation functions that allow for rapid evolutionary optimization requires handling the interaction of the variation operators with the evaluation function. Although there are exceptions, it's often desirable to have the application of small variations to a candidate solution engender small changes to the resulting offspring's quality. This tends to ensure that at least a local stochastic search can be performed. It may also be useful to have the evaluation function be separable into an assortment of independent factors that contribute to overall quality. Under these conditions, special variation operators (e.g., recombination) can be devised to mix, match, or blend useful parts of an overall solution.

Further, you have to consider the interaction of the representation with the search operators in light of the evaluation function. For example, suppose you wanted to devise a finite impulse response filter that would be useful in predicting the output of a transfer function given some input [70]. A natural evaluation function would be to minimize the sum of the squared error between the predicted and actual output of the transfer function over some set of input values. This would allow you to converge closer and closer to the best response filter as you reduce the total squared error. But the "landscape" of this evaluation function depends on the representation. There are several data structures that have been offered to represent these sorts of response filters. Figure 7.3 shows the shape of the sum of squared error evaluation function for each of these representations near the perfect filter for a problem studied in [70]. Depending on the choices you make for varying one solution into another, some of these surfaces will be more "evolution friendly" than others.

In practice, evaluating the quality of the solutions in a population is the most time-consuming part of executing an evolutionary algorithm. This occurs particularly when the evaluation must be conducted over a large sample of cases (e.g., in a pattern recognition problem) or when a simulation must be used to determine the effectiveness of a particular strategy over some extended period of time (e.g., decision making in combat situations). The process can sometimes be accelerated, however, by using incomplete evaluation, where you don't determine an exact evaluation but rather an estimated value or a relative value.

For example, suppose your selection procedure is to maintain the best 10 solutions in your population of 100 candidates as parents for the next generation. Once you have scored the first 10 solutions, you only have to complete the evaluation of any subsequent solution if it has the potential to offer better performance than the worst of these 10. If you can determine early on that this isn't possible then you don't have to complete the evaluation. Say you face a pattern recognition problem where there are 1000 training cases, and the current 10th place solution has 950 correct cases, with your evaluation function simply being the number of correct recognitions. Once any other solution you might test misclassifies its 51st case, it doesn't matter how many of the remaining cases it would get correct, it's not going to be maintained for the next generation. You can move on to the next candidate in the population. Note, however, that for selection methods where even lower-scoring individuals have a probability of entering the next generation, you might still have to complete the evaluation to determine just how good or bad this particular solution was.

Another possible short cut to consider is an evaluation function that merely compares two candidate solutions and indicates which is better. It needn't tell you exactly how much better, just which of the two is better than the other. Consider the problem of designing a controller for an unstable system in a

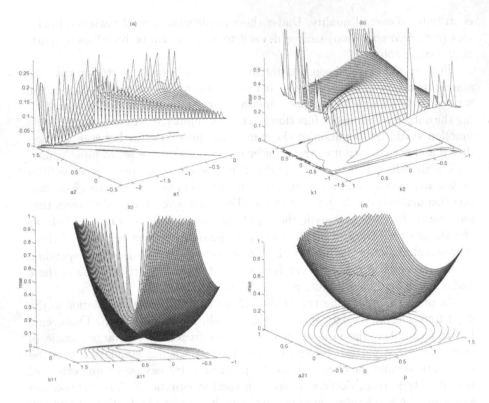

Fig. 7.3. The overall mean squared error surfaces around the optimal solution for the **(a)** direct, **(b)** lattice, **(c)** cascade, and **(d)** parallel forms of a three-pole, three-zero filter model [70]. Each form of filter can be shown to be equivalent to another, yet their error surfaces can be drastically different. Some will be more amenable to evolution than others.

simulation. One possible evaluation function would be the number of time steps in the simulation for which a controller can prevent the system from falling out of tolerance. This requires executing every candidate controller until it fails. In contrast, if you conducted a series of pairwise comparisons between two controllers, you'd only need to run the simulation until one of the two fails and then select the other, repeating this procedure until you have filled up the next population. You don't need to know exactly how good any particular controller is, you just know that it's better than its competitor that failed earlier.

7.3 Variation operators

The practical implementation of variation operators relies on the representation chosen and your intuition of the "landscape" generated by the evaluation function in light of your representation and variation operators. Every landscape requires a distance metric, and here the metric can be defined in terms of the number of variation operators required to traverse from one point (or multiple points) to another or in terms of the probability of transitioning from one solution to another. Although we are trained to think of landscapes as three-dimensional hills and valleys with a Euclidean metric, this is really only a valid metaphor for understanding how an evolutionary search might proceed when using continuous representations and continuous variation operators (e.g., Gaussian perturbations). Envisioning the landscape of a combinatorial optimization problem, such as a TSP, can be quite challenging.

In general, your choice of variation operators should follow naturally from your representation. You might consider various one- or two-parent operators, or even variations on more than two parents. There's no best choice across all problems, so you have to consider the implications of any choices that you make carefully and seek to determine the functional effects they will have in searching for successively better solutions in light of your chosen evaluation function. Let's look at some typical choices for variations in light of the representations mentioned above.

7.3.1 Fixed-length vectors of symbols

When faced with a binary representation, a common choice for an essential variation operator is to assign a probability of flipping a bit from 0 to 1, or vice versa. Often this operator is applied to each bit such that it's possible to transition from any binary string to any other, albeit with low probability. Essentially, this variation operator implies a landscape that is based on the Hamming distance between solutions. The more bits that must be flipped to transform one string into another, the further away these strings are from each other. Although many applications with binary strings use a fixed probability of flipping bits, most nontrivial problems require a variable probability in order to generate an optimal search (i.e., the probability of bit flips must change appropriately over the course of the evolution in light of the current population). Determining how to best effect these changes in the probability is a complex problem that has been solved only for simple cases.

Another common operator in this case relies on combining two or more solutions by crossing over. That is, you line up multiple solutions and then take pieces (or individual bits) from one solution or another in constructing a new solution to test. These operators can be applied to any fixed-length string of symbols, but again remember that their utility will always be problem dependent. Also note that different crossover operators yield different landscapes. The number of crosses that must be applied to two or more solutions in order to generate some other solution varies by the specific operator, and will likely not coincide with the landscape of the bit-flip operator described above.

If the symbols in the string are continuous variables, a very common procedure is to perturb the symbols of a single solution by adding a zero mean Gaussian random variable with a particular standard deviation. The intuitive metric for the landscape here is the Euclidean distance between any two points in the search space, and the standard deviation controls the likelihood of taking relatively big or small jumps on that landscape. Other common perturbations on a single solution rely on adding Cauchy or uniform random variables. The sampling variation of the Cauchy random variable results in longer jumps than the Gaussian, while the uniform variation does not impose any bias for generating solutions that are closer to a parent solution (within the range of the uniform density function being used). Another common variation operator for continuous variables relies on blending components of different solutions, essentially taking their (weighted) average, component by component. This arithmetic operator is essentially like taking the "center of mass" of multiple solutions, and can sometimes accelerate convergence to local optima (this accelerated convergence may be good or bad depending on the quality of the discovered optimum). There are many other possible recombinations that might also be considered (e.g., geometrical crossover rather than arithmetic).

When searching over the integers, it's often useful to vary the value of a symbol by incrementing or decrementing it with equal probability and having the amount of variation follow a Poisson random variable that's based on a mean parameter λ. For example, in the case of evolving the number of lags in an AR model,[5] the "model order" could be increased (up to some maximum limit based on memory constraints) or decreased (down to zero lag terms) with equal probability, with a Poisson number of lags subsequently added or deleted from the model. When the integer values can take on a finite range, particularly a small finite range, it's often convenient to use uniform sampling over the possibilities or a binomial distribution (often as a discrete analog of the Gaussian distribution) instead.

7.3.2 Permutations

The use of permutations often requires the invention of clever variation operators to take advantage of knowledge concerning the interpretation of the permutation. For example, consider the problem of choosing a variation operator if the permutation represents a tour in a TSP. One possibility would be to remove a city chosen at random and replace it randomly in the list. Another possibility would be to choose two positions along the list and reverse the order of the cities between these locations (i.e., much like a 2-opt [284]). For the TSP, the reversal operator turns out to be a better choice because it takes advantage of the landscape associated with the Euclidean distance between vertices of a quadrilateral in Cartesian coordinates. As we saw in chapter 6, given a quadrilateral, the sum of the distances of the diagonals will always be longer than the sum of the distances of either pair of opposite sides (figure 6.2). Thus, any tour that crosses over itself will be suboptimal. Applying the reversal to

---

[5] AR stands for *auto-regressive*, meaning that the value of the process at a given time is a function of the values that occurred at previous times.

two vertices after such a cross removes this imperfection. No such benefit can be obtained from the remove-and-replace operator here, nor would there be any special benefit in the TSP for choosing two positions and swapping just those entries (rather than reversing the entire length of the segment between the positions). Note, however, that for other scheduling problems these kind of operators might be more useful than the reversal; it all depends on the problem. Just as you can imagine implementing the full 2-opt procedure as a variation operator, you can just as well implement a k-opt operator, but certainly this would require additional processing time.

Using two or more permutations to generate an offspring permutation requires some careful planning. Using simple cut-and-splice operations would run the risk of generating lists with multiple copies of some entries and omitting others entirely. There are many possible recombinations for permutations that can be found in the literature. We will mention two that are based on preserving the order of appearance of some of the entries in each parent's permutation.

First take two permutations, P_1 and P_2, and choose two cut points at random along P_1. Copy the entries between the cut points in P_1 into the offspring O_1. Note which entries in P_2 are not included in O_1, then insert them in order starting from the second cut point (figure 7.4) [85].

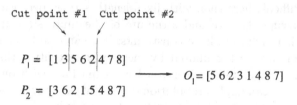

Fig. 7.4. One possible recombination operator for two permutations. The elements between two cut points in the first permutation are copied to the offspring. The remaining elements in the offspring are copied in order from the second permutation.

Compare that with the following: Take two permutations P_1 and P_2, and choose n points at random along P_2. Find the corresponding entries in P_1 for each of these n entries in P_2. Reorder these entries in P_1 according to their order in P_2. Copy the remaining entries in P_1 directly (figure 7.5) [462].

The second procedure does not guarantee that any contiguous section of either parent's permutation will be passed intact to an offspring, but by the same token, you are not constrained to maintain the contiguous sections of either parent. As with uniparent operators, you must consider the functional effects of applying any possible multiparent variation in order to determine which may be suitable for the particular problem at hand.

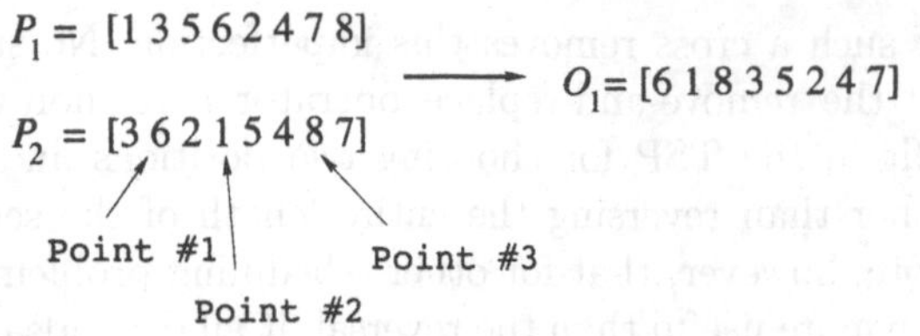

$$P_1 = [1\,3\,5\,6\,2\,4\,7\,8]$$

$$\longrightarrow \quad O_1 = [6\,1\,8\,3\,5\,2\,4\,7]$$

$$P_2 = [3\,6\,2\,1\,5\,4\,8\,7]$$

Point #1   Point #3

Point #2

Fig. 7.5. Another possible recombination operator for two permutations. Here, three points are chosen at random from the second permutation. The elements at these locations are copied to the offspring. The remaining elements are copied in order from the first permutation.

7.3.3 Finite state machines

If you're working with a finite state machine with an output symbol and state transition defined for every possible input in each state, and you choose a matrix representation as offered above, the variation operators to apply follow quite naturally. There are five modes of varying an individual finite state machine: (1) add a state, (2) delete a state, (3) change the start state, (4) change an output symbol, (5) change a next-state transition (figure 7.6). Things become a little more difficult here than with fixed-length strings because you have to account for varying the size and structure of the finite state machine. The possible variation operators have constraints: you can't add more states than some maximum size (either limited by memory or execution time), and you can't delete or change states when you have a machine with only one state. The variations you can apply are solution dependent. Moreover, there are many different ways to implement these operators. For example, when adding a state, you must decide whether or not to mandate that at least some existing state might transition to this new state. When deleting a state, you have to decide some rules for reassigning all the links that previously communicated with this now-removed state. Even when performing something as simple as changing an output symbol, you might impose a random distribution over the possible changes. Rather than picking new symbols uniformly at random, there might be a nearness between the meaning of the available symbols, so you might want to ensure that new symbols closer to the old one have a greater chance of being selected. You also have to decide on the number of variations to apply, which might be a Poisson random number with an adjustable rate.

Recombination operators can be applied to finite state machines too, but again you have to think about the functionality of the resulting machines. If you take three states from one machine that were evolved to operate in the context of that machine, and splice them together with three states from another machine with a different architecture, it's not very likely that the new machine will perform like either parent. It's essentially taking the first half of Hamlet with the second half of King Lear again. Functionally, it's just a "macro-mutation," but sometimes taking what amount to large jumps in the state space of possible solutions can be effective (at least to a limited extent). A different idea is to

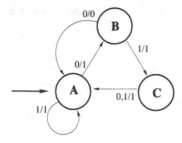

Fig. 7.6. A finite state machine having three states. The machine starts in state A. For each input symbol $(0, 1)$, each state responds with an output symbol $(0, 1)$ and a next-state transition. Input symbols are shown to the left of the slash, while output symbols are given on the right of the slash. Note that the input and output alphabets do not have to be equivalent. The natural modes of varying a finite state machine follow from the diagram.

perform a majority logic operation across multiple machines (more than two). For each input in every combination of states from each machine considered, the outputs are recorded, and that output which receives the majority vote is selected as the output of the majority logic machine. Essentially, this is a way to implement the democratic process across the parents. The disadvantage is that the size of the resultant offspring machine is the product of the sizes of the parent machines, so this can grow very large, very quickly.

7.3.4 Symbolic expressions

For symbolic expressions represented in the form of a parse tree, we again face the situation of using a variable-length structure much like a finite state machine. We can think about growing or shrinking a tree by randomly extending or pruning branches. We can manipulate the order in which operations are conducted within the tree by swapping nodes within a branch or even swapping the order of entire branches (figure 7.7). And we could also modify elements at individual nodes, changing, say, an operator of + to *.

As with finite state machines, we could also swap elements or branches between two or more trees. When swapping branches, if each performs a subroutine that is important for solving a bigger problem then this sort recombination might be helpful. It might be a good way to bring together useful modules from different trees (see figure 7.8). On the other hand, we have to be careful not to blindly think that just because some branch might be useful in one tree that it will be useful in another. The utility of a subroutine or any module is always dependent on the integrated complex systems in which it resides. Imagine swapping CPU mother boards between two arbitrarily chosen personal computers. Both might work fine before the swap, and neither might work at all afterwards, ever again. Even swapping individual resistors from two different computers might have catastrophic effects. However, there's experimental

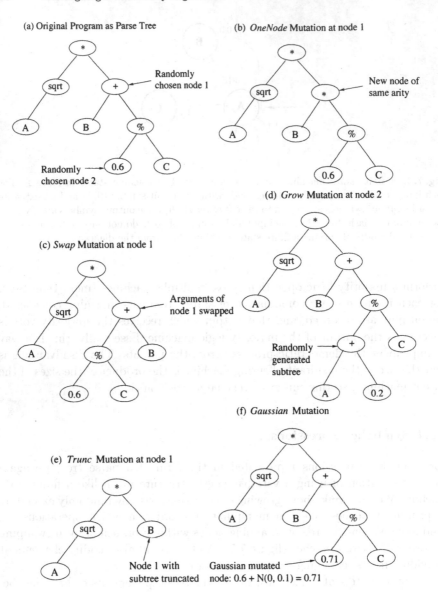

Fig. 7.7. Examples to demonstrate six variation operators for use with a tree representation. The original expression is $\sqrt{A} * (B + 0.6/C)$; **(a)** Original tree; **(b)** One-node variation; **(c)** Swap nodes; **(d)** Grow subtree; **(e)** Truncate subtree; **(f)** Gaussian variation (applied to real-valued numbers).

evidence [267] that such branch swapping between (usually two) trees for symbolic expressions can generate useful results. Again, the effectiveness is problem dependent.

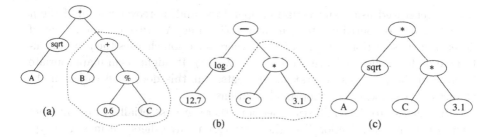

Fig. 7.8. Swapping branches between two trees; areas for swap are marked by *broken lines*: (a) the first tree; (b) the second tree; (c) one of the resulting trees.

7.4 Selection

The selection operator acts on individuals in the current population. Based on each individual's quality, it determines the next population. Selection typically comes in one of two forms: (1) some individuals are eliminated from consideration, while the remaining individuals survive to become parents of the next progeny, or (2) individuals are sampled, with replacement based on their relative quality. In the first form of selection, each individual can contribute at most one copy of themselves as a basis for the next generation; in the second form, any individual can be chosen multiple times. Although there can be different forms for implementing the order of applying selection in an evolutionary algorithm, the main concept for all of these realizations remains the same: selection is the filter by which the algorithm will determine the composition of the next generation. For convenience, let's consider a variety of selection procedures and make the assumption that any of these would be applied after all new offspring have been generated and each individual in the population has been assigned a measure of quality from the evaluation function.

Broadly, selection methods can be classified as either deterministic or stochastic. Given the same population (parents and offspring), deterministic selection will always eliminate the same individuals, whereas stochastic methods generate a probability mass function over the possible compositions of the next iteration. Deterministic selection has the advantage of often being faster to execute than stochastic (or probabilistic) selection, and the common implementations of deterministic selection also lead to faster convergence in the population, although this may or may not be beneficial depending on where the population converges!

The common notation in evolutionary algorithms uses μ to denote the number of parents and λ to denote the number of offspring. This is a bit cryptic, but to remain consistent with prior literature, we'll also employ these symbols. Perhaps the simplest form of selection is termed $(\mu + \lambda)$ in which λ offspring are created from μ parents, all of the $\mu + \lambda$ individuals are evaluated, and selection is applied to keep only the best μ from this collection. Note that the λ offspring can be generated using any variation operators, and moreover not all of the μ parents have to contribute to generating offspring. Another common form of deterministic selection is termed (μ, λ), where selection chooses the μ best solutions only from the λ offspring (of course, $\lambda \geq \mu$). It might seem inappropriate to discard the previous generation's parents, but this does tend to lessen the chances of stagnating at suboptimal solutions.

In contrast to deterministic selection, there are essentially three different forms of stochastic selection (i.e., there are very many others but most can typically be recast as a variant of one of the following three forms). One form that was employed early in the history of evolutionary algorithms but now receives much less attention is proportional selection. In this method, each individual must be evaluated with a positive quantity, and then the number of copies of each individual that propagate into the next generation is made proportional to each individual's relative fitness. That is, given that there will be μ parents in the next population, μ samples are taken from the current population where the probability of selecting the i-th individual is

$$P_i = F_i / \overline{F},$$

where F_i is the "fitness" of the i-th individual and $\overline{F}$ is the mean of all the individuals' fitness in the current population. Those individuals with above average fitness will, on average, receive more attention than those with below average fitness. This biases the evolutionary algorithm to use better solutions as a basis for further exploration, but it doesn't mandate that only the very best solutions be maintained. In fact, it's possible with this method that only the very worst solutions might be maintained just by random sampling, but this would occur with very low probability.

The two other forms of stochastic selection are based on random tournaments. The first samples a collection of the current population, identifies the best individual in that sample, retains that individual, and replaces the remaining individuals for possible resampling. This process is repeated until the appropriate number of parents for the next generation have been identified. The size of the subset being sampled determines the variability of this procedure. It's easy to see that if the collection is made as large as the population itself, this method is essentially equivalent to the deterministic $(\mu + \lambda)$ selection.

The second tournament method assigns a surrogate score to every individual based on comparing it to a random subset of the population. In each comparison, the value of the solution is incremented by one point if it's as least as good as its opponent. The very best solution in the population will always earn points in every comparison, but lesser-quality individuals can earn points too. Even solutions in the lower rank of the population might get lucky and face even lower-ranked individuals, thereby earning more points. In a subset of size q, the maximum number of points earned by any solution is q. Selection then proceeds deterministically by operating on the number of points associated with each individual. Whereas some individuals in the population might be missed using the first tournament method, this alternative method provides a probabilistic evaluation of every individual. Again, as is easily imagined, as the size of the subset q increases, this tournament method becomes equivalent to the $(\mu + \lambda)$ deterministic selection.

It's sometimes convenient to compare these different methods in terms of how stringent their selection pressure is. This can be measured in a variety of ways such as the average reduction in population variance, the change of the mean fitness of the population as a ratio to the population variance, or the average time required for the best solution to takeover the entire population. Unfortunately, it's very difficult to analyze selection operators in light of these criteria when coupled with random variation, so most analyses consider only the repeated application of selection to a population in light of merely copying selected solutions without any modifications. Bäck [20] showed that the most stringent selection procedures are the deterministic ones: $(\mu + \lambda)$ and (μ, λ). The weakest (i.e., slowest to converge) method of those he examined was proportional selection. Tournament selection could be tuned to be weak or strong depending on the number of tournaments (i.e., more competitions lead to more stringent selection).

The term "selection" can be used in two different contexts. It can refer both to the process of selecting parents and to the process of selecting individuals for replacement. Selection methods usually define both of these aspects, but not always. For example, in a *steady-state* evolutionary algorithm only a few (at the limit, just one) offspring are created at each generation before selection. As a consequence, the population undergoes less change at each instance of selection. Here, we need to treat two aspects of selection. Once we've selected the parent(s) and have created an offspring, we have to select an individual to purge from the population. This makes room for the new offspring, presuming of course that we don't instead decide to eliminate this new offspring straightaway. The choice of parents and the choice of which individual to remove could be independent, or in fact made by completely different means altogether. You could use, say, proportional selection to choose parents, and tournament selection to eliminate some individual. You might also apply an *elitist* rule where the best solution(s) in the population are certain to survive to the next generation. This ensures that the fitness of the best solution never deteriorates. The asymptotic convergence properties of evolutionary algorithms rely generally on elitism [399, 145], and elitist evolutionary algorithms often converge faster than those that do not utilize elitism.

7.5 Initialization

Recall that most evolutionary algorithms can be captured by the equation

$$x[t + 1] = s(v(x[t])),$$

where $x[t]$ is the population under a representation x at time t, $v(\cdot)$ is the variation operator(s), and $s(\cdot)$ is the selection operator. This difference equation indicates the probabilistic course of evolution over each generation. Note that you have to determine $x[0]$, the initial population, before starting the process in motion. Comparatively little attention has been given to initializing evolutionary algorithms. The straightforward approach is to sample uniformly at random from the state space of possible solutions, thereby providing an unbiased initial population. This is often useful in benchmarking evolutionary algorithms, but when solving real problems, we often know some information that would help us seed the initial population with useful hints. For example, if we were looking for a linear model with real-valued coefficients that related, say, the height of a tree to its width and its age,

Height $= a_1 \cdot \text{Width} + a_2 \cdot \text{Age}$,

we would probably be well off to constrain the initial values of a_1 and a_2 to be positive. It wouldn't make much sense to think about models where the wider the tree, the smaller it gets — or where trees shrink with age. Biasing the initial population in the direction of better solutions can often save a great deal of time that would otherwise be spent "reinventing the wheel." And why have the evolutionary algorithm learn what we already know?

Population-based evolutionary algorithms offer an additional possible benefit in that we can initialize at least one individual in the population with the best solution(s) obtained by other methods (e.g., a greedy algorithm). This technique can be used as an argument that "evolutionary algorithms can never perform worse than the XYZ system," such as "the evolutionary supply chain optimizer can't do worse than a just-in-time delivery system." A potential user might know and trust the XYZ system, so having the extra security that the performance of the evolutionary algorithm is bounded below by the performance that has already been achieved by other means is reassuring.

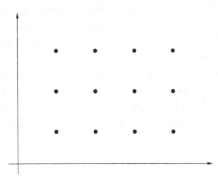

Fig. 7.9. When initializing the population, one way to ensure diversity is to use a grid pattern.

If we really don't know much and can't give the algorithm any hints, there are other ways to initialize the population that need to be considered, rather than sampling randomly. When faced with continuous (or integer) variable optimization problems, one alternative is to lay down a grid or pattern of initial samples (figure 7.9). This precludes the possibility of having a random sample that happens to cluster too much in one area. Another possibility is to mandate that as each individual in the initial population is chosen, each next individual must be at least some minimum distance away from all other individuals. Although this can be time consuming, it again precludes inadvertent clustering of initial candidate solutions while maintaining a randomized procedure.

7.6 Summary

Evolutionary algorithms provide a distinctly different means for solving problems as compared to classical techniques. Their primary advantage is their conceptual simplicity. The basic procedure is really very straightforward: generate a population of potential solutions to your problem, devise some variation operators that can create new solutions from old ones, and apply a selection mechanism for keeping those that have worked the best up to the current point in time.

Of course, you must first have a clear statement of what you are trying to achieve written in mathematical terms so that any candidate solution can be evaluated. You must also think about how to represent the solutions and store them within the computer. For some problems, you might want to use a string of bits. For others, a matrix of real numbers would be more intuitive. For still others, a variable-length structure such as a finite state machine or a neural network encoded in a variable-size matrix of real values might be appropriate. Unfortunately, there's no best representation that will work for every problem, so you have to use your intuition and pick structures that provide insight into how to solve the problem.

If you're running an evolutionary algorithm under a real-time constraint and have to find solutions in a short period of time, then you will most likely want to use a deterministic method of selecting only the best few available solutions to serve as parents for future offspring solutions. If you are facing a problem with multiple local optima, this method might have a greater chance of converging to one of those suboptima, but you may not have the luxury of trying to escape them using probabilistic selection.

If you have knowledge about where to look for good solutions then you can and should incorporate that knowledge into your evolutionary search. You might think about doing that in the initialization of possible solutions as well as in the design of variation operators. In the end, to get the best performance from your evolutionary search, you must consider how all of these facets of the algorithm fit together, for each part cannot be separated and optimized independently of the whole algorithm.

This chapter has provided a first look at the issues involved in designing practical evolutionary algorithms. There are many subtle points that deserve greater attention. For example, how can the evolution be used to find useful parameter settings for the variation operators as a function of what has been learned during the evolutionary process itself? How can we treat problems with constraints, where entire regions of the solution space are infeasible? How can we find multiple solutions to a problem rather than only a single solution? These and other important questions are addressed in subsequent chapters.

VIII. What Is the Shortest Way?

Many challenging problems require finding the shortest way of moving from one point to another while still satisfying some additional constraints. We'll discuss the most famous of these problems in the following chapter. Before we do that, let's consider two other puzzles that involve minimizing the length of a path.

Suppose we have to build a road from city A to city B, but these cities are separated by a river. We'd like to minimize the length of the road between these cities. The bridge must be constructed perpendicular to the banks of the river. Figure VIII.1 illustrates the point.

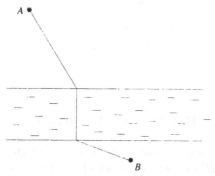

Fig. VIII.1. A river, two cities, A and B, and a road between them

Now, the question is where to build the bridge so as to minimize the total length of the road? There are many methods for solving this problem but some methods require many calculations. For example, we can assume some coordinates (x_a, y_a) and (x_b, y_b) for cities A and B, respectively. We can also assume that the river flows horizontally and is bounded by y_{r_1} and y_{r_2} (where $y_a > y_{r_1} > y_{r_2} > y_b$). Then, we can build a formula for the length of the connection between cities A and B that is a function of an angle α (see figure VIII.2) and find the minimum length.

Things are much simpler, however, if we concentrate on the important parts of the model and reject the "noise" — those elements that obstruct our path to the goal. Let's assume that there's no river. The river is reduced to a line (of width zero) and city B is moved upwards by the distance equal to the original width of the river (figure VIII.3). This problem is extremely easy to solve: a straight line between A and B' gives a solution!

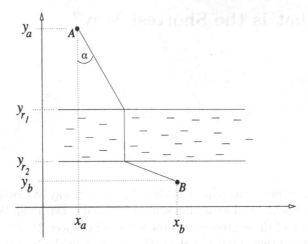

Fig. VIII.2. Calculating the minimum distance between A and B

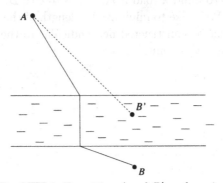

Fig. VIII.3. Two cities, A and B', and no river

This solution also implies the solution to the original problem. The line between city A and B' crosses the bank of the river at some point, say, P. This is the point where the bridge should originate; the terminal point of the bridge (i.e., the point of the other side of the river) is Q (see figure VIII.4). The segments QB and AP are parallel so the total distance

$$AP + PQ + QB$$

is the shortest possible. In other words, for any other connection, say $A - P' - Q' - B$ we have

$$AP' + P'Q' + Q'B = AP' + P'Q' + P'B'$$
$$> AB' + P'Q' = AP + PQ + PB' = AP + PQ + QB,$$

as $P'Q' = PQ$ and $AP' + P'B' > AB'$ (triangle inequality). The best way to solve the problem of how to get across the river is to abstract the river out of existence!

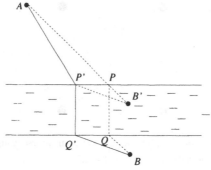

Fig. VIII.4. The shortest road $A - P - Q - B$ between A and B

Let's consider another problem where shortest connections are required. There are four cities located on the vertices of a square (figure VIII.5).

Fig. VIII.5. Location of four cities

The task is to design a network of roads such that (1) every city is connected with every other city and (2) the total length of the roads is minimized. The roads can intersect each other; i.e., it's possible to switch from one road to another while driving between cities. There are many possibilities which satisfy condition (1). Figure VIII.6 displays some of these.

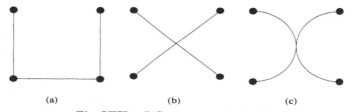

(a)   (b)   (c)

Fig. VIII.6. Different networks of roads

Note, however, that each network has a different total length. Assuming the length of a side of the square is 1, the total lengths of roads in figure VIII.6a is, of course, 3; the total lengths of roads in figure VIII.6b is $2\sqrt{2} = 2.828427$; and the total lengths of roads in figure VIII.6c is $\pi \approx 3.14159$. That's quite a difference!

The optimal solution is not that obvious. Even when revealing the optimal solution (figure VIII.7), the proof of optimality isn't obvious!

Fig. VIII.7. The optimum solution. The angles between intersecting roads are all equal to 120°. The total length of the network is $1 + \sqrt{3} \approx 2.732$

The above problem can be generalized into the following: given n points in the plane, design a network of connections so that the total length of the whole network is as short as possible. This is known as Steiner's problem and is solved only in some special cases. There's no known efficient algorithm that can solve general instances of this problem. At the same time, there are many engineering applications (e.g., telephone networking, routing aircraft) that would benefit from such an algorithm. Can you invent it? If you can, you'll be a billionaire!

8. The Traveling Salesman Problem

> No one realizes how beautiful
> it is to travel until he comes
> home and rests his head on
> his old, familiar pillow.
>
> Lin Yutang, *A trip to Anhwei*

The first problem discussed in section VIII has only a slight connection with the traveling salesman problem (TSP) in that it also requires finding a minimum-distance solution. This is where the similarities end, however, because there aren't any tricks for finding perfect solutions to the TSP. The problem is NP-hard [179]: there are no known algorithms for finding perfect solutions that have a complexity that grows as only a polynomial function of the number of cities. As a consequence, we must rely on generating solutions that are less than perfect, but as close as possible within a reasonable amount of time. This is a significant challenge, and for this reason we will devote this entire chapter to the TSP. Furthermore, the TSP is related to other problems in scheduling and partitioning, and its broad generality makes it an important keystone in combinatorial optimization. Some ideas that might be useful for addressing the TSP might also be useful in a variety of other applications. The TSP also serves to illustrate many of the central concepts of this book: the flexibility of an evolutionary approach to problem solving, the ease of incorporating problem-specific knowledge, and the ease of hybridization.

The traveling salesman problem is a special problem in computer science. The problem itself is conceptually very simple. The traveling salesman must visit every city in his territory exactly once and then return home to his starting point. Given the cost of traveling between each pair of cities, the task is to arrange a complete tour that achieves minimum total cost. One possible search space for the TSP is a set of permutations of n cities. Any single permutation of n cities yields a solution, which is interpreted as a complete tour, with a return home being implied as soon as the last city in the list is visited. The size of this search space is $n!$[1]

The TSP has a long history. It was documented as early as 1759 by Euler, although not with the same name, who was interested in solving the knight's tour problem. A correct solution would have a knight visit each of the 64 squares

---

[1] As we've seen in earlier chapters, due to the symmetry of the problem and the fact that a tour is a complete circuit, the space can be reduced to $(n-1)!/2$ distinct tours. But any search algorithm that operates just on permutations must still treat $n!$ possibilities.

of a chess board exactly once in its tour. The term "traveling salesman" was first used in a German book from 1932 written by a veteran traveling salesman, *The traveling salesman, how and what he should do to get commissions and be successful in his business* [279]. Even though this book didn't focus on the TSP, it was discussed in the last chapter along with scheduling problems.

The TSP was introduced by the RAND Corporation in 1948. The corporation's reputation helped to make the TSP a well-known and popular problem. The TSP also gained attention because of the then new subject of linear programming, and the attempts to use this method to solve combinatorial problems.

The TSP arises in numerous applications, and the size of the problem can be very large [239]:

> Circuit board drilling applications with up to 17,000 cities are mentioned in [287], X-ray crystallography instances with up to 14,000 cities are mentioned in [49], and instances arising in VLSI fabrication have been reported with as many as 1.2 million cities [266]. Moreover, 5 hours on a multi-million dollar computer for an optimal solution may not be cost-effective if one can get within a few percent in seconds on a PC. Thus there remains a need for heuristics.

Within the last few decades, many people have offered algorithms that generate approximate solutions to the TSP. These include nearest neighbor, greedy, nearest insertion, farthest insertion, double minimum spanning trees, strip, space-filling curves, and others offered by Karp, Litke, Christofides, and so forth [239]. Some of these algorithms assume that the cities correspond to points in a plane under some standard distance metric. Another group of algorithms (e.g., 2-opt, 3-opt, Lin-Kernighan) aims at local optimization: improving a tour by imposing local perturbations. The TSP also became a target for the evolutionary computation community in the 1980s and 1990s. Evolutionary approaches were reported in [142], [174], [192], [203], [202], [238], [282], [330], [338], [430], [452], [457], [471], and [494] during the formative years of the field. Furthermore, almost every conference on evolutionary algorithms still includes a few papers on the TSP. Clearly, the quest for the "Holy Grail" — a perfect evolutionary algorithm for the TSP — still remains! Of course, such an algorithm should, hopefully, be competitive with the best available current methods.

It's interesting to compare these approaches, paying particular attention to the representation and variation operators used. We'll trace the evolution of evolutionary algorithms for the TSP in this chapter, but we won't provide the exact results of various experiments for different systems developed over the years. Very few publications provide the computational time required for solving specific instances of the TSP. Instead, the number of function evaluations is reported. This makes it difficult to compare approaches because different operators have different time complexities. Eshelman [129] reported using 2.5 hours on a single-processor Sun SPARCstation to address a 532-city problem. Gorges-Schleuter [195] used between 1 and 2 hours on the same problem but employed a parallel implementation comprising 64 T800 transputers. Braun [56] used about 30 minutes for addressing a 431-city problem on a SUN workstation. These descriptions all leave out the comparison of the quality of the eventual solution, yet they indicate that evolutionary algorithms might be too slow for solving large TSPs (e.g., 10,000+ cities). These quoted performance results depend heavily on many details, including the population size, number of generations, number of cities, and so forth. Furthermore, many results have pertained only to relatively small TSPs (i.e., up to 100 cities). As observed in [239]:

> It does appear that instances as small as 100 cities must now be considered to be well within the state of the global optimization art, and instances must be considerably larger than this for us to be sure that heuristic approaches are really called for.

Such problems are now solved routinely within a few hours [241].

Most of the papers cited in this chapter compare a proposed approach with other approaches. Two main types of test cases are used for these comparisons:

- The cities of the TSP at hand are distributed at random, typically in accordance with a uniform random variable, and a Euclidean distance metric is assumed. Here, an empirical formula for the expected length L^* of a minimal TSP tour is useful:

$$L^* = k\sqrt{n \cdot R},$$

where n is the number of cities, R is the area of the square box within which the cities are placed randomly, and k is an empirical constant. The expected ratio k of the Held-Karp bound to $\sqrt{n}$ for n-city random Euclidean TSP (for $n \geq 100$) is

$$k = 0.70805 + \frac{0.52229}{\sqrt{n}} + \frac{1.31572}{n} - \frac{3.07474}{n\sqrt{n}}.$$

Bonomi and Lutton [51] suggest using $k = 0.749$.

- The cities of the TSP are drawn from a collection of publicly available test cases (TSPLIB [385]) with documented optimal or best-known solutions.[2]

8.1 In search of good variation operators

The TSP has an extremely easy and natural evaluation function: for any potential solution — a permutation of cities — we can determine all of the distances between the ordered cities and, after $n - 1$ addition operations, calculate the total length of the tour. Within a population of tours, we can easily compare any two of them, but the choice of how to represent a tour and the choice of what variation operators to use is not so obvious.

---

[2]See http://www.iwr.uni-heidelberg.de/iwr/comopt/soft/TSPLIB95/TSPLIB.html.

We noted before that one natural representation for a TSP is a permutation. This seems to capture the progressive nature of traveling from one city to the next. Early in the development of evolutionary algorithms, however, it was suggested that binary representations would be advantageous, regardless of the problem at hand [228]. In retrospect, this sounds a bit implausible, and indeed it can be proved to be implausible. No bijective representation is better than any other across all problems. So the question at hand then is does a binary representation provide any apparent advantage when addressing a TSP?

After a little consideration, binary strings seem to be poorly suited for representing solutions to TSPs. It's not too hard to see why, after all, we're interested in the best permutation of cities, i.e.,

$$(i_1, i_2, \ldots, i_n),$$

where $(i_1, i_2, \ldots, i_n)$ is a permutation of $\{1, 2, \ldots, n\}$. A binary encoding of these cities won't make life any easier. In fact, just the opposite is true: A binary representation would require special *repair* operators to fix the errors that would be made when generating illegal (i.e., infeasible) tours. As observed in [494],

> Unfortunately, there is no practical way to encode a TSP as a binary string that does not have ordering dependencies or to which operators can be applied in a meaningful fashion. Simply crossing strings of cities produces duplicates and omissions. Thus, to solve this problem some variation on standard ... crossover must be used. The ideal recombination operator should recombine critical information from the parent structures in a non-destructive, meaningful manner.

Although the sentiment here is correct, note the apparent bias in terms of looking particularly for recombination operators. This is a common mistake in designing evolutionary algorithms. Rather than focus on any one class of variation operators, you should think more broadly about how to construct variation operators that are appropriate for your task at hand. They might include some sort of recombination, or they might not. A variation operator that has a high probability of generating illegal solutions also has a high probability of not being the operator you are looking for.[3]

Three vector representations were considered for the TSP during the 1980s: *adjacency, ordinal,* and *path*. Each of these representations has its own associated variation operators — we'll discuss them in turn. And since it's relatively easy to invent some sort of mutation operator that could impose a small change on a tour, we'll first concentrate on the development of possible recombination operators that might be useful. In all three cases, a tour is described as a list

---

[3]Lidd [282] actually offered some success using a binary representation and simple crossover and mutation operators on a TSP; however, illegal tours were "fixed" using a greedy procedure, so it is unclear how much benefit the evolutionary algorithm was providing. Furthermore, none of the problems involved more than 100 cities.

of cities. We'll use a common example of nine cities, numbered intuitively from one to nine.

Adjacency Representation

The adjacency representation encodes a tour as a list of n cities. City j is listed in the position i if and only if the tour leads from city i to city j. For example, the vector

(2 4 8 3 9 7 1 5 6)

represents the following tour:

$1 - 2 - 4 - 3 - 8 - 5 - 9 - 6 - 7.$

If that's not clear, go back and see if you can identify why each city in the vector is placed in the corresponding order. Each tour has only one adjacency list representation; however, some adjacency lists can represent illegal tours, e.g.,

(2 4 8 1 9 3 5 7 6),

which leads to

$1 - 2 - 4 - 1,$

i.e., a partial tour with a premature cycle.

The adjacency representation doesn't support a simple "cut-and-splice" crossover operator.[4] It might be necessary to impose a subsequent repair procedure to fix illegal solutions. Instead, Grefenstette et al. [203] proposed recombination operators specifically for the adjacency representation: *alternating edges*, *subtour chunks*, and *heuristic* crossovers.

Alternating-edges crossover. This operator builds an offspring by randomly choosing an edge from the first parent, then selecting an appropriate edge from the second parent, etc. — the operator extends the tour by choosing edges from alternating parents. If the new edge from one of the parents introduces a cycle into the current, still partial tour, the operator instead selects a random edge from the remaining edges that does not introduce cycles. For example, the first offspring from the two parents

$p_1 = $ (2 3 8 7 9 1 4 5 6) and
$p_2 = $ (7 5 1 6 9 2 8 4 3)

might be

$o_1 = $ (2 5 8 7 9 1 6 4 3),

---

[4]Even if it did, that wouldn't be a reason for using crossover. There has to be some match between the operator, the representation, and the problem at hand. The simple fact that an operator generates legal solutions isn't sufficient justification to use that operator.

where the process starts from edge (1 2) from parent p_1, and the only random edge introduced during the process of alternating edges is (7 6), in place of (7 8) which would have introduced a premature cycle.

Subtour-chunks crossover. This operator constructs an offspring by choosing a random-length subtour from one of the parents, then choosing a random-length subtour from another parent, etc. — the operator extends the tour by choosing edges from alternating parents. Again, if some edge from one of the parents introduces a cycle into the current, still partial tour, the operator then instead selects a random edge from the remaining edges that does not introduce cycles.

Heuristic crossover. This operator builds an offspring by choosing a random city as the starting point for the offspring's tour. Then it compares the two edges that emanate from this city in both parents and selects the better (i.e., shorter) edge. The city on the other end of the selected edge serves as the starting point in selecting the shorter of the two edges leaving this city, etc. If at some stage, a new edge introduces a cycle into the partial tour, the tour is extended by a random edge from the remaining edges that does not introduce cycles.

This heuristic crossover was modified in [238] by changing two rules: (1) if the shorter edge from a parent introduces a cycle in the offspring tour, check the other longer edge. If the longer edge does not introduce a cycle, accept it; otherwise (2) select the shortest edge from a pool of q randomly selected edges (q is a parameter of the method). The effect of this operator is to glue together short subpaths of the parent tours. However, it may leave undesirable crossings of edges; therefore, this form of heuristic crossover is not appropriate for fine local optimization of potential solutions.

Suh and Gucht [457] introduced an additional heuristic operator based on the 2-opt algorithm [284] that is appropriate for local tuning. The operator randomly selects two edges, $(i\ j)$ and $(k\ m)$, and checks whether or not

$$dist(i, j) + dist(k, m) > dist(i, m) + dist(k, j),$$

where $dist(a, b)$ is a given distance between cities a and b. If this is the case, the edges $(i\ j)$ and $(k\ m)$ in the tour are replaced by the edges $(i\ m)$ and $(k\ j)$.

One possible advantage of the adjacency representation is that it allows us to look for templates that are associated with good solutions. For example, the template

$$(*\ *\ *\ 3\ *\ 7\ *\ *\ *)$$

denotes the set of all tours with edges (4 3) and (6 7). It might be that we could use some sort of branch and bound procedure or other heuristic and focus attention within this narrow subset of all possible solutions. Unfortunately, poor results have been obtained with all standard variation operators when using this representation. The alternating-edges crossover often disrupts good tours due to its own operation by alternating edges from two parents. It's difficult to generate sequences of edges and store them within a single parent with this operator.

Subtour-chunk crossover performs better than alternating-edges crossover on the TSP with an adjacency representation, most likely because its rate of disrupting templates is lower. But even then, empirical results with this operator have not been encouraging. In contrast, for the TSP, heuristic crossover is a better choice because the first two crossovers are blind, i.e., they don't take into account the actual lengths of the edges. On the other hand, heuristic crossover selects the better edge from two possible edges. In some respects, it uses local information to improve the likelihood of generating better solutions. In fairness, however, even heuristic crossover's performance has not been outstanding. In three experiments on 50-, 100-, and 200-city TSPs, an evolutionary algorithm found tours within 25, 16, and 27 percent of the optimum, in approximately 15000, 20000, and 25000 generations, respectively [203].

Ordinal Representation:

The ordinal representation encodes a tour as a list of n cities; the i-th element of the list is a number in the range from 1 to $n - i + 1$. The idea that underlies this representation is as follows. There is some ordered list of cities C, which serves as a reference point for lists in ordinal representations. Assume, for example, that such an ordered list (reference point) is

$$C = (1\ 2\ 3\ 4\ 5\ 6\ 7\ 8\ 9).$$

A tour

$$1 - 2 - 4 - 3 - 8 - 5 - 9 - 6 - 7$$

is represented as a list l of references,

$$l = (1\ 1\ 2\ 1\ 4\ 1\ 3\ 1\ 1),$$

and should be interpreted as follows:

The first number on the list l is 1, so take the first city from the list C as the first city of the tour (city number 1), and remove it from C. The resulting partial tour is

1.

The next number on the list l is also 1, so take the first city from the current list C as the next city of the tour (city number 2), and remove it from C. The resulting partial tour is then

$$1 - 2.$$

The next number on the list l is 2, so take the second city from the current list C as the next city of the tour (city number 4), and remove it from C. The new partial tour is

1 – 2 – 4.

The next number on the list l is 1, so take the first city from the current list C as the next city of the tour (city number 3), and remove it from C. The partial tour is

1 – 2 – 4 –3.

The next number on the list l is 4, so take the fourth city from the current list C as the next city of the tour (city number 8), and remove it from C. The partial tour is

1 – 2 – 4 – 3 – 8.

The next number on the list l is again 1, so take the first city from the current list C as the next city of the tour (city number 5), and remove it from C. The partial tour is

1 – 2 – 4 – 3 – 8 – 5.

The next number on the list l is 3, so take the third city from the current list C as the next city of the tour (city number 9), and remove it from C. The partial tour is

1 – 2 – 4 – 3 – 8 – 5 – 9.

The next number on the list l is 1, so take the first city from the current list C as the next city of the tour (city number 6), and remove it from C. The partial tour is

1 – 2 – 4 – 3 – 8 – 5 – 9 – 6.

The last number on the list l is 1, so take the first city from the current list C as the next city of the tour (city number 7, the last available city), and remove it from C. The final tour is

1 – 2 – 4 – 3 – 8 – 5 – 9 – 6 – 7.

The main (possibly only) advantage of the ordinal representation is that the classic cut-and-splice crossover actually works! Given any two tours in ordinal representation, if you make a cut after some position and take the first part of one tour with the second part of the other (and vice versa), you'll generate a legal tour. For example, the two parents

$$p_1 = (1\ 1\ 2\ 1\ |\ 4\ 1\ 3\ 1\ 1)\ \text{and}$$
$$p_2 = (5\ 1\ 5\ 5\ |\ 5\ 3\ 3\ 2\ 1),$$

which correspond to the tours

$$1 - 2 - 4 - 3 - 8 - 5 - 9 - 6 - 7 \text{ and}$$
$$5 - 1 - 7 - 8 - 9 - 4 - 6 - 3 - 2,$$

with the crossover point marked by "|", produce the following offspring:

$$o_1 = (1\ 1\ 2\ 1\ 5\ 3\ 3\ 2\ 1)\ \text{and}$$
$$o_2 = (5\ 1\ 5\ 5\ 4\ 1\ 3\ 1\ 1).$$

These offspring correspond to

$$1 - 2 - 4 - 3 - 9 - 7 - 8 - 6 - 5 \text{ and}$$
$$5 - 1 - 7 - 8 - 6 - 2 - 9 - 3 - 4.$$

It's easy to see that partial tours to the left of the crossover point don't change, whereas partial tours to the right of the crossover point are disrupted essentially at random. That's bad. Hopefully, when you read above that the application of a simple crossover would actually work on this representation, you thought "so what?" Again, it's not enough to just generate legal tours, there has to be an inheritance from parent(s) to offspring that supports a better effort than can be obtained simply by blind random search. Not surprisingly, the results for coupling this representation with simple one-point crossover have been poor [203].

Path Representation:

The path representation is perhaps the most natural representation of a tour. For example, a tour

$$5 - 1 - 7 - 8 - 9 - 4 - 6 - 2 - 3$$

is represented as

$$(5\ 1\ 7\ 8\ 9\ 4\ 6\ 2\ 3).$$

The best-known crossovers defined for the path representation are: *partially-mapped* (PMX), *order* (OX), and *cycle* (CX) crossovers.

PMX. This crossover builds an offspring by choosing a subsequence of a tour from one parent and preserving the order and position of as many cities as possible from the other parent [192]. A subsequence of a tour is selected by choosing two random cut points, which serve as boundaries for the swapping operations. For example, the two parents (with two cut points marked by "|")

$$p_1 = (1\ 2\ 3\ |\ 4\ 5\ 6\ 7\ |\ 8\ 9)\ \text{and}$$
$$p_2 = (4\ 5\ 2\ |\ 1\ 8\ 7\ 6\ |\ 9\ 3)$$

would produce offspring as follows. First, the segments between cut points are swapped (the symbol "x" can be interpreted as "at present unknown"):

$$o_1 = (\text{x x x} \mid 1\ 8\ 7\ 6 \mid \text{x x}) \text{ and}$$
$$o_2 = (\text{x x x} \mid 4\ 5\ 6\ 7 \mid \text{x x}).$$

This swap also defines a series of mappings:

$$1 \leftrightarrow 4,\ 8 \leftrightarrow 5,\ 7 \leftrightarrow 6,\ \text{and } 6 \leftrightarrow 7.$$

Then, we can fill in additional cities from the original parents for which there's no conflict:

$$o_1 = (\text{x}\ 2\ 3 \mid 1\ 8\ 7\ 6 \mid \text{x}\ 9) \text{ and}$$
$$o_2 = (\text{x x}\ 2 \mid 4\ 5\ 6\ 7 \mid 9\ 3).$$

Finally, the first x in the offspring o_1 (which should be 1, but there was a conflict) is replaced by 4 because of the mapping $1 \leftrightarrow 4$. Similarly, the second x in offspring o_1 is replaced by 5, and the respective x and x in offspring o_2 are 1 and 8. The offspring are

$$o_1 = (4\ 2\ 3 \mid 1\ 8\ 7\ 6 \mid 5\ 9) \text{ and}$$
$$o_2 = (1\ 8\ 2 \mid 4\ 5\ 6\ 7 \mid 9\ 3).$$

The PMX operator exploits similarities in the value and ordering simultaneously when used with an appropriate reproductive plan [192]. Nevertheless, an evolutionary algorithm in [142] offered better results in fewer function evaluations simply using a one-parent "remove-and-replace" variation operator on an ordinal representation for 100-city TSPs.

OX. This crossover builds offspring by choosing a subsequence of a tour from one parent and preserving the relative order of cities from the other parent [86]. For example, two parents (with two cut points marked by '|')

$$p_1 = (1\ 2\ 3 \mid 4\ 5\ 6\ 7 \mid 8\ 9) \text{ and}$$
$$p_2 = (4\ 5\ 2 \mid 1\ 8\ 7\ 6 \mid 9\ 3)$$

would produce offspring as follows. First, the segments between cut points are copied into offspring:

$$o_1 = (\text{x x x} \mid 4\ 5\ 6\ 7 \mid \text{x x}) \text{ and}$$
$$o_2 = (\text{x x x} \mid 1\ 8\ 7\ 6 \mid \text{x x}).$$

Next, starting from the second cut point of one parent, the cities from the other parent are copied in the same order, omitting symbols that are already present. Reaching the end of the string, we continue from the first place of the string. The sequence of the cities in the second parent (from the second cut point) is

$$9 - 3 - 4 - 5 - 2 - 1 - 8 - 7 - 6;$$

after removing cities 4, 5, 6, and 7, which are already in the first offspring, we obtain

$$9 - 3 - 2 - 1 - 8.$$

This sequence is placed in the first offspring (starting from the second cut point):

$$o_1 = (2\ 1\ 8\ |\ 4\ 5\ 6\ 7\ |\ 9\ 3).$$

Similarly we obtain the other offspring:

$$o_2 = (3\ 4\ 5\ |\ 1\ 8\ 7\ 6\ |\ 9\ 2).$$

The OX operator exploits a property of the path representation that the relative order of the cities (as opposed to their specific positions) is important, i.e., the two tours

$$9 - 3 - 4 - 5 - 2 - 1 - 8 - 7 - 6 \text{ and}$$
$$4 - 5 - 2 - 1 - 8 - 7 - 6 - 9 - 3$$

are in fact identical.

CX. This crossover builds offspring in such a way that each city and its position comes from one of the parents [338]. Cycle crossover works as follows. Two parents, say:

$$p_1 = (1\ 2\ 3\ 4\ 5\ 6\ 7\ 8\ 9) \text{ and}$$
$$p_2 = (4\ 1\ 2\ 8\ 7\ 6\ 9\ 3\ 5),$$

produce the first offspring by taking the first city from the first parent:

$$o_1 = (1\ \text{x x x x x x x x}).$$

Since every city in the offspring should be taken from one of its parents (from the same position), we don't have any choice at this point. The next city to be considered must be city 4, as it's the city from parent p_2 that's just "below" the selected city 1. In p_1 this city is at position "4", thus

$$o_1 = (1\ \text{x x}\ 4\ \text{x x x x x}).$$

This, in turn, implies city 8, as it's the city from parent p_2 that's just "below" the selected city 4. Thus

$$o_1 = (1\ \text{x x}\ 4\ \text{x x x}\ 8\ \text{x}).$$

Following this rule, the next cities to be included in the first offspring are 3 and 2. Note, however, that the selection of city 2 requires selecting city 1, which is already on the list. Thus, we have completed a cycle

$$o_1 = (1\ 2\ 3\ 4\ \text{x x x}\ 8\ \text{x}).$$

The remaining cities are filled in from the other parent:

$$o_1 = (1\ 2\ 3\ 4\ 7\ 6\ 9\ 8\ 5).$$

Similarly,

$$o_2 = (4\ 1\ 2\ 8\ 5\ 6\ 7\ 3\ 9).$$

CX preserves the absolute position of the elements in the parent sequence.

It's possible to define other operators for the path representation. For example, Syswerda [461] defined two modified versions of the order crossover operator. The first modification (called order-based crossover) selects several positions in a vector randomly, and the order of the cities in the selected positions in one parent is imposed on the corresponding cities in the other parent. For example, consider two parents

$$p_1 = (1\ 2\ 3\ 4\ 5\ 6\ 7\ 8\ 9) \text{ and}$$
$$p_2 = (4\ 1\ 2\ 8\ 7\ 6\ 9\ 3\ 5).$$

Assume that the selected positions are 3rd, 4th, 6th, and 9th. The ordering of the cities in these positions from parent p_2 will be imposed on parent p_1. The cities at these positions, in the given order, in p_2 are 2, 8, 6, and 5. In parent p_1 these cities are present at positions 2, 5, 6, and 8. In the offspring, the elements on these positions are reordered to match the order of the same elements from p_2 (the order is 2 – 8 – 6 – 5). The first offspring is a copy of p_1 at all positions except 2, 5, 6, and 8:

$$o_1 = (1\ \text{x}\ 3\ 4\ \text{x}\ \text{x}\ 7\ \text{x}\ 9).$$

All other elements are filled in the order given in parent p_2, i.e., 2, 8, 6, 5, so finally,

$$o_1 = (1\ 2\ 3\ 4\ 8\ 6\ 7\ 5\ 9).$$

Similarly, we can construct the second offspring,

$$o_2 = (3\ 1\ 2\ 8\ 7\ 4\ 6\ 9\ 5).$$

The second modification (called position-based crossover) is more similar to the original order crossover. The only difference is that in position-based crossover, instead of selecting one subsequence of cities to be copied, several cities are (randomly) selected for that purpose.

It's interesting to note that these two operators (order-based crossover and position based crossover) are in some sense equivalent. An order-based crossover with some number of positions selected as crossover points, and a position-based crossover with complimentary positions as its crossover points will always produce the same result. This means that if the average number of crossover points is $n/2$, where n is the total number of cities, these two operators should provide the same performance. However, if the average number of crossover points is, say, $n/10$, then the two operators will display different characteristics.

For more information on these operators and some theoretical and empirical comparisons, see [174], [338], [452], and [460].

In surveying the different reordering operators that have emerged in recent years, we should mention the inversion operator as well. Simple inversion selects two points along the length of the permutation, which is then cut at these points, and the substring between these points is reversed. Thus the operator works on only a single parent at a time. For example, a permutation

$$(1\ 2\ |\ 3\ 4\ 5\ 6\ |\ 7\ 8\ 9)$$

with two cut points marked by '|', is changed into

$$(1\ 2\ |\ 6\ 5\ 4\ 3\ |\ 7\ 8\ 9).$$

Such simple inversion guarantees that the resulting offspring is a legal tour. It also has the potential to uncross a tour that travels over itself, but of course it can also introduce such crossings into tours that don't. Some experiments with inversion on a 50-city TSP showed that it outperformed an evolutionary algorithm that used a "cross and correct" operator [494]. Increasing the number of cut points decreased the performance. Other examples with inversion are offered in [147]. Toward the end of this chapter we'll return to the inversion operator and discuss its newest version, the so-called *inver-over* operator.

At this point we should mention other efforts that rely on either a single parent tour to create an offspring, or even more than two parents. Herdy [213] experimented with four different variation operators:

- *Inversion* — as described above.

- *Insertion* — selects a city and inserts it in a random place.

- *Displacement* — selects a subtour and inserts it in a random place.

- *Reciprocal exchange* — swaps two cities.

Herdy also used a version of heuristic crossover where several parents contribute to producing offspring. After selecting the first city of the offspring tour at random, all left and right neighbors of that city (from all parents) are examined. The city that yields the shortest distance is selected. The process continues until the tour is completed.

Another way to encode and operate on tours is to actually use a vector of n floating-point numbers, where n corresponds to the number of cities. Components of the vector are sorted and their order determines the tour. For example, the vector

$$\mathbf{v} = (2.34, -1.09, 1.91, 0.87, -0.12, 0.99, 2.13, 1.23, 0.55)$$

corresponds to the tour

$$2 - 5 - 9 - 4 - 6 - 8 - 3 - 7 - 1,$$

since the smallest number, -1.09, is the second component of the vector $\mathbf{v}$, the second smallest number, -0.12, is the fifth component of the vector $\mathbf{v}$, etc. New solutions could be generated by a variety of methods that apply to real-valued vectors, but again, the connection (or at least the intuition) between the representation and the problem may be lost here.

Most of the operators discussed so far take into account cities (i.e., their positions and order) as opposed to edges — links between cities. What might be important isn't the particular position of a city in a tour, but rather the linkage of that city with other cities. Homaifar and Guan [230] offered

> Considering the problem carefully, we can argue that the basic building blocks for TSP are edges as opposed to the position representation of cities. A city or short path in a given position without adjacent or surrounding information has little meaning for constructing a good tour. However, it is hard to argue that injecting city a in position 2 is better than injecting it in position 5. Although this is the extreme case, the underlying assumption is that a good operator should extract edge information from parents as much as possible. This assumption can be partially explained from the experimental results in Oliver's paper [338] that OX does 11% better than PMX, and 15% better than the cycle crossover.

Grefenstette [202] developed a class of heuristic operators that emphasizes edges. They work along the following lines:

1. Randomly select a city to be the current city c of the offspring.

2. Select four edges (two from each parent) incident to the current city c.

3. Define a probability distribution over the selected edges based on their cost. The probability for the edge associated with a previously visited city is 0.

4. Select an edge. If at least one edge has a nonzero probability, selection is based on the above distribution; otherwise, selection is made at random from the unvisited cities.

5. The city on "the other end" of the selected edge becomes the current city c.

6. If the tour is complete, stop; otherwise, go to step 2.

As reported in [202], however, such operators transfer around 60 percent of the edges from parents — which means that 40 percent of the edges are selected randomly. Again, the procedure tends to look more like a blind search than an evolutionary search.

In contrast, the *edge recombination* (ER) operator [494] transfers more than 95 percent of the edges from the parents to a single offspring. The ER operator explores the information on edges in a tour, e.g., for the tour

(3 1 2 8 7 4 6 9 5),

the edges are (3 1), (1 2), (2 8), (8 7), (7 4), (4 6), (6 9), (9 5), and (5 3). After all, edges — not cities — carry values (distances) in the TSP. The evaluation function to be minimized is the total cost of the edges that constitute a legal tour. The position of a city in a tour isn't important — tours are circular. Also, the direction of an edge isn't important: both edges (3 1) and (1 3) only signal that cities 1 and 3 are connected.

The general idea behind the ER crossover is that an offspring should be built exclusively from the edges present in both parents. This is done with the help of the edge list created from both parent tours. The edge list provides, for each city c, all of the other cities that are connected to city c in at least one of the parents. For each city c there are at least two and at most four cities on the list. For example, for the two parents

$p_1 = (1\ 2\ 3\ 4\ 5\ 6\ 7\ 8\ 9)$ and
$p_2 = (4\ 1\ 2\ 8\ 7\ 6\ 9\ 3\ 5)$,

the edge list is

City 1: edges to other cities: 9 2 4
City 2: edges to other cities: 1 3 8
City 3: edges to other cities: 2 4 9 5
City 4: edges to other cities: 3 5 1
City 5: edges to other cities: 4 6 3
City 6: edges to other cities: 5 7 9
City 7: edges to other cities: 6 8
City 8: edges to other cities: 7 9 2
City 9: edges to other cities: 8 1 6 3.

Constructing an offspring starts with a selection of an initial city from one of the parents. In [494] the authors selected one of the initial cities (e.g., 1 or 4 in the example above). The city with the smallest number of edges in the edge list is selected. If these numbers are equal, a random choice is made. This increases the chance that we'll complete a tour with all of the edges selected from the parents. With a random selection, the chance of having an "edge failure," i.e., being left with a city that doesn't have a continuing edge, would be much higher. Assume that we've selected city 1. This city is directly connected to three other cities: 9, 2, and 4. The next city is selected from these three. In our example, cities 4 and 2 have three edges, and city 9 has four. A random choice is made between cities 4 and 2. Assume that city 4 was selected. Again, the candidates for the next city in the constructed tour are 3 and 5, since they are directly connected to the last city, 4. Again, city 5 is selected, since it has only three edges as opposed to the four edges of city 3. So far, the offspring has the following entries:

(1 4 5 x x x x x x).

Continuing this procedure we finish with the offspring

(1 4 5 6 7 8 2 3 9),

which is composed entirely of edges taken from the two parents. From a series of experiments, edge failure occurred with a very low rate (1–1.5 percent) [494].

Edge recombination crossover was further enhanced in [452]. The idea was that the "common subsequences" weren't preserved in the ER crossover. For example, if the edge list contains the row with three edges

City 4: edges to other cities: 3 5 1,

one of these edges repeats itself. Referring to the previous example, it's the edge (4 5). This edge is present in both parents. However, it's listed as other edges, e.g., (4 3) and (4 1), which are present only in one parent. The proposed solution [452] modifies the edge list by storing "flagged" cities:

City 4: edges to other cities: 3 -5 1,

where the character '-' means simply that the flagged city 5 should be listed twice. In the previous example of two parents

$p_1 = (1\ 2\ 3\ 4\ 5\ 6\ 7\ 8\ 9)$ and
$p_2 = (4\ 1\ 2\ 8\ 7\ 6\ 9\ 3\ 5)$,

the enhanced edge list is:

City 1: edges to other cities: 9 -2 4
City 2: edges to other cities: -1 3 8
City 3: edges to other cities: 2 4 9 5
City 4: edges to other cities: 3 -5 1
City 5: edges to other cities: -4 6 3
City 6: edges to other cities: 5 -7 9
City 7: edges to other cities: -6 -8
City 8: edges to other cities: -7 9 2
City 9: edges to other cities: 8 1 6 3.

The algorithm for constructing a new offspring gives priority to flagged entries. This is important only in the cases where three edges are listed — in the two other cases either there are no flagged cities, or both cities are flagged. This enhancement, coupled with a modification for making better choices for when random edge selection is necessary, further improved the performance of the system [452].

The edge recombination operators indicate that the path representation might not be sufficient to represent important properties of a tour — this is why it was complemented by the edge list. Are there other representations that are more suitable for the traveling salesman problem? The question demands that we also ask about the variation operators that are applied to that representation. Nevertheless, it's worthwhile to experiment with other, possibly nonvector, representations if only to see what we might discover.

There have been at least three independent attempts to evolve solutions for a TSP where solutions were represented in a matrix [174, 430, 230]. Fox and McMahon [174] represented a tour as a precedence binary matrix M. The matrix element m_{ij} in row i and column j contains a 1 if and only if the city i occurs before city j in the tour. For example, a tour

$(3\ 1\ 2\ 8\ 7\ 4\ 6\ 9\ 5)$,

is represented in matrix form in figure 8.1.

	1	2	3	4	5	6	7	8	9
1	0	1	0	1	1	1	1	1	1
2	0	0	0	1	1	1	1	1	1
3	1	1	0	1	1	1	1	1	1
4	0	0	0	0	1	1	0	0	1
5	0	0	0	0	0	0	0	0	0
6	0	0	0	0	1	0	0	0	1
7	0	0	0	1	1	1	0	0	1
8	0	0	0	1	1	1	1	0	1
9	0	0	0	0	1	0	0	0	0

Fig. 8.1. Matrix representation of a tour. The matrix element m_{ij} in row i and column j contains a 1 if and only if the city i occurs before city j in the tour

In this representation, the $n \times n$ matrix M that represents a tour (i.e., the total order of cities) has the following properties:

1. The number of 1s is exactly $\frac{n(n-1)}{2}$.

2. $m_{ii} = 0$ for all $1 \le i \le n$.

3. If $m_{ij} = 1$ and $m_{jk} = 1$ then $m_{ik} = 1$.

If the number of 1s in the matrix is less than $\frac{n(n-1)}{2}$, and the two other previous requirements are satisfied, the cities are partially ordered. This means that we can complete such a matrix (in at least one way) to get a legal tour. Fox and McMahon [174] offered

> The Boolean matrix representation of a sequence encapsulates all of the information about the sequence, including both the micro-topology of individual city-to-city connections and the macro-topology of predecessors and successors. The Boolean matrix representation can be used to understand existing operators and to develop new operators that can be applied to sequences to produce desired effects while preserving the necessary properties of the sequence.

The two new operators developed in [174] were *intersection* and *union*. The intersection operator is based on the observation that the intersection of bits from both matrices results in a matrix where (1) the number of 1s is not greater than $\frac{n(n-1)}{2}$, and (2) the two other requirements are satisfied. Thus, we can complete such a matrix to get a legal tour.

For example, two parents

$$p_1 = (1\ 2\ 3\ 4\ 5\ 6\ 7\ 8\ 9) \text{ and } p_2 = (4\ 1\ 2\ 8\ 7\ 6\ 9\ 3\ 5)$$

are represented by two matrices (figure 8.2).

	1	2	3	4	5	6	7	8	9
1	0	1	1	1	1	1	1	1	1
2	0	0	1	1	1	1	1	1	1
3	0	0	0	1	1	1	1	1	1
4	0	0	0	0	1	1	1	1	1
5	0	0	0	0	0	1	1	1	1
6	0	0	0	0	0	0	1	1	1
7	0	0	0	0	0	0	0	1	1
8	0	0	0	0	0	0	0	0	1
9	0	0	0	0	0	0	0	0	0

	1	2	3	4	5	6	7	8	9
1	0	1	1	0	1	1	1	1	1
2	0	0	1	0	1	1	1	1	1
3	0	0	0	0	1	0	0	0	0
4	1	1	1	0	1	1	1	1	1
5	0	0	0	0	0	0	0	0	0
6	0	0	1	0	1	0	0	0	1
7	0	0	1	0	1	1	0	0	1
8	0	0	1	0	1	1	1	0	1
9	0	0	1	0	1	0	0	0	0

Fig. 8.2. Two parents in the Boolean matrix representation

The first stage of the intersection of these two matrices gives the matrix displayed in figure 8.3.

	1	2	3	4	5	6	7	8	9
1	0	1	1	0	1	1	1	1	1
2	0	0	1	0	1	1	1	1	1
3	0	0	0	0	1	0	0	0	0
4	0	0	0	0	1	1	1	1	1
5	0	0	0	0	0	0	0	0	0
6	0	0	0	0	0	0	0	0	1
7	0	0	0	0	0	0	0	0	1
8	0	0	0	0	0	0	0	0	1
9	0	0	0	0	0	0	0	0	0

Fig. 8.3. First phase of the intersection operator

The partial order imposed by the result of intersection requires that city 1 precedes cities 2, 3, 5, 6, 7, 8, and 9. City 2 precedes cities 3, 5, 6, 7, 8, and

9. City 3 precedes city 5. City 4 precedes cities 5, 6, 7, 8, and 9. Cities 6, 7, and 8 precede city 9. During the next stage of the intersection operator, one of the parents is selected; some 1s that are unique to this parent are added, and the matrix is completed into a sequence through an analysis of the sums of the rows and columns. For example, the matrix from figure 8.4 is a possible result after completing the second stage. It represents the tour (1 2 4 8 7 6 3 5 9).

	1	2	3	4	5	6	7	8	9
1	0	1	1	1	1	1	1	1	1
2	0	0	1	1	1	1	1	1	1
3	0	0	0	0	1	0	0	0	1
4	0	0	1	0	1	1	1	1	1
5	0	0	0	0	0	0	0	0	1
6	0	0	1	0	1	0	0	0	1
7	0	0	1	0	1	1	0	0	1
8	0	0	1	0	1	1	1	0	1
9	0	0	0	0	0	0	0	0	0

Fig. 8.4. Final result of the intersection

The union operator is based on the observation that the subset of bits from one matrix can be safely combined with a subset of bits from the other matrix, provided that these two subsets have an empty intersection. The operator partitions the set of cities into two disjoint groups (a special method was used to make this partition in [174]). For the first group of cities, it copies the bits from the first matrix, and for the second group of cities, it copies the bits from the second matrix. Finally, it completes the matrix into a sequence through an analysis of the sums of the rows and columns (similar to the intersection operator). For example, the two parents p_1 and p_2 and the partition of cities $\{1, 2, 3, 4\}$ and $\{5, 6, 7, 8, 9\}$ produce the matrix (figure 8.5), which is completed in the same manner as the intersection operator.

The reported experimental results on different arrangements of cities (e.g., random, clusters, concentric circles) revealed that the combination of union and intersection operators was able to improve solutions over successive generations even when elitist selection was not applied. (Recall that elitist selection always preserves the best solution found so far.) The same behavior was not evidenced for either the ER or PMX operators. A more detailed comparison of several binary and unary (swap, slice, and invert) operators in terms of performance, complexity, and execution time is provided in [174].

The second approach in using a matrix representation was offered by Seniw [430]. The matrix element m_{ij} in the row i and column j contains a 1 if and only if the tour goes from city i directly to city j. This means that there is only one nonzero entry for each row and each column in the matrix (for each city i

	1	2	3	4	5	6	7	8	9
1	0	1	1	1	x	x	x	x	x
2	0	0	1	1	x	x	x	x	x
3	0	0	0	1	x	x	x	x	x
4	0	0	0	0	x	x	x	x	x
5	x	x	x	x	0	0	0	0	0
6	x	x	x	x	1	0	0	0	1
7	x	x	x	x	1	1	0	0	1
8	x	x	x	x	1	1	1	0	1
9	x	x	x	x	1	0	0	0	0

Fig. 8.5. First phase of the union operator

there is exactly one city visited prior to i, and exactly one city visited next to i). For example, the matrix in figure 8.6a represents a tour that visits the cities (1, 2, 4, 3, 8, 6, 5, 7, 9) in this order. Note also that this representation avoids the problem of specifying the starting city, i.e., figure 8.6a also represents the tours (2, 4, 3, 8, 6, 5, 7, 9, 1), (4, 3, 8, 6, 5, 7, 9, 1, 2), etc.

	1	2	3	4	5	6	7	8	9
1	0	1	0	0	0	0	0	0	0
2	0	0	0	1	0	0	0	0	0
3	0	0	0	0	0	0	0	1	0
4	0	0	1	0	0	0	0	0	0
5	0	0	0	0	0	0	1	0	0
6	0	0	0	0	1	0	0	0	0
7	0	0	0	0	0	0	0	0	1
8	0	0	0	0	0	1	0	0	0
9	1	0	0	0	0	0	0	0	0

(a)

	1	2	3	4	5	6	7	8	9
1	0	1	0	0	0	0	0	0	0
2	0	0	0	1	0	0	0	0	0
3	0	0	0	0	0	0	0	1	0
4	0	0	0	0	1	0	0	0	0
5	0	0	0	0	0	0	1	0	0
6	0	0	0	0	0	0	0	0	1
7	1	0	0	0	0	0	0	0	0
8	0	0	0	0	0	1	0	0	0
9	0	0	1	0	0	0	0	0	0

(b)

Fig. 8.6. Binary matrix tours

It's interesting to note that each complete tour is represented as a binary matrix with only one bit in each row and one bit in each column set to one. However, not every matrix with these properties would represent a single tour. Binary matrices can represent multiple subtours. Each subtour will eventually loop back onto itself, without connecting to any other subtour in the matrix. For example, a matrix from figure 8.6b represents two subtours: (1, 2, 4, 5, 7) and (3, 8, 6, 9).

The subtours were allowed in the hope that natural clustering would take place. After the evolutionary algorithm terminates, the best matrix is reduced to a single tour by successively combining pairs of subtours using a deterministic algorithm. Subtours that comprised just one city (a tour leaving a city to travel right back to itself), having a distance cost of zero, were not allowed. A lower limit of $q = 3$ cities in a subtour was set in an attempt to prevent the evolutionary algorithm from reducing a TSP to a large number of subtours, each having very few cities.

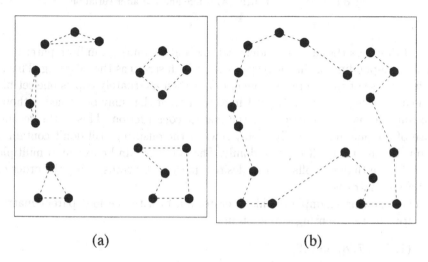

(a)   (b)

Fig. 8.7. (a) The subtours resulting from a sample run of the algorithm on a number of cities intentionally placed in clusters. As expected, the algorithm developed isolated subtours. (b) The tour after the subtours have been combined.

Two variation operators were defined. One took a matrix, randomly selected several rows and columns, and then removed the set bits in the intersections of those rows and columns and replaced them randomly, possibly in a different configuration. For example, consider the tour from figure 8.6a represented by

$$(1, 2, 4, 3, 8, 6, 5, 7, 9).$$

Assume that rows 4, 6, 7, 9 and columns 1, 3, 5, 8, and 9 are randomly selected to participate in the variation operation. The marginal sums for these rows and columns are calculated. The bits at the intersections of these rows and columns are removed and replaced randomly, although they must agree with the marginal sums. In other words, the submatrix corresponding to rows 4, 6, 7, and 9, and columns 1, 3, 5, 8, and 9 from the original matrix (figure 8.8a), is replaced by another submatrix (figure 8.8b).

The resulting matrix represents a solution with two subtours:

$$(1, 2, 4, 5, 7) \text{ and } (3, 8, 6, 9)$$

and is represented in figure 8.6b.

The other operator begins with a matrix that has all of its bits set to zero. The operator first examines the two selected matrices from the population and

0	1	0	0	0
0	0	1	0	0
0	0	0	0	1
1	0	0	0	0

(a)

0	0	1	0	0
0	0	0	0	1
1	0	0	0	0
0	1	0	0	0

(b)

Fig. 8.8. Part of the matrix (a) before and (b) after variation

when it discovers the same bit (identical row and column) set in both parents, it sets a corresponding bit in the new matrix, which serves as the offspring. This is the first phase of the operator. The operator then alternately copies one set bit from each parent until no bits exist in either parent that may be copied without violating the basic restrictions of the matrix construction. This is the second phase of the operator. Finally, if any rows in the offspring still don't contain a set bit, the matrix is filled in randomly. The operator can be executed multiple times if more than one offspring is desired, perhaps by transposing the order of the parent matrices.

The following example of this operator starts with the first parent matrix in figure 8.9a representing two subtours

(1, 5, 3, 7, 8) and (2, 4, 9, 6),

and the second parent matrix (figure 8.9b) representing a single tour

(1, 5, 6, 2, 7, 8, 3, 4, 9).

	1	2	3	4	5	6	7	8	9
1	0	0	0	0	1	0	0	0	0
2	0	0	0	1	0	0	0	0	0
3	0	0	0	0	0	0	1	0	0
4	0	0	0	0	0	0	0	0	1
5	0	0	1	0	0	0	0	0	0
6	0	1	0	0	0	0	0	0	0
7	0	0	0	0	0	0	0	1	0
8	1	0	0	0	0	0	0	0	0
9	0	0	0	0	0	1	0	0	0

(a)

	1	2	3	4	5	6	7	8	9
1	0	0	0	0	1	0	0	0	0
2	0	0	0	0	0	0	1	0	0
3	0	0	0	1	0	0	0	0	0
4	0	0	0	0	0	0	0	0	1
5	0	0	0	0	0	1	0	0	0
6	0	1	0	0	0	0	0	0	0
7	0	0	0	0	0	0	0	1	0
8	0	0	1	0	0	0	0	0	0
9	1	0	0	0	0	0	0	0	0

(b)

Fig. 8.9. (a) First and (b) second parents

The first two phases of constructing the first offspring are displayed in figure 8.10.

The first offspring from the operator, after the final phase, is displayed in figure 8.11. It represents a subtour

	1	2	3	4	5	6	7	8	9
1	0	0	0	0	1	0	0	0	0
2	0	0	0	0	0	0	0	0	0
3	0	0	0	0	0	0	0	0	0
4	0	0	0	0	0	0	0	0	1
5	0	0	0	0	0	0	0	0	0
6	0	1	0	0	0	0	0	0	0
7	0	0	0	0	0	0	0	1	0
8	0	0	0	0	0	0	0	0	0
9	0	0	0	0	0	0	0	0	0

(a)

	1	2	3	4	5	6	7	8	9
1	0	0	0	0	1	0	0	0	0
2	0	0	1	0	0	0	0	0	0
3	0	0	0	0	0	0	0	0	0
4	0	0	0	0	0	0	0	0	1
5	0	0	0	0	0	0	1	0	0
6	0	1	0	0	0	0	0	0	0
7	0	0	0	0	0	0	0	1	0
8	1	0	0	0	0	0	0	0	0
9	0	0	0	0	0	0	0	0	0

(b)

Fig. 8.10. Offspring for crossover, **(a)** after phase 1 and **(b)** phase 2

	1	2	3	4	5	6	7	8	9
1	0	0	0	0	1	0	0	0	0
2	0	0	1	0	0	0	0	0	0
3	0	0	0	1	0	0	0	0	0
4	0	0	0	0	0	0	0	0	1
5	0	0	0	0	0	1	0	0	0
6	0	1	0	0	0	0	0	0	0
7	0	0	0	0	0	0	0	1	0
8	1	0	0	0	0	0	0	0	0
9	0	0	0	0	0	0	1	0	0

Fig. 8.11. Offspring from the two-parent operator after the final phase

(1, 5, 6, 2, 3, 4, 9, 7, 8).

The second offspring represents

(1, 5, 3, 4, 9) and (2, 7, 8, 6).

Note that there are common segments of the parent matrices in both offspring.

This procedure evidenced reasonable performance on several test cases from 30 cities to 512 cities, but the effect of the parameter q, the minimum number of cities in a subtour, remains unclear. In addition, the procedures for combining subtours into a single overall tour aren't at all obvious. On the other hand, the method has some similarities with Litke's recursive clustering algorithm [287], which recursively replaces clusters of size B by single representative cities until fewer than B cities remain. The smaller problem is then solved optimally. All of the clusters are expanded one-by-one and the algorithm sequences the expanded set between the two neighbors in the current tour. Also, the approach might be useful for solving the multiple traveling salesmen problem, where several salesman must complete separate nonoverlapping tours.

The third approach for using matrices to represent solutions for the TSP was offered in [230]. As with the previous approach, the element m_{ij} of the binary matrix M is set to 1 if and only if there is an edge from city i to city j, however, different variation operators were developed.

Homaifar and Guan [230] defined two matrix crossover (MX) operators. These operators exchange all of the entries of the two parent matrices that fall either after a single crossover point (one-point crossover) or between two crossover points (two-point crossover). An additional "repair algorithm" is run to (1) remove duplications, i.e., to ensure that each row and each column has precisely one 1, and (2) cut and connect cycles (if any) to produce a legal tour.

A two-point crossover is illustrated by the following example. Two parent matrices are given in figure 8.12. They represent two legal tours:

(1 2 4 3 8 6 5 7 9) and (1 4 3 6 5 7 2 8 9).

Two crossovers points are selected. These are points between columns 2 and 3 (first point), and between columns 6 and 7 (second point). The crossover points cut the matrices vertically. For each matrix, the first two columns constitute the first part of the division, columns 3, 4, 5, and 6 constitute the middle part, and the last three columns constitute the third part.

	1	2	3	4	5	6	7	8	9
1	0	1	0	0	0	0	0	0	0
2	0	0	0	1	0	0	0	0	0
3	0	0	0	0	0	0	0	1	0
4	0	0	1	0	0	0	0	0	0
5	0	0	0	0	0	0	1	0	0
6	0	0	0	0	1	0	0	0	0
7	0	0	0	0	0	0	0	0	1
8	0	0	0	0	0	1	0	0	0
9	1	0	0	0	0	0	0	0	0

(a)

	1	2	3	4	5	6	7	8	9
1	0	0	0	1	0	0	0	0	0
2	0	0	0	0	0	0	0	1	0
3	0	0	0	0	0	1	0	0	0
4	0	0	1	0	0	0	0	0	0
5	0	0	0	0	0	0	1	0	0
6	0	0	0	0	1	0	0	0	0
7	0	1	0	0	0	0	0	0	0
8	0	0	0	0	0	0	0	0	1
9	1	0	0	0	0	0	0	0	0

(b)

Fig. 8.12. Binary matrices with crossover points marked *(double lines)*

After the first step of the two-point MX operator, entries of both matrices are exchanged between the crossover points (i.e., entries in columns 3, 4, 5, and 6). The intermediate result is given in figure 8.13.

Both offspring, (a) and (b), are illegal; however, the total number of 1s in each intermediate matrix is correct (i.e., nine). The first step of the "repair

	1	2	3	4	5	6	7	8	9
1	0	1	0	1	0	0	0	0	0
2	0	0	0	0	0	0	0	0	0
3	0	0	0	0	0	1	0	1	0
4	0	0	1	0	0	0	0	0	0
5	0	0	0	0	0	0	1	0	0
6	0	0	0	0	1	0	0	0	0
7	0	0	0	0	0	0	0	0	1
8	0	0	0	0	0	0	0	0	0
9	1	0	0	0	0	0	0	0	0

(a)

	1	2	3	4	5	6	7	8	9
1	0	0	0	0	0	0	0	0	0
2	0	0	0	1	0	0	0	1	0
3	0	0	0	0	0	0	0	0	0
4	0	0	1	0	0	0	0	0	0
5	0	0	0	0	0	0	1	0	0
6	0	0	0	0	1	0	0	0	0
7	0	1	0	0	0	0	0	0	0
8	0	0	0	0	0	1	0	0	1
9	1	0	0	0	0	0	0	0	0

(b)

Fig. 8.13. Two intermediate offspring after the first step of MX operator

algorithm" moves some 1s in the matrices in such a way that each row and each column has precisely one 1. For example, in the offspring shown in figure 8.13a, the duplicate 1s occur in rows 1 and 3. The algorithm may move the entry $m_{14} = 1$ into m_{84}, and the entry $m_{38} = 1$ into m_{28}. Similarly, in the other offspring (figure 8.13b) the duplicate 1s occur in rows 2 and 8. The algorithm may move the entry $m_{24} = 1$ into m_{34}, and the entry $m_{86} = 1$ into m_{16}. After the completion of the first step of the repair algorithm, the first offspring represents a legal tour,

(1 2 8 4 3 6 5 7 9),

and the second offspring represents a tour that consists of two subtours,

(1 6 5 7 2 8 9) and (3 4).

The second step of the repair algorithm should be applied only to the second offspring. During this stage, the procedure cuts and connects subtours to produce a legal tour. The cut and connect phase takes into account the existing edges in the original parents. For example, the edge (2 4) is selected to connect these two subtours because this edge is present in one of the parents. Thus the complete tour (a legal second offspring) is

(1 6 5 7 2 4 3 8 9).

A heuristic inversion operator was also used in [230] to complement the MX crossover. The operator reverses the order of cities between two cut points just as with simple inversion (see above). If the distance between the two cut points is large (high-order inversion), the operator explores connections between "good" paths, otherwise the operator performs a local search (low-order inversion). There are two differences between the classical and proposed inversion operators. The first difference is that the resulting offspring is accepted only if the new tour is better than the original. The second difference is that the inversion procedure selects a single city in a tour and checks for an improvement for inversions of the lowest possible order: two. The first inversion that results in an improvement is accepted and the inversion procedure terminates. Otherwise, inversions of order three are considered, and so on.

The results reported in [230] indicate that the evolutionary algorithm using two-point MX and inversion performed well on 30 100-city TSPs. In one experiment, the result of this algorithm for a 318-city problem was only 0.6 percent worse than the optimal solution.

8.2 Incorporating local search methods

The main effort of researchers in the evolutionary computation community who address the TSP has been directed toward inventing suitable representations coupled with a recombination operator that preserves partial tours from the parents. One-parent (i.e., unary) operators have not received comparable attention. None of these efforts so far, however, can really compete with other heuristics (e.g., Lin-Kernighan) both with respect to the quality of the solution that was generated or the time required to generate it. The approaches that have been based on recombination have been computationally intensive, and the approaches based solely on mutation (e.g., simple inversions or swapping operators) do not appear to escape local optima efficiently in the absence of other probabilistic mechanisms of selection.

These obstacles have led many to incorporate local search operators into an evolutionary framework. Such hybridization was examined even in the early stages of evolutionary algorithms [202].[5] When designing solutions for real-world problems, it's often useful to examine how to combine deterministic heuristics with stochastic evolutionary search. Indeed, this can sometimes lead to significant improvements in performance. Figure 8.14 displays the possible structure of an evolutionary algorithm that's extended by a local optimization routine (we added an ∗ to the term "evolutionary algorithm").[6]

The difference between the generic evolutionary algorithm (see figure 6.5) and its extension with a local optimizer is small, but significant. Each individual in the population, whether just initialized or created by one or more parents, undergoes a local improvement before being evaluated. The local optima thus found replace the original individuals. (You might think of this as a form of Lamarckian evolution.) The search space for the evolutionary algorithm

---

[5] Actually, hybridizing local and evolutionary search was offered as early as 1967 by Howard Kaufman [253].

[6] The term *memetic algorithm* refers to an evolutionary algorithm where local optimization is applied to every solution before evaluation. This can be thought of as an evolutionary algorithm that's applied in the subspace of local optima with local optimization acting as a repair mechanism for offspring that reside outside this subspace (i.e., they aren't locally optimal).

```
procedure evolutionary algorithm*
begin
    t ← 0
    initialize P(t)
    apply local optimizer to P(t)
    evaluate P(t)
    while (not termination-condition) do
    begin
        t ← t + 1
        select P(t) from P(t − 1)
        alter P(t)
        apply local optimizer to P(t)
        evaluate P(t)
    end
end
```

Fig. 8.14. Structure of an evolutionary algorithm with a local optimizer

is reduced to only those solutions that are locally optimal with respect to the evaluation function and the particular heuristic used for local improvements.

There are many possibilities for developing evolutionary algorithms that incorporate local search operators. With respect to the TSP, you could rely on Lin-Kernighan, 2-opt, 3-opt, and so forth. In addition, you could develop other single- or multiple-parent operators and couple them with different selection methods. There are many open avenues to explore.

Some efforts have been directed at applying a variety of crossover operators (e.g., MPX operator [329]) to locally optimal individuals [56, 195, 329, 302, 471, 129], i.e., solutions that are obtained after being improved by a local search. In some of these cases, when evolutionary techniques have been extended with local operators, they have performed better than multistart local search algorithms alone and have returned near-optimum solutions for test cases of 442, 532, and 666 cities, where the deviation from the optimal tour length was less than one percent.

Let's illustrate some of these approaches. Mühlenbein et al. [330], encouraged an "intelligent evolution" of individuals. Their algorithm

- Uses a 2-opt procedure to replace each tour in the current population with a locally optimal tour.

- Places more emphasis on higher-quality solutions by allowing them to generate more offspring.

- Uses recombination and mutation.

- Searches for the minimum by using a local search with each individual.

- Repeats the last three steps until some halting condition is met.

They use a version of the order crossover (OX) where two parents (with two cut points marked by '|'), say,

$p_1 = (1\ 2\ 3\ |\ 4\ 5\ 6\ 7\ |\ 8\ 9)$ and
$p_2 = (4\ 5\ 2\ |\ 1\ 8\ 7\ 6\ |\ 9\ 3),$

produce the offspring as follows. First, the segments between cut points are copied into the offspring:

$o_1 = (x\ x\ x\ |\ 4\ 5\ 6\ 7\ |\ x\ x)$ and
$o_2 = (x\ x\ x\ |\ 1\ 8\ 7\ 6\ |\ x\ x).$

Next, instead of starting from the second cut point of one parent, as was the case for OX, the cities from the other parent are copied in the same order from the beginning of the string, omitting those symbols that are already present:

$o_1 = (2\ 1\ 8\ |\ 4\ 5\ 6\ 7\ |\ 9\ 3)$ and
$o_2 = (2\ 3\ 4\ |\ 1\ 8\ 7\ 6\ |\ 5\ 9).$

The experimental results on a 532-city problem have been encouraging, as they found a solution that is within 0.06 percent of the optimum solution (found by Padberg and Rinaldi [342]).

Another approach is worthy of mention. Craighurst and Martin [81] concentrated on exploring a connection between "incest" prevention and the performance of an evolutionary algorithm for the TSP. In addition to incorporating 2-opt as a local operator, they introduced the notion that a solution cannot recombine with other solutions that it is "related" to. The k-th incest law prohibited mating an individual with $k - 1$ relations. That is, for $k = 0$, there are no restrictions. For $k = 1$ an individual can't mate with itself. For $k = 2$, it can't mate with itself, with its parents, its children, its siblings, and so forth. Several experiments were conducted using six test problems from TSPLIB ranging in size from 48 to 101 cities. The results indicated a strong and interesting interdependence between incest prohibition laws and the mutation rate. For low mutation rates, incest prevention improved the results; however, for larger mutation rates, the significance of incest prevention mechanism decreased until it actually impaired the final performance. Different laws for incest prevention were also tried, where the similarity between two individuals was measured as a ratio of the difference between the total number of edges in a solution and the number of common edges shared between solution taken over the total number of edges. Interestingly, these rules didn't affect the population's diversity significantly. Sometimes results are counterintuitive. For more discussion, see [81].

8.3 Other possibilities

Valenzuela and Jones [473] proposed an interesting approach to extending evolutionary algorithms based on the divide and conquer technique of Karp-Steele algorithms for the TSP [251, 453]. Their evolutionary divide and conquer (EDAC) algorithm can be applied to any problem in which some knowledge of good solutions of subproblems is useful in constructing a global solution. They applied this technique to the geometric TSP (i.e., the triangle inequality of distances between cities is satisfied: $dist(A, B) + dist(B, C) \geq dist(A, C)$ for any cities A, B, and C).

Several bisection methods can be considered. These methods cut a rectangle with n cities into two smaller rectangles. For example, one of these methods partitions the problem by exactly bisecting the area of the rectangle parallel to the shorter side. Another method intersects the $\lfloor n/2 \rfloor$ closest city to the shorter edge of the rectangle, thus providing a "shared" city between two subrectangles. Final subproblems are quite small (typically between five and eight cities) and relatively easy to solve. The 2-opt procedure was chosen for its speed and simplicity. A patching algorithm (i.e., repair algorithm) replaces some edges in two separate tours to get one larger tour. The major role of the evolutionary algorithm is to determine the direction of bisection (horizontal or vertical) used at each stage.

There are other possibilities as well. In [474] an evolutionary system is used to improve a simple heuristic algorithm for the TSP by perturbing city coordinates; results for problem sizes up to 500 cities were reported. In [306] a new local search operator was discussed, and in [297] a new selection method was introduced as well as new crossovers (edge exchange crossover EEX and subtour exchange crossover SXX).

We conclude this chapter by describing two recent evolutionary approaches for the TSP. The first, proposed by Nagata et al. [334], incorporates local search in an interesting way that's not as explicit as in other systems. The second approach represents a "pure" evolutionary algorithm (i.e., no local search method is involved) and is based on a new operator that combines inversion and crossover. Both systems also share some common features.

8.3.1 Edge assembly crossover

Nagata et al. [334] offered a new edge assembly crossover operator (EAX). The process of offspring construction is quite complex and requires a few steps. Assume that two parents have been selected: parent A and B (figure 8.15). First, a graph G is constructed that contains edges of both parents (figure 8.16). From the graph G, a set of so-called AB-cycles is constructed. An AB-cycle is an even-length subcycle of G with edges that come alternately from A and B. Note that an AB-cycle can repeat cities, but not edges. After adding a single edge, we check to see if an AB-cycle is present as a subset of selected edges. Note also that some initial edges selected for a construction of an AB-cycle may not be included in the final AB-cycle. The continuation of our example illustrates these points.

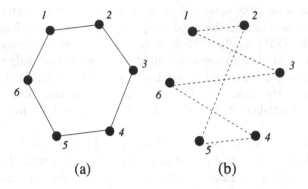

Fig. 8.15. (a) Parents A and (b) B for a six-city TSP

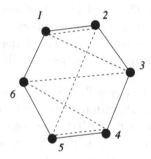

Fig. 8.16. Graph G: The edges of A and B are included.

Assume that we start at city 6 and that the first edge of an AB-cycle is (6 1). This edge comes from parent A so the next selected edge should come from parent B and originate at city 1. Let's say that we select edge (1 3). Further selected edges (in order) are (3 4) from parent A, (4 5) from parent B, (5 6) from parent A, and (6 3) from parent B. At this point we have the first AB-cycle (figure 8.17a). Note that the first two edges, (6 1) and (1 3), aren't included. The edges of the AB-cycle are removed from the graph G and the search for the next AB-cycle is started. Note as well that during this process we can get "ineffective" AB-cycles, which consist of only two cities (figure 8.17b).

The next step involves selecting a subset of AB-cycles (called an E-set). Nagata et al. [334] considered two selection mechanisms: (1) a particular deterministic heuristic method, and (2) a random method, where each AB-cycle is selected with a probability of 0.5.

Once the E-set is fixed, an intermediate offspring C is constructed as follows:

set $C \leftarrow A$
for each edge $e \in E$-set **do**

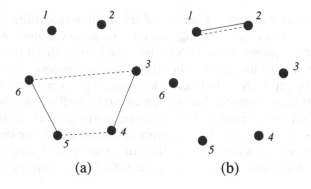

Fig. 8.17. (a) An AB-cycle and (b) an ineffective AB-cycle that consists of only two cities

> **if** $e \in A$ **then** $C \leftarrow C - \{e\}$
> **if** $e \in B$ **then** $C \leftarrow C \cup \{e\}$

The intermediate offspring C represents a set of disjoint subtours that cover all cities (as was the case with one of the methods discussed earlier, see figure 8.7). At this stage an intermediate child is transformed into a single legal tour using a simple greedy algorithm. This greedy procedure incorporates "local search," as the lengths of several links are considered and the best connection is selected. In particular, it starts from a subtour T with the smallest number of edges. Then it selects two edges: one from the selected subtour T, the other from all edges in the remaining subtours. Assume that

$$(v_q\, v_{q+1}) \in T \text{ and } (v'_r\, v'_{r+1}) \notin T$$

(the increments $q+1$ and $r+1$ are computed modulo the number of nodes in corresponding subtours). Taking these two edges out, we reduce the cost by

$$\text{cut}(q,r) = L(v_q, v_{q+1}) + L(v'_r, v'_{r+1}),$$

where $L(a,b)$ denotes the length of the edge (a,b). Then we replace them by a shorter connection:

$$\text{link}(q,r) = \min\{L(v_q, v'_r) + L(v_{q+1}, v'_{r+1}),\ L(v_q, v'_{r+1}) + L(v_{q+1}, v'_r)\}.$$

Thus, we seek

$$\min_{r,q}\{\ \text{link}(q,r) - \text{cut}(q,r)\ \}.$$

An additional heuristic was used to reduce the number of considered edge pairs in [334]. After making a connection, the number of subtours is decreased by one and the process is repeated (i.e., a subtour with the smallest number of edges is selected). Finally, a legal tour is produced. This is a legitimate offspring of parents A and B. The system with EAX produced results that are a small fraction of one percent above the optimum on a variety of test cases from 100 to over 3000 cities.

There are several interesting features of the proposed algorithm [334]. First, it's important to note that a very high selective pressure was used. An offspring replaces one of the parents only if it's better than both of them! If the offspring isn't better than both its parents, the edge assembly crossover operator is applied repeatedly (up to $N = 100$ times) in an attempt to produce an offspring that's better than its parents. The EAX may produce different offspring each time, as the first one is produced by a deterministic heuristic method for selecting the E-set, and for further attempts (if necessary), a random selection of E-set components is made. This method of iterative child generation (ICG) increases the probability of finding an improved offspring from the same set of parents.

The algorithm doesn't use a local search method like 2-opt in a direct manner, but the EAX operator does merge subtours by exploiting local information in choosing which edges should be used for cutting and linking. Moreover, the EAX operator incorporates an implicit mutation by introducing new edges into an offspring during the process of connecting subtours. An attempt was made [485] to evaluate the performance contribution of the various components of the algorithm:

> Edges not in the parents, or perhaps not even in the population, are introduced into offspring. We are now convinced that a good operator for the TSP must be able to introduce new edges into the offspring.

This is a significant departure from the usual design of constructing variation operators. All of the initial attempts (section 8.1) tried to preserve as many edges from the parents as possible!

In the following subsection we describe another algorithm for the TSP. It preserves many features of the above algorithm like competition between an offspring and its parent and the introduction of new edges, but it doesn't use any local information.

8.3.2 Inver-over operator

Suppose that we'd like to construct a "pure" evolutionary algorithm for tackling the TSP. That is, our algorithm won't use any local search at all. How could we design such a procedure so that it can be competitive with local methods? After examining many attempts to evolve solutions, we might conclude that the TSP should be addressed using a relatively strong selection pressure. We might also want to include a computationally efficient variation operator that produces offspring with the possibility for escaping local optima. Certainly, unary operators require less time than two-parent operators, but we might want to combine the advantages of both.

With all that in mind, consider the following evolutionary algorithm developed for the TSP [463] that has the following characteristics:

- Each individual competes only with its offspring.

- There's only one variation operator but this *inver-over* operator is adaptive.

- The number of times the operator is applied to an individual during a single generation is variable.

Such an algorithm can be perceived as a set of parallel hill-climbing procedures that preserve the spirit of the Lin-Kernighan algorithm. Each hill-climber performs a variable number of edge swaps. However, the inver-over operator has adaptive components: (1) the number of inversions applied to a single individual and (2) the segment to be inverted is determined by another randomly selected individual. So it's also possible to view this method as an evolutionary algorithm with a strong selective pressure and an adaptive variation operator.

procedure inver-over
random initialization of the population P
while (not satisfied termination-condition) **do**
begin
  for each individual $S_i \in P$ **do**
  begin
    $S' \leftarrow S_i$
    select (randomly) a city c from S'
    repeat
      if $(rand() \leq p)$
        select the city c' from the remaining cities in S'
      else
        select (randomly) an individual from P
        assign to c' the city 'next' to the city c in the selected individual
      if (the next city or the previous city of city c in S' is c')
        exit from repeat loop
      invert the section from the next city of city c to the city c' in S'
      $c \leftarrow c'$
    if $(eval(S') \leq eval(S_i))$
      $S_i \leftarrow S'$
  end
end

Fig. 8.18. The outline of the inver-over algorithm

Figure 8.18 provides a more detailed description of the whole inver-over algorithm in general and of the proposed operator in particular. With a low probability,[7] p, the second city for inversion is selected randomly from within

---

[7]Interestingly, experimental results indicated that the value of this parameter was independent of the number of cities in a test case. The function $rand()$ in figure 8.18 generates a random floating-point value from the range $[0..1]$.

the same individual. This is necessary. Without a possibility of generating new connections, the algorithm would only search among connections between cities that are present in the initial population. If $rand() > p$, a randomly selected mate provides information required to accomplish the inversion. In this way, the inversion operator actually resembles crossover because part of the pattern (at least two cities) of the second individual appears in the offspring.

Let's illustrate a single iteration of this operator on the following example. Assume that the current individual S' is

$$S' = (2, 3, 9, 4, 1, 5, 8, 6, 7),$$

and the current city c is 3. If the generated random number $rand()$ doesn't exceed p, another city c' from the same individual S' is selected (say, c' is 8), and the appropriate segment is inverted, producing the offspring

$$S' \leftarrow (2, 3, 8, 5, 1, 4, 9, 6, 7).$$

The position of the cutting points for the selected segment are located after cities 3 and 8. Otherwise (i.e., $rand() > p$), another individual is selected at random from the population. Suppose that it's $(1, 6, 4, 3, 5, 7, 9, 2, 8)$. We search this individual for the city c' that is "next" to city 3 (which is city 5), thus the segment for inversion in S' starts after city 3 and terminates after city 5. Consequently, the new offspring is

$$S' \leftarrow (2, 3, 5, 1, 4, 9, 8, 6, 7).$$

Note that a substring 3 – 5 arrived from the "second parent." Note also that in either case, the resulting string is intermediate in the sense that the above inversion operator is applied several times before an offspring is evaluated. This process terminates when, c', the city next to the current city c in a randomly selected individual is also the "next city" in the original individual. For example, say that after a few inversions, the current individual S' is

$$S' = (9, 3, 6, 8, 5, 1, 4, 2, 7),$$

and the current city c is 6. If $rand() > p$, a city "next" to city 6 is recovered from a randomly selected individual from the population. Suppose that it's city 8 (if $rand() \leq p$, a random city is selected, so it may also happen that city 8 was chosen). Since city 8 already follows city 3, the sequence of inversions terminates.

The results presented in [463] demonstrate the efficiency of the algorithm: (1) for a test case with 144 cities the average solution was only 0.04 percent above the optimum, (2) for a test case with 442 cities it was 0.63 percent above the optimum, and (3) for the test case with 2392 cities it was 2.66 percent worse than optimum. Moreover, for a random test case with 10,000 cities, the average solution stayed within 3.56 percent of the Held-Karp lower bound, whereas the best solution found in these 10 runs was less than 3 percent above this lower bound. The execution time for the algorithm was also reasonable: a few seconds for problems with up to 105 cities, under 3 minutes for the test case of 442 cities, and under 90 minutes for the test case with 2392 cities.

There are a few interesting observations that can be made on the basis of the experiments:

- The proposed system is probably the quickest evolutionary algorithm for the TSP developed so far. All of the other algorithms that are based on crossover operators provide much worse results in a much longer time.

- The proposed system has only three parameters: the population size, the probability p of generating random inversion, and the number of iterations in the termination condition. Most of the other evolutionary systems have many additional parameters and each requires some thought about how to set that parameter.

- The precision and stability of the evolutionary algorithm is quite good for relatively small test cases (almost 100 percent accuracy for every test case considered, up to 105 cities), and the computational time was also acceptable (3–4 seconds).

- The system introduces a new, interesting operator that combines features of inversion (or mutation) and crossover. Results of experiments reported in the previous section indicate that the inver-over evolutionary algorithm has generated results that are significantly better than those evolutionary algorithms that rely on random inversion.

- The probability parameter p (which was held constant at 0.02 in all experiments) determines a proportion of blind inversions and guided (adaptive) inversions.

8.4 Summary

The traveling salesman problem is a prototypical example of how an apparently simple problem can blossom with many unexpected challenges. It just doesn't seem that difficult to find an ordered list of cities such that the total path length is minimized. Why should that be so hard? And yet, the TSP, like many other problems in computer science is NP-hard. We have no polynomial-time algorithms that can generate perfect solutions. We can rely on enumeration, but only for small, trivial cases. Sometimes we can employ heuristics that are similar to branch and bound techniques, but even then, we have yet to be able to find any method that will find a perfect solution to the TSP that has a complexity that grows as a polynomial function of the number of cities. If you can develop such an algorithm, not only will you receive many awards from different computer science and mathematics societies, your name will be permanently etched in the history books! And you might make a little money from it on the side, too.

In the absence of perfect algorithms that can generate answers fast enough to be useful, a great deal of effort has been spent trying to generate nearly optimum solutions quickly. Evolutionary algorithms have been explored in depth in this regard. The TSP presents a straightforward encoding problem and scoring function. Once you've chosen your representation, your main obstacle is the choice of variation operators.

As we've seen in this review, most of the attention has been devoted to two-parent recombination operators. The imaginative efforts that have been aimed in this direction are truly impressive. This is one of the great aspects of evolutionary computation. It's so flexible that it can treat any representation and evaluation function. You are free to determine what seems right for your problem. You're also free then to invent the variation operators that will transform parent solutions into offspring. Without having to be concerned with evaluation functions that are smooth and continuous, or even doubly differentiable, and without having to encode solutions into some fixed data structure, the opportunities for you to display your creativity are nearly endless.

Why have so many efforts been directed at the use of two-parent recombination? Actually, this is a historical handicap. The emphasis on two-parent variation operators seems mostly to have been a result of having previous researchers also emphasize these operators. There is something "sexy" about having solutions that reside in the computer have "sex," but that doesn't mean such variation operators will be the best choice. In the massive effort to invent ways to crossover (just) two parent tours in the TSP, other useful variation operators have until recently been overlooked (e.g., inver-over). The possibility for using more than two parents when generating offspring has received relatively little attention [115, 119].[8] Experiments have shown that using more than two parents can offer advantages in some problems, and that even forgoing recombination altogether can also be beneficial [112].

The central point, and this is a critically important point, is that as a problem solver, you have to take responsibility for your own approach to solving the problem at hand. It's not enough for you to say, "Well, so-and-so used the xyx-algorithm, so that's what I'll do too." Or worse, "Everyone is using the xyz-algorithm — if I don't use it, I'll look dumb." In fact, these are truly cardinal sins. You have a duty to search for better ways to solve problems and to question the assumptions that others have made before you.

It's by this means that you will invent novel techniques for addressing problems. Think about ways to incorporate knowledge about the problem into the search for a solution. Perhaps there are local operators that can be assimilated. Perhaps an evolutionary algorithm can assist in overcoming local optima. Perhaps you can invent some new operator, like inver-over. Perhaps no searching is necessary because there's a closed-form analytical solution! Perhaps . . .

---

[8]The idea of using more than two parents in recombination actually arose in the 1960s [253, 58, 161].

IX. Who Owns the Zebra?

The real world often poses what are called *constraint satisfaction problems*, where the task is to find a feasible solution. In other words, we don't have to optimize anything, but we have to satisfy all of the problem's constraints instead.

One of the best-known examples of a constraint satisfaction problem is the famous "Zebra" problem. It goes like this: There are five houses, each of a different color and inhabited by men of different nationalities, with different pets, favorite drinks, and cars. Moreover,

1. The Englishman lives in the red house.

2. The Spaniard owns the dog.

3. The man in the green house drinks cocoa.

4. The Ukrainian drinks eggnog.

5. The green house is immediately to the right (your right) of the ivory house.

6. The owner of the Oldsmobile also owns snails.

7. The owner of the Ford lives in the yellow house.

8. The man in the middle house drinks milk.

9. The Norwegian lives in the first house on the left.

10. The man who owns the Chevrolet lives in the house next to the house where the man owns a fox.

11. The Ford owner's house is next to the house where the horse is kept.

12. The Mercedes-Benz owner drinks orange juice.

13. The Japanese drives a Volkswagen.

14. The Norwegian lives next to the blue house.

The question is ... who owns the zebra? Furthermore, who drinks water?

Several sentences provide direct information. For example, sentence 8 says that milk is consumed in house 3 (i.e., the middle house). Sentence 9 tells us that the Norwegian lives in house 1 (i.e., the first house on the left). Sentence 14 indicates that house 2 is blue (as this is the only house next to the Norwegian). So, the current knowledge we have is

House	1	2	3	4	5
Color		blue			
Drink			milk		
Country	Norwegian				
Car					
Pet					

Two other sentences are of immediate interest. Sentence 1 says that the Englishman lives in the red house, and sentence 5 says that the green house is immediately to the right of the ivory house. The conclusions are that

- House 1 is not red (because it's occupied by the Norwegian).

- Either houses 3 and 4 or houses 4 and 5 are ivory and green, respectively.

Thus house 1 must be yellow. Because of that, the Norwegian owns the Ford (sentence 7) and the horse is kept in house 2 (sentence 11). Thus, our current knowledge is

House	1	2	3	4	5
Color	yellow	blue			
Drink			milk		
Country	Norwegian				
Car	Ford				
Pet		horse			

From here, we can consider two cases, as there are only two possible sequences of colors for houses 3, 4, and 5:

- Case 1: (ivory, green, red),

- Case 2: (red, ivory, green).

Let's consider these two cases in turn. In case 1, we can infer additional information. The Englishman lives in house 5, as it is red (sentence 1), cocoa is drunk in house 4, because it's green (sentence 3), the Ukrainian drinks eggnog (sentence 4), so he must live in house 2 (as the Norwegian and Englishman live in houses 1 and 5, and milk and cocoa are consumed in houses 3 and 4), the Englishman owns a Mercedes and drinks orange juice (sentence 12, as orange juice is drunk either in house 1 or 5, but the Norwegian from house 1 owns a Ford). Thus, in case 1 our current knowledge is

House	1	2	3	4	5
Color	yellow	blue	ivory	green	red
Drink		eggnog	milk	cocoa	orange
Country	Norwegian	Ukrainian			Englishman
Car	Ford				Mercedes
Pet		horse			

And now it's clear that this case doesn't lead to a solution! To see this, let's investigate who the owner of the Oldsmobile is? It's not the Norwegian nor the Englishman, as we already know what they own. It is not the Japanese, as sentence 13 says that he drives a Volkswagen. It is not the Ukrainian, as the Oldsmobile owner owns snails (sentence 6) and the Ukrainian owns a horse. And it can't be the Spaniard because he owns a dog (sentence 2). Thus, we've reached a contradiction. In case 1, it's impossible to satisfy the remaining constraints of the problem. Thus, case 2 must hold true. Now we know that the colors of houses 3, 4, and 5 are red, ivory, and green, respectively. Again, sentences 1 and 3 give us additional pieces of certain information and our current knowledge is

House	1	2	3	4	5
Color	yellow	blue	red	ivory	green
Drink			milk		cocoa
Country	Norwegian		Englishman		
Car	Ford				
Pet		horse			

At this stage we can reason as follows. As the Ukrainian drinks eggnog (sentence 4), he must live either in house 2 or 4 (cocoa's drunk in house 5). But if the Ukrainian lives in house 4, then the Spaniard (who owns a dog by sentence 2) must live in house 5, and the Japanese must live in house 2. Also, orange juice must be drunk in house 2 (sentence 12) whose inhabitant drives a Mercedes. This is a contradiction because the Japanese owns a Volkswagen! We conclude then that the Ukrainian must live in house 2 and the current information we have is

House	1	2	3	4	5
Color	yellow	blue	red	ivory	green
Drink		eggnog	milk		cocoa
Country	Norwegian	Ukrainian	Englishman		
Car	Ford				
Pet		horse			

From this stage it's now easy to find the solution. As the owner of Mercedes-Benz drinks orange juice (sentence 12), he must live in house 4. As the Japanese owns a Volkswagen, he must live in house 5. Thus the Spaniard (who also owns a dog) lives in house 4. At this stage we know that

House	1	2	3	4	5
Color	yellow	blue	red	ivory	green
Drink		eggnog	milk	orange	cocoa
Country	Norwegian	Ukrainian	Englishman	Spaniard	Japanese
Car	Ford			Mercedes	Volkswagen
Pet		horse		dog	

Sentence 6 says that the Oldsmobile owner also owns snails, so he must live in house 3. Sentence 10 allows us to place the Chevrolet and the fox. It's clear now that the Japanese owns the zebra and the Norwegian drinks water:

House	1	2	3	4	5
Color	yellow	blue	red	ivory	green
Drink	water	eggnog	milk	orange	cocoa
Country	Norwegian	Ukrainian	Englishman	Spaniard	Japanese
Car	Ford	Chevrolet	Oldsmobile	Mercedes	Volkswagen
Pet	fox	horse	snails	dog	zebra

And you thought we'd never get there!

Another way of looking at this puzzle is a bit more formal. There are 15 variables of the problem:

a_i — the color of house i,
b_i — the drink consumed in house i,
c_i — the nationality of the person living in house i,
d_i — the car that's owned by the tenant of house i, and
e_i — the pet of the tenant of house i.

Each of these variables has a domain:

$$dom(a_i) = \{yellow(Y), blue(B), red(R), ivory(I), green(G)\},$$
$$dom(b_i) = \{water(W), eggnog(E), milk(M), orangejuice(O), cocoa(C)\},$$
$$dom(c_i) = \{Norwegian(N), Ukrainian(U), Englishman(E),$$
$$Spaniard(S), Japanese(J)\},$$
$$dom(d_i) = \{Ford(F), Chevrolet(C), Oldsmobile(O), Mercedes(M),$$
$$Volkswagen(V)\}, \text{ and}$$
$$dom(e_i) = \{fox(F), horse(H), snails(S), dog(D), zebra(Z)\}.$$

We can represent our knowledge as a table with possible domains for each variable. For example, after analyzing sentences 8, 9, 14, 1, and 5 we know[9]

House	1	2	3	4	5
Color	$\{Y\}$	$\{B\}$	$\{R, I\}$	$\{I, G\}$	$\{R, G\}$
Drink	$\{W\}$	$\{O, E\}$	$\{M\}$	$\{C, O, E\}$	$\{C, O, E\}$
Country	$\{N\}$	$\{U, J\}$	$\{E, S, J\}$	$\{S, U, J\}$	$\{E, S, U, J\}$
Car	$\{F\}$	$\{M, V, C\}$	$\{O, V, C\}$	$\{M, O, V, C\}$	$\{M, O, V, C\}$
Pet	$\{F, Z\}$	$\{H\}$	$\{D, S, F, Z\}$	$\{D, S, F, Z\}$	$\{D, S, F, Z\}$

[9]Note that by filling the table with the domains of the appropriate variables, we know immediately that the Norwegian drinks water. Finding the owner of zebra is more difficult.

The remaining constraints are interpreted as constraints for values of the appropriate variables. For example, sentence 2 becomes

if $c_i = S$ then $e_i = D$ (for $i = 2, 3, 4, 5$).

So the task is to select appropriate values for each attribute (color, drink, country, car, and pet) and each house, so that all of the constraints are satisfied.

It's good to practice on constraint satisfaction problems like this because it forces you to think logically, considering what is possible and what is impossible. These problems also seem to show up on graduate school entrance exams, so remember the old boy scout motto: Be prepared! And if you've already finished school, then hey, aren't you glad that's over with?!

9. Constraint-Handling Techniques

> Misery acquaints a man with strange bedfellows.
>
> Shakespeare, *The Tempest*

Every real-world problem poses constraints. You can't get away from them. It's only the textbooks that allow you to solve problems in the absence of constraints. Dhar and Ranganathan [102] wrote:

> Virtually all decision making situations involve constraints. What distinguishes various types of problems is the form of these constraints. Depending on how the problem is visualized, they can arise as rules, data dependencies, algebraic expressions, or other forms.

We would only amend this by removing the word "virtually." In the most extreme case, if a real-world problem is worth solving, then it is worth solving within your lifetime, or at least the foreseeable life span of the human race. The eventual nova of our sun puts a constraint on everything, and a very final one at that.

Back closer to home, we can face everyday problems with very pesky constraints. We acknowledged the importance of including constraints in problem-solving methods right from the start in chapter 1. Here we now concentrate on a variety of constraint-handling techniques that you might incorporate into evolutionary algorithms. In this case we have the potential of treating both feasible and infeasible solutions simultaneously within a single evolving population.

In evolutionary computation methods the evaluation function serves as the main link between the problem and the algorithm by rating individual solutions in the population. Superior individuals are usually given higher probabilities for survival and reproduction. It's crucial that we define the evaluation function to capture and characterize the problem in a reasonable way. This can be a significant challenge when facing the possibility of having infeasible solutions. Our final answer must be a *feasible* solution, otherwise it's really no *solution* at all. It might be useful, however, to operate on infeasible solutions while searching for better feasible solutions. Finding a proper evaluation measure for feasible and infeasible individuals is of paramount importance. It can mean the difference between success or failure.

Constraints are often subtle. For example, very early in our mathematical education, we're exposed to a problem that reads

> A snail is climbing a wooden post that is ten meters high. During the day the snail climbs up five meters, but during the night it falls asleep and slides down four meters. How many days will it take the snail to climb to the top of the post?

Most children (and even some adults!) reason that in 24 hours (i.e., day and night) the total "gain" of snail is precisely one meter. Since the post is ten meters high, it takes ten days to reach the top.

The above reasoning is valid, but only for the first five days because one of the constraints is violated if this line of reasoning is applied after that. The constraint is that the post is *only* ten meters high! If the snail sits at the six-meter mark at the beginning of the day, there's no room to climb five additional meters. So, after five days and nights the snail is at the five-meter mark on the post and it will reach the top during the sixth day.

This simple constraint causes some difficulties in otherwise simple reasoning, but it needn't always be the case. Sometimes constraints are helpful and can guide you in the right direction. We'll explore some possibilities for taking advantage of constraints later in the chapter. First, let's tackle some of the general issues that are connected with handling constraints in evolutionary algorithms. Section 9.2 illustrates many of these issues in the domain of nonlinear programming problems (NLPs).

9.1 General considerations

When facing constrained optimization problems using evolutionary algorithms, it's very important to process infeasible individuals [311]. Particularly in real-world problems, you'll find it difficult to design operators that avoid them entirely while still being effective in locating useful feasible solutions. In general, a search space S consists of two disjoint subsets of feasible and infeasible subspaces, $\mathcal{F}$ and $\mathcal{U}$, respectively (see figure 9.1).

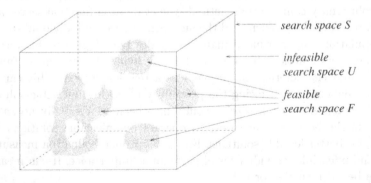

search space S

infeasible search space U

feasible search space F

Fig. 9.1. A search space and its feasible and infeasible parts

Here, we're not making any assumptions about these subspaces. In particular, they don't have to be convex or even connected (e.g., as shown in figure 9.1 where the feasible part $\mathcal{F}$ of the search space consists of four disjoint subsets). In solving optimization problems we search for a *feasible* optimum. During the search process we have to deal with various feasible and infeasible individuals. For example (see figure 9.2), at some stage of the evolution, a population might contain some feasible individuals (b, c, d, e, j) and infeasible individuals $(a, f, g, h, i, k, l, m, n, o)$, while the (global) optimum solution is marked by "X".

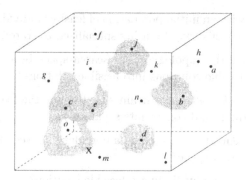

Fig. 9.2. A population of 15 individuals, $a - o$

Having to operate on both feasible and infeasible solutions can affect how we design various aspects of an evolutionary algorithm. Suppose we were using some form of elitist selection. Should we simply maintain the best *feasible* individual, or should we perhaps maintain an *infeasible* individual that happens to score better by our evaluation function? Questions sometimes arise in designing variation operators as well. Some operators might only be applicable to feasible individuals. But without a doubt, the major concern is the design of a suitable evaluation function to treat both feasible and infeasible solutions. This is far from trivial.

In general, we'll have to design two evaluation functions, $eval_f$ and $eval_u$, for the feasible and infeasible domains, respectively. There are many important questions to be addressed:

1. How should we compare two feasible individuals, e.g., "c" and "j" from figure 9.2? In other words, how should we design the function $eval_f$?

2. How should we compare two infeasible individuals, e.g., "a" and "n"? In other words, how should we design the function $eval_u$?

3. How are the functions $eval_f$ and $eval_u$ related? Should we assume, for example, that $eval_f(s) \succ eval_u(r)$ for any $s \in \mathcal{F}$ and any $r \in \mathcal{U}$? (The symbol $\succ$ is interpreted as "is better than," i.e., "greater than" for maximization and "less than" for minimization problems.)

4. Should we consider infeasible individuals harmful and eliminate them from the population?

5. Should we "repair" infeasible solutions by moving them into the closest point of the feasible space (e.g., the repaired version of "m" might be the optimum "X," figure 9.2)?

6. If we repair infeasible individuals, should we replace an infeasible individual by its repaired version in the population or should we instead only use a repair procedure for evaluations?

7. Since our aim is to find a feasible optimum solution, should we choose to penalize infeasible individuals?

8. Should we start with an initial population of feasible individuals and maintain the feasibility of their offspring by using specialized operators?

9. Should we change the topology of the search space by using decoders that might translate infeasible solutions into feasible solutions?

10. Should we extract a set of constraints that define the feasible search space and process individuals and constraints separately?

11. Should we concentrate on searching a boundary between feasible and infeasible parts of the search space?

12. How should we go about finding a feasible solution?

Several trends for handling infeasible solutions have emerged in evolutionary computation; most of these have only come about quite recently, making efforts from a decade or more ago almost obsolete (e.g., [391]). Even when using penalty functions to degrade the quality of infeasible solutions, this area of application now consists of several methods that differ in many important respects. Other newer methods maintain the feasibility of the individuals in the population by means of specialized operators or decoders, impose a restriction that any feasible solution is "better" than any infeasible solution, consider constraints one at the time in a particular order, repair infeasible solutions, use multiobjective optimization techniques, are based on cultural algorithms, or rate solutions using a particular coevolutionary model. We'll discuss these techniques by addressing issues 1 – 12 in turn.

9.1.1 Designing $eval_f$

For textbook problems, designing the evaluation function f is usually easy: it's usually given to you. For example, when treating most operations research problems, such as knapsack problems, the TSP, set covering, and so forth, the evaluation function comes part-and-parcel along with the problem. But when dealing with the real world, things aren't always so obvious. For example, in building an evolutionary system to control a mobile robot you'll need to evaluate the robot's paths. It's unclear, however, in figure 9.3, whether path 1 or path 2 should be favored taking into account their total distance, clearance from obstacles, and smoothness. Path 1 is shorter, but path 2 is smoother. When facing these sorts of problems, there's a need to incorporate heuristic measures into the evaluation function. Note that even the subtask of measuring the smoothness or clearance of a path isn't that simple.

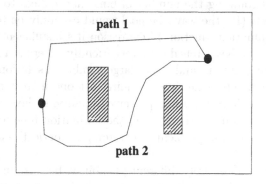

Fig. 9.3. Paths in an environment. Which path is better, the shorter one or the smoother one?

This is also the case in many design problems, where there are no clear formulas for comparing two feasible designs. Some problem-dependent heuristics are necessary in these cases, which should provide a numerical measure $eval_f(x)$ of the quality of a feasible individual x.

One of the best examples that illustrates the difficulties in evaluating even feasible individuals comes in the SAT problem. For a given formula in conjunctive normal form, say

$$F(\mathbf{x}) = (\overline{x_1} \vee x_2 \vee x_3) \wedge (\overline{x_1} \vee \overline{x_3}) \wedge (x_2 \vee x_3),$$

it's hard to compare the two feasible individuals $\mathbf{p} = (0,0,0)$ and $\mathbf{q} = (1,0,0)$ (in both cases $F(\mathbf{p}) = F(\mathbf{q}) = 0$). De Jong and Spears [95] examined a few possibilities. For example, you could define $eval_f$ to be a ratio of the number of conjuncts that evaluate to true. In that case,

$$eval_f(\mathbf{p}) = 0.666 \text{ and } eval_f(\mathbf{q}) = 0.333.$$

It's also possible [347] to change the Boolean variables x_i into floating-point numbers y_i and to assign

$$eval_f(\mathbf{y}) = |y_1 - 1||y_2 + 1||y_3 - 1| + |y_1 + 1||y_3 + 1| + |y_2 - 1||y_3 - 1|,$$

or

$$eval'_f(\mathbf{y}) = (y_1 - 1)^2(y_2 + 1)^2(y_3 - 1)^2 + (y_1 + 1)^2(y_3 + 1)^2 + (y_2 - 1)^2(y_3 - 1)^2.$$

In the above cases the solution to the SAT problem corresponds to a set of global minimum points of the evaluation function: the TRUE value of $F(\mathbf{x})$ is equivalent to the global minimum value 0 of $eval_f(\mathbf{y})$.

Similar difficulties arise in other combinatorial optimization problems such as bin packing. As noted in [134], where the task was to pack some maximum number of potentially different items into bins of various size, the obvious evaluation function of counting the number of bins that suffices to pack all of the items is insufficient. (By the way, the goal would certainly be to minimize the value from the evaluation function.) The reason it's insufficient is that the resulting "landscape" to be searched isn't very friendly. There is a relatively small number of optimum solutions and a very large number of solutions that evaluate to just one higher number (i.e., they require just one more bin). All of these suboptimal solutions have the same perceived quality, so how can we traverse the space when there's no guidance from the evaluation function on which directions to go? Clearly, the problem of designing a "perfect" $eval_f$ is far from trivial.

Actually, there's an entirely different possibility, because in many cases we don't have to define the evaluation function $eval_f$ to be a mapping to the real numbers. In a sense, we don't have to define it at all! It's really only mandatory if we're using a selection method that acts on the solutions' values, such as proportional selection. For other types of selection, however, all that might be required is an ordering relation that says that one solution is better than another. If an ordering relation $\succ$ handles decisions of the type "is a feasible individual x better than a feasible individual y?" then such a relation $\succ$ is sufficient for tournament and ranking selection methods, which require either selecting the best individual out of some number of individuals, or a rank ordering of all individuals, respectively.

Of course, we might have to use some heuristics to build such an ordering relation $\succ$. For example, for multiobjective optimization problems it's relatively easy to establish a partial ordering between individual solutions. Additional heuristics might be needed to order individuals that aren't comparable by the partial relation.

In summary, it seems that tournament and ranking selection offer some additional flexibility. Sometimes it's easier to rank order two solutions than to calculate a numeric quality for each and then compare those values. But with rank ordering methods, we still have to resolve the additional problems of comparing two infeasible individuals (see section 9.1.2) as well as comparing feasible and infeasible individuals (see section 9.1.3).

9.1.2 Design of $eval_u$

Designing the evaluation function for treating infeasible individuals is really very difficult. It's tempting to avoid it altogether by simply rejecting infeasible solutions (see section 9.1.4). Alternatively, we can extend the domain of the function $eval_f$ in order to handle infeasible individuals, i.e., $eval_u(x) = eval_f(x) \pm Q(x)$, where $Q(x)$ represents either a penalty for the infeasible individual x, or a cost for repairing such an individual (i.e., converting it to a feasible solution, see section 9.1.7). Another option is to design a separate evaluation function $eval_u$

that's independent of $eval_f$; however, we then have to establish some relationship between these two functions (see section 9.1.3).

Evaluating infeasible individuals presents quite a challenge. Consider a knapsack problem, where your goal is to get as many items as possible into a knapsack of some particular size. The amount by which you violate the capacity of the knapsack might not be a very good measure of that particular solution's "fitness" (see section 9.1.7). This also holds true for many scheduling and timetabling problems, and even for path planning problems. It's unclear whether path 1 or path 2 is better (figure 9.4) since path 2 has more intersection points with obstacles and is longer than path 1 but, using the above criteria, most infeasible paths are "worse" than the straight line (path 1).

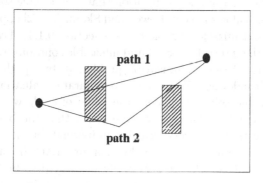

Fig. 9.4. Infeasible paths in an environment. Which path is better?

As with feasible solutions (section 9.1.1), it's possible to develop an ordering relation for infeasible individuals (as opposed to constructing $eval_u$). In both cases it's necessary to establish a relationship between the evaluations of feasible and infeasible individuals (section 9.1.3).

9.1.3 Relationship between $eval_f$ and $eval_u$

Let's say that we've decided to treat both feasible and infeasible individuals and evaluate them using two evaluation functions, $eval_f$ and $eval_u$, respectively. That is, a feasible individual x is evaluated using $eval_f(x)$ and an infeasible individual y is evaluated using $eval_u(y)$. Now we have to establish a relationship between these two evaluation functions.

As mentioned previously, one possibility is to design $eval_u$ based on $eval_f$, $eval_u(y) = eval_f(y) \pm Q(y)$, where $Q(y)$ represents either a penalty for the infeasible individual y, or a cost for repairing such an individual (see section 9.1.7). Alternatively, we could construct a global evaluation function $eval$ as

$$eval(p) = \begin{cases} q_1 \cdot eval_f(p) & if \ p \in \mathcal{F} \\ q_2 \cdot eval_u(p) & if \ p \in \mathcal{U}. \end{cases}$$

In other words, we have two weights, q_1 and q_2, that are used to scale the relative importance of $eval_f$ and $eval_u$.

Note that both of the methods provide the possibility for an infeasible solution to be "better" than a feasible solution. That is, it's possible to have a feasible individual x and an infeasible individual y such that $eval(y) \succ eval(x)$. This may lead the algorithm to converge to an infeasible solution. This phenomenon has led many researchers to experiment with dynamic penalty functions Q (see section 9.1.7) that increase the pressure on infeasible solutions as the evolutionary search proceeds. The problem of selecting $Q(x)$ (or weights q_1 and q_2), however, can be as difficult as solving the original problem itself.

On the other hand, some researchers [362, 501] have reported good results when working under the simple assumption that any feasible solution is always better than any infeasible solution. Powell and Skolnick [362] applied this heuristic for numerical optimization problems (see section 9.1.4). Feasible solutions were measured in the interval $(-\infty, 1)$ and infeasible solutions were measured in the the interval $(1, \infty)$ (for minimization problems). In a path planning problem, Xiao and Michalewicz [501] used two separate evaluation functions for feasible and infeasible solutions, where the value of $eval_u$ was increased (i.e., made less attractive) by adding a constant such that the best infeasible individual was worse than the worst feasible individual, but it's really doubtful that these sorts of categorical rules are always appropriate. In particular, should the feasible individual "b" (figure 9.2) really be judged to be better than the infeasible individual "m", which is "just next" to the optimal solution? A similar example can be drawn from the path planning problem. It's not at all clear that the feasible path 2 (see figure 9.5) deserves a better evaluation than path 1, which is infeasible! Unfortunately, there really are no generally useful heuristics for establishing relationships between the evaluation of feasible and infeasible solutions.

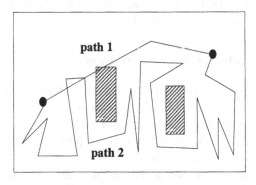

Fig. 9.5. Infeasible and feasible paths in an environment. Which path is better?

9.1.4 Rejecting infeasible solutions

The "death penalty" heuristic is a popular option in many evolutionary algorithms. Simply kill off all infeasible solutions at every step. Note that rejecting infeasible individuals does simplify things. For example, there's no need to design $eval_u$ and to compare it with $eval_f$.

Eliminating infeasible solutions may work well when the feasible search space is convex and constitutes a reasonable part of the whole search space. Otherwise this approach has serious limitations. For example, there are many search problems where a random sampling of solutions may generate an initial population that's entirely infeasible. It would therefore be essential to improve these solutions instead of reject them outright. Wiping the slate clean doesn't help here because you're right back where you started. Moreover, for many variation operators, it's quite often the case that the evolutionary algorithm can reach the optimum solution more quickly if it can "cross" an infeasible region (this is especially true in nonconvex feasible search spaces).

9.1.5 Repairing infeasible individuals

The idea of repairing infeasible solutions enjoys a particular popularity in the evolutionary computation community, and especially so for certain combinatorial optimization problems (e.g., TSP, knapsack problem, set covering problem). In these cases, it's relatively easy to repair an infeasible solution and make it feasible. Such a repaired version can be used either for evaluation, i.e.,

$$eval_u(y) = eval_f(x),$$

where x is a repaired (i.e., feasible) version of y, or it can replace the original individual in the population (perhaps with some probability, see section 9.1.6). Note that the repaired version of solution "m" (figure 9.2) might be the optimum "X".

The process of repairing infeasible individuals is related to a combination of learning and evolution (the so-called *Baldwin effect* [491]). Learning (as local search in general, and local search for the closest feasible solution, in particular) and evolution interact with each other. The fitness value of the local improvement is transferred to the individual. In that way a local search is analogous to the learning that occurs during a single generation.

The weakness of these methods lies in their problem dependence. Different repair procedures have to be designed for each particular problem. There are no standard heuristics for accomplishing this. It's sometimes possible to use a greedy method for repairing solutions, and at other times you might use some random procedure, or a variety of alternative choices. And then sometimes repairing infeasible solutions might become as complex a task as solving the original problem. This is often the case in nonlinear transportation problems, scheduling, and timetabling. Nevertheless, one evolutionary system for handling constrained problems, GENOCOP III (described in section 9.2) is based on repair algorithms.

9.1.6 Replacing individuals by their repaired versions

The idea of replacing repaired individuals is related to what's called *Lamarckian evolution* [491], which assumes that an individual improves during its lifetime and that the resulting improvements are coded back into the genetics of that individual. This is, of course, not the way nature works, but remember that we're designing evolutionary algorithms as computational procedures for solving problems, and we don't have to be constrained by the way nature works.

Orvosh and Davis [339] reported a so-called 5-percent-rule, which states that in many combinatorial optimization problems, when coupling a repair procedure with an evolutionary algorithm, the best results are achieved when 5 percent of repaired individuals replace their infeasible original versions. We know that this rule can't work in all problem domains, but this is at least a starting point to try when facing some new problem. In continuous domains, a new replacement rule is emerging. The GENOCOP III system (section 9.2) that uses a repair function appears to work well on certain problems when repairing 15 percent of the individuals. Higher and lower percentages have yielded inferior performance. Again, these types of settings are problem dependent and might even vary while a problem is being solved (indeed, you might think about how to tune these sorts of parameters; see chapter 10).

9.1.7 Penalizing infeasible individuals

The most common approach to handling infeasible solutions is to provide a penalty for their infeasibility by extending the domain of $eval_f$, and assuming that

$$eval_u(p) = eval_f(p) \pm Q(p),$$

where $Q(p)$ represents either a penalty for an infeasible individual p, or a cost for repairing such an individual. The primary question then concerns how we should design such a penalty function $Q(p)$. Intuition suggests that the penalty should be kept as low as possible, just above the threshold below which infeasible solutions are optimal (the so-called *minimal penalty rule* [280]), But it's often difficult to implement this rule effectively.

The relationship between an infeasible individual "p" and the feasible part $\mathcal{F}$ of the search space $\mathcal{S}$ plays a significant role in penalizing such individuals. An individual might be penalized just for being infeasible, for the "amount" of its infeasibility, or for the effort required to repair the individual. For example, for a knapsack problem with a weight capacity of 99 kilograms we may have two infeasible solutions that yield the same profit (which is calculated based on the items you can fit in the knapsack), where the total weight of all items taken is 100 and 105 kilograms, respectively. It's difficult to argue that the first individual with the total weight of 100 is better than the other one with a total weight of 105, despite the fact that for this individual the violation of the capacity constraint is much smaller than for the other one. The reason is that the first solution may involve five items each weighing 20 kilograms, and the second solution may contain (among other items) an item of low profit and a weight of six kilograms: Removing this item would yield a feasible solution, one that might be much better than any repaired version of the first individual. In these cases, the appropriate penalty function should consider the ease of repairing an individual, as well as the quality of the repaired version. Again, this is problem dependent.

It seems that the appropriate choice of the penalty method may depend on (1) the ratio between the sizes of the feasible search space and the whole search space, (2) the topological properties of the feasible search space, (3) the type of the evaluation function, (4) the number of variables, (5) the number of constraints, (6) the types of constraints, and (7) the number of active constraints at the optimum. Thus the use of penalty functions is not trivial and only some partial analysis of their properties is available. Also, a promising direction for applying penalty functions is the use of self-adaptive penalties: Penalty factors can be incorporated in the individual solutions and varied along with the components that comprise a solution (see chapter 10).

9.1.8 Maintaining a feasible population using special representations and variation operators

It seems that one of the most reasonable heuristics for dealing with the issue of feasibility is to use specialized representations and variation operators to maintain the feasibility of individuals in the population.

Several specialized systems have been developed for particular optimization problems. These evolutionary algorithms rely on unique representations and specialized variation operators. Some examples were described in [91] and many others are described here. For example, GENOCOP (section 9.2) assumes that the problem you face has only linear constraints and a feasible starting point (or a feasible initial population). A closed set of variation operators maintains the feasibility of solutions. There's no need to ever treat infeasible solutions when these conditions hold.

Very often such systems are much more reliable than other evolutionary techniques based on a penalty approach. Many practitioners have used problem-specific representations and specialized variation operators in numerical optimization, machine learning, optimal control, cognitive modeling, classical operation research problems (TSP, knapsack problems, transportation problems, assignment problems, bin packing, scheduling, partitioning, etc.), engineering design, system integration, iterated games, robotics, signal processing, and many others. The variation operators are often tailored to the representation (e.g., [167, 267]).

9.1.9 Using decoders

Using some form of decoder offers an interesting option when designing an evolutionary algorithm. In these methods, the data structure that represents an individual doesn't encode for a solution directly, but instead provides the instruction for how to build a feasible solution. For example, a sequence of items for the knapsack problem can be interpreted as "take an item if possible" — such an interpretation would always lead to feasible solutions. Consider the following scenario. We have to solve the 0–1 knapsack problem (the binary condition here applies to each item you might take — either you take it or you don't) with n items. The profit and weight of the i-th item are p_i and w_i, respectively. We can sort all of the items in decreasing order of p_i/w_is and interpret the binary string

$$\langle 11001100010011101010010101110101101\ldots0010 \rangle$$

as follows. Take the first item from the list (i.e., the item with the largest ratio of profit-to-weight) if the item fits in the knapsack. Continue with second, fifth, sixth, tenth, etc. items from the sorted list until the knapsack is full or there are no more items available. Note that the sequence of all 1s corresponds to a greedy solution. Any sequence of bits would translate into a feasible solution, and every feasible solution may have many possible codes. We can apply classical binary operators (crossover and mutation); any offspring is clearly feasible. The effect of the classic binary operators, however, isn't that clear, and we shouldn't necessarily expect their functionality to be the same on this binary problem as we would on some other binary problem.

It's important to point out several factors that should be taken into account while using decoders. Each decoder imposes a relationship T between a feasible solution and a decoded solution (see figure 9.6).

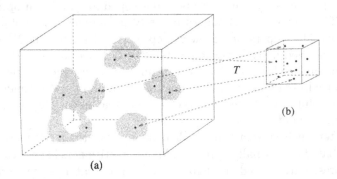

(a)

(b)

Fig. 9.6. Transformation T between solutions in (a) original and (b) decoder's space

It's important that several conditions are satisfied: (1) for each solution $s \in \mathcal{F}$ there is an encoded solution d, (2) each encoded solution d corresponds to a feasible solution s, and (3) all solutions in $\mathcal{F}$ should be represented by the same number of encodings d.[1] Additionally, it's reasonable to request that (4) the transformation T is computationally fast and (5) it has a locality feature in the sense that small changes in the encoded solution result in small changes in the solution itself. An interesting study on coding trees in evolutionary algorithms was reported by Palmer and Kershenbaum [343], where the above conditions were formulated.

9.1.10 Separating individuals and constraints

Separating out individuals and constraints is a general and interesting heuristic. The first possibility includes the use of multiobjective optimization methods, where the evaluation function f and constraint violation measures f_j (for m constraints) constitute an $(m+1)$-dimensional vector $\mathbf{v}$:

$$\mathbf{v} = (f, f_1, \ldots, f_m).$$

Using some multiobjective optimization method, we can attempt to minimize its components. An ideal solution x would have $f_j(x) = 0$ for $1 \leq j \leq m$ and $f(x) \leq f(y)$ for all feasible y (minimization problems). Surry [458] presented a successful implementation of this approach.

Another heuristic is based on the idea of handling constraints in a particular order. Schoenauer and Xanthakis [416] called this method a "behavioral memory" approach. We discuss this method in section 9.2.3.

9.1.11 Exploring boundaries between feasible and infeasible parts of the search space

One of the most recently developed approaches for constrained optimization is called *strategic oscillation*. This was originally proposed in conjunction with the evolutionary strategy of *scatter search*, and more recently has been applied to a variety of problem settings in combinatorial and nonlinear optimization (see, for example, the review of Glover [184]). The approach, which is a bit tricky to describe, is based on identifying a *critical level*, which for our purposes represents a boundary between feasibility and infeasibility, but which also can include such elements as a stage of construction or a chosen interval of values for a functional. In the feasibility–infeasibility context, the basic strategy is to approach and cross the feasibility boundary by a design that is implemented either by adaptive penalties and "inducements" (which are progressively relaxed

---

[1] As observed by Davis [90], however, the requirement that all solutions in $\mathcal{F}$ should be represented by the same number of decodings seems overly strong. There are cases in which this requirement might be suboptimal. For example, suppose we have a decoding and encoding procedure that makes it impossible to represent suboptimal solutions, and which encodes the optimal one. This might be a good thing. An example would be a graph coloring order-based structure with a decoding procedure that gives each node its first legal color. This representation couldn't encode solutions where some nodes that could be colored in fact weren't colored, but this is a good thing!

or tightened according to whether the current direction of search is to move deeper into a particular region or to move back toward the boundary) or by simply employing modified gradients or subgradients to progress in the desired direction. Within the context of neighborhood search, the rules for selecting moves are typically amended to take account of the region traversed and the direction of traversal. During the process of repeatedly approaching and crossing the feasibility frontier from different directions, the possibility of retracing a prior trajectory is avoided by mechanisms of memory and probability.

The application of different rules — according to region and direction — is generally accompanied by crossing a boundary to different depths on different sides. An option is to approach and retreat from the boundary while remaining on a single side, without crossing. One-sided oscillations are especially relevant in a variety of scheduling and graph theory settings, where a useful structure can be maintained up to a certain point and then is lost (by running out of jobs to assign, or by going beyond the conditions that define a tree, tour, etc.). In these cases, a constructive process for building to the critical level is accompanied by a destructive process for dismantling the structure.

It's frequently important in strategic oscillation to spend additional time searching regions that are close to the boundary. This may be done by inducing a sequence of tight oscillations about the boundary as a prelude to each larger oscillation to a greater depth. If greater effort is allowed for executing each move, the method may use more elaborate moves (such as various forms of "exchanges") to stay at the boundary for longer periods. For example, such moves can be used to proceed to a local optimum each time a critical proximity to the boundary is reached. A strategy of applying such moves at additional levels is suggested by a *proximate optimality principle*, which states roughly that good constructions at one level are likely to be close to good constructions at another. Approaches that embody such ideas may be found, for example, in [176, 256, 483, 185].

A similar approach was proposed in evolutionary algorithms where variation operators can be designed to search the boundary (and only the boundary) between feasible and infeasible solutions [318]. The general concept for boundary operators is that all individuals of the initial population lie on the boundary between feasible and infeasible areas of the search space and the operators are closed with respect to the boundary. Then all the offspring are at the boundary points as well. As these operators were proposed in the context of constrained parameter optimization, we discuss this approach in section 9.2.

9.1.12 Finding feasible solutions

There are problems for which any feasible solution would be of value. These are tough problems! Here, we really aren't concerned with any optimization issues (finding *the best* feasible solution) but instead we have to find *any* point in the feasible search space $\mathcal{F}$. These problems are called *constraint satisfaction problems*. A classical example of a problem in this category is the well-known

N-queens problem, where the task is to position N queens on a chess board with N rows and N columns in such a way that no two queens attack each other. Either you can do it, or you can't. There's no inbetween.

Some evolutionary algorithms have been designed to tackle these problems, typically by keeping track of the number of constraints that are violated and using that to assay the quality of the evolved solutions (e.g., [52]). In addition, some have tried so-called *scanning* recombination operators that can use pieces from any solution in the population to construct a new solution, thereby potentially using all of the available building blocks [116].

Paredis experimented with two different approaches to constraint satisfaction problems. The first approach [348, 349] was based on a clever representation of individuals where each component was allowed to take on values from the search domain, as well as an additional value of '?,' which represented choices that remained undecided. The initial population consisted of strings that possessed all ?s. A selection-assignment-propagation cycle then replaced some ? symbols by values from the appropriate domain (the assignment is checked for consistency with the component in question). The quality of such partially-defined individuals was defined as the value of the evaluation function of the best complete solution found when starting the search from a given partial individual. Variation operators were extended to incorporate a repair process (a constraint-checking step). This system was implemented and executed on several N-queens problems [349] as well as some scheduling problems [348].

In the second approach, Paredis [350] investigated a coevolutionary model, where a population of potential solutions coevolves with a population of constraints. Fitter solutions satisfy more constraints, whereas fitter constraints are violated by more solutions. This means that individuals from the population of solutions are considered from the whole search space, and that there's no distinction between feasible and infeasible individuals. The evaluation of an individual is determined on the basis of constraint violation measures f_js; however, better f_js (e.g., active constraints) contribute more towards the evaluation of solutions. The approach was tested on the N-queens problem and compared with other single-population approaches [351, 350].

9.2 Numerical optimization

There have been many efforts to use evolutionary algorithms for constrained numerical optimization [310, 313]. They can be grouped into five basic categories:

1. Methods based on preserving the feasibility of solutions.

2. Methods based on penalty functions.

3. Methods that make a clear distinction between feasible and infeasible solutions.

4. Methods based on decoders.

5. Other hybrid methods.

We've just explored many of the issues involved in each of these categories. Now let's revisit and take a look at methods in more detail. Furthermore, for most of the methods enumerated here, we'll provide a test case and the result of the method on that case.

9.2.1 Methods based on preserving the feasibility of solutions

There are two methods that fall in this category. Let's discuss them in turn.

Use of specialized operators. The idea behind the GENOCOP system[2] [316, 307] is based on specialized variation operators that transform feasible individuals into other feasible individuals. These operators are closed on the feasible part $\mathcal{F}$ of the search space. As noted earlier, the method assumes that we are facing only linear constraints and that we have a feasible starting point (or population). Linear equations are used to eliminate some variables, which are replaced as a linear combination of the remaining variables. Linear inequalities are updated accordingly. For example, when a particular component x_i of a solution vector $\mathbf{x}$ is varied, the evolutionary algorithm determines its current domain $dom(x_i)$ (which is a function of linear constraints and the remaining values of the solution vector $\mathbf{x}$) and the new value of x_i is taken from this domain (either with a uniform probability distribution or other probability distributions for nonuniform and boundary-type variations). Regardless of the specific distribution, it's chosen such that the offspring solution vector is always feasible. Similarly, arithmetic crossover, $a\mathbf{x} + (1-a)\mathbf{y}$, on two feasible solution vectors $\mathbf{x}$ and $\mathbf{y}$ always yields a feasible solution (for $0 \leq a \leq 1$) in convex search spaces. Since this evolutionary system assumes only linear constraints, this implies the convexity of the feasible search space $\mathcal{F}$.

GENOCOP gave surprisingly good results on many linearly constrained test functions [307]. For example, consider the following problem [137] to minimize the function

$$G1(\mathbf{x}) = 5x_1 + 5x_2 + 5x_3 + 5x_4 - 5\sum_{i=1}^{4} x_i^2 - \sum_{i=5}^{13} x_i,$$

subject to the constraints,

$$2x_1 + 2x_2 + x_{10} + x_{11} \leq 10,$$
$$2x_1 + 2x_3 + x_{10} + x_{12} \leq 10,$$
$$2x_2 + 2x_3 + x_{11} + x_{12} \leq 10,$$
$$-8x_1 + x_{10} \leq 0,$$

---

[2]GENOCOP stands for GEnetic algorithm for Numerical Optimization of COnstrained Problems and was developed before the different branches of evolutionary algorithms became functionally similar.

$$-8x_2 + x_{11} \leq 0,$$
$$-8x_3 + x_{12} \leq 0,$$
$$-2x_4 - x_5 + x_{10} \leq 0,$$
$$-2x_6 - x_7 + x_{11} \leq 0,$$
$$-2x_8 - x_9 + x_{12} \leq 0,$$

and bounds, $0 \leq x_i \leq 1$, $i = 1, \ldots, 9$, $0 \leq x_i \leq 100$, $i = 10, 11, 12$, $0 \leq x_{13} \leq 1$. The problem has 13 variables and 9 linear constraints. The function $G1$ is quadratic, with its global minimum at

$$\mathbf{x}^* = (1,1,1,1,1,1,1,1,1,3,3,3,1),$$

where $G1(\mathbf{x}^*) = -15$. Out of the 9 constraints, 6 are active at the global optimum (i.e., all except $-8x_1 + x_{10} \leq 0$, $-8x_2 + x_{11} \leq 0$, and $-8x_3 + x_{12} \leq 0$).

GENOCOP required fewer than 1,000 generations to arrive at the global solution.[3]

When facing linear constraints, and therefore a convex feasible space, the search effort is often reduced as compared with nonconvex spaces, where the feasible parts of the search space can be disjoint and irregular. Note as well that the method can be generalized to handle nonlinear constraints provided that the resulting feasible search space $\mathcal{F}$ is convex. The weakness of the method lies in its inability to deal with nonconvex search spaces (i.e., to deal with general nonlinear constraints).

Searching the boundary of the feasible region. Searching along the boundaries of the feasible–infeasible regions can be very important when facing optimization problems with nonlinear equality constraints or with active nonlinear constraints at the target optimum. Within evolutionary algorithms, there's a significant potential for incorporating specialized operators that can search such boundaries efficiently. We provide two examples of such an approach; for more details see [415].

Consider the numerical optimization problem introduced in chapter 1 (see section 1.1 and figure 1.2). The problem is to maximize the function

$$G2(\mathbf{x}) = |\frac{\sum_{i=1}^n \cos^4(x_i) - 2\prod_{i=1}^n \cos^2(x_i)}{\sqrt{\sum_{i=1}^n i x_i^2}}|,$$

subject to

$\prod_{i=1}^n x_i \geq 0.75$, $\sum_{i=1}^n x_i \leq 7.5n$, and bounds $0 \leq x_i \leq 10$ for $1 \leq i \leq n$.

Function $G2$ is nonlinear and its global maximum is unknown, lying somewhere near the origin. The problem has one nonlinear constraint and one linear constraint. The latter is inactive around the origin and will be omitted.

---

[3]For details on GENOCOP system including its operators, selection method, initialization routine, etc., see [307].

Some potential difficulties of solving this test case are illustrated in figure 1.2, where infeasible points were assigned a value of zero. The boundary between feasible and infeasible regions is defined by the equation $\prod x_i = 0.75$. The problem is difficult and no standard methods (deterministic or evolutionary) gave satisfactory results. This function was the first for which the idea of searching only the boundary was tested [318]. Specific initialization procedures and variation operators could be tailored to the problem owing to the simple analytical formulation of the constraint.

- **Initialization.** Randomly choose a positive value for x_i, and use its inverse as a value for x_{i+1}. The last variable is either 0.75 (when n is odd), or is multiplied by 0.75 (if n is even), so that the point lies on the boundary surface.

- **Crossover.** The variation operator of *geometrical crossover* is defined by $(x_i)(y_i) \rightarrow (x_i^\alpha y_i^{1-\alpha})$, with α randomly chosen in $[0,1]$. Figure 9.7 illustrates the possible offspring from two parents for all values of α.

- **Mutation.** Pick two variables randomly, multiply one by a random factor $q > 0$ and the other by $\frac{1}{q}$ (restrict q to respect the bounds on the variables).

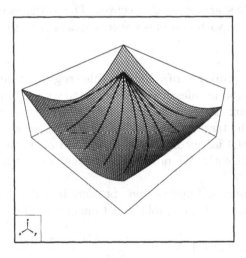

Fig. 9.7. Geometrical crossover on the hyperboloid around its center. When the parameter α goes from 0 to 1, the offspring of two parents defines the line joining them on the plot.

The simple evolutionary algorithm described above (with standard proportional selection extended by an elitist rule) gave outstanding results. For the case $n = 20$ the system reached the value of 0.80 in fewer than 4,000 generations (with a population size of 30, probability of crossover $p_c = 1.0$, and a probability of mutation $p_m = 0.06$) in all runs. The best value found, 0.803553, was better than the best values of any method discussed earlier, whereas the worst value found was 0.802964. Similarly, for $n = 50$, all results (in 30,000 generations) were better than 0.83, with the best being 0.8331937. It's interesting to note the importance of geometrical crossover. When using a fixed population size of 30 solutions, better performance was observed from higher (geometrical) crossover probabilities and lower rates of mutation.

Since evolutionary methods for searching boundaries of feasible regions constitute a new development in constrained optimization, let's illustrate this approach on another test case. The test problem suggests the use of a *sphere crossover* [318, 415]. The task is to maximize

$$G3(\mathbf{x}) = (\sqrt{n})^n \cdot \prod_{i=1}^n x_i,$$

where

$$\sum_{i=1}^n x_i^2 = 1 \quad \text{and} \quad 0 \le x_i \le 1 \quad \text{for} \quad 1 \le i \le n.$$

The function $G3$ has a global solution at $(x_1, \ldots, x_n) = (\frac{1}{\sqrt{n}}, \ldots, \frac{1}{\sqrt{n}})$ and the value of the function at this point is 1. The evolutionary algorithm uses the following components:

- **Initialization.** Randomly generate n variables y_i, calculate $s = \sum_{i=1}^n y_i^2$, and initialize an individual (x_i) by $x_i = y_i/s$ for $i \in [1, n]$.

- **Crossover.** The variation operator *sphere crossover* produces one offspring (z_i) from two parents (x_i) and (y_i) by

$$z_i = \sqrt{\alpha x_i^2 + (1 - \alpha) y_i^2} \quad i \in [1, n], \text{ with } \alpha \text{ randomly chosen in } [0, 1].$$

- **Mutation.** Similarly, the problem-specific mutation transforms (x_i) by selecting two indices $i \ne j$ and a random number p in $[0, 1]$, and setting

$$x_i \to p \cdot x_i \text{ and } x_j \to q \cdot x_j, \text{where } q = \sqrt{(x_i/x_j)^2(1 - p^2) + 1}.$$

The simple evolutionary algorithm (again with proportional selection coupled with elitism) described above gave very good results. For the case where $n = 20$, the system reached the value of 0.99 in fewer than 6,000 generations (with a population size of 30, probability of crossover $p_c = 1.0$, and a probability of mutation $p_m = 0.06$) in every trial. The best value found in 10,000 generations was 0.999866.

9.2.2 Methods based on penalty functions

The main efforts to treat constrained evolutionary optimization have involved the use of (extrinsic) penalty functions that degrade the quality of an infeasible solution. In this manner, the constrained problem is made unconstrained by using the modified evaluation function

$$eval(\mathbf{x}) = \begin{cases} f(\mathbf{x}), & \text{if } \mathbf{x} \in \mathcal{F} \\ f(\mathbf{x}) + penalty(\mathbf{x}), & \text{otherwise,} \end{cases}$$

where $penalty(\mathbf{x})$ is zero if no violation occurs, and is positive, otherwise (assuming the goal is minimization). The *penalty* function is usually based on some form of distance that measures how far the infeasible solution is from the feasible region $\mathcal{F}$, or on the effort to "repair" the solution, i.e., to transform it into $\mathcal{F}$. Many methods rely on a set of functions f_j $(1 \leq j \leq m)$ to construct the penalty, where the function f_j measures the violation of the j-th constraint:

$$f_j(\mathbf{x}) = \begin{cases} \max\{0, g_j(\mathbf{x})\}, & \text{if } 1 \leq j \leq q \\ |h_j(\mathbf{x})|, & \text{if } q + 1 \leq j \leq m, \end{cases}$$

(see section 3.2.1 for definitions of g_j and h_j). But these methods differ in many important details with respect to how the penalty function is designed and applied to infeasible solutions; some details are provided below.

Method of static penalties. Homaifar et al. [231] proposed that a family of intervals be constructed for every constraint that we face, where the intervals determine the appropriate penalty coefficient. The idea works as follows:

- For each constraint, create several (ℓ) levels of violation.

- For each level of violation and for each constraint, create a penalty coefficient R_{ij} $(i = 1, 2, \ldots, \ell, j = 1, 2, \ldots, m)$. Higher levels of violation require larger values of this coefficient (again, we assume minimization).

- Start with a random population of individuals (feasible or infeasible).

- Evolve the population. Evaluate individuals using

$$eval(\mathbf{x}) = f(\mathbf{x}) + \sum_{j=1}^{m} R_{ij} f_j^2(\mathbf{x}).$$

The weakness of the method arises in the number of parameters. For m constraints the method requires $m(2\ell + 1)$ parameters in total: m parameters to establish the number of intervals for each constraint, ℓ parameters for each constraint thereby defining the boundaries of the intervals (the levels of violation), and ℓ parameters for each constraint representing the penalty coefficients R_{ij}. In particular, for $m = 5$ constraints and $\ell = 4$ levels of violation, we need to set 45 parameters! And the results are certainly parameter dependent. For any particular problem, there might exist an optimal set of parameters for which we'd generate feasible near-optimum solutions, but this set might be very difficult to find.

A limited set of experiments reported in [308] indicates that the method can provide good results if the violation levels and penalty coefficients R_{ij} are tuned to the problem. For example, consider the problem [219] of minimizing a function of five variables:

$$G4(\mathbf{x}) = 5.3578547x_3^2 + 0.8356891x_1x_5 + 37.293239x_1 - 40792.141,$$

subject to three double inequalities,

$$0 \leq 85.334407 + 0.0056858x_2x_5 + 0.0006262x_1x_4 - 0.0022053x_3x_5 \leq 92$$
$$90 \leq 80.51249 + 0.0071317x_2x_5 + 0.0029955x_1x_2 + 0.0021813x_3^2 \leq 110$$
$$20 \leq 9.300961 + 0.0047026x_3x_5 + 0.0012547x_1x_3 + 0.0019085x_3x_4 \leq 25,$$

and bounds,

$$78 \leq x_1 \leq 102,\ 33 \leq x_2 \leq 45,\ 27 \leq x_i \leq 45 \text{ for } i = 3, 4, 5.$$

The best solution obtained in ten trials by Homaifar et al. [231] was

$$\mathbf{x} = (80.49, 35.07, 32.05, 40.33, 33.34)$$

with $G4(\mathbf{x}) = -30005.7$, whereas the optimum solution [219] is

$$\mathbf{x}^* = (78.0, 33.0, 29.995, 45.0, 36.776),$$

with $G4(\mathbf{x}^*) = -30665.5$. Two constraints (the upper bound of the first inequality and the lower bound of the third inequality) are active at the optimum.

Homaifar et al. [231] reported a value of 50 for the penalty coefficient that modified the violation of the lower boundary of the first constraint and a value of 2.5 for the violation of the upper boundary of the first constraint. It's reasonable to expect better performance if the penalty coefficients could be adapted to the problem, or if they could be adapted online while the problem is being solved (see chapter 10).

Method of dynamic penalties. In contrast to [231], Joines and Houck [247] applied a method that used dynamic penalties. Individuals are evaluated at each iteration (generation) t, by the formula

$$eval(\mathbf{x}) = f(\mathbf{x}) + (C \times t)^\alpha \sum_{j=1}^m f_j^\beta(\mathbf{x}),$$

where C, α, and β are constants. A reasonable choice for these parameters is $C = 0.5$, $\alpha = \beta = 2$ [247]. This method doesn't require as many parameters as the static method described earlier, and instead of defining several violation levels, the selection pressure on infeasible solutions increases due to the $(C \times t)^\alpha$ component of the penalty term: as t grows larger, this component also grows larger.

Joines and Houck [247] experimented with a test case [227] that involved minimizing the function

$$G5(\mathbf{x}) = 3x_1 + 0.000001x_1^3 + 2x_2 + 0.000002/3x_2^3$$

subject to

$$x_4 - x_3 + 0.55 \geq 0,\ x_3 - x_4 + 0.55 \geq 0,$$
$$1000\sin(-x_3 - 0.25) + 1000\sin(-x_4 - 0.25) + 894.8 - x_1 = 0$$
$$1000\sin(x_3 - 0.25) + 1000\sin(x_3 - x_4 - 0.25) + 894.8 - x_2 = 0$$
$$1000\sin(x_4 - 0.25) + 1000\sin(x_4 - x_3 - 0.25) + 1294.8 = 0$$
$$0 \leq x_i \leq 1200,\ i = 1, 2,\ \text{and}\ -0.55 \leq x_i \leq 0.55,\ i = 3, 4.$$

The best-known solution is

$$\mathbf{x}^* = (679.9453, 1026.067, 0.1188764, -0.3962336),$$

and $G5(\mathbf{x}^*) = 5126.4981$.

The best result reported in [247] (over many trials and various values of α and β) was evaluated at 5126.6653. Interestingly, no solution was fully feasible due to the three equality constraints, but the sum of the violated constraints was quite small (10^{-4}). It seems that sometimes this method penalizes solutions too much. The factor $(C \times t)^\alpha$ often grows too quickly to be useful. As a result, the evolution can stall out in local optima. Experiments with this method in [308] indicated that the best individual was often discovered in the early generations. Nevertheless, the method gave very good results for test cases that involved quadratic evaluation functions.

Method of annealing penalties. A different means for dynamically adjusting penalties can take a clue from the annealing algorithm: Perhaps we can anneal the penalty values using a parameter that is analogous to "temperature." This procedure was incorporated into the second version of GENOCOP (GENOCOP II) [312, 307]:

- Divide all the constraints into four subsets: linear equations, linear inequalities, nonlinear equations, and nonlinear inequalities.

- Select a random single point as a starting point. (The initial population consists of copies of this single individual.) This initial point satisfies the linear constraints.

- Set the initial temperature $\tau = \tau_0$.

- Evolve the population using

  $$eval(\mathbf{x}, \tau) = f(\mathbf{x}) + \frac{1}{2\tau}\sum_{j=1}^m f_j^2(\mathbf{x}),$$

- If $\tau < \tau_f$, stop; otherwise,

  - Decrease the temperature τ.

  - Use the best available solution to serve as a starting point for the next iteration.

  - Repeat the previous step of the algorithm.

This method distinguishes between linear and nonlinear constraints. The algorithm maintains the feasibility of all linear constraints using a set of closed operators that convert a feasible solution (feasible, that is, only in terms of linear constraints) into another feasible solution. The algorithm only considers active constraints at every generation and the selective pressure on infeasible solutions increases due to the decreasing values of the temperature τ.

The method has an additional unique feature: it starts from a single point.[4] It's relatively easy, therefore, to compare this method with other classical optimization methods whose performances are tested for a given problem from some starting point. The method requires starting and "freezing" temperatures, τ_0 and τ_f, respectively, and a cooling scheme to decrease the temperature τ. Standard values [312] are $\tau_0 = 1$, $\tau_{i+1} = 0.1 \cdot \tau_i$, with $\tau_f = 0.000001$. One of the test problems [137] explored by Michalewicz and Attia [312] was

$$\text{minimize } G6(\mathbf{x}) = (x_1 - 10)^3 + (x_2 - 20)^3,$$

subject to nonlinear constraints,

$$\begin{aligned} c1: \quad & (x_1 - 5)^2 + (x_2 - 5)^2 - 100 \geq 0, \\ c2: \quad & -(x_1 - 6)^2 - (x_2 - 5)^2 + 82.81 \geq 0, \end{aligned}$$

and bounds,

$$13 \leq x_1 \leq 100 \text{ and } 0 \leq x_2 \leq 100.$$

The known global solution is $\mathbf{x}^* = (14.095, 0.84296)$, and $G6(\mathbf{x}^*) = -6961.81381$ (see figure 9.8). Both constraints are active at the optimum. The starting point, which is not feasible, was $\mathbf{x}_0 = (20.1, 5.84)$.

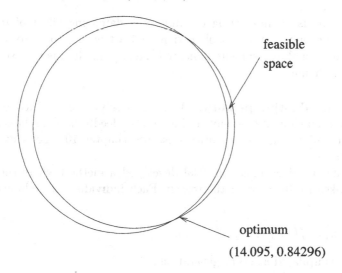

feasible space

optimum
(14.095, 0.84296)

Fig. 9.8. The feasible space for test case $G6$

GENOCOP II approached the optimum very closely at the 12th iteration. The progress of the system is reported in table 9.1.

Some experimental evidence is presented in [308] to the effect that linear constraints of a problem may prevent the system from moving closer to the

[4]This feature, however, isn't essential. The only important requirement is that the next population contains the best individual from the previous population.

Table 9.1. Progress of GENOCOP II on test case of function $G6$. For iteration 0, the best point is the starting point

Iteration number	The best point	Active constraints
0	(20.1, 5.84)	c_1, c_2
1	(13.0, 0.0)	c_1, c_2
2	(13.63, 0.0)	c_1, c_2
3	(13.63, 0.0)	c_1, c_2
4	(13.73, 0.16)	c_1, c_2
5	(13.92, 0.50)	c_1, c_2
6	(14.05, 0.75)	c_1, c_2
7	(14.05, 0.76)	c_1, c_2
8	(14.05, 0.76)	c_1, c_2
9	(14.10, 0.87)	c_1, c_2
10	(14.10, 0.86)	c_1, c_2
11	(14.10, 0.85)	c_1, c_2
12	(14.098, 0.849)	c_1, c_2

optimum. This is an interesting example of a damaging effect of restricting the population only to the feasible region, at least with respect to linear constraints. Additional experiments indicated that the method is very sensitive to the cooling scheme.

Methods of adaptive penalties. Instead of relying on a fixed decreasing temperature schedule, it's possible to incorporate feedback from the search into a means for adjusting the penalties (also see chapter 10). Let's discuss two possibilities.

Bean and Hadj-Alouane [36, 205] developed a method where the penalty function takes feedback from the search. Each individual is evaluated by the formula

$$eval(\mathbf{x}) = f(\mathbf{x}) + \lambda(t) \sum_{j=1}^{m} f_j^2(\mathbf{x}),$$

where $\lambda(t)$ is updated at every generation t:

$$\lambda(t+1) = \begin{cases} (1/\beta_1) \cdot \lambda(t), & \text{if } \mathbf{b}^i \in \mathcal{F} \text{ for all } t-k+1 \le i \le t \\ \beta_2 \cdot \lambda(t), & \text{if } \mathbf{b}^i \in \mathcal{S} - \mathcal{F} \text{ for all } t-k+1 \le i \le t \\ \lambda(t), & \text{otherwise,} \end{cases}$$

where $\mathbf{b}^i$ denotes the best individual in terms of function $eval$ at generation i, $\beta_1, \beta_2 > 1$ and $\beta_1 \neq \beta_2$ to avoid cycling. In other words, (1) if all of the best individuals in the last k generations were feasible the method decreases the penalty component $\lambda(t+1)$ for generation $t+1$, and (2) if all of the best individuals in the last k generations were infeasible then the method increases the penalties.

If there are some feasible and infeasible individuals as best individuals in the last k generations, $\lambda(t+1)$ remains without change.

The method introduces three additional parameters, β_1, β_2, and a time horizon, k. The intuitive reasoning behind Bean and Hadj-Alouane's method suggests that if constraints don't pose a problem, the search should continue with decreased penalties; otherwise, the penalties should be increased. The presence of both feasible and infeasible individuals in the set of best individuals in the last k generations means that the current value of the penalty component $\lambda(t)$ is set appropriately. At least, that's the hope.

A different method was offered by Smith and Tate [435] that uses a "near-feasible" threshold q_j for each constraint $1 \leq j \leq m$. These thresholds indicate the distances from the feasible region $\mathcal{F}$ that are considered "reasonable" (or, in other words, that qualify as "interesting" infeasible solutions because they are close to the feasible region). Thus the evaluation function is defined as

$$eval(\mathbf{x}, t) = f(\mathbf{x}) + F_{feas}(t) - F_{all}(t) \sum_{j=1}^{m} (f_j(\mathbf{x})/q_j(t))^k,$$

where $F_{all}(t)$ denotes the unpenalized value of the best solution found so far (up to generation t), $F_{feas}(t)$ denotes the value of the best feasible solution found so far, and k is a constant. Note, that the near-feasible thresholds $q_j(t)$ are dynamic. They are adjusted during the search based on feedback from the search. For example, it's possible to define $q_j(t) = q_j(0)/(1+\beta_j t)$ thus increasing the penalty component over time. To the best of our knowledge, neither of the adaptive methods described in this subsection has been applied to continuous nonlinear programming problems.

Death penalty method. The death penalty method simply rejects infeasible solutions, killing them immediately (see above). This simple method can provide quality results for some problems. For example, one test case in [308] included the following problem [227]. Minimize a function

$$G7(\mathbf{x}) = x_1^2 + x_2^2 + x_1 x_2 - 14x_1 - 16x_2 + (x_3 - 10)^2 + 4(x_4 - 5)^2 + (x_5 - 3)^2 + 2(x_6 - 1)^2$$
$$+ 5x_7^2 + 7(x_8 - 11)^2 + 2(x_9 - 10)^2 + (x_{10} - 7)^2 + 45,$$

subject to the constraints,

$$105 - 4x_1 - 5x_2 + 3x_7 - 9x_8 \geq 0,$$
$$-3(x_1 - 2)^2 - 4(x_2 - 3)^2 - 2x_3^2 + 7x_4 + 120 \geq 0,$$
$$-10x_1 + 8x_2 + 17x_7 - 2x_8 \geq 0,$$
$$-x_1^2 - 2(x_2 - 2)^2 + 2x_1 x_2 - 14x_5 + 6x_6 \geq 0,$$
$$8x_1 - 2x_2 - 5x_9 + 2x_{10} + 12 \geq 0,$$
$$-5x_1^2 - 8x_2 - (x_3 - 6)^2 + 2x_4 + 40 \geq 0,$$
$$3x_1 - 6x_2 - 12(x_9 - 8)^2 + 7x_{10} \geq 0,$$
$$-0.5(x_1 - 8)^2 - 2(x_2 - 4)^2 - 3x_5^2 + x_6 + 30 \geq 0,$$

and bounds,

$$-10.0 \leq x_i \leq 10.0, \quad i = 1, \ldots, 10.$$

The problem has three linear and five nonlinear constraints. The function $G7$ is quadratic and has its global minimum at

$$\mathbf{x}^* = (2.171996, 2.363683, 8.773926, 5.095984, 0.9906548,$$
$$1.430574, 1.321644, 9.828726, 8.280092, 8.375927),$$

where $G7(\mathbf{x}^*) = 24.3062091$.

Six out of the eight constraints are active at the global optimum (all except the last two). The GENOCOP system extended by the "death penalty" method generated respectable solutions, the best of which had the value of 25.653.

This method requires initialization with feasible solutions so comparisons to other methods can be tricky, but an interesting pattern emerged from the experiments analyzed in [308]. Simply rejecting infeasible methods performed quite poorly and wasn't as robust as other techniques (i.e., the standard deviation of the solution values was relatively high).

Segregated evolutionary algorithms. When tuning penalty coefficients, too small a penalty level leads to solutions that are infeasible because some penalized solutions will still exhibit an apparently higher quality than some feasible solutions. On the other hand, too high a penalty level restricts the search to the feasible region and thereby foregoes any short cut across the infeasible region. This often leads to premature stagnation at viable solutions of lesser value. One method for overcoming this concern was offered in [280].

The idea is to design two different penalized evaluation functions with static penalty terms p_1 and p_2. Penalty p_1 is purposely too small, while penalty p_2 is hopefully too high. Every individual in the current population undergoes recombination and mutation (or some form of variation). The values of the two evaluation functions $f_i(\mathbf{x}) = f(\mathbf{x}) + p_i(\mathbf{x})$, $i = 1, 2$, are computed for each resulting offspring (at no extra cost in terms of evaluation function calls), and two ranked lists are created according to the evaluated worth of all parents and offspring for each evaluation function.

Parents are selected for the next generation by choosing the best individual from each list, alternately, and then removing that individual from the list. The main idea is that this selection scheme will maintain two subpopulations: those individuals selected on the basis of f_1 will be more likely to lie in the infeasible region while those selected on the basis of f_2 will be more likely to lie in the feasible region. Overall, the evolutionary search can reach the optimum feasible solutions from both sides of the boundary between feasible and infeasible regions. This method gave excellent results in the domain of laminated design optimization [280], but has not yet, again to the best of our knowledge, been applied to continuous nonlinear programming problems.

9.2.3 Methods based on a search for feasible solutions

Some evolutionary methods emphasize a distinction between feasible and infeasible solutions. One method considers the problem constraints in sequence. Once a sufficient number of feasible solutions is found in the presence of one constraint, the next constraint is considered. Another method assumes that any feasible solution is better than any infeasible solution (as we discussed above). Yet another method repairs infeasible individuals. Let's take each of these examples in turn.

Behavioral memory method. Schoenauer and Xanthakis [416] proposed what they call a "behavioral memory" approach:

- Start with a random population of individuals (feasible or infeasible).

- Set $j = 1$ (j is a constraint counter).

- Evolve this population with $eval(\mathbf{x}) = f_j(\mathbf{x})$ until a given percentage of the population (a so-called flip threshold ϕ) is feasible for this constraint.[5]

- Set $j = j + 1$.

- The current population is the starting point for the next phase of the evolution, where $eval(\mathbf{x}) = f_j(\mathbf{x})$ (defined in the section 9.2.2). During this phase, points that don't satisfy one of the first, second, ..., or $(j-1)$-th constraints are eliminated from the population. The halting criterion is again the satisfaction of the j-th constraint using the flip threshold percentage ϕ of the population.

- If $j < m$, repeat the last two steps, otherwise ($j = m$) optimize the evaluation function, i.e., $eval(\mathbf{x}) = f(\mathbf{x})$, while rejecting infeasible individuals.

The method requires a sequential ordering of all constraints that are processed in turn. The influence of the order of constraints on the end results isn't very clear, and different orderings can generate different results both in the sense of total running time and accuracy. In all, the method requires three parameters: the sharing factor σ, the flip threshold ϕ, and a particular order of constraints. The method's very different from many others, and really is quite different from other penalty approaches since it only considers one constraint at a time. Also, the method concludes by optimizing using the "death penalty" approach, so it can't be neatly parceled into one or another category.

Schoenauer and Xanthakis [416] tried the procedure on the following problem. Maximize the function

$$G8(\mathbf{x}) = \frac{\sin^3(2\pi x_1) \cdot \sin(2\pi x_2)}{x_1^3 \cdot (x_1 + x_2)},$$

subject to the constraints,

---

[5] The method suggests using a so-called sharing scheme (see chapter 11) to maintain population diversity.

$$c1(\mathbf{x}) = x_1^2 - x_2 + 1 \le 0,$$
$$c2(\mathbf{x}) = 1 - x_1 + (x_2 - 4)^2 \le 0,$$

and bounds,

$$0 \le x_1 \le 10 \text{ and } 0 \le x_2 \le 10.$$

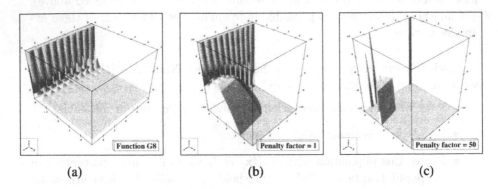

(a)   (b)   (c)

Fig. 9.9. Function $G8$ **(a)** together with two fitness landscapes that correspond to two different penalization parameters α; **(b)** a small value of α leaves the highest peaks outside the feasible region; **(c)** a large value of α makes the feasible region look like a needle in a haystack (all plots are truncated between -10 and 10 for the sake of visibility).

Function $G8$ has many local optima; the highest peaks are located along the x-axis, e.g., $G8(0.00015, 0.0225) > 1540$. In the feasible region, however, $G8$ has two maxima of almost equal fitness, 0.1. Figure 9.9 shows the penalized fitness landscape for different penalty parameters, i.e., plots of the function $G8(\mathbf{x}) - \alpha(c1(\mathbf{x})^+ + c2(\mathbf{x})^+)$ for different values of α. All of the values of the penalized function are truncated at -10 and 10 so that we can distinguish the values in the feasible region around 0. This situation makes any attempt to use penalty parameters a difficult task. Small penalty parameters leave the feasible region hidden among much higher peaks of the penalized fitness (figure 9.9b) while penalty parameters that are large enough to allow the feasible region to really stand out from the penalized fitness landscape imply vertical slopes of that region (figure 9.9c). It's difficult to envision an evolutionary algorithm climbing up the slopes of a vertical cliff. Finding the global maximum of the penalized fitness, therefore, requires a lucky move away from the huge and flat low-fitness region.

On the other hand, the behavioral memory method has no difficulty in localizing 80 percent of the population first into the feasible region for constraint $c1$ (step 1), and then into the feasible region for both constraints (step 2), as can be seen on figure 9.10. The optimization of $G8$ itself in that region is then straightforward (see the smooth bimodal fitness landscape in that region in figure 9.9). Throughout all the steps, however, you have to preserve the diversity of the population. The resulting population will be used as the basis for another

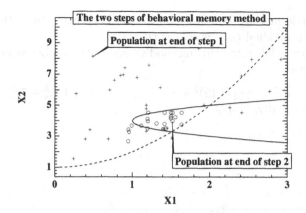

Fig. 9.10. Illustrating a sample run of behavioral memory method on $G8$: zoom around the feasible region. The *dashed curve* is constraint $c1$ and the *continuous curve* is constraint $c2$. The population at the end of the first step samples the feasible region for $c1$ while the population at the end of step 2 samples the feasible region for both $c1$ and $c2$.

evolution so it should sample the target feasible region as uniformly as possible. Since the evaluation function is multimodal in that feasible region, we have to take care to locate both peaks. This is achieved using a "sharing" scheme (see chapter 10), and it's even more of a concern when the feasible region comprises several distinct connected components, as shown in [416] on a similar problem with a slightly different constraint $c2$.

Method of superiority of feasible points. Powell and Skolnick [362] used a method that's based on a classic penalty approach, but with one notable exception. Each individual is evaluated by the formula

$$eval(\mathbf{x}) = f(\mathbf{x}) + r \sum_{j=1}^{m} f_j(\mathbf{x}) + \theta(t, \mathbf{x}),$$

where r is a constant; however, the original component $\theta(t, \mathbf{x})$ is an additional iteration-dependent function that influences the evaluations of infeasible solutions. The point is that the method distinguishes between feasible and infeasible individuals by adopting an additional heuristic (suggested earlier [391]). For any feasible individual $\mathbf{x}$ and any infeasible individual $\mathbf{y}$: $eval(\mathbf{x}) < eval(\mathbf{y})$, i.e., any feasible solution is better than any infeasible one. This can be achieved in many ways. One possibility is to set

$$\theta(t, \mathbf{x}) = \begin{cases} 0, & \text{if } \mathbf{x} \in \mathcal{F} \\ \max\{0, \max_{\mathbf{x} \in \mathcal{F}}\{f(\mathbf{x})\} \\ \quad - \min_{\mathbf{x} \in \mathcal{S} - \mathcal{F}}\{f(\mathbf{x}) + r \sum_{j=1}^{m} f_j(\mathbf{x})\}\}, & \text{otherwise.} \end{cases}$$

Infeasible individuals are penalized such that they can't be better than the worst feasible individual (i.e., $\max_{\mathbf{x} \in \mathcal{F}}\{f(\mathbf{x})\}$).[6]

Michalewicz [308] tested this method on one of the problems in [227]. Minimize the function

$$G9(\mathbf{x}) = (x_1 - 10)^2 + 5(x_2 - 12)^2 + x_3^4 + 3(x_4 - 11)^2 + 10x_5^6 + 7x_6^2 + x_7^4$$
$$-4x_6 x_7 - 10x_6 - 8x_7,$$

subject to the constraints,

$$127 - 2x_1^2 - 3x_2^4 - x_3 - 4x_4^2 - 5x_5 \geq 0,$$
$$282 - 7x_1 - 3x_2 - 10x_3^2 - x_4 + x_5 \geq 0,$$
$$196 - 23x_1 - x_2^2 - 6x_6^2 + 8x_7 \geq 0,$$
$$-4x_1^2 - x_2^2 + 3x_1 x_2 - 2x_3^2 - 5x_6 + 11x_7 \geq 0,$$

and bounds,

$$-10.0 \leq x_i \leq 10.0, \; i = 1, \ldots, 7.$$

The problem has four nonlinear constraints. The function $G9$ is nonlinear and has its global minimum at

$$\mathbf{x}^* = (2.330499, 1.951372, -0.4775414, 4.365726, -0.6244870,$$
$$1.038131, 1.594227),$$

where $G9(\mathbf{x}^*) = 680.6300573$. Two out of the four constraints are active at the global optimum (the first and last). The method of Powell and Skolnick showed reasonable performance. The best solution found was 680.934. The method may have some difficulties, however, when locating feasible solutions in other test problems in [308].

In a recent study [97], this approach was modified using tournament selection coupled with the evaluation function

$$eval(\mathbf{x}) = \begin{cases} f(\mathbf{x}), & \text{if } \mathbf{x} \text{ is feasible,} \\ f_{\max} + \sum_{j=1}^{m} f_j(\mathbf{x}), & \text{otherwise,} \end{cases}$$

where $f_{\max}$ is the function value of the worst feasible solution in the population. The main difference between this approach and Powell and Skolnick's approach is that here the evaluation function value is not considered in evaluating an infeasible solution. Additionally, a niching scheme (chapter 10) is introduced to maintain diversity among feasible solutions. Thus, the search focuses initially on finding feasible solutions and then, when an adequate sampling of feasible solutions have been found, the algorithm finds better feasible solutions by maintaining a diverse set of solutions in the feasible region. There's no need for penalty coefficients here because the feasible solutions are always evaluated to be better than infeasible solutions, and infeasible solutions are compared purely on the basis of their constraint violations. Normalizing the constraints $f_j(\mathbf{x})$ is suggested.

---

[6]Powell and Skolnick [362] achieved the same result by mapping evaluations of feasible solutions into the interval $(-\infty, 1)$ and infeasible solutions into the interval $(1, \infty)$. This difference in implementation isn't important for ranking and tournament selection methods.

Repairing infeasible individuals. GENOCOP III (the successor to the previous GENOCOP systems), incorporates the ability to repair infeasible solutions, as well as some of the concepts of coevolution [317]. As with the original GENOCOP (section 9.2.1), linear equations are eliminated, the number of variables is reduced, and linear inequalities are modified accordingly. All points included in the initial population satisfy linear constraints. Specialized operators maintain their feasibility in the sense of linear constraints from one generation to the next. We denote the set of points that satisfy the linear constraints by $\mathcal{F}_l \subseteq \mathcal{S}$.

Nonlinear equations require an additional parameter (γ) to define the precision of the system. All nonlinear equations $h_j(\mathbf{x}) = 0$ (for $j = q + 1, \ldots, m$) are replaced by a pair of inequalities:

$$-\gamma \leq h_j(\mathbf{x}) \leq \gamma.$$

Thus, we only deal with nonlinear inequalities. These nonlinear inequalities further restrict the set $\mathcal{F}_l$. They define the fully feasible part $\mathcal{F} \subseteq \mathcal{F}_l$ of the search space $\mathcal{S}$.

GENOCOP III extends GENOCOP by maintaining two separate populations where a development in one population influences the evaluations of individuals in the other. The first population P_s consists of so-called search points from $\mathcal{F}_l$ that satisfy the linear constraints of the problem. As mentioned earlier, the feasibility of these points, in the sense of linear constraints, is maintained by specialized operators (see section 9.2.1).

The second population P_r consists of so-called reference points from $\mathcal{F}$; these points are fully feasible, i.e., they satisfy *all* the constraints.[7] Reference points $\mathbf{r}$ from P_r, being feasible, are evaluated directly by the evaluation function, i.e., $eval(\mathbf{r}) = f(\mathbf{r})$. On the other hand, search points from P_s are "repaired" for evaluation and the repair process works as follows. Assume there's a search point $\mathbf{s} \in P_s$. If $\mathbf{s} \in \mathcal{F}$, then $eval(\mathbf{s}) = f(\mathbf{s})$, since $\mathbf{s}$ is fully feasible. Otherwise (i.e., $\mathbf{s} \notin \mathcal{F}$), the system selects[8] one of the reference points, say $\mathbf{r}$ from P_r, and creates a sequence of random points $\mathbf{z}$ from a segment between $\mathbf{s}$ and $\mathbf{r}$ by generating random numbers a from the range $\langle 0, 1 \rangle$: $\mathbf{z} = a\mathbf{s} + (1 - a)\mathbf{r}$.[9] Once a fully feasible $\mathbf{z}$ is found, $eval(\mathbf{s}) = eval(\mathbf{z}) = f(\mathbf{z})$.[10]

Additionally, if $f(\mathbf{z})$ is better than $f(\mathbf{r})$, then the point $\mathbf{z}$ replaces $\mathbf{r}$ as a new reference point in the population of reference points P_r. Also, $\mathbf{z}$ replaces $\mathbf{s}$ in the population of search points P_s with some probability of replacement p_r.

---

[7]If GENOCOP III has difficulties in locating such a reference point for the purpose of initialization, it prompts the user for it. In cases where the ratio $|\mathcal{F}|/|\mathcal{S}|$ of feasible points in the search space is very small, it may happen that the initial set of reference points consists of multiple copies of a single feasible point.

[8]Better reference points have greater chances of being selected. A nonlinear ranking selection method was used.

[9]Note that all such generated points $\mathbf{z}$ belong to $\mathcal{F}_l$.

[10]The same search point $\mathcal{S}$ can evaluate to different values in different generations due to the random nature of the repair process.

The structure of GENOCOP III is shown in figure 9.11 and the procedure for evaluating search points (which aren't necessarily fully feasible) from population P_s is given in figure 9.12.

procedure GENOCOP III
begin
$\quad t \leftarrow 0$
$\quad$ initialize $P_s(t)$
$\quad$ initialize $P_r(t)$
$\quad$ evaluate $P_s(t)$
$\quad$ evaluate $P_r(t)$
$\quad$ **while** (**not** termination-condition) **do**
$\quad$ **begin**
$\quad\quad t \leftarrow t + 1$
$\quad\quad$ select $P_s(t)$ from $P_s(t-1)$
$\quad\quad$ alter $P_s(t)$
$\quad\quad$ evaluate $P_s(t)$
$\quad\quad$ **if** $t \bmod k = 0$ **then**
$\quad\quad$ **begin**
$\quad\quad\quad$ alter $P_r(t)$
$\quad\quad\quad$ select $P_r(t)$ from $P_r(t-1)$
$\quad\quad\quad$ evaluate $P_r(t)$
$\quad\quad$ **end**
$\quad$ **end**
end

Fig. 9.11. The structure of GENOCOP III

Note that there's some asymmetry between processing a population of search points P_s and a population of reference points P_r. While we apply selection and variation operators to P_s every generation, population P_r is modified every k generations (some additional changes in P_r are possible during the evaluation of search points, see figure 9.12). The main reason behind this arrangement is the efficiency of the system. Searching within the feasible part of the search space $\mathcal{F}$ is treated as a background event. Note also, that the "selection" and "alternation" steps are reversed in the evolution loop for P_r. Since there's often a low probability of generating feasible offspring, at first the parent individuals reproduce and later the best feasible individuals (both parents and offspring) are selected for survival.

GENOCOP III uses the evaluation function only for judging the quality of fully feasible individuals, so the evaluation function isn't adjusted or altered, as we've seen with methods that are based on penalty functions. Only a few additional parameters are introduced (e.g., the population size of the reference points, the probability of replacement, the frequency of application of variation operators to the population of reference points, the precision γ). It always returns a feasible solution. A feasible search space $\mathcal{F}$ is searched (population P_r)

```
    procedure evaluate Ps(t)
    begin
        for each s ∈ Ps(t) do
            if s ∈ F
            then evaluate s (as f(s)) else
            begin
                select r ∈ Pr(t)
                generate z ∈ F
                evaluate s (as f(z))
                if f(r) > f(z) then replace r by z in Pr
                replace s by z in Ps with probability pr
            end
    end
```

Fig. 9.12. Evaluation of population P_s

by making references from the search points and by applying variation operators every k generations (see figure 9.11). The neighborhoods of better reference points are explored more frequently. Some fully feasible points are moved into the population of search points P_s (the replacement process), where they undergo additional transformation by specialized operators.

One of the most interesting parameters of the developed system is the probability of replacement p_r (replacement of s by z in the population of search points P_s, see figure 9.12). Experiments in [317] suggested that a replacement probability of $p_r = 0.2$ offered the best results on a selected test suite of problems. For example, the following problem from [227] proved very difficult for all of the prior methods that were tested in the literature. The problem is to minimize

$$G10(\mathbf{x}) = x_1 + x_2 + x_3,$$

subject to the constraints,

$$1 - 0.0025(x_4 + x_6) \geq 0,$$
$$1 - 0.0025(x_5 + x_7 - x_4) \geq 0,$$
$$1 - 0.01(x_8 - x_5) \geq 0,$$
$$x_1 x_6 - 833.33252 x_4 - 100 x_1 + 83333.333 \geq 0,$$
$$x_2 x_7 - 1250 x_5 - x_2 x_4 + 1250 x_4 \geq 0,$$
$$x_3 x_8 - 1250000 - x_3 x_5 + 2500 x_5 \geq 0,$$

and bounds,

$$100 \leq x_1 \leq 10000, \quad 1000 \leq x_i \leq 10000, \text{ for } i = 2, 3,$$
$$10 \leq x_i \leq 1000, \text{ for } i = 4, \ldots, 8.$$

The problem has three linear and three nonlinear constraints. The function $G10$ is linear and has its global minimum at

$$\mathbf{x}^* = (579.3167, 1359.943, 5110.071, 182.0174, 295.5985,$$
$$217.9799, 286.4162, 395.5979),$$

where $G10(\mathbf{x}^*) = 7049.330923$. All six constraints are active at the global optimum. For the above problem, GENOCOP III's best result was 7286.650 — better than the best result of the best system from those discussed in [308].

Similar performance was observed on two other problems, $G9$ (with 680.640) and $G7$ (with 25.883). Interestingly, the standard deviation of the results generated by GENOCOP III was very low. For example, for problem $G9$, all of the results were between 680.640 and 680.889. In contrast, other systems have produced a variety of results (between 680.642 and 689.660; see [308]). Of course, all of the resulting points $\mathbf{x}$ were feasible, which wasn't always the case with other systems (e.g., GENOCOP II produced a value of 18.917 for the problem $G7$ — the systems based on the methods of Homaifar et al. [231] and Powell and Skolnick [362] gave results of 2282.723 and 2101.367, respectively, for the problem $G10$, and these solutions weren't feasible).

9.2.4 Methods based on decoders

Decoders offer an interesting alternative for designing evolutionary algorithms, but they've only been applied in continuous domains recently [272, 273]. It's relatively easy to establish a one-to-one mapping between an arbitrarily convex feasible search space $\mathcal{F}$ and the n-dimensional cube $[-1,1]^n$ (see figure 9.13).

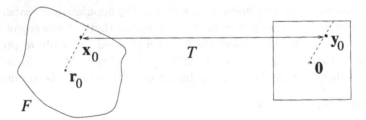

Fig. 9.13. A mapping T from a space $\mathcal{F}$ into a cube $[-1,1]^n$ (two-dimensional case)

Note that an arbitrary point (other than $\mathbf{0}$) $\mathbf{y}_0 = (y_{0,1}, \ldots, y_{0,n}) \in [-1,1]^n$ defines a line segment from the $\mathbf{0}$ to the boundary of the cube. This segment is described by

$$y_i = y_{0,i} \cdot t, \text{ for } i = 1, \ldots, n,$$

where t varies from 0 to $t_{max} = 1/\max\{|y_{0,1}|, \ldots, |y_{0,n}|\}$. For $t = 0$, $\mathbf{y} = \mathbf{0}$, and for $t = t_{max}$, $\mathbf{y} = (y_{0,1}t_{max}, \ldots, y_{0,n}t_{max})$ — a boundary point of the $[-1,1]^n$ cube. Consequently, the corresponding feasible point (to $\mathbf{y}_0 \in [-1,1]^n$) $\mathbf{x}_0 \in \mathcal{F}$ (with respect to some reference point[11] $\mathbf{r}_0$) is defined as

---

[11] A reference point $\mathbf{r}_0$ is an arbitrary internal point of the convex set $\mathcal{F}$. Note, that it's not necessary for the feasible search space $\mathcal{F}$ to be convex. Assuming the existence of the

$$\mathbf{x}_0 = \mathbf{r}_0 + \mathbf{y}_0 \cdot \tau,$$

where $\tau = \tau_{max}/t_{max}$, and τ_{max} is determined with arbitrary precision by a binary search procedure such that

$$\mathbf{r}_0 + \mathbf{y}_0 \cdot \tau_{max}$$

is a boundary point of the feasible search space $\mathcal{F}$.

The above mapping satisfies all the requirements of a "good" decoder. Apart from being one-to-one, the transformation is fast and has a "locality" feature (i.e., points that are close before being mapped are close after being mapped). Additionally, there are a few other features that make the proposed method interesting, including

- As opposed to most of the other constraint-handling methods, there's no need for any additional parameters (e.g., the frequency of seven operators in GENOCOP [307], penalty coefficients). Only the basic parameters of an evolutionary algorithm (e.g., population size, manner of applying variation operators) are required.

- In contrast with some other constraint-handling methods, there's no need for any specialized operators to maintain the feasibility of solutions (e.g., operators of GENOCOP [307] to maintain linear constraints and specialized boundary operators to search the boundary between feasible and infeasible parts of the search space). Any evolutionary algorithm can be used together with the proposed mapping.

- In contrast with most other constraint-handling methods (i.e., all the methods that don't reject infeasible solutions), there's no need to evaluate infeasible solutions (in particular, there's no need to penalize them, tune penalty coefficients, to repair).

- In contrast to other constraint-handling methods (e.g., methods based on penalty functions, or hybrid methods), the proposed method always returns a feasible solution.

We can extend the proposed approach with an additional method for iteratively improving solutions that's based on the relationship between the location of the reference point and the efficiency of the proposed approach. The location of the reference point $\mathbf{r}_0$ has an influence on the "deformation" of the domain of the optimized function. The evolutionary algorithm doesn't optimize the evaluation function, but rather some other function that's topologically equivalent to the original. For example, consider the case when the reference point is located nearby the edge of the feasible region $\mathcal{F}$ — it's easy to notice a strong irregularity of transformation T. The part of the cube $[-1, 1]^2$ that's on the left side of the vertical line is transformed into a much smaller part of the set $\mathcal{F}$ than the part on the right side of this line (see figure 9.14).

---

reference point $\mathbf{r}_0$, such that every line segment originating in $\mathbf{r}_0$ intersects the boundary of $\mathcal{F}$ in precisely one point, is sufficient. This requirement is satisfied, of course, for any convex set $\mathcal{F}$.

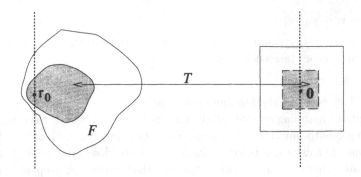

Fig. 9.14. The influence of the location of the reference point on the transformation T

Given these considerations, it seems intuitive to localize the reference point in the neighborhood of the expected optimum if this optimum is close to the edge of the set $\mathcal{F}$. In such a case, the area between the edge of $\mathcal{F}$ and the reference point $\mathbf{r}_0$ is explored more precisely. If we lack information about the approximate location of a solution, then we should place the reference point close to the geometrical center of the set $\mathcal{F}$. We can do this by sampling the set $\mathcal{F}$ and setting

$$\mathbf{r}_0 = 1/k \sum_{i=1}^{k} \mathbf{x}_i,$$

where $\mathbf{x}_i$ are samples from $\mathcal{F}$. It's also possible to take advantage of the mentioned effect for the purpose of improving the best-found solution iteratively. To obtain this effect we have to repeat the optimization process with a new reference point $\mathbf{r}_0'$ that's located on a line segment between the current reference point $\mathbf{r}_0$ and the best solution $\mathbf{b}$ found so far:

$$\mathbf{r}_0' = t \cdot \mathbf{r}_0 + (1 - t) \cdot \mathbf{b},$$

where $t \in (0, 1]$ should be close to zero. Changing the location of the reference point like this means that the neighborhood of the discovered optimum will be explored in more detail as compared with the remaining part of the feasible region in the next iteration. There's evidence that this method can give good results for problems where the optimal solutions are localized on the edge of the feasible region [272, 273]. The approach can also be extended to handle nonconvex search spaces [272, 273].

9.2.5 Hybrid methods

It's easy to develop hybrid methods that combine evolutionary algorithms with deterministic procedures for numerical optimization problems.[12] For example,

---

[12]This was accomplished as early as the 1960s [253].

Waagen et al. [484] combined a floating-point representation and Gaussian variation with the direction set method of Hooke-Jeeves. This hybrid was tested on three unconstrained continuous-valued test functions. Myung et al. [332] considered a similar approach but experimented with constrained problems in the continuous domain. Again, they used floating-point representations and Gaussian variation, but combined this with a Lagrange method developed by Maa and Shanblatt [294] into a two-phased search. During the first phase, the evolutionary algorithm optimizes the function

$$eval(\mathbf{x}) = f(\mathbf{x}) + \frac{s}{2}\left(\sum_{j=1}^{m} f_j^2(\mathbf{x})\right),$$

where s is a constant. After this phase terminates, the second phase applies the Maa and Shanblatt procedure to the best solution found during the first phase. This second phase iterates until the system

$$\mathbf{x}' = -\nabla f(\mathbf{x}) - \left[\sum_{j=1}^{m} \nabla f_j(\mathbf{x})(sf_j(\mathbf{x}) + \lambda_j)\right],$$

is in equilibrium, where the Lagrange multipliers are updated as $\lambda_j' = \epsilon s f_j$ for a small positive constant ϵ.

This hybrid method was applied successfully to a few test cases. For example, minimize

$$G11(\mathbf{x}) = x_1^2 + (x_2 - 1)^2,$$

subject to a nonlinear constraint,

$$x_2 - x_1^2 = 0,$$

and bounds,

$$-1 \leq x_i \leq 1, i = 1, 2.$$

The known global solutions are $\mathbf{x}^* = (\pm 0.70711, 0.5)$, and $G11(\mathbf{x}^*) = 0.75000455$. For this problem, the first phase of the evolutionary process converged in 100 generations to $(\pm 0.70711, 0.5)$, and the second phase drove the system to the global solution. The method wasn't tested, however, on problems of higher dimension.

Several other constraint-handling methods also deserve attention. For example, some methods use the values of the evaluation function f and penalties f_j $(j = 1, \ldots, m)$ as elements of a vector and apply multiobjective techniques to minimize all the components of the vector. For example, Schaffer [410] selects $1/(m+1)$ of the population based on each of the objectives. Such an approach was incorporated by Parmee and Purchase [353] in the development of techniques for constrained design spaces. On the other hand, Surry et al. [458], ranked all the individuals in the population on the basis of their constraint violations. This rank, r, together with the value of the evaluation function f, leads to the two-objective optimization problem. This approach gave a good performance on optimization of gas supply networks [458].

Hinterding and Michalewicz [222] used a vector of constraint violations, to assist in the process of parent selection, where the length of the vector corresponded to the number of constraints. For example, if the first parent satisfies constraints 1, 2, and 4 (say out of 5), it's mated preferably with an individual that satisfies constraints 3 and 5. It's also possible to incorporate knowledge of the problem constraints into the belief space of cultural algorithms [388]. These algorithms provide a possibility for conducting an efficient search of the feasible search space [389].

Recently, Kelly and Laguna [257] developed a new procedure for difficult constrained optimization problems that

> ...has succeeded in generating optimal and near-optimal solutions in minutes to complex nonlinear problems that otherwise have required days of computer time in order to verify optimal solutions. These problems include nonconvex, nondifferentiable and mixed discrete functions over constrained regions. The feasible space explicitly incorporates linear and integer programming constraints and implicitly incorporates other nonlinear constraints by penalty functions. The objective functions to be optimized may have no 'mathematically specifiable' representation, and can even require simulations to generate. Due to this flexibility, the approach applies also to optimization under uncertainty (which increases its relevance to practical applications).

Sounds pretty good! The method combines evolutionary algorithms with (1) scatter search, (2) procedures of mixed-integer optimization, and (3) adaptive memory strategies of tabu search. No doubt you can imagine many more ways to construct hybrid methods for handling constraints.

9.3 Summary

We've covered a great many facets of constrained optimization problems and provided a survey of some of the attempts to treat these problems. Even though it might seem that this survey was lengthy, in fact, it was overly brief. It only skimmed the surface of what has been done with evolutionary algorithms and what remains to be done. Nevertheless, we have to admit that the characteristics that make a constrained problem difficult for an evolutionary algorithm (or for that matter, other methods) aren't clear. Problems can be characterized by various parameters: the number of linear constraints, the number of nonlinear constraints, the number of equality constraints, the number of active constraints, the ratio $\rho = |\mathcal{F}|/|\mathcal{S}|$ of the size of feasible search space to the whole, and the type of evaluation function in terms of the number of variables, the number of local optima, the existence of derivatives, and so forth.

We discussed 11 test cases ($G1$–$G11$) for constrained numerical optimization problems [319] in this chapter. These test cases include evaluation functions of various types (linear, quadratic, cubic, polynomial, nonlinear) with various numbers of variables and different types (linear inequalities, nonlinear equalities and inequalities) and numbers of constraints. The ratio ρ between the size of the feasible search space $\mathcal{F}$ and the size of the whole search space $\mathcal{S}$ for these test cases varies from zero to almost 100 percent. The topologies of feasible search spaces are also quite different. These test cases are summarized in table 9.2. The number n of variables, the type of the function f, the relative size of the feasible region in the search space given by the ratio ρ, the number of constraints of each category (linear inequalities LI, nonlinear equations NE and inequalities NI), and the number a of active constraints at the optimum (including equality constraints) are listed for each test case.

Table 9.2. Summary of 11 test cases. The ratio $\rho = |\mathcal{F}|/|\mathcal{S}|$ was determined experimentally by generating 1,000,000 random points from $\mathcal{S}$ and checking to see if they belong to $\mathcal{F}$ (for $G2$ and $G3$ we assumed $k = 50$). LI, NE, and NI represent the number of linear inequalities, and nonlinear equations and inequalities, respectively

Function	n	Type of f	ρ	LI	NE	NI	a
$G1$	13	quadratic	0.0111%	9	0	0	6
$G2$	k	nonlinear	99.8474%	0	0	2	1
$G3$	k	polynomial	0.0000%	0	1	0	1
$G4$	5	quadratic	52.1230%	0	0	6	2
$G5$	4	cubic	0.0000%	2	3	0	3
$G6$	2	cubic	0.0066%	0	0	2	2
$G7$	10	quadratic	0.0003%	3	0	5	6
$G8$	2	nonlinear	0.8560%	0	0	2	0
$G9$	7	polynomial	0.5121%	0	0	4	2
$G10$	8	linear	0.0010%	3	0	3	6
$G11$	2	quadratic	0.0000%	0	1	0	1

Even though there were many reported successes, the results of many tests haven't provided meaningful patterns to predict the difficulty of problems. No single parameter, such as the number of linear, nonlinear, active constraints, and so forth, can suffice to describe the problem difficulty. Many evolutionary methods approached the optimum quite closely for test cases $G1$ and $G7$ (with $\rho = 0.0111\%$ and $\rho = 0.0003\%$, respectively), whereas most of the methods had trouble with test case $G10$ (with $\rho = 0.0010\%$). Two quadratic functions (test cases $G1$ and $G7$) with a similar number of constraints (nine and eight, respectively) and an identical number (six) of active constraints at the optimum, posed a significant challenge to most of these methods.

Several methods were also quite sensitive to the presence of a feasible solution in the initial population. There's no doubt that more extensive testing and analysis is required. This is definitely an area where you can help contribute to the knowledge of the field. Not surprisingly, the experimental results of [319] suggest that the question of how to make an appropriate choice of an evolutionary method for a nonlinear optimization problem *a priori* remains open. It seems that more complex properties of the problem (e.g., the characteristic of the evaluation function together with the topology of the feasible region) may constitute quite significant measures of the difficulty of the problem. Also, some additional measures of the problem characteristics due to the constraints might be helpful. So far, we don't have this sort of information at hand.

Michalewicz and Schoenauer [319] offered:

> It seems that the most promising approach at this stage of research is experimental, involving the design of a scalable test suite of constrained optimization problems, in which many [...] features could be easily tuned. Then it should be possible to test new methods with respect to the corpus of all available methods.

There's a clear need for a parameterized test-case generator that can be used for analyzing various methods in a systematic way instead of testing them on a few selected cases. Furthermore, it's not clear if the addition of a few extra specific test cases is really of any help. Such a test-case generator has been just constructed [314, 315]. Time will tell if we can fathom out the salient properties of constrained problems so that we can better tune our evolutionary algorithms to the task at hand.

X. Can You Tune to the Problem?

Sometimes it's difficult to apply a general problem-solving strategy to a particular problem, as the problem at hand is simply unique in some sense. When this happens, you have to analyze the problem and develop a method that's just right for the particular instance you face. Many mathematical puzzles have this quality in the sense that the solution rarely generalizes to include a variety of problems. Consider, for example, the following problem:

Show, that $\lfloor (2+\sqrt{3})^n \rfloor$ is odd for all natural n.

It's hard to apply any general problem-solving method here. For example, mathematical induction doesn't seem appropriate. We can easily show that for $n = 1$, the given expression (before cutting off the fractional part) has a value of 3.732, that for $n = 2$ the corresponding value is 19.928, and for $n = 3$, it's 51.981. But it might be quite difficult to prove the induction step. If $\lfloor (2+\sqrt{3})^n \rfloor$ is odd, then $\lfloor (2+\sqrt{3})^{n+1} \rfloor$ must be odd as well. What's so special about $\sqrt{3} = 1.7320508075688772935$ such that any further powers of $(2+\sqrt{3})$ result in an odd integer plus some fraction?

Another possibility for applying a general mechanism for this problem would be to calculate the n-th power and to write down the resulting expression:

$$(2+\sqrt{3})^n = 2^n + n \cdot 2^{n-1}\sqrt{3} + \ldots + \sqrt{3}^n,$$

but it's of little help. Some components of this expression are even, but the others are multiplied by $\sqrt{3}$ in some power, and suddenly the reasoning becomes very murky. This approach doesn't seem to make life any easier.

What can we do? It seems that we have to discover something special about this problem, something that would help us move forward. This is probably the most difficult part of problem solving and there are no rules for how to do it. The only hint we can give here is that such discoveries usually come from experimentation and experience.

Just out of frustration, we can check some other powers for n, to see what would happen. Indeed, something very interesting is going on. We've already observed that

$$(2+\sqrt{3})^1 = 3.732,$$
$$(2+\sqrt{3})^2 = 13.928, \text{ and}$$
$$(2+\sqrt{3})^3 = 51.981.$$

Further,

$$(2 + \sqrt{3})^4 = 199.995,$$
$$(2 + \sqrt{3})^5 = 723.9986, \text{ and}$$
$$(2 + \sqrt{3})^6 = 2701.9996.$$

It seems that higher values of n result in larger fractional parts of the number, which approach one, while the integer part of the number remains odd. The resulting number is even, less some fractional part that approaches zero for larger values of n. Indeed, we can write

$$(2 + \sqrt{3})^1 = 4 - 0.268,$$
$$(2 + \sqrt{3})^2 = 14 - 0.072,$$
$$(2 + \sqrt{3})^3 = 52 - 0.019,$$
$$(2 + \sqrt{3})^4 = 200 - 0.005,$$
$$(2 + \sqrt{3})^5 = 724 - 0.0014, \text{ and}$$
$$(2 + \sqrt{3})^6 = 2702 - 0.0004.$$

If we could show that $(2 + \sqrt{3})^n$ is always equal to an even number minus a fraction, then our proof would be complete. For $n = 1$, the fraction being subtracted is $0.268 = 2 - \sqrt{3}$. This gives us a hint. Is it true that $0.072 = (2 - \sqrt{3})^2$? Yes, it is! It's also clear that $(2 - \sqrt{3})^n$ is always a fraction from the range $[0..1]$.

At this stage it's sufficient to prove that

$$(2 + \sqrt{3})^n + (2 - \sqrt{3})^n$$

is even for all n. This is elementary, however, because all of the components with $\sqrt{3}$ raised to an odd power will cancel out (having opposite signs in the two expressions), and even powers of $\sqrt{3}$ are either multiplied by 2 (in some power), or added together (e.g., $3^n + 3^n = 2 \cdot 3^n$ for even n). This concludes the proof.

Actually, only the last paragraph constitutes the proof; everything before that only demonstrates the reasoning that leads to the proof. The proof itself is very short and elegant, but it's not that easy to get there. The method developed was certainly problem specific. It's unlikely that we would ever encounter a similar problem where this method might be of some use.

Having to develop problem-specific methods is a little bit frustrating. It would be ideal to have a method that could tune itself to the problem at hand. Actually, there are some hints at how to do this with evolutionary algorithms that we'll see in the next chapter, but before we proceed, try your hand at the following problems and see how well you do.

One problem is simply stated:

Find all of the numbers that consist of six digits such that the result of multiplying that number with any number in the range from two to six, inclusive, is a number that consists of the same six digits as the original number but in a different order.

Thus the search space consists of all six-digit numbers, i.e., from 100,000 to 999,999. Which of them have the above property? See if you can solve this without our help first.

The required property constrains the search space. Only some six-digit numbers (if any) satisfy the requirement, i.e., are feasible. Altogether, there are five constraints, one for each multiplication (by two, by three, and so forth). How can these constraints help us in solving the problem? How can we narrow down the search?

Let's look at the first (most significant) digit of the number we're searching for. The last constraint implies that the first digit has to be a 1 because if it's 2 or greater, then after multiplying this number by 6 we'd get a seven-digit number! So, this single constraint allows us to narrow the search space down to the set of all numbers from 100,000 to 199,999.

But we can reason further. What are the consequences of having digit 1 in the first position? If the number we're searching for is x, then the number $2x$ starts from a digit that's either 2 or 3. In general, the first digit of $6x$ is greater than the first digit of $5x$, which is greater than the first digit of $4x$, which is in turn greater than the first digit of $3x$, which is greater than the first digit of $2x$, which is greater than the first digit of x. From this simple observation we can draw a few conclusions:

- Since the six digits for x, $2x$, ..., $6x$ are the same, and each of these numbers starts with a digit greater than the first digit of the previous number, then all six digits are different!

- The digit 0 is not included.

Still, we can reason further. What's the last (least significant) digit of the number x? It can't be even because $5x$ would have 0 as its last digit, which is impossible. It can't be 5 because the last digit of $2x$ would be 0, which is also impossible. It can't be 1 because the first digit is 1, and all of the digits are distinct. So the only possible choices for the last digit are 3, 7, and 9.

If the last digit is 3, then the last digits of $2x$, ..., $6x$ are 6, 9, 2, 5, and 8, respectively. This is impossible because the six-digit number would contain digits $\{3, 6, 9, 2, 5, 8\}$, and we already know that the digit 1 must be included in such a set (as x starts with 1). If the last digit is 9 then the last digits of $2x$, ..., $6x$ are: 8, 7, 6, 5, and 4, respectively. Again, this is impossible for the same reasons as before.

So the last digit of the number x must be 7. We are searching then for numbers from 100,007 to 199,997, and only every tenth number from this set has to be considered.

Knowing the last digit to be 7, we also know that the digits we're searching for are $\{1, 2, 4, 5, 7, 8\}$ because the last digits of $2x$, ..., $6x$ are 4, 1, 8, 5, and 2, respectively. This allows us to find the first digits of $2x$, ..., $6x$, as they must constitute a growing sequence. Everything we've discovered so far is summarized in figure X.1.

$$
\begin{aligned}
x &= 1 - - - - 7 \\
2x &= 2 - - - - 4 \\
3x &= 4 - - - - 1 \\
4x &= 5 - - - - 8 \\
5x &= 7 - - - - 5 \\
6x &= 8 - - - - 2
\end{aligned}
$$

Fig. X.1. The current information while solving the problem. The dashed lines represent the digits that we must still determine

To find x it's now sufficient to search for a permutation of just four numbers, 2, 4, 5, and 8, since x starts with 1 and the last digit is 7. The problem is easy now because all we have to do is check $4! = 24$ possibilities. Note too that the difference between, say, $5x$ and $2x$ is $3x$, the difference between $6x$ and $4x$ is $2x$, etc. In short, any difference between two different numbers from $x, \ldots, 6x$ is a number from this set.

How can this observation help us? We can make the following claim: the set of digits of particular significance (i.e., whether the most significant, the second-most significant, etc., up to the least significant) of $x, \ldots, 6x$ is always $\{1, 2, 4, 5, 7, 8\}$. In other words, each "vertical" column of digits (see figure X.1) is also a permutation of our six digits $\{1, 2, 4, 5, 7, 8\}$. The reason for that is simple. If a digit appears twice in a particular column, say, for $5x$ and $2x$, then their difference (which is $3x$) would have either 0 or 9 in that column, which is impossible because these digits do not belong the considered set.

This simple observation allows us to sum six equations from figure X.1. Since we know that $7 + 4 + 1 + 8 + 5 + 2 = 27$, we can sum these equations, column after column, as each column sums up to 27. The left-hand sides sum to $x + 2x + 3x + 4x + 5x + 6x = 21x$, so

$$21x = 2,999,997.$$

Consequently, $x = 142857$. We've narrowed the search space to a single point! Not bad! How well did you do?

Here's another one for you. There are many *practical* problems that can be solved easily if you use the right model. Consider the following scenario. There are two rolls of toilet paper. The dimensions of the outer and inner circles are given in figure X.2.

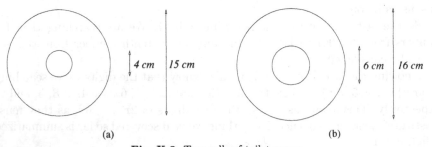

(a)   (b)

Fig. X.2. Two rolls of toilet paper

Each roll has 480 tissues, but the length of a single tissue in roll (a) is 16 cm, whereas the length of a tissue in roll (b) is 17 cm. The question is which roll has thicker tissues?

Note that the question is very important (!) and the solution quite useful. Many manufacturers provide only partial information on the number of tissues. If you're instead interested in the thickness of a tissue, you have to do all the calculations! Can you do it?

There are a few ways to determine the solution. Probably the simplest way is to calculate the side area of the rolls:

Area of roll (a) $= \pi(7.5^2 - 2^2) = 164.15$, and
Area of roll (b) $= \pi(8^2 - 3^2) = 172.79$,

and the total length of their tissues:

Length of roll (a) $= 480 \times 16 = 7680$, and
Length of roll (b) $= 480 \times 17 = 8160$.

Then the tissue thickness for each of the rolls is

$164.15/7680 = 0.0214$ cm and $172.79/8160 = 0.0212$ cm,

respectively, so roll (a) has slightly thicker tissues!

Note that we can equally answer an additional question: what is the number of turns of tissue on each roll? The thickness of all the turns of tissue on the whole roll (a) is 5.5 cm, so the number of turns is

$5.5/0.0855 \approx 64$.

(The number of turns on the roll (b) is just 59.)

Okay, new problem: Suppose someone gives you a sack of gold that weighs between 1 and 40 oz. In fact, to make things more interesting, suppose that the sack can only take on integer values, and suppose that in front of you is a balance scale. Your problem is to find a minimum cardinality set of weights for determining the weight of the sack by bringing the scale into equilibrium.

If we think about the problem in terms of powers of two, we might consider the weights

1, 2, 4, 8, 16, and 32.

Any integer weight for the sack between $1 \leq x \leq 40$ can be determined by placing some weights on one side of the balance and the sack on the other side, for example,

$x = 35 = 1 + 2 + 32$.

It's like expressing an integer in a binary notation. But to be able to balance against any integer from 1 to 40 we'd need 6 weights (see above).

If we instead think in terms of powers of three, we can do much better. The weights are

1, 3, 9, and 27.

Any number $1 \leq x \leq 40$ can be expressed as a unique sum of the above weights, provided that these weighs can take on a negative sign. This represents the case where a weight is placed on the same side as the sack of gold, for example,

$$35 = -1 + 9 + 27,$$

which means that on one side of the scale we should place the sack of gold together with a weight of 1, whereas on the other side we have the weights 9 and 27. So we can weigh any item using only four weights!

If powers of three are good, what about powers of four?

The final puzzle of this section is as follows. It's interesting to note that the digits $1, \ldots, 9$ can be arranged to form two numbers, whose ratio is $\frac{1}{2}, \frac{1}{3}, \ldots, \frac{1}{9}$. For example,

$$\frac{7293}{14586} = \frac{1}{2}$$

and

$$\frac{6381}{57429} = \frac{1}{9}.$$

Can you find all the other fractions?

10. Tuning the Algorithm to the Problem

> He who wishes to be rich in a day
> will be hanged in a year.
>
> Leonardo da Vinci, *Notebooks*

Almost every practical heuristic search algorithm is controlled by some set of parameters. In simulated annealing, for example, there's a temperature parameter, and what's more, you have to decide the schedule for reducing the temperature over time. In hill-climbing, there's a parameter that controls the size of the local neighborhood in which you'll look for improvements. In tabu search, you must determine how to implement the rules for the memory structure. None of these algorithms comes neatly wrapped in a gift box where all you have to do is open the box and receive your nice surprise!

It's no different with evolutionary algorithms. In fact, evolutionary algorithms often present more parameters to control than other methods. This is both (1) a source of their robustness to different problems as well as (2) a source of frustration for the person who has to design these algorithms. Right, that's you. You have to consider many facets: (1) the representation, (2) the evaluation function, (3) the variation operators, (4) the population size, (5) the halting criterion, and so forth. Considerable effort has been expended in the search for parameter settings that will offer reasonable performance across a range of problems. In some cases, these efforts have actually met with success, but in general the quest for a magic set of parameter values that will give the best possible results is a fool's errand.

What's more, the trial-and-error method for finding useful parameter values is a very tedious and time-consuming adventure. It would be nice if there were a theory to guide how to set some of the parameters of evolutionary algorithms. In some cases such a theory actually exists, but these cases are of limited scope. In the absence of a theory, it would be nice if we could find some automated way for optimizing the parameters that control the evolutionary search for improved solutions. As we'll see, there are a variety of such methods, some which even involve the use of the evolutionary algorithm itself (i.e., the evolution of evolution).

10.1 Parameter control in evolutionary algorithms

After you've defined a representation and the evaluation function, which really constitute the bridge between your original problem and the framework for the evolutionary algorithm, you're still faced with numerous challenges. How are you going to transform parent solutions into offspring? That is, what form of variation operators are you going to apply? No doubt, whatever choice you make, you'll be faced with a decision about how to parameterize those operators. For example, if you believe that some form of recombination might be helpful in your problem, then you must decide whether this will be one-point crossover (i.e., pick a point at random and then splice together the data that come before that point in the first parent and after that point in the second parent), or analogously, two-point crossover, n-point crossover, uniform crossover (i.e., choose each element from either parent uniformly at random), arithmetic crossover (which is a blending or averaging of components), majority logic (which compares three or more parents and sets each component to agree with the majority of the parents selected), and so forth. Let's say that you choose to use a majority logic operator. Now you're faced with a new parameter control problem. How many parents will be involved in the decision? The minimum reasonable number is three, but maybe you'd consider involving the entire population, regardless of how big it is. Let's say that you choose to have 20 parents, selected at random for each offspring, participate in the majority logic operator. Now you're faced with yet another parameter control problem. Should you set the values of the offspring based on the majority vote for each component, or should you perhaps set each value based on a probability that reflects the fraction of the population that possesses each different value for that component? That is, say you were working with a binary representation, and 11 of the parents had a 1 at the first value, while 9 of the parents had a 0. Would you set the offspring's first entry to 1 because this is the majority vote, or would you say that there's a 11/20 probability of setting it to a 1, and a corresponding 9/20 probability of setting it to a 0? No matter which variation operator you choose, there are a raft of possible parameterizations that you must decide.

Each decision is important. If you make a poor choice for your parameters you can generate truly disastrous results. We've seen some people give up on evolutionary algorithms for just this reason. They tried something they read in some book on the subject, found that it didn't "work," or maybe it suggested that they should use a binary encoding on a problem that was inherently nonbinary, which just seemed like extra effort, and then they lost interest. So you should know going in that whenever you use any search algorithm, evolutionary or otherwise, you should always expect to consider how best to tune that algorithm for your problem, and you should never simply accept that simply because some set of parameters worked well for someone else on their particular problem that those same parameter values will work well for you.

Here's a classic example. Early in the development of evolutionary algorithms, De Jong [93] studied a set of five NLPs and tried to find suitable parameter values for the population size, the probability of recombination, the probability of mutation, the selection strategy, and so forth. Each of these problems involved continuous values, but De Jong transformed them into a binary encoding and used a one-point crossover and simple bit-flipping mutation for variation. After executing many experiments, De Jong concluded that a good set of parameters for these five functions was: population size of 50, a crossover probability of 0.6, a mutation probability of 0.001, and an elitist selection strategy that always preserved the best solution found up to the current generation.

A decade later, Grefenstette [201] used an evolutionary algorithm to search over the parameter values that De Jong had tuned on the same functions. The performance of the primary evolutionary algorithm was quantified in both *online* and *offline* measures, where the former is based on the quality of every solution generated in the evolutionary experiment, and the latter is based only on the quality of the best solution at each time step. The evolutionary search for the best parameters for the evolutionary search suggested the following values, with the offline performance values shown in parentheses: a population size of 30 (80), crossover probability of 0.95 (0.45), mutation probability of 0.01 (0.01), and a selection that was elitist (nonelitist).

Here are some interesting things to consider that followed these studies. First, most of the publications in evolutionary algorithms in the 1980s and early 1990s used values that were very close to one of these two studies. Certainly, these publications had a significant effect on other applications. But was this appropriate? After all, most of these other applications had nothing to do with the five functions that were studied in these two investigations. Second, five years later, Davis [89] showed that a simple hill-climbing algorithm that used random initialization outperformed the evolutionary algorithms on these same problems:

> A central tenet of our field has been that genetic algorithms do well because they combine schemata, multiple-member subcomponents of good solutions. Yet [random bit climbing] is an algorithm that manipulates one bit at a time and finds solutions from 3 to 23 times faster than the genetic algorithm, on the very problems that we have used to tune and compare genetic algorithms.

There is no doubt that many of the early efforts, and even some recent efforts, in evolutionary computation were overly restrictive, accepting one particular way to encode problems and then leaving the only remaining problem as the search for the best parameters for the available operators. We know now that you must allow for the possibility of tuning the available parameters, and even the relationship between the variation operators, selection operator, and the representation for each problem that you face. The scope for "optimal" parameter settings is necessarily narrow. Any quest for generally (near-)optimal parameter settings is lost from the start.

One alternative to tuning parameters by hand or based on several other trials of an evolutionary algorithm is to rely on mathematical analysis. This can be effective in some restricted domains. For example, in the early years of evolutionary algorithms, Rechenberg [383] showed that if you are using a population comprising a single parent that generates only one offspring, and if the representation you're using is continuous (i.e., real-valued), and if you are trying to find the minimum of a strongly convex (e.g., quadratic) bowl or a planar corridor function, and if the dimensionality of these functions is large (i.e., $n \to \infty$), and if you are using a Gaussian mutation with zero mean and a tunable standard deviation, then the best settings for the standard deviation will generate improved solutions with a probability of 0.27 and 0.18 for these two functions, respectively. As a compromise, Rechenberg recommended a so-called one-fifth-success-rule: If the frequency of generating solutions that are better than the parent is greater than one-fifth then increase the standard deviation; if it is less than one-fifth then decrease the standard deviation. The idea is that if you are generating too many improved offspring, then you are being too conservative and taking steps that are too small. On the other hand, if your success rate is less than one-fifth you are being too aggressive and need to scale back your step size. This analysis suggests an online procedure for adapting the standard deviation to generate the maximum expected rate of improvement.

Strictly speaking, this only works on the functions that Rechenberg studied, and even then only in the limit as the number of dimensions grows large. Can we apply the one-fifth-rule to, say, a traveling salesman problem? Fogel and Ghozeil [157] showed that the results of doing so could be very poor, as in some cases the maximum expected improvement occurred for a probability of improvement of 0.008, which isn't very close to one-fifth. Nevertheless, the idea that you can feed back information from how the search is progressing to the evolutionary algorithm and use that to adjust the strategy for exploration is enticing and useful.

In contrast, tuning by hand is fraught with problems. Typically, only one parameter is tuned at a time, which may result in some less-than-optimal choices because parameters interact in complex ways. The simultaneous tuning of more parameters, however, leads to an enormous amount of experimentation. The technical drawbacks of parameter tuning based on experimentation are

- The parameters aren't independent, but trying all possible combinations is practically impossible.

- The process of parameter tuning is time consuming, even if parameters are optimized one and time.

- The results are often disappointing, even when a significant expenditure of time is made.

Hand tuning and systematic experimentation is truly a Mt. Everest of challenges: you might make it, or you might die trying.

Another option for choosing appropriate parameter settings for an evolutionary algorithm is to rely on an apparent analogy between your problem and approach and someone else's problem and their approach. With any luck, what worked for them will also work for you. It's not clear, however, how to measure the similarity between problems so as to have any good idea that there really is any analogy between two problems or methods for their solution. Certainly, it's not enough that two approaches rely on the same cardinality of representation, for example. Just because two approaches use, say, binary strings for representation doesn't imply anything in the way of guidance for how their parameters should be set.

Another option is to rely on theory to set the parameters, but this will be a very short discussion because there is really very little theory to be used beyond Rechenberg's result. Later in the chapter, we'll discuss some work that determined the optimum probability of mutation in linearly separable problems, but realistically these are not the problems for which you'd want to use evolutionary algorithms. There have been some theoretical investigations into the optimal population size [466, 207, 189] and optimal operator probabilities [191, 467, 19, 413], but these were based on simple problems and their practical value is limited.

Returning to the idea of finding parameter values that hold across multiple problems, regardless of how the parameters are tuned, we must recognize that any run of an evolutionary algorithm is an intrinsically dynamic, adaptive process. The use of fixed parameter values runs contrary to this spirit. More practically, it's obvious that different values of parameters might be optimal at different stages of the evolutionary process [88, 461, 16, 17, 19, 215, 442]. This is particularly true because the population changes its composition at each generation. What works for one generation and one composition shouldn't necessarily work for every other generation and composition. It might be that large mutation steps are good in the early generations of some application of evolutionary algorithms because it would help the exploration of the search space. Smaller mutations might be needed later to help in fine tuning. (Note this is a principle that is similar to cooling the temperature in simulated annealing.) The use of static parameters can itself lead to inferior performance.

The straightforward way to overcome this problem is to use parameters that change values over time, that is, by replacing a parameter p by a function $p(t)$, where t is the generation counter, but just as the problem of finding optimal *static* parameters for a particular problem can be quite difficult, designing an optimal function $p(t)$ may be even more difficult. Another possible drawback to this approach is that if the parameter value $p(t)$ changes solely as a function of time t, without taking any notion of the actual progress made in solving the problem then there is a disconnect between the algorithm and the problem. Yet many researchers (see section 10.4) have improved their evolutionary algorithms (i.e., they improved the quality of results returned by their algorithms while working on particular problems) by using simple deterministic rules to change parameters. A suboptimal choice of $p(t)$ can often lead to better results than a suboptimal choice of p.

To this end, let's admit that finding good parameter values for an evolutionary algorithm is a poorly structured, ill-defined, complex problem. But these are the kinds of problems for which evolutionary algorithms are themselves quite apt! It's natural to think about using an evolutionary algorithm not only for finding solutions to a problem, but also for tuning that same algorithm to the particular problem. Technically speaking, this amounts to modifying the values of parameters during the run of the algorithm by taking the actual search process into account. There are essentially two ways to accomplish this. You can either use some heuristic rule that takes feedback from the current state of the search and modifies the parameter values accordingly, or you can incorporate parameters into the data structure that represents a solution, thereby making those parameters subject to evolution along with the solution itself.

10.2 Illustrating the case with an NLP

Let's assume that we face a numerical optimization problem:

optimize $f(\mathbf{x}) = f(x_1, \ldots, x_n)$,

subject to some inequality and equality constraints,

$g_i(\mathbf{x}) \leq 0$ $(i = 1, \ldots, q)$ and $h_j(\mathbf{x}) = 0$ $(j = q+1, \ldots, m)$,

and bounds, $l_i \leq x_i \leq u_i$ for $1 \leq i \leq n$, defining the domain of each variable.

For such a numerical optimization problem we may consider an evolutionary algorithm based on a floating-point representation. Each individual $\mathbf{x}$ in the population is represented as a vector of floating-point numbers

$\mathbf{x} = \langle x_1, \ldots, x_n \rangle$.

For the sake of illustration, let's assume that we're using Gaussian mutation to produce offspring for the next generation. Recall that a Gaussian mutation operator requires two parameters: the mean, which we take here to be zero so that on average each offspring is no different from its parent, and the standard deviation σ, which can be interpreted as the mutation step size. Mutations are then realized by replacing components of the vector $\mathbf{x}$ by

$x_i' = x_i + N(0, \sigma)$,

where $N(0, \sigma)$ is a random Gaussian number with mean zero and standard deviation σ. The simplest method to specify the mutation mechanism is to use the same σ for all vectors in the population, for all variables of each vector, and for the whole evolutionary process, for instance, $x_i' = x_i + N(0, 1)$. Intuitively, however, it might be beneficial to vary the mutation step size.[1]

First, we could replace the static parameter σ by a dynamic parameter, i.e., a function $\sigma(t)$. This function could be defined by some heuristic that assigns different values depending on the number of generations. For example, the mutation step size might be defined as

$\sigma(t) = 1 - 0.9 \cdot \frac{t}{T}$,

---

[1]There are formal arguments supporting this in specific cases, e.g., [16, 17, 19, 215].

where t is the current generation number, 0 to T, which is the maximum generation number. Here, the mutation step size $\sigma(t)$, which is used for every vector in the population and for every component of every vector, will decrease slowly from 1 at the beginning of the run ($t = 0$) to 0.1 as the number of generations t approaches T. This might aid in fine tuning solutions as the evolutionary algorithm proceeds. In this approach, the value of the given parameter changes according to a fully deterministic scheme, and you have full control of the parameter and its value at a given time t because it's completely determined and predictable.

Second, it's possible to incorporate feedback from the search, still using the same σ for every vector in the population and for every component of every vector. Rechenberg's one-fifth-rule is an example. An online implementation of the rule is

if ($t \bmod n = 0$) **then**

$$\sigma(t) \leftarrow \begin{cases} \sigma(t-n)/c, & \text{if } p_s > 1/5 \\ \sigma(t-n) \cdot c, & \text{if } p_s < 1/5 \\ \sigma(t-n), & \text{if } p_s = 1/5 \end{cases}$$

else

$$\sigma(t) \leftarrow \sigma(t-1);$$

where p_s is the relative frequency of successful mutations, measured over some number of generations and $0.817 \le c \le 1$ [21]. (Schwefel [418] suggests $c = 0.82$.) Changes in the parameter values are now based on feedback from the search, and σ-adaptation takes place every n generations. Your influence on the parameter values is much less here than in the deterministic scheme above. Of course, the mechanism that embodies the link between the search process and parameter values is still a man-made heuristic that dictates how the changes should be made, but the values of $\sigma(t)$ are not predictable.

Third, it's possible to assign an individual mutation step size to each vector (i.e., each solution). The representation for an individual can be extended to be of length $n + 1$ as

$$\langle x_1, \ldots, x_n, \sigma \rangle.$$

The value σ governs how the values x_i will be mutated and is itself subject to variation. A typical variation would be:

$$\sigma' = \sigma \cdot e^{N(0,\tau_0)}$$
$$x_i' = x_i + N(0, \sigma'),$$

where τ_0 is a parameter of the method that is often set to $1/\sqrt{n}$. This mechanism is commonly called *self-adapting* the mutation step sizes.

In the above scheme, the value of σ is applied to a single individual. Each individual has its own adaptable step size parameter. It learns how to search the space of potential solutions as it is searching the space of potential solutions. This can be easily extended to each separate component of each solution vector.

A separate mutation step size can be applied to each x_i. If an individual is represented as

$$\langle x_1, \ldots, x_n, \sigma_1, \ldots, \sigma_n \rangle,$$

then mutations can be realized by replacing the above vector according to a similar formula as discussed above:

$$\sigma_i' = \sigma_i \cdot e^{N(0,\tau_0)}$$
$$x_i' = x_i + N(0, \sigma_i'),$$

where τ_0 is again a parameter of the method. Each component x_i now has its own mutation step size σ_i that is self-adapted. This mechanism implies a larger degree of freedom for adapting the search strategy to the topology of the fitness landscape.[2]

We've illustrated how you can control (adapt) the mutation operator during the evolutionary process. Certainly, there are many components and parameters that can also be changed and tuned for optimal algorithm performance. In general, the three options we sketched for the mutation operator are valid for any parameter of an evolutionary algorithm, whether it's the population size, mutation step, the penalty coefficient, selection pressure, and so on. In some cases, however, you might need to implement this approach in a hierarchical fashion. For example, to adapt the population size you might choose to compete multiple populations, each with its own size parameter.

It's natural to try to construct a taxonomy that can usefully describe the effect of different parameter tuning and control procedures. The mutation step size, for example, can have different domains of influence, which we call its *scope*. Using the $\langle x_1, \ldots, x_n, \sigma_1, \ldots, \sigma_n \rangle$ model, a particular mutation step size applies only to one variable of one individual. Thus, the parameter σ_i acts on a subindividual level. In the $\langle x_1, \ldots, x_n, \sigma \rangle$ representation, the scope of σ is one individual, whereas the dynamic parameter $\sigma(t)$ was defined to affect all individuals and thus has the whole population as its scope. Let's turn to a more complete taxonomy of approaches to controlling parameters.

10.3 Taxonomy of control techniques

There are many aspects that we can consider when classifying different parameter control techniques:

1. *What* exactly is being changed (e.g., representation, evaluation function, operators, selection process, mutation rate)?

2. *How* is that change being made (e.g., by deterministic heuristic, feedback-based heuristic, or in a self-adaptive manner)?

---

[2]If you're familiar with probability and statistics, then note that this procedure does not allow for correlated mutations. Additional mechanisms are required to control generating offspring when steps in one dimension are correlated with steps in other dimensions.

3. What is the *scope/level* of the change (e.g., population-level, individual-level)?

4. What statistic or *evidence* is used to affect the change (e.g., monitoring the performance of operators, the diversity of the population)?

Each of these aspects deserves attention.

To classify parameter control techniques from the perspective of what's being changed, it's first necessary to agree on a list of *all* of the components of an evolutionary algorithm. This is a difficult task in itself but we can consider the following facets:

- The representation of individuals.

- The evaluation function.

- The variation operators and their associated probabilities.

- The selection (or replacement) operator and any associated rules.

- The population in terms of size, topology, etc.

Additionally, each component can be parameterized, and the number of parameters is not clearly defined. For example, an offspring produced by an *arithmetic crossover* or averaging of k parents $\mathbf{x}_1, \ldots, \mathbf{x}_k$ can be defined by the following formula

$$\mathbf{v} = a_1 \mathbf{x}_1 + \ldots + a_k \mathbf{x}_k,$$

where $a_1, \ldots, a_k$, and k can be considered as parameters of this operator. Parameters for a population can include the number and sizes of subpopulations, migration rates between subpopulations, etc. (this is for a general case; when more than one population is involved, see chapter 16).

In spite of these limitations we can maintain the focus on the aspect of what is being changed. This allows us to locate where a specific mechanism has a direct effect (it may indirectly affect any number of other aspects of the evolutionary algorithm). It also is the way many people search for ways to improve their own algorithm, e.g., "I want to experiment with changing mutation rates. Let me see how others have done this."[3]

As we saw above, each method for changing the value of a parameter can be classified into one of three categories:

---

[3]We'd be remiss, however, if we didn't say that this sort of bottom-up approach to problem solving is almost always the wrong way to go about it. This is akin to saying, "Hey, I've got a hammer here, I wonder what other people have done with hammers?" This completely misses the point of problem solving in the first place. "What is your purpose? What are you trying to achieve?" The hammer is a means for achieving something, not an end to itself.

- *Deterministic* parameter control. This takes place when the value of a strategy parameter (i.e., a parameter that controls how the evolutionary search is made, such as σ in the previous example) is altered by some deterministic rule. This rule modifies the strategy parameter deterministically without using any feedback from the search. A time-varying schedule is often used, where the rule will be applied when a set number of generations have elapsed since the last time the rule was activated.

- *Adaptive* parameter control. This takes place when there is some form of feedback from the search that is used to determine the direction and/or magnitude of the change to the strategy parameter. The assignment of the value of the strategy parameter may involve credit assignment, and the action of the evolutionary algorithm may determine whether or not the new value persists or propagates throughout the population.

- *Self-adaptive* parameter control. The idea of the evolution of evolution can be used to implement the self-adaptation of parameters. Here the parameters to be adapted are encoded into the data structure(s) of the individual and undergo variation (mutation and recombination). The "better" values of these encoded individuals lead to "better" individuals, which in turn are more likely to survive and produce offspring and hence propagate these "better" parameter values.

This terminology leads to the taxonomy illustrated in figure 10.1. Other taxonomies and terminology have been offered [6] but we believe the approach offered here has advantages.

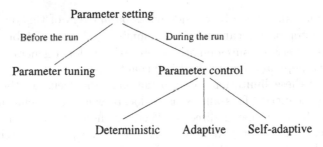

Fig. 10.1. Global taxonomy of parameter setting in evolutionary algorithms

Any change may affect a component, an entire individual, the whole population, or even the evaluation function. This is the aspect of the scope or level of adaptation [6, 223, 439, 436]. Note, however, that the scope/level usually depends on the component of the evolutionary algorithm where the change takes place, e.g., a change of the mutation step size may affect a component, an individual, or the whole population, depending on the particular implementation used. On the other hand, a change in the penalty coefficients for constraint violations always affects the whole population. So, the scope/level feature is secondary, usually depending on the given component and its actual implementation.

The issue of the scope of the parameter is really more complicated than indicated in section 10.2, however. First, the scope depends on how the given parameters are interpreted. For example, an individual might be represented as

$$\langle x_1, \ldots, x_n, \sigma_1, \ldots, \sigma_n, \alpha_1, \ldots, \alpha_{n(n-1)/2} \rangle,$$

where the vector α denotes the covariances between the variables $\sigma_1, \ldots, \sigma_n$. In this case the scope of the strategy parameters in α is the whole individual, although the notation might suggest that they act on a subindividual level.

The next example illustrates that the same parameter encoded in an individual can be interpreted in different ways, leading to different algorithmic variants with different scopes of this parameter. Spears [443] experimented with individuals that contained an extra bit to determine whether one-point crossover or uniform crossover was to be used (the bit 1/0 stood for one-point/uniform crossover, respectively). Two interpretations were considered. The first interpretation was based on a pairwise operator choice. If both parental bits were the same, the corresponding operator was used; otherwise, a random choice was made. Thus, under this interpretation, this parameter acted at an individual level. The second interpretation was based on the bit-distribution over the whole population. If, for example, 73 percent of the population had bit 1, then the probability of one-point crossover for a selected individual was 0.73. Under this interpretation, the parameter acted on the level of the population. Note, that these two interpretations can be combined easily. For instance, similar to the first interpretation, if both parental bits were the same, the corresponding operator would be used. However, if they differed, the operator would be selected according to the bit-distribution, just as in the second interpretation. The scope/level of this parameter in this interpretation would be neither individual nor population, but rather both. This example shows that the notion of scope can be ill-defined and arbitrarily complex.

Another possible criterion for classifying a particular approach is the type of evidence or statistics used for determining the change of parameter value [439, 436]. Most often, the progress of the search is what's monitored (e.g., the performance of variation operators). It's also possible to look at other measures, such as the diversity of the population. The information gathered by such monitoring is used as feedback for adjusting the parameters. Although this is a meaningful distinction, it only appears in adaptive parameter control. A similar distinction can be made in deterministic control, which might be based on any counter that's not related to search progress. One option is the number of fitness evaluations, as the description of deterministic control above indicates — there's nothing in the number of times you call the evaluation function that relates directly to the progress of the search. There are other possibilities, changing the probability of mutation on the basis of the number of executed mutations, for instance, but we won't give further consideration to that here.

Thus, the main criteria for classifying methods that change the values of the strategy parameters of an algorithm during its execution are

1. *What* is changed?

2. *How* is the change made?

The classification is thus two-dimensional. It takes into account the type of control and the component of the evolutionary algorithm that incorporates the parameter. The *type* and *component* are orthogonal and encompass typical forms of parameter control within evolutionary algorithms. The *type* of parameter change consists of three categories: deterministic, adaptive, and self-adaptive mechanisms. The *component* of parameter change consists of five categories: representation, evaluation function, variation operators (mutation and recombination), selection (replacement), and population.

10.4 The possibilities for parameter control

The following section provides a survey of many experimental efforts to control parameters of an evolutionary algorithm. The discourse groups the work based on what is being adapted. As a consequence, the following subsections correspond to the above list of five components of an evolutionary algorithm, with the efforts in controlling variation operators further divided into mutation and recombination.

10.4.1 Representation

Most of the effort in adapting representations has been made within binary data structures. This is likely due to the many problems that were encountered historically when trying to encode diverse problems into bit strings. The populations often prematurely converged and there were so-called *Hamming cliffs* where values that are next to each other functionally often had few components in common (e.g., the numbers 7 and 8 are represented as 0111 and 1000, so even though they are only 1 unit apart in real values, they are maximally separated in binary). Keep in mind that there is no reason to restrict the application of adapting a representation solely to binary strings. Three main approaches to adaptive representation are considered here and they all use adaptive parameter control.

Shaefer [431] offered a strategy that used a flexible mapping of the variables in a function to the components in an individual (one component per variable). His mechanism allowed not only the number of bits per parameter to vary, thereby providing an adaptive resolution to the problem, it also had the potential to vary the range of each parameter by contraction or expansion, and also by shifting the center of the interval. Remember that when translating from a continuous domain to a binary domain, you have to fix some level of precision within an interval (or collection of intervals), so the possibility for extending, shrinking, and shifting those intervals during the search was attractive.[4] The representation was changed on-the-fly, based on rules that governed the degree of convergence of components across the population, the variance of the components, and how closely the component values approached the range boundaries for each variable.

Mathias and Whitley [492, 301] offered a technique that also modifies the representation, but it does this over multiple restarts. The first trial is used to locate an *interim solution*, while subsequent trials interpret the components as distances (so-called *delta values*) from the last interim solution. Each restart forms a new hypercube with the interim solution at its origin. The resolution of the delta values can also be modified at each restart to expand or contract the search space. Restarts are triggered when the Hamming distance between the best and worst individuals of the population is one or less.

Schraudolph and Belew [417] offered a method that alters the interpretation of bits in a string. Given an initial interval for each variable in the evaluation function, as the population converges in a variable's interval, the defined range of that variable is narrowed, thereby allowing the evolutionary algorithm to zoom in on the interval but still have the same bit-length encoding. No method for zooming out was offered, however.

Some other efforts that deserve mention include the technique in [190] where the data structure is of variable length and may contain too few or too many bits for defining an individual (i.e., it may be overspecified in some variables and underspecified in others, in which case they are filled in using additional rules). The interpretations of an individual therefore vary because for variables that are overspecified, whatever specification is encountered first is the interpretation that's acted upon. As the data structures undergo variation and selection, the interpretation for each component of a solution can be turned on or off depending on the other components in that solution. In addition, some of the earliest use of adaptive representations involved self-adaptive control for the dominance mechanism of a simulated diploid encoding structure. In this more complicated set up, there are two copies of every structure in each individual. The extra copy encodes alternative values and the dominance rules determine which of the solutions will be expressed. Considerable work was done within this framework in the 1960s and 1970s [28, 395, 229, 59], with more recent efforts in [193, 196, 197].

10.4.2 Evaluation function

If you incorporate penalty functions into the evaluation of a potential solution, you also have the potential to adapt the evaluation function during the evolutionary search. Alternative mechanisms for varying penalties according to

---

[4]Note that this is really only attractive when mapping a problem from the continuous domain to a representation with significantly lower cardinality.

predefined deterministic schedules are offered in [247] (which was discussed in section 10.2) and [312], which uses the following idea. The evaluation function *eval* incorporates an additional parameter τ,

$$eval(\mathbf{x}, \tau) = f(\mathbf{x}) + \frac{1}{2\tau} \sum_J f_j^2(\mathbf{x}),$$

that is decreased every time the evolutionary algorithm converges (usually, $\tau \leftarrow \tau/10$). Copies of the best solution found after convergence are taken as the initial population of the next iteration with the new decreased value of τ. Thus, there are several cycles within a single run of the system. The evaluation function is fixed for each particular cycle ($J \subseteq \{1, \ldots, m\}$ is a set of active constraints at the end of a cycle) and the penalty pressure increases thereby changing the evaluation function when the evolutionary algorithm switches from one cycle to another.

The method of Eiben and Ruttkay [118] falls somewhere between tuning and adaptive control of the fitness function. They apply a method for solving constraint satisfaction problems that changes the evaluation function based on the performance of a run. The penalties (weights) of the constraints that are violated by the best individual after termination are increased, and the new weights are used in the next run. In this manner, the evolutionary search is forced to put more emphasis on satisfying the constraints that it has failed to satisfy so far (also see [123, 122]).

A technical report [36] from 1992 offers what may be the earliest example of adaptive evaluation functions for constraint satisfaction, where the penalties for constraint violations in a constrained optimization problem were adapted during a run (see section 10.2). Since then, many other efforts have been made to adjust the penalty values associated with constraints that are violated. The penalty can vary depending on the number of violated constraints [435], or by identifying when the evolutionary algorithm appears to be stuck in a local optimum [108, 109] (following [327]). No doubt there are many other possibilities.

10.4.3 Mutation operators and their probabilities

There has been considerable effort in finding optimal values for mutation rates and parameters for mutation distributions. Let's first focus on their tuned optimal rates before moving on to attempt to control them online.

As we saw before, across five different cases, De Jong [93] and Grefenstette [201] offered recommended mutation rates, p_m, on binary-encoded continuous-valued optimization problems that differed by an order of magnitude of 0.001 and 0.01, respectively. Schaffer et al. also suggested a probability of bit mutation that was in the range $[0.005, 0.01]$ [412]. These recommendations were based solely on experimental evidence so their generalizability was limited.

Much earlier, Bremermann [58, 57] proved that the optimum mutation rate for binary function optimization problems that are linearly separable is $1/L$, where L is the number of binary variables. This yields the maximum expected rate of convergence toward the best solution. Furthermore, Bremermann [57] indicated that the mutation rate should depend on the number of binary variables that were set correctly, thereby giving early evidence that the mutation rate should not be held constant. A similar result was later derived by Mühlenbein [328]. Smith and Fogarty compared this rate with several fixed rates and found it to be the best possible fixed value for p_m [437]. Bäck [21] also found $1/L$ to be a good value for p_m when using Gray-coding (which is an alternative form of binary encoding).

Fogarty [141] used deterministic control schemes that decrease p_m over time and over the components (see [22] for the exact formulas). As mentioned before in another context, this idea is similar to reducing the temperature in simulated annealing. The observed improvement with this scheme makes it an important contribution. Hesser and Männer [215] derived theoretically optimal schedules for changing p_m deterministically when facing the counting-ones problem.[5] They suggested

$$p_m(t) = \sqrt{\frac{\alpha}{\beta}} \times \frac{\exp\left(\frac{-\gamma t}{2}\right)}{\lambda\sqrt{L}}$$

where α, β, γ are constants, λ is the population size, and t is the time (generation counter) (cf. [57]).

Bäck also presented an optimal schedule for decreasing the mutation rate as a function of the distance to the optimum in the counting-ones problem (as opposed to a function of time) [16], being

$$p_m(f(\mathbf{x})) \approx \frac{1}{2(f(\mathbf{x}) + 1) - L}.$$

Bäck and Schütz [27] constrained a function to control the probability of mutation to decrease starting from $p_m(0) = 0.5$ and then using $p_m(T) = \frac{1}{L}$ if a maximum of T evaluations are used:

$$p_m(t) = \left(2 + \frac{L - 2}{T} \cdot t\right)^{-1} \text{ if } 0 \le t \le T.$$

Janikow and Michalewicz [237] experimented with a *non-uniform mutation*, where

$$x_k^{t+1} = \begin{cases} x_k^t + \Delta(t, r(k) - x_k) & \text{if a random binary digit is 0} \\ x_k^t - \Delta(t, x_k - l(k)) & \text{if a random binary digit is 1} \end{cases}$$

for $k = 1, \ldots, n$. The function $\Delta(t, y)$ returns a value in the range $[0, y]$ such that the probability of $\Delta(t, y)$ being close to 0 increases as t increases (t is the generation number). This causes the operator to search the space uniformly at first (when t is small), and very locally at later stages. Janikow and Michalewicz [237] used the function

---

[5]This is the problem of finding the string of 0s and 1s such that the sum of the bits is maximized.

$$\triangle(t, y) = y \cdot r \cdot (1 - \frac{t}{T})^b,$$

where r is a random number from $[0..1]$, T is the maximal generation number, and b is a system parameter determining the degree of non-uniformity.

At this point you should stop and consider the similarity that has been discovered repeatedly in these separate investigations. It's often useful to have larger variations initially and then focus on smaller variations later. Across all possible problems, by the no free lunch theorems [499] we know that this won't make any difference. That tells us two important things: (1) it's possible to focus on a subset of all possible problems and find patterns that are effective in controlling the values of the variation parameters, and (2) the typical problems used for testing and analyzing variation operators have fundamental similarities. This latter property may not be advantageous. We may be focusing on too narrow a subset of all possible problems and generating methods that may not be as effective as we would like on real-world problems. An important point then is to always test your procedures on problems that are as close to the real world as you can get.

We already mentioned two other means for controlling the parameters of a continuous mutation operator: (1) Rechenberg's one-fifth-rule, and (2) the self-adaptation of the σ value in a Gaussian mutation. The latter can also be extended to be a control variable for scaling other random mutation distributions, such as a Cauchy random variable, or a mixture of Cauchy and Gaussians [65]. Furthermore, the update of the σ parameter doesn't have to be performed using a lognormal distribution, i.e., $e^N(0, \sigma)$. Fogel et al. [155] offered the approach of using another Gaussian random variable to update the value of σ for each component of an individual. Comparisons of these methods have been offered in [408, 7]. As you might expect, each method has particular instances when it appears to offer an advantage, but there has been no definite pattern identified as yet to suggest when either method should be applied. Self-adaptive control of mutation step sizes is discussed at length in [21, 419]. There are very many applications and variations of self-adapting search parameters in continuous domains (e.g., [220, 17, 16, 437])

Self-adaptation of mutation has also been used for non-numeric problems. Fogel et al. [164] used self-adaptation to control the relative probabilities of five mutation operators for the components of a finite state machine. Hinterding [221] used a multiple-structure representation to implement self-adaptation in the cutting stock problem where self-adaptation varied the probability of using one of the two available mutation operators. Chellapilla and Fogel [66] used self-adaptation to vary the length of an inversion operator in the TSP. The notion of using evolution to optimize itself (i.e., self-adaptation) is widely applicable in both continuous and discrete problem domains.

10.4.4 Crossover operators and their probabilities

When working with binary representations, in contrast to a mutation rate p_m that's applied to each bit, the crossover rate p_c acts on a pair of solutions, giving the probability that the selected pair undergoes recombination. The efforts of De Jong [93], Grefenstette [201], and Schaffer [412] suggested values for p_c that range from 0.6 to 0.95. Given that the mutation rates in these efforts were always very low, it's not surprising that high rates of recombination were required. For alternative coding schemes that don't rely on binary strings, however, you have to be cautious about applying the same settings because the effect of recombination within various representations can be entirely different.

In what follows here, we'll treat mechanisms for controlling recombination probabilities and mechanisms for controlling the recombination mechanism itself separately. Let's begin with efforts to control the probability of the operator.

Davis [91] adapted the rate of variation operators (e.g., crossover) by rewarding those that show success in creating better offspring. This is a form of reinforcement learning where the reward is propagated back to operators that acted a few generations in the past, in the hope that these operators helped set up the current rewards. Rewards are diminished as they are propagated back in time to operators used previously. In a similar vein, Julstrom [249] adapted the rates of crossover and mutation by comparing the respective improvements that were obtained using these operators in the past generation (or longer). The probability of crossover was set to the ratio of the credit assigned for the improvements made by crossover divided by the number of instances of crossover, taken as a ratio to that same quotient summed with the analogous quotient for mutation

$$p_c = \frac{Credit(Cross)/N(Cross)}{Credit(Cross)/N(Cross) + Credit(Mut)/N(Mut)},$$

where *Cross* and *Mut* refer to crossover and mutation, respectively. When crossover is generating more successes, its probability of being used in the future is amplified, and similarly for mutation. Note that there's no convincing rationale for making the probability of crossover the complement of the probability of mutation. You could just as well adapt these probabilities independently. Tuson and Ross [470] offered an extensive study of various "cost-based" operator rate adaptation mechanisms and found that this general procedure was effective on timetabling problems, but was also detrimental on numerous other problems. Certainly, there are many ways to implement this general idea and each will have advantages and limitations. For other efforts along these lines, see [286, 486].

Spears [443] offered a means for self-adapting the choice between two different crossovers, two-point crossover, and uniform crossover, by adding one extra bit to each individual (see section 10.3). This extra bit determined which type of crossover was used for that individual. The offspring inherited the choice for its type of crossover from its parents as a result of the variation operator. Earlier,

Fogel and Atmar [151] adapted the use of crossover and mutation in a population of individuals by tagging each one but without allowing for an individual to change its form of variation. The population was initialized randomly with respect to the variation operators and then the evolutionary dynamics took over as the population converged ultimately on one strategy for generating offspring. The method was implemented to find solutions to systems of linear equations, and the population converged to the mutation strategy nearly ten times more often than the strategy for recombination.

Schaffer and Morishima [413] used self-adaptation to determine the number and locations of crossover points. Special markers were introduced into the representation that kept track of the sites where crossover was to occur. Their experiments indicated that this self-adaptive approach performed as well as or better than a nonadaptive approach on a set of test problems.

One of the earliest efforts in the self-adaptation of crossover (and mutation) probabilities was offered in [384] where individuals represented strategies for playing a simple card game. The individuals had parameters for playing the game that governed rules such as the probability of "betting high" when having a "high card." They also had parameters for adjusting the probabilities of mutation and for turning on and off the possibility of crossing over with other individuals. Furthermore, this effort used co-evolution to determine the worth of each strategy, as all of the strategies competed directly with the others that were present in the population. Keep in mind that this was done in 1967![6]

Up to this point, we've only discussed controlling the recombination of two parents. A new parameter is introduced when extending the recombination to multiple parents [111], namely, the number of parents applied in recombination. Eiben et al. [121] used an adaptive mechanism to adjust the arity of recombination based on competing subpopulations. In particular, the population was divided into disjoint subpopulations, each using a crossover of different arity. Subpopulations evolved independently for a period of time and exchanged information by allowing migration after each period (for more on this concept of subpopulations, see chapter 16). Quite naturally, migration was arranged in such a way that those populations that evidenced greater progress in the given period grew in size, while those populations with small progress decreased in size. There's also a mechanism for preventing subpopulations (and thus crossover operators) from complete extinction. This method yielded an algorithm showing comparable performance with the traditional (one-population, one-crossover) version using a variant that employed six-parent crossover. The mechanism failed to clearly identify the better operators by making the corresponding subpopulations larger. This is, in fact, in accordance with the findings of Spears [443] in a self-adaptive framework.

---

[6]The idea that the operator rules could be adapted online actually goes back to Bledsoe in 1961 [50]: "One of the difficulties in using such a technique on the computer is the necessity of selecting the mating rules. One interesting prospect is to let the mating rule itself evolve — i.e. by a selection process the mating rule is chosen which provided for the best survival of the whole population."

10.4.5 Parent selection

The family of the so-called Boltzmann selection mechanisms embodies a method that varies the selection pressure along the course of the evolution according to a predefined "cooling schedule" [298], again much like we've seen in simulated annealing. The name originates from the Boltzmann trial from condensed matter physics, where a minimal energy level is sought by state transitions. Being in a state i, the chance of accepting state j is

$$P[\text{accept } j] = \exp\left(\frac{E_i - E_j}{K_b \cdot T}\right),$$

where E_i, E_j are the energy levels, K_b is a parameter called the Boltzmann constant, and T is the temperature. This acceptance rule is called the Metropolis criterion. The mechanism proposed by de la Maza and Tidor [96] applies the Metropolis criterion for defining the evaluation of a solution. In this way, the selection criterion changes over time. A related effort using multiple populations is offered in [402].

As compared with adapting variation operators or representation, methods for adaptively selecting parents have received relatively little attention, but some selection methods do have parameters that can be easily adapted. For example, consider linear ranking, which assigns to each individual a selection probability that is proportional to the individual's rank i,[7]

$$p(i) = \frac{2 - b + 2i(b - 1)/(pop\_size - 1)}{pop\_size},$$

where the parameter b represents the expected number of offspring to be allocated to the best individual. By changing this parameter within the range of [1..2] we can vary the selective pressure of the procedure. Similar possibilities exist for other ranking methods and tournament selection.

10.4.6 Population

There have been many efforts to determine the appropriate population size in an evolutionary algorithm, but most of the theoretical analysis has not been very useful [188, 189, 441]. As a result, empirical trials have been used to determine optimal settings for particular problems. We mentioned De Jong's early study [93] as well as Grefenstette [201] before, where they searched for good population sizes on a set of five test problems. Grefenstette essentially adapted the population sizes by his use of a meta-level evolutionary algorithm that searched for the best setting. Many experiments have tested evolutionary algorithms at different population sizes to determine if one setting is better than another (e.g., [238, 63, 181]), but these did not adapt the population size during evolution.

---

[7]The rank of the worst individual is zero, whereas the rank of the best individual is $pop\_size - 1$.

Smith [440] offered a brief abstract regarding one possibility for adapting population size as a function of the variability of the quality of templates within individual solutions. The proposed method estimated variability by sampling alternative solutions within the population and then resized the population as a result of the estimated variability. Unfortunately, the length of the abstract (one printed page) didn't permit a fully detailed explanation of the theory or mechanism that supports this idea, but it does offer some possibilities for further work.

Arabas et al. [12] implemented a rule for adapting the size of the population by tagging individuals with a parameter that controlled its own life span. The life span parameter was defined in terms of the maximum number of generations that the particular individual could survive. An individual was automatically removed from the population once its life span was exceeded, regardless of its quality. However, the life span assigned upon creation of a new solution was made a function of the solution's quality (i.e., higher quality implied longer life span). The population size could thereby grow or shrink based on the performance of the individuals at each generation.

10.5 Combining forms of parameter control

By far, most efforts to study parameter control in evolutionary algorithms have focused on controlling only one aspect of the algorithm at a time. This is probably because (1) the exploration of capabilities of adaptation was performed experimentally, and (2) it's easier to report positive results in simpler cases. Combining forms of control is much more difficult because the interactions of even static parameter settings for different components of an evolutionary algorithm are not well understood. They often depend on the evaluation function [209], representation [464], and many other factors. Several empirical studies have been performed to investigate the interactions between various parameters of an evolutionary algorithm [131, 412, 500]; some stochastic models based on Markov chains have also been developed and analyzed to understand these interactions [64, 336, 459, 482]. Nevertheless, there is a dearth of work on adapting combinations of control parameters in evolutionary algorithms.

In combining forms of control, the most common method is related to mutation. For example, when treating floating-point representations and applying Gaussian mutation, there are several possibilities for controlling its operation. We can distinguish the setting of the standard deviation of the mutations (mutation step size) at a global level, for each solution, or for components within a solution. We can also control the preferred direction of mutation by incorporating rotational parameters that define the correlation between steps in each dimension.

Other examples of combining the adaptation of different mutation parameters are given in Yao et al. [504] and Ghozeil and Fogel [183]. Yao et al. [504] combined the adaptation of the step size with the mixing of Cauchy and Gaussian mutation. Recall that the Cauchy random variable looks a lot like a Gaussian random variable, except that it has much fatter tails, so it generates offspring that are more distant from their parents with a much higher probability. Here the scaling step-size parameter is self-adapted, and the step size is used to generate two new individuals from one parent: one using Cauchy mutation and the other using Gaussian mutation. The worse of these two individuals is discarded. Empirical evidence suggested that this method was at least as good as using either just Gaussian or Cauchy mutations alone, even though the population size was halved to compensate for generating two individuals from each parent. Ghozeil and Fogel compared the use of polar coordinates for the mutation step size and direction over the generally used Cartesian representation. While these results were preliminary, they indicate a possibility for not only controlling multiple facets of variation but also a shift in representational basis.

It's much more rare to find efforts that adapt multiple control parameters that govern different components of the evolutionary algorithm. Hinterding et al. [224] combined self-adaptation of the mutation step size with a feedback-based adaptation of the population size. Here feedback from a cluster of three evolutionary algorithms, each with different population sizes, was used to adjust the population size of one or more of the evolutionary algorithms at "epochs" of 1,000 evaluations. Self-adaptive Gaussian mutation was used in each of the evolutionary algorithms. The procedure adapted different strategies for different types of test functions. For unimodal functions it adapted to small population sizes for all of the evolutionary algorithms. In contrast, for multimodal functions it adapted one of the evolutionary algorithms to a large but oscillating population size, apparently to assist in escaping from local optima.

Smith and Fogarty [438] self-adapted both the mutation step size and the preferred crossover points in a evolutionary algorithm. Each component in an individual included: (1) the encoded variable for the overall solution, (2) a mutation rate for the associated component, and (3) two linkage flags, one at each end of the component that are used to link components into larger *blocks* when two adjacent components have their adjacent linkage flags set. Here, crossover acted on multiple parents and occurred at block boundaries, whereas mutation could affect all the components of a block and the rate was the average of the mutation rates in a block. It's difficult to describe such complex algorithms in a short paragraph, but the main idea to convey is that it's possible to encode the adaptive mechanisms for controlling various parameters all within a single individual, or even within a single component.

The most comprehensive combination of forms of control has been offered by Lis and Lis [286], as they combined the adaptation of the mutation probability, crossover rate, and population size. A statistical experimental design (Latin Squares) was used over a number of generations to assess the performance of different parameter settings. The results then implied how to change parameters for the next set of generations, and so forth.

It's interesting that all but one of the efforts that combine various forms of control use self-adaptation. Hinterding et al. [224] used feedback-based adaptation rather than self-adaptation to control the population size in order to minimize the number of separate populations. While the interactions of static parameter settings for the various components of an evolutionary algorithm are complex, the interactions of the dynamics of adapting parameters using either deterministic or feedback-based adaptation is even more complex and therefore much more difficult to determine. Self-adaptation appears to be the most promising way of combining forms of control as we leave it to evolution itself to determine the beneficial interactions among various components while finding a near-optimum solution to the problem.

10.6 Summary

The effectiveness of an evolutionary algorithm depends on many of its components (e.g., representation, variation operators) and their interactions. The variety of parameters included in these components, the many possible choices (e.g., to change or not to change), and the complexity of the interactions between various components and parameters make the selection of a "perfect" evolutionary algorithm for a given problem very difficult, if not impossible.

How can we find the "best" evolutionary algorithm for a given problem? As discussed earlier, we can make some effort to tune the parameters, trying to find good values before the run of the algorithm. But even if we assume for a moment that there is a perfect configuration, finding it is really a hopeless task. Figure 10.2 illustrates this point. The search space S_{EA} of all possible evolutionary algorithms is huge, much larger than the search space S_P of the given problem P, so our chances of *guessing* the right configuration (again, if one exists!) for an evolutionary algorithm are rather slim (e.g., much smaller than the chances of guessing the optimum permutation of cities for a large instance of the traveling salesman problem). Even if we restrict our attention to a relatively narrow subclass of evolutionary algorithms, say, S_{GA}, which represents the classical "genetic" algorithm, the number of possibilities is still prohibitive.[8] Note, that within this (relatively small) class there are many possible algorithms with different population sizes, different frequencies of the two variation operators (whether static or dynamic), and so forth. What's more, guessing the right values of the parameters might be of limited value anyway. Any set of static parameters is likely to be inappropriate because an evolutionary algorithm is an intrinsically dynamic, adaptive process. So the use of rigid parameters that don't change during evolution is unlikely to be optimal. Different values for the parameters may be appropriate at different stages of the evolutionary process.

Adaptation provides the opportunity to customize the evolutionary algorithm to the problem and to modify the configuration and the strategy param-

---

[8] A subspace of *classical* genetic algorithms, $S_{GA} \subset S_{EA}$, consists of evolutionary algorithms where individuals are represented by binary-coded fixed-length strings and has two variation operators: one-point crossover and a bit-flip mutation. It also uses proportional selection.

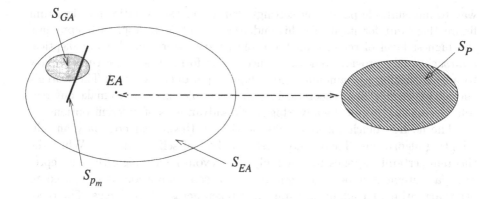

Fig. 10.2. An evolutionary algorithm EA for problem P as a single point in the search space S_{EA} of all possible evolutionary algorithms. EA searches (*broken line*) the solution space S_P of the problem P. S_{GA} represents a subspace of classical genetic algorithms, whereas S_{p_m} represents a subspace that consists of evolutionary algorithms that are identical except for their mutation rate p_m.

eters that control the evolutionary search while the solution to the problem is sought. We can incorporate domain-specific information and multiple variation operators into the evolutionary algorithm more easily by using adaptation, and can also allow the algorithm to itself select those values and operators that provide better results. When we allow some degree of adaptation within an evolutionary algorithm we are essentially searching two different spaces simultaneously. While solving the problem P (searching the space of solutions S_P), a part of S_{EA} is also searched for the best evolutionary algorithm EA at that stage of the search of S_P. So far, the investigations (see section 10.4) have examined only a tiny part of the search space S_{EA}. For example, when adapting the mutation rate, p_m, we consider only a subspace S_{p_m} (see figure 10.2), which consists of all evolutionary algorithms that have all of their control parameters fixed except for the mutation rate. Even more, the early experiments of Grefenstette [201] were further restricted just to the subspace S_{GA}.

One of the main obstacles of optimizing parameter settings of evolutionary algorithms arises from the nonlinear (so-called *epistatic*) interactions between these parameters. The mutual influence of different parameters and the combined influence of parameters together on the behavior of an evolutionary algorithm is very complex. A pessimistic conclusion would be that such an approach is not appropriate, since the ability of evolutionary algorithms to cope with epistasis is limited. But then again, the ability to handle arbitrary nonlinear interactions between parameters is always limited, for any search algorithm. Parameter optimization falls in the category of ill-defined, poorly structured (at least poorly understood) problems that prevents an analytical approach. This is a class of problem for which evolutionary algorithms provide a potentially useful alternative to other methods. Roughly speaking, we might not have a better way to find suitable parameter settings than to let the evolutionary algorithm figure them out for itself. To this end, the self-adaptive approach represents the highest level of reliance on the evolutionary algorithm. A more skeptical approach would provide some assistance in the form of heuristic rules on how to adjust parameters, amounting to adaptive parameter control. There aren't enough experimental or theoretical results available currently to make any reasonable conclusions on the advantages or disadvantages of different options.

The no free lunch theorem [499] provides a theoretical boundary on self-adapting algorithms. There's no "best" method of self-adaptation. While the theorem certainly applies to a self-adjusting evolutionary algorithm, it represents a statement about the performance of self-adaptation as compared to other algorithms for adapting parameters taken across all problems. The theorem isn't relevant in the practical sense here because these "other" algorithms hardly exist. The appropriate comparison to make is between the self-adaptive methods for controlling parameters and the use of human "oracles," the latter being the more common and painful practice.

It could be argued that relying on human intelligence and expertise is the best way to design an evolutionary algorithm, including the parameter settings. After all, the "intelligence" of an evolutionary algorithm would always be limited by the small fraction of the predefined problem space it encounters during the search, while human designers may have global insight of the problem to be solved. This, however, does not imply that the human insight leads to better parameter settings (see the discussion of the approaches called parameter tuning and parameter setting by analogy in section 10.1). Furthermore, human expertise is costly and might not be available for the problem at hand, so relying on computer power is often the most practical option. The domain of applicability of the evolutionary problem-solving technology as a whole could be significantly extended by evolutionary algorithms that are able to configure themselves, at least partially.

Some researchers [37] have suggested that adaptive control substantially complicates the task of an evolutionary algorithm and that the rewards in solution quality don't justify the cost. Certainly, there is some learning cost involved in adaptive and self-adaptive control mechanisms. Either some statistics are collected during the run, or additional operations are performed on extended individuals. Comparing the efficiency of algorithms both with and without (self-)adaptive mechanisms might be misleading, however, since it disregards the time needed for tuning. A more fair comparison would be based on a model that includes the time needed to set up (i.e., to tune) and to run the algorithm. We're not aware of any such comparisons at the moment.

Online parameter control mechanisms may have a particular significance in nonstationary environments, where we often have to modify the current solution in light of various changes in the environment (e.g., machine breakdowns, sick employees). The potential for an evolutionary algorithm to handle such changes and to track the optimum efficiently have been studied by a number of researchers [8, 24, 477, 478]. As we discussed above, several mechanisms have been considered. Some used (self-)adaptation of various parameters of the algorithm, while other mechanisms were based on maintaining diversity in the population (sometimes using other adaptive schemes based on diploidy and dominance).

There are several exciting research issues connected with controlling parameters in evolutionary algorithms, including

- Developing models for comparing algorithms with and without (self-) adaptive mechanisms. These models should include stationary and dynamic environments.

- Understanding the utility of changing parameter values as well as the behavior of the algorithm that results from their interactions using simple deterministic control rules. For example, you might consider an evolutionary algorithm with a constant population size versus an evolutionary algorithm where the population size decreases or increases at a predefined rate such that the total number of function evaluations in both algorithms remain the same.

- Justifying or rebuking popular heuristics for adaptive control.

- Trying to find the general conditions under which adaptive control works. There are some guidelines for self-adaptive mutation step sizes concerning the number of offspring that should be created per parent [419], but there has been no verification that these or other guidelines are correct or broadly useful.

- Understanding the interactive effects of adaptively controlled parameters.

- Investigating the merits and drawbacks of self-adaptation of several (possibly all) parameters of an evolutionary algorithm.

- Developing a formal mathematical basis for the proposed taxonomy for parameter control in evolutionary algorithms in terms of functionals that transform the operators and variables they require.

There are myriad possibilities for you to adjust and mold an evolutionary algorithm for your particular problem. When you don't have sufficient knowledge to set parameters appropriately, it's often useful to have heuristics that can guide things as the evolution proceeds, and you can even use evolution itself to give you this guidance.

XI. Can You Mate in Two Moves?

There's a beautiful chess puzzle which illustrates that problems shouldn't be considered as being static. The concept of "time" might be quite essential! After several moves of a chess game, the board situation is as follows:

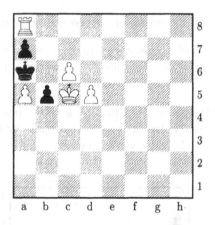

It's now white's move (white moves upwards). Is it possible to check-mate black in two moves?

At first glance, the problem doesn't seem so difficult. There are only a few pieces on the board, so a search tree won't be too wide. And white has to check-mate black in only two moves, so the search tree can't be too deep. So, let's try!

White has four possibilities for the first move: move the pawn on c6, the pawn on d5, the king, or the rook. Let's discuss these possibilities one by one.

- White advances the pawn: c6 – c7. This move fails to check-mate in the following move because black can respond by advancing pawn: b5 – b4. This makes it impossible to check-mate (the resulting board is given at the top of the next page).

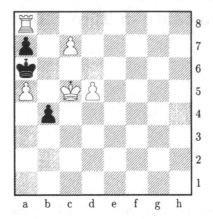

Note that white has two possibilities to check:

- a8 × a7, or

- c7 – c8 with a promotion of the pawn to a queen or bishop.

Neither of these checking moves gives a check-mate. In the former case
the black king can retake the white rook, whereas in the latter case, the
black king may take the white pawn on a5.[1]

- White advances the pawn: d5 – d6. This is a very peaceful move for which
  no defense is necessary. Black can respond again by advancing the pawn
  b5 – b4 yielding the board given below.

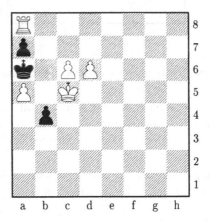

In the second move, white can check only by a8 × a7, but as we've seen
previously, this doesn't lead to a check-mate (the black king takes the
rook).

---

[1] White can then check-mate in the following move, a8 × a7, but it's white's third move
so that's not good enough.

- White can move the king. However, the move c5 – b4 results in a *pat* (which yields a draw) because black can't make any legal move. Two other moves (c5 – d6 and c5 – d4) also fail to achieve the goal: the black king can safely take the white pawn on a5. If white checks the black king — the only possibility to do so is again by the move a8 × a7 — then black can respond with a5 – b4 or a5 – b6. Thus, moving the white king doesn't lead to the solution.

- White can move the rook. Again, there are three possibilities to consider:

  - a8 × a7. But the black king can retake the white rook. After that, in the second move, white can't even check the black king.

  - a8 – b8. Black can advance the pawn b5 – b4. At this stage the only possibility to check the black king is by using the rook: b8 – b6+. But this isn't a check-mate because black can take the rook: a7 × b6.

  - a8 – ?8 which means that white moves the rook to either c8, d8, e8, f8, g8, or h8. In this case black can advance the pawn, b5 – b4, and again, white can't even place the black king in check on the second move.

We can wrap up the analysis of the board with a simple conclusion: it is *impossible* for white to check-mate black in only two moves!

So, where's the trick? The only way to solve the problem is to take into account the time factor! The diagram shown at the start of the section shouldn't be considered as a static snapshot, but rather as a particular stage of some dynamic development. White is about to move at this stage of the game, so the previous move was made by black. This much is trivial, but which move did black make? There are only three black pieces on the board so we can investigate all of the possibilities:

- The pawn on a7 is still on its original position from the beginning of the game, so we can eliminate that possibility.

- The king could have arrived at a6 from either b6 or b7. But upon further inspection, this is impossible. The black king couldn't occupy b6 on the previous move because of the presence of the white king on c5. Also, the black king couldn't occupy b7 on the previous move because of the white pawn on c6 (which couldn't move on the previous move).

Thus, the conclusion is that the last move was made by the black pawn on b5. What move was that? Either b7 – b5, or b6 – b5. The pawn on b5 couldn't have occupied b6, however, because of the presence of the white king on c5, so the previous move had to be b7 – b5.

With this knowledge, the solution of the puzzle is easy! The required move by white is simply

a5 × b6 (*en passant*)

and the resulting board is given below:

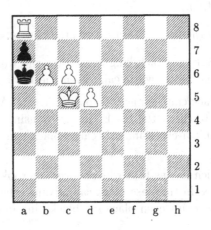

Now black is forced to respond with a6 − a5. It's the only possible move, and white check-mates in the second move:

a8 × a7++.

Beautiful, isn't it?

11. Time-Varying Environments and Noise

Change is the law of life.
And those who look only to the past
or the present are certain
to miss the future.

John F. Kennedy, speech, June 25, 1963

As we have seen, real-world problems are usually associated with large sample spaces and nonlinear, multimodal evaluation functions. In many cases, there are no closed-form mathematical procedures for generating solutions, and even when these techniques exist, they are often too computationally intensive to be practical. Their computational complexity increases at such a rapid rate that when we try to apply them to anything but trivial problems they are practically useless. In order to make them more useful, we accept simplified assumptions about the real world, perhaps mandating the application of a linear evaluation function, with linear constraints, integer values, or other common mathematical devices. We sacrifice the right answer just to get any answer at all.

Actually, the situation is worse than this because in real-world problem solving it isn't enough to solve a problem. The problem must be solved continually, over and over again. This is necessary because the world is always changing around us. The solution for the problem at hand quickly becomes the solution for yesterday's problem. Being content with deriving a single solution to any significant real-world challenge is tantamount to accepting defeat.

11.1 Life presents a dynamic landscape

To give an idea of the pervasive nature of how problems change let's consider the simple task of flying from New York to Los Angeles. Before going even one step further, the first question to ask is "what is the objective?" That is, what are the parameters of concern in evaluating alternative solutions to the problem? One obvious objective is to arrive in Los Angeles on time, or better yet, arrive early. Arriving late is undesirable, but certainly there is more to the flight than just a timely arrival. What about safety? That's an easy one to treat: the flight should certainly be safe. But what level of risk are you willing to accept? While you are pondering this parameter, let's add another dimension of concern. Let's presume that you own the plane that will be used for the flight. Then you'll also need to worry about cost. For example, the cost of fuel burned during the flight. A faster flight might give a better result for the timeliness parameter, but it would cost more. And of course you'll also want a smooth ride. To get it, you may have to adjust your altitude of flight to avoid turbulence, with each adjustment likely costing you more fuel. The optimal flight requires trading off each of the parameters in a most appropriate manner.

Although this is a difficult challenge in its own right, let's give you the benefit of having already defined the appropriate normalizing function to handle the various levels of importance of each of these parameters and let's presume that you have actually discovered the best possible route to fly. You take off. The flight plan indicates that you'll pass over Denver, Colorado in order to take advantage of some of the predicted winds aloft, thereby reducing your fuel costs while still arriving on time. Along the way, bad weather comes up before you get to Denver and you must deviate from your plan. Now what? How are you going to replan? You're stuck. You're stuck because you haven't considered this possibility and incorporated it into your prior planning. The best hope might seem to be starting over as if this were a new problem. How can I now get from where I am (somewhere over, say, the state of Ohio) to Los Angeles in an optimal manner in terms of my evaluation function? The problem is that there is insufficient time to find the best solution in this manner. The required computation is too great. You must adopt some alternative technique.

While debating with the pilot over what to do next, the weather forecasters report that the winds aloft are no longer as predicted, and neither are the reported areas of turbulence. In response, you are forced to do *something* so you pick a new path that "seems to work" and press on, knowing that you have quite likely missed many opportunities for optimizing your path already.

When you arrive in Los Angeles, they are reporting fog, with a cloud ceiling 220 feet above the ground. This is just barely above the minimum conditions (a 200 foot ceiling) that are required to land within regulations. Will the weather be this way when you arrive over the end of the runway? How can you determine this? You ask what the weather was like 20 minutes ago, and an hour ago. The trend shows the weather is getting worse: the fog is lowering. Will it stay above the minimum ceiling? How can you determine this? The best you can do is guess. The impact of your guess is important. If you decide to go for the approach to the airport and find that the weather is below the minimum safe standard you'll be forced to "go around" and waste at least 30 minutes being routed by air traffic control before trying for another approach – alternatively you could make an illegal and possibly unsafe landing in less than minimum conditions. (We now see that we left out one possible parameter: conducting a legal flight.) Rather than opt for landing at Los Angeles, you might be better off diverting to a nearby airport that has better weather conditions and then renting a car to get to your final destination. You'll be a little late and it will cost more than you had originally planned, but it might be worth it.

Certainly, we could make the dynamics even more interesting by imposing an engine failure, a fire in flight, an inoperative instrument, or any of a number of other conditions on the situation.[1] But by now the point has been made: solving problems requires planning for contingencies. Not only that, it also requires being able to handle contingencies that weren't anticipated in a timely manner.

Let's make the example more relevant to industry so that we can see the effects on the bottom line of profit and loss. You run a factory that manufactures clothing garments. It requires 100 employees to operate the ten machines that are needed to process each of the five different types of clothes. A company has made a simulation of the factory for you so that alternative schedules can be tried. Before taking the first step, the question is "What is the objective?" A possible answer is to produce the most clothes, but that's probably not the complete answer. You probably want to put more emphasis on producing more of the clothes that make you more money. You don't want clothes sitting in a warehouse not being sold. You also want the least cost of operations. And you also want good employee morale, which means that you can't change their schedules too quickly; otherwise, the workers won't feel comfortable with their job assignments. What's more, each worker has different talents so it is important both for morale and efficiency to ensure that the tasks assigned to each employee reflect their job skills. Further, there are constraints here because not all of the workers can operate each different piece of machinery. Can we go further? How about allowing for the proper maintenance of equipment? You can't run machines 24 hours per day unless you want to face unexpected breakdowns.

Suppose you properly normalize all of these parameters of concern and somehow manage to find the best schedule in light of the evaluation function. As you proceed to post the schedule for the employees you learn that: (1) one of the pieces of equipment is no longer working and the estimated time to repair the machine is six to eight hours, but it might be "significantly longer," and (2) three people are sick today and four others are late to work because of an accident on the freeway. How should you now modify your plans? The first piece of information dictates that you must make some modification because one of the machines you were going to use is no longer available. Your plan doesn't meet the new constraints. The second piece of information doesn't necessarily mandate that you change your plans, but in light of one less machine being available perhaps you should take the available workforce into consideration.

You then find that one of your supervisors calls and says that they have a rush order for an important customer and they want you to suspend all of your operations and process the incoming order until it is finished. How does this affect your schedule? Later in the day, one of your suppliers phones and says that they cannot make their promised delivery date for ten percent of

---

[1]One of the authors (D.F.) holds a commercial pilot's license. In over 500 hours of flying, on separate occasions, he has encountered (1) a complete radio failure (on his first solo cross-country flight), (2) the baggage door opening in a flight at night over clouds that prevented visual contact with the ground, (3) the engine sputtering while flying over mountains, (4) the main cabin door failing to latch and coming open in flight, and (5) numerous instances where in-flight navigational aids on the ground failed to function. Although the description of the fictional flight from New York to Los Angeles contains many significant unexpected occurrences, these are actually not atypical.

your garments. How do you reschedule in light of this information? As you are executing your plan, some of the workers forget to maintain an accurate log of the materials used. As a result, you believe that you have more stock on hand than you actually have. How will you adjust your schedule to compensate for this problem when you discover it three or four days from now? Without a proper sequence of decisions, this situation could rapidly deteriorate into an unworkable setting where every action is taken simply to overcome the most recent anomaly. This is such a common occurrence in normal business practice that it's described by the phrase "putting out fires." It shouldn't be this way.

Clearly, problem solving is more than finding the one best solution. It is at once (1) defining the problem, (2) representing the possible candidate solutions, (3) anticipating how the problem may change over time, (4) estimating the probabilities of the possible changes, and (5) searching for solutions that are robust to those changes. Moreover, it is the action of re-solving the problem as it changes based on the most recent available information. Problem solving is a never-ending process.

It is at this point that some of the heuristics that we have discussed can be compared in terms of their utility in adapting to changing circumstances. The classic methods of generating solutions to planning, scheduling, and other combinatorial problems are mathematical. These include dynamic and linear programming. These methods are notably poor for adapting to changing circumstances. The methods compute only a single solution and when conditions change, the new best solution must be determined from scratch. The work that has already been devoted to solving the problem is lost.

In contrast, the methods that *search* for solutions, rather than calculate them, often appear better suited to adapting to changing circumstances. The appropriate analogy to have in mind is one where the evaluation function appears as a landscape that is the result of mapping alternative solutions to their corresponding functional values. Each possible solution corresponds to a point on that landscape. When the conditions of the problem change there are two alternatives: (1) the landscape changes, or (2) the constraints on the feasible region of possible solutions change. Of course, it might be that both change simultaneously. Let's consider each of these possibilities further.

A change in the landscape will be encountered when the evaluation function is altered. Remember that the evaluation function is properly a subjective statement of what is to be achieved (and what is to be avoided). If the task at hand changes, so will the evaluation function. For example, if the relative importance of a timely arrival in Los Angeles changes in relation to the cost for fuel, the evaluation function will in turn need to be revised to reflect this change. The effect will be seen as a new fitness (or error) landscape. This effect might also be seen if the cost for fuel itself increases or decreases, because this will affect the degree of achievement of each possible flight path with respective to cost efficiency. Every time the price of oil changes, so does the landscape.

The constraints on the feasible region can change as a function of the available resources. In the factory setting above, when one of the 10 machines breaks down, what changes is the feasible region. There is now a hard constraint indicating that all schedules must use up to nine machines. Previously feasible schedules that utilized all of the available machines are now infeasible. (The manner in which to treat infeasible solutions was considered in chapter 9).

When the purpose or conditions for solving the problem change, so too does the landscape and/or the feasible region change, but in some cases the change may not be very great. If the price of oil changes by $0.01 per barrel then it's likely that there will be a strong similarity between the evaluated worth of any previous solution and the new worth in light of the change. Our previous best-discovered solution would still likely be very good, and what's more, it's intuitive that those solutions that offer an improvement will be found in the local neighborhood of our previous best solution. Of course, it doesn't have to turn out this way. It might be that slight change in, say, the winds aloft, might mandate an entirely different path from New York to Los Angeles. That's possible, but quite often that's not the way the real world works.

Not all changes are small or slight, however, and when larger changes are encountered, the similarity between previously discovered solutions and new opportunities may be slim. Our previous solution(s) may be of little or no value at all. We have seen this in the course of natural evolution. There have been several mass extinctions in our history, the most well known being the extinction of the dinosaurs (as well as over 60 percent of all species on the planet) when a comet impacted the earth 65 million years ago [379]. The environmental change that was felt from the impact was severe. The majority of species were both unprepared for and unable to adapt to the new conditions. Evolution is not a guarantee for successful adaptation in the face of changing circumstances.

Nevertheless, evolutionary computation provides several advantages over other heuristics when trying to solve problems as conditions change. The primary advantage is that evolutionary algorithms maintain a population of solutions at any point in time, rather than just a single solution. This provides the potential for a diversity of approaches to problem solving. When the problem changes, to cite a cliché, we don't have all of our "eggs in one basket." If the constraints change and make one solution infeasible, perhaps another reasonable solution in the population will still be feasible. We can move from solution to solution within the population to determine if any of the other available alternatives that have been maintained up to this point are of value.

A second and related advantage is that we can begin searching for new solutions using the diversity in the population. Each available candidate solution offers a starting point for discovering new ways to treat the problem given whatever change has occurred. We don't have to rely on only a single starting point, and we certainly don't have to recompute a new solution starting from *tabula rasa*. If there are any similarities between the old problem and the new problem, it's possible that these will be reflected in the solutions that are present in the population.

To see an example of the potential for using evolutionary computation to adapt to problems as they change, let's consider the alternatives of either using an evolutionary algorithm or a greedy nearest-neighbor algorithm to solve a TSP. Suppose that we needed to solve a problem with 70 cities. For the example, we have simply randomized the locations of these cities uniformly in a square of 100 units on a side. The positions are shown in figure 11.1. Just judging from this figure, there aren't any obvious solutions. The problem of finding a suitable tour for these cities seems to be a significant challenge.

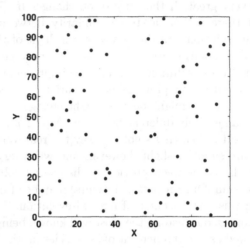

Fig. 11.1. A set of 70 cities for the traveling salesman to visit, including home base at (6,2)

Let's start by applying the greedy approach. Suppose that the city at location (6,2) is our home base (that's the city in the lower left-hand corner). Our first decision is to determine which of the 69 other cities is nearest. If we were looking at the plot of the points we could easily discard some of the possibilities, but since we are considering the application of an algorithm here we must make 69 comparisons for the distance between each other city and our home at (6,2). In doing this we find that location (14,7) is closest. Next, we must determine which of the remaining 68 cities is closest to (14,7). The answer is (11,18). And so forth. Figure 11.2 shows the resulting tour. It doesn't seem very good. It has some very long paths and it crosses over itself numerous times. The total length of the tour is 843.95 units.[2]

Having found this solution, let's now compare it to a solution that we evolve. Suppose that we use a permutation representation, numbering the cities from 1 to 70 in an ordered list. To keep things simple, our initial population consists of only four parent tours. Each one is constructed at random. The variation operator we'll apply is the reversal procedure described earlier in chapter 7. Here we pick two locations along the permutation and reverse the order of appearance of each city in this segment. Again, for simplicity we'll restrict our variation operator to only generate one offspring per parent. Then selection will compare

---

[2]Using the formula given in chapter 8 for the expected length of a minimal TSP tour we find that the greedy solution is about 35 percent worse than the expected best tour.

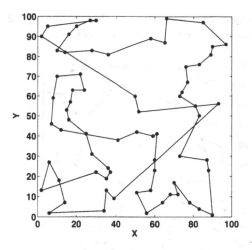

Fig. 11.2. The greedy solution that starts at (6,2) and takes each next nearest neighboring city. Being greedy in the short term results in some very long links between cities overall.

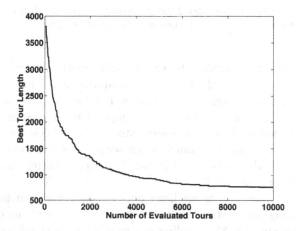

Fig. 11.3. The rate of evolutionary optimization of the length of the best tour found as a function of the number of evaluated tours. With four parents at each generation, 10,000 tours is equivalent to 2,500 generations.

the distances of each tour in the population and keep the best four tours to be parents for the next generation. Figure 11.3 shows the rate of evolutionary optimization in one trial. The best tour length at the first generation was 3,816.32, which is much worse than the greedy solution we found above. But after 2,500 generations, evolution discovered the tour shown in figure 11.4, which has a length of 758.54 units. This is 21.04 percent worse than the expected best solution, but we only used a population of size four, so certainly we could anticipate doing better if we implemented a larger population. Nevertheless, the solution in figure 11.4 does appear much better than that in figure 11.2, and in fact it's a 10.12 percent improvement over that greedy solution.

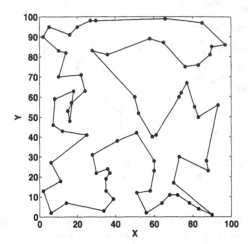

Fig. 11.4. The best evolved tour after 2,500 generation. Even though this solution is not perfect, it is about a 10 percent improvement over the greedy solution.

To be fair, we must consider the amount of computation required for both procedures. For the greedy algorithm, we computed 2,415 distances in order to perform all of the necessary comparisons. For the evolutionary algorithm, discounting the computations required to implement the variation operator, we tested 10,000 tours, which translates into about 700,000 distances. So we performed about 290 times as much work with the evolutionary algorithm, but we ended up with a solution that was 10 percent better and much more aesthetically pleasing as well.[3]

Now, let's say that just as we head out the door from our home at (6,2) to start performing our business in each city we receive a phone call from the customer at location (11,18) canceling our visit with them for today. Now we must replan.

Using the greedy algorithm we must start over with a new 69-city TSP. There are now 68 comparisons to make on the first choice, and so forth. After computing 2,346 distances, we create the new solution shown in figure 11.5. The length of this new tour is 962.64 units. Note that this is *larger* than the previous greedy solution. Even though we have *removed* a city, the length of our tour has become *longer*. This is not very encouraging. Our new solution is about 55 percent worse than the expected best solution for 69-city uniform TSPs.

---

[3]This is actually a serious consideration. In real-world conditions, people will often favor a solution that is measurably *worse* than another but is more aesthetically pleasing. Also in fairness, the evolutionary algorithm first outperformed the greedy solution after 96,250 tours were evaluated, which is about 160 times as much work as was required by the greedy solution. Still, this is a considerably greater effort.

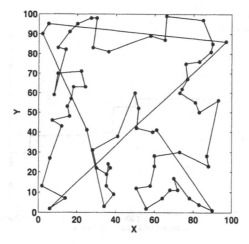

Fig. 11.5. After removing the customer at (11,18), the greedy solution deteriorates, even in the face of a smaller problem. Note that the last two legs of the tour are extremely long, covering almost the entire range of the salesman's area.

Instead, let's compare how we would replan using the evolutionary algorithm. We already have a solution to the previous problem. If we take that solution and eliminate the city that corresponds to the location (11,18) the resulting new tour can serve as a starting point for our population. Let's make three additional copies of this solution so we again have four initial parents. If we then apply the evolutionary algorithm for 500 generations we find the optimization curve shown in figure 11.6. Our initial solution at the first generation had a length of 752.51 units, and we improved it to a total length of 729.01 units. The new best-evolved solution is shown in figure 11.7.

Note that when we started with the best-evolved solution to the 70-city TSP, our initial population for the new 69-city TSP was already better than the new solution produced by the greedy algorithm. The extra processing that was done with the evolutionary algorithm only served to improve the tour even further. In contrast to the greedy algorithm, which generated a solution that was actually worse after a city was removed, the evolutionary algorithm was able to use the information that it had already built up in the ordering of the cities and reoptimize for the new problem very quickly. Further, the evolutionary algorithm in this example was really very limited because we only used four solutions as parents. If the population size were larger, we could expect better performance in adapting to changes in the problem.

In addition to this ability to readapt to dynamic environments, evolutionary algorithms offer the potential for incorporating possible changes in the conditions directly into the evaluation of a solution. If we can limit the possible outcomes and assign them reasonable probabilities then we can optimize solutions with respect to several different criteria: (1) a weighted expectation of the

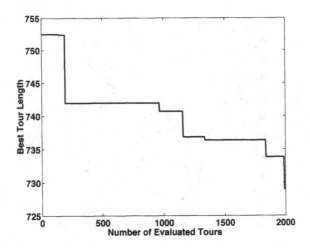

Fig. 11.6. Evolutionary optimization after the problem is changed with the removal of the customer at (11,18). The evolutionary algorithm picks up where it left off.

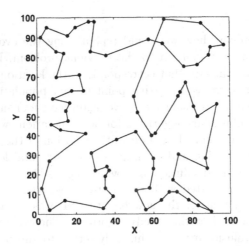

Fig. 11.7. The best-evolved solution to the new 69-city TSP starting with the population that addressed the previous 70-city TSP. Again, after evaluating 2000 additional tours, the best-found solution is not perfect (it crosses over itself), but it is a significant improvement over the greedy solution. Furthermore, it's easy to envision creating heuristics to "clean up" a tour such as this, which suggests a possible synergy from hybridizing approaches (see chapter 16).

outcomes based on their likelihood, (2) the maximum likelihood outcome, (3) the worst-case outcome, and so forth. As an example, if we were designing a communications network with many connected nodes, we could evaluate any candidate solution not just in terms of its functionality when the network performs perfectly, but also when connections fail. The performance of a solution under these possible breakdowns in the system can be incorporated into the solution's overall worth. In the end, the evolved solution will have been tested on many different possible "failures" and we can have confidence that it will be able to respond appropriately [423].

You'll recall that in the last paragraph of chapter 9, we suggested the need for a parametrized test-case generator for analyzing various contraint-handling methods in a systematic way. There's a similar need for treating dynamic environments and the first attempts to construct such test-case generators have already appeared [472].

11.2 The real world is noisy

Another source of difficulty in problem solving is that our observations are always in error. We can measure the same thing over and over and find different results for each measurement. In part, this occurs because our instruments are not sufficiently precise. In addition, there are extraneous factors that may influence the observations that we make. If we measure the amount of grain that can be harvested in May from, say, a specific acre of land in British Columbia, we could record this value year after year and find different numbers of bushels each time. The grain yield is affected by the winter temperature, the precipitation, and numerous other factors that are beyond our control. The typical manner to treat these exogenous variables is to abstract them into a general "noise" variable and then model that noise, or perhaps simply make some assumptions about the noise (e.g., it is Gaussian with zero mean). If we cannot make these assumptions, we are often left without the ability to calculate answers directly and must instead use a search algorithm for a set of parameters that are optimal in some sense.

Thus, we return to the concept of searching a fitness or error landscape for a maximum or minimum value, but the value that we obtain for every set of adjustable parameters varies each time we evaluate them. Even if we were searching a parameter space like that shown in figure 11.8 (a quadratic bowl) under the imposition of noise, the underlying structure of the function can become lost or, at the very least, difficult to discern (figure 11.9).

This poses a considerable challenge to a gradient method because unless the underlying structure of the function is given beforehand, approximating the gradient at any point becomes highly unreliable. It is simply a function of the noise. Similarly, even a hill-climbing (or equivalent descent) method can have trouble because steps will be accepted as a result of random noise that should have been rejected.

Indeed, this is true for all sampling techniques. It is just as true for hill-climbing as it is for, say, simulated annealing or evolutionary algorithms. The difference is that in the latter case, there is a population of contending solutions to call upon, rather than a single point. In a sense, the population contains more

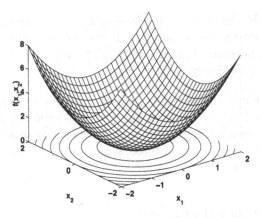

Fig. 11.8. A quadratic bowl

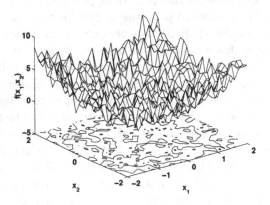

Fig. 11.9. A quadratic bowl where every sampled point has standard Gaussian noise superimposed on it. Even though gradient methods can be very effective at locating the extrema of functions such as that shown in figure 11.8, when noise is added to the evaluation of candidate solutions, these methods can fail.

information about the shape of the landscape than can be held in just a single point. Over successive generations, the mean of the population can reflect the statistical average fitness acquired over many samples. Another advantage of evolutionary algorithms is that for a solution to be maintained it must compete effectively, not just with a previous single best solution, but rather with a collection of solutions. This often means that a lucky point that evaluates well one time will be discarded in the next generation. In contrast, when a hill-climbing method locates a point that evaluates well simply because of lucky random noise, it may be difficult to displace that point by searching within its local neighborhood unless additional heuristics are employed or that point is resampled and scored again.

The degree of difficulty faced depends in part on the signal-to-noise ratio (SNR). As the SNR decreases, the surface's underlying structure becomes less and less evident. This can pose significant challenges for some classic methods of optimization, particularly if the function is multimodal or has deep grooves. To illustrate the problem, consider the simple function shown in figure 11.10,

$$x^4 + 2x^3 e^x,$$

and let's use Brent's method (see chapter 3), which relies on successive function evaluations in order to locate a minimum. If we begin with a bracketing region of $[-5, 5]$ with a starting midpoint of 3, the minimum can be located at -0.630591 with an error tolerance of 10^{-6}. But what happens when we add standard Gaussian noise to the function every time the function evaluation routine in Brent's method is called? Averaged over 100 independent trials, the method locates the minimum at -0.396902 with a standard deviation of 0.391084. The addition of noise has a marked effect on the final outcome.

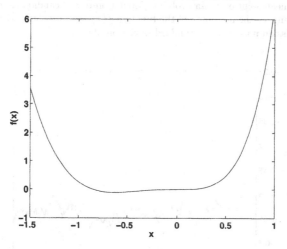

Fig. 11.10. The function $f(x) = x^4 + 2x^3 e^x$

For comparison, let's apply an evolutionary algorithm with a population size of 100 parents, each parent initialized in the range $[-5, 5]$, one offspring per parent generated by adding a zero mean Gaussian random variable with a fixed standard deviation of 0.01, and selection maintaining the best 100 solutions at each generation. Figure 11.11 shows the mean solution in the population as a function of the number of generations. The mean over 500 generations is -0.506095 with a standard deviation of 0.152638. Figure 11.12 shows the mean solution in the population over successive generations when the population size is increased to 1,000 parents. The mean over 500 generations is -0.528244 with a standard deviation of 0.051627. The introduction of random noise to the function biases the evolution away from the true minimum, but the effect is not as great as is observed using Brent's method.

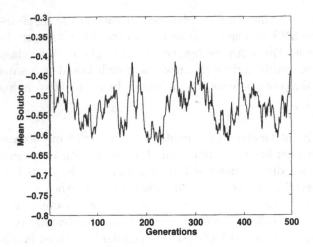

Fig. 11.11. The mean solution in an evolving population of 100 candidates over 500 generations when searching for the minimum of the function $f(x) = x^4 + 2x^3 e^x$ under the imposition of zero mean Gaussian noise with a standard deviation of 0.01

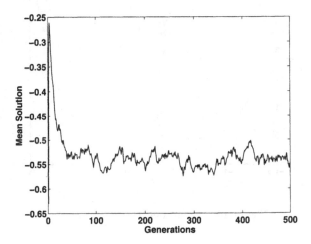

Fig. 11.12. The mean solution in an evolving population of 1000 candidates over 500 generations when searching for the minimum of $f(x) = x^4 + 2x^3 e^x$ under the same imposition of noise

Even though Brent's method was affected negatively by the addition of noise to the function, here the effects weren't tragic. The technique was still able to generate points that might be regarded as being close to the minimum, and the average solution over 100 trials was within one estimated standard deviation of the true optimum. The effects of noise on classic optimization methods, however, can be devastating when a gradient of the function must be approximated. This is often required because we cannot usually write the evaluation function in a form that is suitable for calculating the derivative at any point directly. Techniques that use estimated gradients based on sampled points essentially rely on the difference in the height of the evaluation function taken over a distance between two points:

$$\frac{g(x_1) - g(x_2)}{|x_1 - x_2|},$$

where $|x_1 - x_2|$ is hopefully small. Under the imposition of noise, both $g(x_1)$ and $g(x_2)$ could be rewritten as

$$g(x_1) = f(x_1) + h(x_1) \text{ and } g(x_2) = f(x_2) + h(x_2),$$

where $f(\cdot)$ is the true evaluation of the solution and $h(\cdot)$ is the noise that is added. If the noise terms added to each function evaluation are statistically independent, then the variance of the noise on the difference $g(x_1) - g(x_2)$ is the *sum* of the variances of $h(x_1)$ and $h(x_2)$. If $h(\cdot)$ is, say, standard Gaussian random variation with zero mean and unit variance, then the random variable $G = g(x_1) - g(x_2)$ will have a variance of 2.0. If the standard deviation of the G is greater than the difference between $f(x_1)$ and $f(x_2)$ then it will become difficult to obtain reliable estimates of even the correct sign of the gradient. Methods that then rely on the estimated gradient for a new direction to search may be sent on a "wild goose chase."

11.2.1 Ensuring diversity

The diversity that can be offered in a collection of potential solutions is often valuable. Nevertheless, there is nothing in a typical evolutionary algorithm that guarantees that the population will remain diverse. Selection tends to cull out the least-fit solutions. The result is that the population may converge to a single point, or a collection of points that are all within some small neighborhood of the best solution found so far. The initial wide coverage of the fitness landscape that is obtained from a random distribution of trials decreases rapidly, and after several generations all of the solutions in the population may be essentially similar. The advantages of a population of solutions diminish greatly in the face of dynamic, noisy environments when all of the solutions are the same. Thus, it is natural to develop means for enforcing diversity within a population.

One possibility is called *niching*. In ecology, a niche refers to the sum total of an organism's use of the living and nonliving resources of its environment. For example, the temperature range, the location of its dwelling, and its food source, are all parts of its niche. In evolutionary computation, niching refers to the notion that competing species cannot coexist in the same niche (the competitive exclusion principle). One way to enforce this concept into an evolutionary optimization algorithm is to demand that each solution share its fitness with other solutions that are within the same region of the search space. This so-called *fitness sharing* works by assigning a fitness to each individual solution

i

$$F_i' = \frac{F(i)}{\sum_{j=1}^{n} sh(d(i,j))},$$

where there are n solutions in the population, $F(i)$ is the result obtained from the evaluation function (where the goal is maximization), $d(i,j)$ is a "distance" from the i-th solution to the j-th solution, and $sh(d(i,j))$ is a sharing function, commonly assigned as

$$sh(d) = \begin{cases} 1 - (d/\sigma_{share})^{\alpha} & \text{if } d < \sigma_{share} \\ 0 & \text{otherwise,} \end{cases}$$

where σ_{share} defines the size of the neighborhood around the i-th solution and α is a scaling parameter. The closer two solutions are to each other, either in terms of their structure or their function, the more each is penalized for occupying the adjacent locations. This imposition forces the population to remain spread out across multiple extrema, with each peak being able to support a number of individuals that is in proportion to its height.

Although this method sounds enticing, it has some associated drawbacks. The primary drawback is that the effect of fitness sharing is determined by adjusting the parameters σ_{share} and α. Alternative choices will result in alternative clustering of solutions in the search space. Which choice is best depends on the problem and in turn the evaluation function. The value for σ_{share} should be sufficiently small to differentiate between peaks, but how can we know how small this should be before we start to solve our problem? The common result is a series of ad hoc changes to the parameters until the diversity simply appears acceptable. From the practitioner's point of view, applying such a procedure can be challenging because the results depend very strongly on the choices for parameters, and making good decisions regarding these parameters can be quite difficult. Another secondary drawback is the computational expense required to compare solutions, determine their relative distance, and compute the shared fitness. For large population sizes, the procedure may be computationally prohibitive.

A different method for enforcing diversity is called *crowding*. When applying selection to determine which solutions to maintain, new solutions that are similar to old ones simply replace the old ones. In this manner, the population doesn't tend to build up an excess of similar solutions. Again, a distance measure is required to determine how similar two solutions are, and the appropriate setting for this criterion is problem dependent.

Each of the niching and crowding methods requires some measure of distance between any new solutions. When treating problems where solutions lie in the real numbers, this seems straightforward because we can rely on familiar norms for this measure (e.g., a Euclidean measure). When treating combinatorial problems in discrete space, however, things are not as clear. If solutions are represented as binary strings, we might be inclined to use the Hamming distance, but what if the solutions are permutations? What is the distance between [1 3 2 4 5 7 6] and [3 6 7 5 1 2 4]? There's no obvious answer here. More to the point, the "distance" between two solutions might be better reflected by how they are interpreted in terms of their functionality rather than their coding. For example, if we were solving a TSP, then the two solutions [1 2 3 4 5 6 7] and [4 5 6 7 1 2 3] which represent ordered lists of cities to be visited, with a return from the last city to the first city in the list being implicit, are equivalent functionally. They both describe exactly the same tour. Intuitively, the distance between these solutions is effectively zero. This situation really presents an important obstacle. We could no doubt encode both of these permutations in some binary language. If we did that, we'd be tempted to use a Hamming distance to describe their similarity, but this would be an entirely artificial measure of their relatedness that would very likely have nothing to do with how similar the two solutions are functionally. The same argument holds for any other representation that we might choose. Sometimes the obvious straightforward distance measure is the wrong measure to use!

Yet another method for maintaining population diversity is to set up the evolutionary algorithm to compete entire populations against each other, each being judged not only on the fitness scores of their survivors but also on their diversity. One possible measure for a population's quality might be

$$Q(\mathbf{x}) = \alpha \overline{f} + (1 - \alpha)\sigma_{\mathbf{x}}^2,$$

where $\mathbf{x}$ is the vector of solutions in the population, $\overline{f}$ is the mean worth of all solutions in $\mathbf{x}$, $\sigma_{\mathbf{x}}^2$ is the variance of the solutions in $\mathbf{x}$, and α is a scaling term to adjust the weighted contribution of fitness and diversity. Each population of solutions takes the place of a typical individual in a single population. Selection is used to eliminate entire populations of solutions, and variation is used to generate completely new populations based on random alterations of existing populations. Of course, the normal application of an evolutionary algorithm for each of the populations is also ongoing, so the design here is a hierarchical structure where, at the lower level, individuals within a population undergo variation and selection, and at a higher level complete populations undergo variation and selection. The concept of a hierarchical series of evolutionary algorithms is intriguing, but the evident drawback for the approach is its computational complexity and memory requirements for maintaining a population of populations.

Still another way to enforce diversity in a population is to have it remember where it's been before. There are really two ways to do this by using either *explicit* or *implicit* memory structures. The idea behind explicit memory is that specific information is stored during an evolutionary trial that can be re-introduced into the population at a later stage. The important questions to be addressed are

- What is remembered and when?
- What is recalled and when?
- What are the memory structures and how large are they?
- What is removed from the memory (if necessary)?

Several experiments with explicit memory have been published. For example, Louis and Xu [292] considered the case where only the best individual is stored at regular intervals (i.e., the best individual in the population is added to the memory at the end of *each* interval). After a change in the environment, a (small) percentage of r-initialized individuals is taken from the memory and all of the other individuals are re-initialized randomly. Of course, this still leaves the problem of detecting a "change" in the environment. Ramsey and Grefenstette [376] developed a system that also stored the best individual in the population but it was stored along with some environmental variables, thus establishing a connection between a state of the environment and the solution for this state. After a change in the environment, around 50 percent of re-initialized individuals were taken from the memory. Attention was given to those individuals that were stored during environments that were *similar* to the new environment. Again, how to measure the change that necessitates recalling stored individuals and the similarity between environments is an open question.

Mori et al. [326] experimented with a system in which the best individual is stored in memory *every* generation. Memory is always limited, however, so an individual is also removed from memory every generation. The decision of which to remove is based on the number of generations that the individual has resided in memory and on their contribution to the diversity of that memory. Individuals from the memory might be selected as new parents in evolutionary process. Trojanowski et al. [468, 469] experimented with individual memory structures assigned to each solution in the population. An individual "remembers" its most recent ancestors due to the limited size of the available memory. After a change in the environment, each individual is evaluated along with all the solutions in its memory and the best solution is selected as the new current solution.

In contrast to explicit memory, implicit memory leaves the control of what to remember, when to recall it, and so forth, to the evolutionary algorithm itself. One method for achieving this is through self-adaptation, which we saw in chapter 10. Those strategies for generating new solutions that have worked well in the past are then "memorized" in a sense by their offspring. Another method is based on the ideas of diploidy and dominance that arise in genetics. In this model, individuals maintain two or more solutions (or data structures) and a dominance function selects pieces from one or more structures to build a final overall solution. The key issue in this approach is how to design this function so that meaningful results can emerge.

How do to this isn't that clear. For example, Goldberg and Smith [193] worked on a binary representation with the twist of having three values: 0, recessive 1, and dominant 1. The value 0 is selected if and only if the corresponding value in the other structure (where the individual comprised two such structures) was also a 0 or a recessive 1. Otherwise, a 1 would be selected. Ng and Wong [335] extended this to include recessive and dominant 0s and 1s. Ryan [404] proposed an additive dominance scheme where the corresponding values (which could be numeric, not just binary) were added between each data structure, with the overall outcome being set to 0 if the sum was below a threshold value of 10.

Every one of these approaches is an ad hoc procedure. That's perhaps a kind way of saying that there's no theory for choosing what dominance function to design, how many structures to include in an individual, and indeed whether or not to have the dominance function change over time. Lewis et al. [281] observed that changing the dominance function is essential for nonstationary environments, but the observation is too general to be practical. At present, the area of diploid or polyploid structures and dominance functions is a wide open area of research with little theory to guide the practitioner.

Each of these methods — niching, sharing, memory — can be applied usefully to ensure population diversity as a hedge against a dynamic and/or noisy environment. Each also has a few parameters that must be set and each parameter increases the computational complexity of the procedures. The best use of these or other methods for generating diverse solutions requires considerable judgment. There really are no effective heuristics to guide the choices to be made that will work in general, regardless of the problem to be faced.

11.3 Modeling of ship trajectory

In chapter 9, we briefly discussed the difficulties in designing evaluation functions for feasible and infeasible individuals in the context of a path planning problem (see section 9.1.3), i.e., in an environment where several obstacles were present. It's even more interesting to consider the approach in path planning problems when some obstacles are dynamic, moving in different directions with variable speeds.

This particular scenario was considered in [434] in the context of modeling ship trajectories in collision situations. The paper presents experiments with a modified version of the Evolutionary Planner/Navigator (EP/N) developed earlier for static environments [501]. This new extension of EP/N computes a "safe-optimum" path of a ship operating in environments with static and dynamic obstacles. The new system also incorporates time, the variable speed of the ship, and time-varying constraints that represent movable ships (also known as "strange-ships" or "targets," which are other vessels in the environment that must be avoided based on a safe distance between passing targets, their speed ratio, and their bearing).

The new version of EP/N uses a different path representation, where each segment of the path has an accompanying value for the associated own-ship's speed, and a different evaluation function that incorporates the time needed to traverse the path, as well as the standard smoothness and clearance-from-other-vessel parameters. The papers [501] and [434] provide details on the original EP/N system and its extension to dynamic environments.

Let's consider an interesting navigation problem across a strait and see how EP/N handles it. Let's assume that the environment has six static navigational constraints (see figure 11.13) and two targets. The own-ship starts close to the upper-left corner of the environment and must reach a location on the opposite

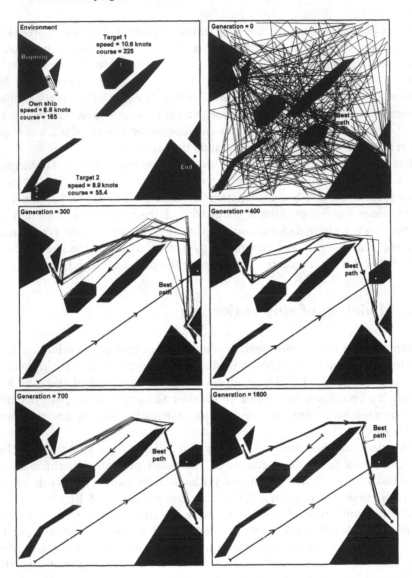

Fig. 11.13. Path evolution when approaching two moving targets in the presence of static navigation constraints. The own-ship travels at $\vartheta = 8.6$ knots.

side of the strait (close to the lower-right corner). There's a passage between two narrow islands located in the middle of the strait, but the movement of targets 1 and 2 makes this passage difficult. The own-ship travels at $\vartheta = 8.6$ knots.

Since targets 1 and 2 are blocking the passage of the own-ship between two islands, the system develops a trajectory (figure 11.13) that "goes around" the islands rather than between them. Target 1 doesn't pose a collision threat in

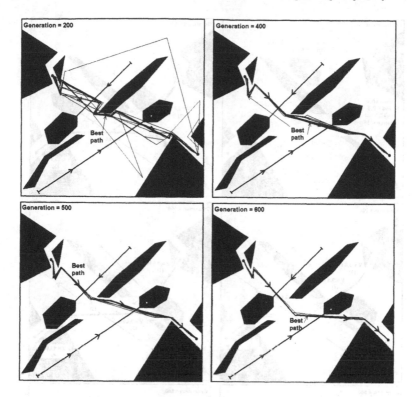

Fig. 11.14. Path evolution when approaching two moving targets in the presence of static navigation constraints. The own-ship travels at $\vartheta = 5.6$ knots.

this case, but it's still necessary to avoid target 2. This task is accomplished by a tiny maneuver at the upper tip of the upper island: by making a sharp right turn and then a left turn shortly afterwards, the own-ship can avoid the hexagon of target 2. (Note that the locations of the dynamic areas are shown (black hexagons) with respect to the best path. These locations depend on the time determined by the first crossing point between the own-ship's path and the trajectory of the target.)

It's interesting to see what happens when the experiment is repeated, but with a slower speed of $\vartheta = 5.6$ knots for the own-ship. Since the own-ship approaches the gap between the islands at a slower rate, there's a benefit for performing some "waiting maneuvers" to let target 1 pass the gap area. This is exactly what EP/N does (see figure 11.14) where the best trajectory, after several tiny maneuvers at the beginning of the path, finds its way between the islands and avoids both targets: it passes behind the sterns of both targets 1 and 2.

We can see that the own-ship's speed can greatly influence the final best trajectory. So what happens when the own-ship can change its speed during the maneuver? In reality, such changes are quite limited because changing the tra-

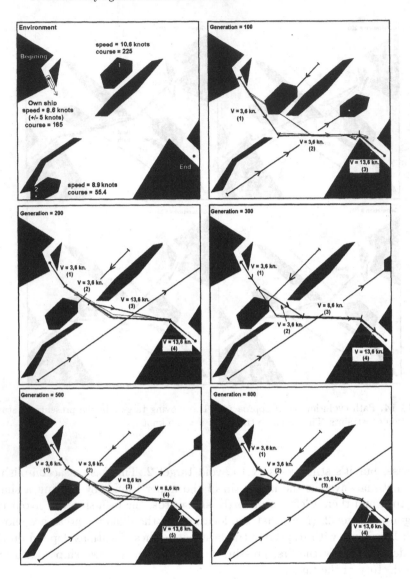

Fig. 11.15. Path evolution when approaching two moving targets in the presence of static navigation constraints. The own-ship has a variable speed.

jectory is considered primarily as an anti-collision maneuver. Such maneuvers can include changing speed, but only when such a change decreases the probability of collision. (The unwillingness to change the *Titanic*'s speed is considered to be one of the reasons for its famous disaster [291].)

Let's assume that the own-ship's speed can be 3.6, 8.6, or 13.6 knots (described more commonly as slow ahead, half ahead, or full ahead). This simple change triggers many important changes in the code of the evolutionary system. First, the speed of the ship is represented as a variable associated with every segment of a path. It's also necessary to develop operators that change the speed values. If speed is represented by a small set of values, then a binary representation and simple bit-flip operations are appropriate. If the speed is instead allowed to vary continuously, the floating-point representation and Gaussian or other similar variation is more appropriate.

Figure 11.15 displays the result of an experiment where the ship has the three speeds indicated above. The environment is the same as in the previous experiments (reported in figures 11.13 and 11.14). In accordance with the "common sense" result for such a situation, the initial segments of the best trajectory are covered with the minimum speed of 3.6 knots, whereas the final two segments are covered with the maximum speed of 13.6 knots.

These simple examples do not do the full justice to the system. The own-ship is only "aware" of some (but not all) potential targets — those visible on its radar screen. Also, the own-ship only "knows" the current speed and direction of the visible targets (and in real life these are estimates). So the "landscape" can change any time: Any strange ship can change its direction and/or speed (reacting, for example, to movements of other targets or of the own-ship or of another ship that was undetected previously), and targets can appear or disappear from radar contact with the own-ship. In such situations, the modified EP/N has to come up with a new solution quickly, and just as with the TSP example earlier, the current population at any time provides a basis for looking for new solutions on-the-fly, or, in this case, on-the-wave.

11.4 Summary

In real-world problem solving, it's not enough to find one solution and stick with it. You have to continually find solutions as the problem changes. The objective in solving the problem may change, and consequently, so should our solution. The available resources or the constraints on those resources may change, and again, so should our solution. Furthermore, the information that we obtain on how well we are currently solving a problem may not be completely accurate. Thus we face the challenge of iteratively improving our solutions to a series of problems that change over time, and we must do this in light of unanticipated events and incomplete knowledge. The fact that uncertainty and inaccuracy will enter into our models, projections, and processes must be accepted and even used to the greatest possible benefit.

Even when our own purpose remains essentially the same over time, in a competitive situation the manner in which our purpose interacts with our potential competitors' objectives may change. This again requires rethinking the solutions to the problems that we face. Say you're trying to develop the best sports team for your city. What does "best" mean? What's your purpose? First, win the most games. Second, attract the most fans. And of course you do that by winning, by having a convenient well-designed stadium, by having television broadcast your games nationally, and so forth. Third, minimize payroll and other expenses. Fourth, develop the most profitable marketing of associated merchandise (caps, jerseys, trading cards, etc.) and broadcast rights to the games. Fifth, develop good community relations: you don't want stories about your players getting into trouble with the law. Sixth, minimize the degree to which you compete directly with other sports teams in the same market, and these might not even be in the same game (for example, the Chicago Bulls basketball team competes with the Chicago Bears football team and both the Chicago Cubs and White Sox baseball teams).

Proceeding more formally, you'd first want to devise your objectives to be preferentially independent: maximize profitability, ensure customer relations, ensure public relations, and minimize conflicts with other team owners in the area. Each of these parameters can be broken down into lower-level concerns. Profitability involves attracting fans, setting the prices for tickets appropriately based on the demographics of the fans, selling merchandise, the cost of company payroll, and so forth. Let's presume we could explicate the degree of achievement with regard to each parameter, as well as their relative importance.

None of these parameters, however, takes into account the relationship you have with your competitors: the other sports team owners in your game. If you were simply to attempt to maximize the evaluation function derived from the preceding effort you would undoubtedly do quite poorly because there are other team owners who are trying to achieve success as well, and their success may come only as a result of your failure!

Suppose that you represent a smaller city with no large television stations, such as San Diego, California or Charlotte, North Carolina. You face a situation where the owners of teams in large-market cities like New York, Chicago, and Los Angeles have more money to dispose on higher-quality athletes, more people in the vicinity to fill their stadiums, and larger revenues from television rights so that they can maintain or even increase the amount of money they have to spend on players every year. In contrast, you only have enough money to attract a few of the highest-quality performers and must scout and trade for other players that you need. You appear doomed to failure because the situation is "stacked" against you.

Rather than accept this situation, you might decide to form a coalition with other small-market team owners and have more bargaining clout against the large-market owners. You might be able to force them to share more of their revenues with you (after all, without all the small-market teams there'd be nobody to play against). You might be more inclined to make deals with owners of other small-market teams and boycott interactions with the "big guys." You might find that you are less interested in defeating some teams as you are in defeating others. You might propose a "salary cap" to put a limit on the amount of money that any one team can spend on players. Of course, the big-market owners would fight against this move. Depending on the outcome of these political challenges, your available resources might change thereby providing you with an entirely different set of constraints on your problem, and an entirely different corresponding set of solutions.

Note that your interactions aren't limited to being just with the owners. While you're fighting and cooperating with different owners, you must also negotiate with a players' union that will seek to maximize player revenues, travel accommodations, pensions, and other benefits. And you must deal with city managers who may want to use your stadium for other events, unless you happen to own your own stadium.

Every time you interact with someone else, you must take into account the joint purpose: not just what you are trying to achieve, but what they are trying to achieve and whether or not you are favorable or antagonistic to that purpose, and vice versa. That means that every time you identify another "player in the game" you have to redetermine whether or not your previous allocation of resources, your previous *solution*, is still suitable in light of this new player.

A classic example of having to deal with a changing set of opponents emerged out of a famous game called the iterated prisoner's dilemma. The game models the following situation. Suppose you and your partner committed a crime. You were both caught and placed in separate cells. The prosecutor comes to your cell and says that he's giving both of you the same deal. If you and your partner both admit to the crime then the prosecutor will put you away for four years in prison. But if you admit your culpability and agree to give evidence against your partner, and if your partner refuses to help the prosecutor, then the prosecutor will set you free and your partner will get five years in prison. If you both instead deny responsibility, the prosecutor still has enough evidence to send you "up the river" for two years. Written in the form of a mathematical game, the payoffs (in this case they are penalties) are

	Sell Out	Keep Quiet
Sell Out	(4,4)	(0,5)
Keep Quiet	(5,0)	(2,2)

where you get to choose the row and your partner chooses the column. The catch is that you don't get to talk to your partner before you make the decision.

You look at the matrix and decide that no matter what your partner does, it's better for you to admit guilt and "sell him out." If he keeps quiet, then you can either also keep quiet and get two years, or sell out and go free! If he sells you out, then you can either keep quiet and go to prison for five years, or sell him out too and only go away for four years. Regardless of your partner's choice, admitting guilt and helping the prosecution seems to be the way to go. The problem is that this rationale is also foremost in your partner's mind. He is figuring the same things you are figuring. The result is that the logical play for both of you is to sell out, resulting in both of you getting four-year sentences when you could have reduced your time in prison to just two years if you'd both kept quiet. What a dilemma indeed!

We can just as well turn the problem around and talk about rewards instead of penalties. Suppose you have two options: *cooperate* or *defect*. Cooperation implies working to increase both your own reward as well as that of the other player. Defecting, on the other hand, implies working to increase your reward at the expense of the other player. In this case, a set of payoffs might be

	Cooperate	Defect
Cooperate	(3,3)	(0,5)
Defect	(5,0)	(1,1)

Again, you have the same dilemma. Defecting seems better than cooperating, but mutual cooperation pays off three times better than mutual defection.

This sort of game can be used to model many real-world circumstances. For example, the countries that participate in OPEC play a prisoner's dilemma. Each country agrees to limit its oil production so that the price can go up, but each country can also violate the agreement and make more money than they would otherwise. The two main differences between the OPEC model and the situation with the two prisoners above is that there are more than two players in OPEC, and more importantly, they must continue to play the game over and over. The players in the game have a memory. They remember who defected last time and will want to take that into account when they encounter that player next time. Similarly, they remember who cooperated as well.

When the prisoner's dilemma is iterated over many time steps, it turns out that mutual cooperation can emerge as a rational course of action. One of the more robust strategies in this game is called tit-for-tat: cooperate first, but then do whatever your opponent(s) do. If they defect, you defect. If they cooperate, you cooperate. This way, you never get taken advantage of for more than one play. But it's clear that the best course of action depends on the other players in the game, and they are often constantly changing in the real world. Against a player that always defects, there's no point in cooperating. Against a player that plays tit-for-tat, all you want to do is cooperate. Against a player that always cooperates, you should defect because that player is a sucker. In the real world, there is often some sort of selection where suckers and constant bullies tend to be eliminated by various means. The result is that you have to continually update your strategy in light of what you know about the other players in the game.

The penultimate example comes in the area of military combat. Every new entity must be identified as friend, foe, or neutral. Every engagement changes as a function of the joint purpose: your objective and how it interacts with everyone else's. The best course of action changes as a function of the anticipated reactions of the enemy and the expected outcomes. Mission planning requires constant adaptation. The phrase "adapt or die" could never be more appropriate.

In a sense, real-world problems never get solved. Each solution leads to the next problem. The trick is to anticipate that problem and adapt your solution so as to best handle the possibilities that you might face. If you can incorporate anticipated situations into the evaluation function you may be able to discover a robust solution that will be able to treat those settings. If you can't anticipate the eventual outcomes with sufficient fidelity then evolving a population of diverse but worthy solutions may provide a hedge against a dynamic environment. It may also serve to ameliorate the inevitable effects of noise that degrade your ability to assess your situation and determine the effectiveness of your strategies.

Having now brought up the prisoner's dilemma, it's difficult to resist the temptation to talk about another version of the problem which is also known by the same name. It's a tough problem to digest though, so maybe we should have tried to resist a little harder. Anyway, the problem is as follows.

> There are three prisoners in a cell. They know that two of them will be executed in the morning and one of them will be set free, but they don't know who. The decision has already been made and the prison officers know the details. As it happens, one of these prisoners is a statistician. He calculates the probability of his survival as 1/3. That evening, which is the last evening for two of the three prisoners, the warden comes to the cell with their dinners. Our hero, the statistician, says to the warden,
>
> "Look, as two of us will be executed in the morning, I know that at least one of the other two prisoners will be executed. So there's no harm if you tell me who of the other two prisoners will die — after all, I'm not asking about myself!"
>
> The warden thinks for a moment and replies, "OK, I'll tell you. He will die," pointing to one of the other men in the cell.
>
> "Great!" exclaimed the statistician, "Now my chances of survival have increased to 1/2!"

The question for you is whether or not the statistician is mistaken in his reasoning.

It seems that with the new knowledge he received, his chances for survival indeed increased to 1/2 as it's now between him and the other prisoner. Think about it this way. The prison officials are conducting a statistical experiment. There are three outcomes:

	A	B	S
1	D	D	L
2	L	D	D
3	D	L	D

where S is the statistician, A and B are the other men, D indicates death, and the L indicates life. If we assume, as the statistician did, that each of these outcomes is equally likely, then once the warden points at either one of the two other men, the sample space is reduced. If the warden points at A then we know that either outcome 1 or outcome 3 are the possibilities. Of these, one indicates that the statistician lives, the other that he dies. Since all of the possibilities are equal going in, the statistician can reason that his chances of survival are now 1 in 2. The same reasoning would apply if the warden picked B instead of A.

On the other hand, you might think about this problem in a different way. The experimental outcome space isn't defined primarily in terms of the three outcomes above, but rather in terms of who the warden will point at. Since we know the warden won't point at S, there are only two possibilities: A or B. But here the outcomes have different implications. We know that if A and S will die, then the warden will point at A. That happens with a 1/3 probability. Likewise, we know that if B and S will die, then the warden will point at B, again, this will happen with a 1/3 probability. The final case is when A and B will die, in which case the warden can be assumed to point at either A or B with equal likelihood. Since the outcome of A and B dying has a probability of 1/3, the chance of pointing to A is half of that, namely, 1/6, as it is for B.

Now let's say that the warden points at B. We've now reduced the sample space to the possibilities of B and S dying, or A and B dying. But, and this is where the catch lies, the outcomes are defined in terms of the warden's fickle finger! So the outcomes are "B and S will die and the warden points at B" and "A and B will die and the warden points at B." The probability for the first case was 1/3 while the probability for the second case, which is the case that makes our statistician happy because he gets to live, is 1/6. Therefore, the likelihood of the outcome where the statistician lives, given that the warden points at B, is

$$\frac{\frac{1}{6}}{\frac{1}{3} + \frac{1}{6}} = \frac{1}{3}.$$

The same reasoning would hold if the warden picks A instead of B.

Gulp! Which one is correct? Maybe they should call this problem the statistician's dilemma! Can you figure it out? Try to write a computer simulation that models the situation and see what happens!

XII. Day of the Week of January 1st

Which day of the week appears more often as the first day of a year, Saturday or Sunday?

Most people would answer: neither. It's intuitive that in the long run the frequency of Saturdays and Sundays that fall on January 1st should be the same. But that's not the case.

Before we go any further, let's remind ourselves of the definition of a leap year. A particular year is a leap year if its number is divisible by 4, but not by 100, unless it's divisible by 400. For example, years 1892 and 1896 were leap years, 1900 was not, 1904, 1908, etc. were leap years, and 2000 is.

Thus, within any period of 400 years we have exactly 97 leap years (i.e., all of the years that are divisible by 4 — and there are 100 of them — except three years whose numbers are divisible by 100 but not by 400). Within any period of 400 years we have exactly

$$97 \cdot 366 + 303 \cdot 365 = 146,097 \text{ days.}$$

The number 146,097 is divisible by 7. This means that any period of 400 years consists of an integer number of whole weeks. It also means that 400 years constitute a cycle. Since 1 January 2001 is a Monday, then 1 January 2401 is without a doubt a Monday as well.

Since 400 is not divisible by 7, different days of the week may have different frequencies. It's impossible to have the same number of Mondays, Tuesdays, Wednesdays, etc. within each 400 years, so it's sufficient to find out whether Saturday or Sunday appears more often as January 1st within an *arbitrary* period of 400 years.

Since we have the freedom of selecting *any* 400-year period, let's do some counting for the period of 2001 – 2400. Within this period we have four long, regular intervals where a leap year occurs precisely every four years. The only exceptions are 2100, 2200, and 2300, which are *not* leap years. Note also that any 28-year period that's within one of these regular intervals (i.e., every period of 28 years that doesn't contain the years 2100, 2200, or 2300) has four instances of each day as the first day of the year. For example, within the period 1 January 2034 – 31 December 2061, there are precisely four Mondays, four Tuesdays, four Wednesdays, etc. as days for January 1st. We needn't be concerned with these periods. During these 28-year intervals there are 7 leap years, and

$$7 \cdot 366 + 21 \cdot 365 = 10,227,$$

which is divisible by 7.

Therefore, let's consider the following set of periods, which amount to 400 consecutive years:

(a) 2001 – 2028   (e) 2101 – 2128   (i) 2201 – 2228   (m) 2301 – 2328
(b) 2029 – 2056   (f) 2129 – 2156   (j) 2229 – 2256   (n) 2329 – 2356
(c) 2057 – 2084   (g) 2157 – 2184   (k) 2257 – 2284   (o) 2357 – 2384
(d) 2085 – 2100   (h) 2185 – 2200   (l) 2285 – 2300   (p) 2385 – 2400

There are four repetitions of each day of the week (as January 1st) in each of the periods: (a), (b), (c), (e), (f), (g), (i), (j), (k), (m), (n), and (o), so the only real counting is needed for the remaining periods: (d), (h), (l), and (p). We have to check (e.g., using a universal calendar) that

1 January 2085 is a Monday,
1 January 2185 is a Saturday,
1 January 2285 is a Thursday, and
1 January 2385 is a Tuesday.

Then it's easy to find out the exact number of days of the week in each of these four periods:[4]

	20—	21—	22—	23—
—85	Mon	Sat	Thu	Tue
—86	Tue	Sun	Fri	Wed
—87	Wed	Mon	Sat	Thu
—88	Thu	Tue	Sun	Fri
—89	Sat	Thu	Tue	Sun
—90	Sun	Fri	Wed	Mon
—91	Mon	Sat	Thu	Tue
—92	Tue	Sun	Fri	Wed
—93	Thu	Tue	Sun	Fri
—94	Fri	Wed	Mon	Sat
—95	Sat	Thu	Tue	Sun
—96	Sun	Fri	Wed	Mon
—97	Tue	Sun	Fri	Wed
—98	Wed	Mon	Sat	Thu
—99	Thu	Tue	Sun	Fri
—00	Fri	Wed	Mon	Sat

Thus, the count, for periods (d), (h), (l), and (p), is

---

[4]The last row of the table gives the day of the week for the first day of years 2100, 2200, 2300, and 2400, respectively.

Monday:	8
Tuesday:	10
Wednesday:	9
Thursday:	9
Friday:	10
Saturday:	8
Sunday:	10

and the total count for 400 years is

Monday:	56
Tuesday:	58
Wednesday:	57
Thursday:	57
Friday:	58
Saturday:	56
Sunday:	58

Thus, the probability that an arbitrary January 1st is a Sunday is 0.145 (i.e., 58/400), whereas the same probability for a Saturday is only 0.14 (i.e., 56/400).

How can you make some money from that knowledge? Now there's a real-world problem!

12. Neural Networks

If I only had a brain.

The Scarecrow, from *The Wizard of Oz*

If you spend any significant amount of time trying to solve problems, "racking your brain," you will inevitably contemplate the prospects of automating your thinking process. The dream is to make a computer simulation that simulates the way your brain functions so that it can solve problems the same way you do. But brains don't function with the same mechanisms as serial digital computers. Biological processing is inherently and massively parallel in character, whereas traditional computing is sequential. Each step in an algorithm is conducted in turn until a final halting condition is reached. Further, although there are conceptual similarities between neurons in living brains and logic gates in computers, the firing rates of biological neurons are much slower than computer logic gates (on the order of milliseconds for neurons versus nanoseconds for computers). These and other differences in design lead to an intuition that these alternative input–output devices should each be able to tackle different problems with efficiency.

Computers are excellent devices for quickly determining arithmetic results. If you need to know what 412.14823 multiplied by 519.442 is, you would much rather have a handy calculator than an eager friend armed with only paper and a pencil. In contrast, computers are not intrinsically good at generalizing, handling conditions that fall outside the prescribed domain of possibilities. If one of our friends shaves his beard, we are still likely to be able to recognize him. In contrast, a computer that relies on a sequence of if–then conditions that correspond to the identification of specific features of that person's face, might have considerably more trouble.

Does it have to always be this way? Is this a fundamental restriction of computer processing? Or is it possible to design computers to function much as biological neural networks do? After all, a neural network is an input–output device, just as is a finite state machine or other representation for a computer. It should be possible to create models of how neural networks perform their input–output behavior and capture those behaviors in a computer. The resulting artificial neural network (ANN) might yield some of the processing capabilities of living brains, while still providing the speed of computation that can be attained in silicon.

12.1 Threshold neurons and linear discriminant functions

Attempts to do just this go back very many years, even before the advent of modern digital computing. Warren McCulloch and Walter Pitts offered a famous model of neural activity in 1943 [303]. It was a simple but important early step toward generalizing the input–output behavior of single neurons. Their model was essentially an all-or-none threshold description. If the inputs to a neuron, defined by their total "activity level," or β, were sufficiently high, the neuron would respond by firing, otherwise it would remain inactive. Mathematically, if $f(\cdot)$ is the function that represents the neuron's behavior, then

$$f(\beta) = \begin{cases} 1 & \text{if } \beta \geq 0 \\ 0 & \text{if } \beta < 0 \end{cases}$$

The neuron fires with an output activity of 1, otherwise it remains at rest with an activity of 0.

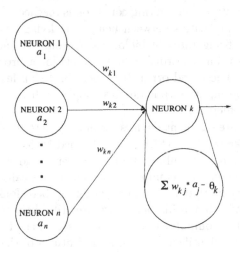

Fig. 12.1. A simple neural network where n input neurons connect to a single output neuron k. There is a weighted connection between each input neuron and the output. The output neuron performs the function $\sum_{j=1}^{n} w_{kj} \cdot a_j - \theta_k$ and then compares the computed value to a threshold (e.g., zero). The network "fires" if the output is not less than the threshold.

The input activity level to a neuron can be described as a weighted sum of the output activities passed along from other neurons (this was proposed by Rosenblatt [396] in a model called the perceptron). In essence, each neuron is connected to other neurons by "synapses" that can amplify or diminish the signal that is being forwarded. The situation is shown in figure 12.1. Describing this with symbols,

$$\beta_k = \sum_{j=1}^{n}(w_{kj}a_j) - \theta_k,$$

where β_k is the activity of the k-th neuron, w_{kj} is the weighted connection from neuron j to neuron k, a_j is the output activity of the j-th neuron that is being forwarded to neuron k, θ_k is a threshold for neuron k, and there are n neurons that are communicating with neuron k. Note that θ_k could just as well be added as subtracted from the sum simply by changing its sign.

It's easy to now envision the case where hundreds, if not millions or billions of these neurons are connected, processing information from the environment, and sending out signals that might control various aspects of a living organism. That is, it's easy to envision or at least imagine this in its potential, but actually identifying the computational properties of these sorts of interlinked mathematical neurons requires analysis rather than conjecture.

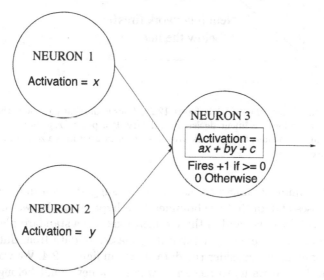

Fig. 12.2. A neural network with two input nodes, each having activations of x and y, respectively. The weighted connections to the output node are a and b, respectively. Finally, the output of the network is computed by calculating $ax + by + c$ and comparing that total to zero. The network returns the value $+1$ if $ax + by + c >= 0$ and returns a 0 otherwise.

A good place to begin to understand how a network of these neurons could function is to examine how a single neuron functions. Here, this is quite elementary. A neuron is a linear combination of the weighted inputs offset by a threshold term. If the total activity level is greater than or equal to zero, then the neuron fires. To simplify things further, let's consider the case where there are only two neurons that are connected to a third neuron (figure 12.2). The output of the first neuron is x and it is weighted by a. The output of the second neuron is y and it is weighted by b. Finally the threshold value of the third neuron is given by c. Thus we have:

If $(ax + by + c \geq 0)$ then (Neuron 3 fires and outputs a 1).

A quick recollection from grade school mathematics will remind us that $ax + by + c = 0$ is the equation for a line. Therefore, what the third neuron is essentially doing is firing if the inputs being passed along from the prior neurons fall on one side of a line defined by a, b, and c (see figure 12.3). This is a rudimentary pattern recognition system: a linear discriminant function.

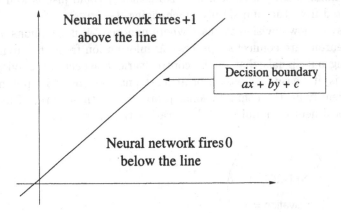

Fig. 12.3. The simple neural network in figure 12.2 defines a line that partitions the xy-plane. The line $ax + by + c$ forms a linear decision boundary. If a point (x, y) is presented to the network that is above the line, the network fires a $+1$. Otherwise it fires a zero (or equivalently is said not to fire).

Linear discriminant functions, as the name suggests, can differentiate between two classes of data that are bounded by a hyperplane. In two dimensions, a hyperplane is just a line, and in three dimensions it is a standard plane. These functions are often convenient for separating classes of data from different distributions. For example, consider the data shown in figure 12.4. We want to find a simple rule that allows us to recognize whether a new point belongs to class 1 or class 2. If the data can be separated easily by a line then this line provides a rule for making the determination. If a new point falls above the line, the formula $ax + by + c$ will be greater than zero, and the rule will be satisfied — equivalently, our neuron would output a 1, indicating a detection. Similarly, the obverse behavior of outputing a 0 would occur if the new point fell below the line. This now begs the question of how to determine the appropriate values for a, b, and c so that we can best separate some data by a linear discriminant function.

To make the discussion more general, let's consider the case in figure 12.5. There are n inputs to a neuron, and these are represented by the vector $\mathbf{x}$. As is typically offered in survey literature, we can also include the neuron's threshold term as one of the inputs that is constantly set at -1 (or equivalently at $+1$). This, then, is the $(n + 1)$-th input. The weights on the inputs (including the threshold) are represented in a vector $\mathbf{w} = [\theta, w_1, \ldots, w_n]$, where θ is the weight for the threshold and the remaining weights indicate the associated amplification on the associated input elements in $\mathbf{x}$.

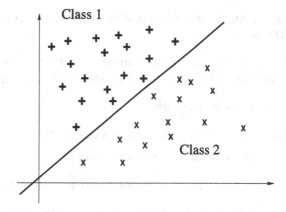

Fig. 12.4. In this case, pluses represent examples from class 1, whereas *crosses* represent examples from class 2. All of the *pluses* and *crosses* can be separated by a linear boundary, indicating that the neural network in figure 12.2 could correctly classify all of the examples.

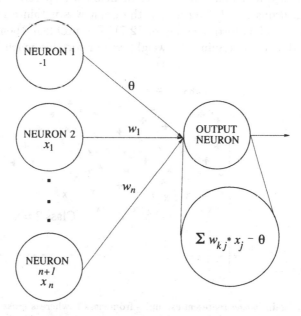

Fig. 12.5. An alternative representation for a neural network where the threshold term is shown as an input node with an activation of -1 and a connection weight of θ

A perfect set of weights $\mathbf{w}$ would yield $\mathbf{w}^T\mathbf{x} \geq 0$ if $\mathbf{x} \in$ class 1 and $\mathbf{w}^T\mathbf{x} < 0$ if $\mathbf{x} \in$ class 2. The neuron would fire every time a point in class 1 was presented, and would fail to fire if the point was from class 2. Suppose there are p input patterns $\mathbf{x}_1, \ldots, \mathbf{x}_p$. A simple algorithm that can achieve the desired condition above ([210], p.109) is

1. For each vector $\mathbf{x}_i$ in turn, if the current weight vector $\mathbf{w}$ generates a neuron output that correctly classifies $\mathbf{x}_i$, then do not change $\mathbf{w}$.

2. Otherwise, update $\mathbf{w}$ by the rule:

   (a) $\mathbf{w} \leftarrow \mathbf{w} - \eta \mathbf{x}_i$, if $\mathbf{w}^T \mathbf{x}_i \geq 0$ and $\mathbf{x}_i \in$ class 2, else

   (b) $\mathbf{w} \leftarrow \mathbf{w} + \eta \mathbf{x}_i$, if $\mathbf{w}^T \mathbf{x}_i < 0$ and $\mathbf{x}_i \in$ class 1.

   where η is a constant step size parameter that is greater than zero.

This simple rule can be shown to be able to separate any linearly separable data in some number of steps, say, s, given an initial condition that $\mathbf{w} = \mathbf{0}$.

Suppose we were interested in classifying data from two classes that were not linearly separable, as shown in figure 12.6. Our single threshold neuron and algorithm for adjusting the weights would not be sufficient. We would need more neurons, perhaps arranged hierarchically, so that some early neurons could determine if a new point were above or below a respective line, and then subsequent neurons could determine if the point was within a region bounded by these lines, and so forth (see figure 12.7). The problem, then, is how to devise an algorithm to determine the weight vector for this much larger network of neurons.

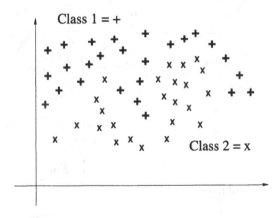

Fig. 12.6. Again, *pluses* represent examples from class 1, whereas *crosses* represent examples from class 2. In this case, the examples are not linearly separable. The neural network in figure 12.2 cannot classify all of these points without making errors.

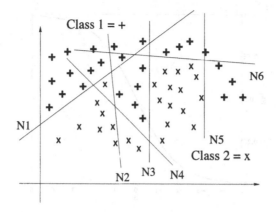

Fig. 12.7. Although a single linear decision boundary cannot separate these data, it is possible to use multiple linear decision boundaries. Here, each boundary is represented as $N_1, \ldots, N_6$, signifying that the line corresponds to the firing threshold of a hypothetical neuron. By combining the outputs of multiple neurons, it is possible to correctly identify all of the examples. Elements can be identified using the following rules: (1) if the point is above N_1, then it is class 1; (2) if the point is below N_1 and below N_2 then it is class 2; (3) if the point is below N_1 and above N_2 and below N_3 and above N_4 then it is class 1; (4) if the point is below N_1 and below N_4 then it is class 2; (5) if the point is below N_1 and above N_5 then it is class 1; (6) if the point is above N_3 and below N_5 and below N_6 then it is class 2; (7) if the point is above N_6 then it is class 1. Note that each of these rules could be encoded into neurons that process the output of $N_1, \ldots, N_6$.

12.2 Back propagation for feed forward multilayer perceptrons

One possibility is to make an approximation to the thresholding (all or nothing) output function of each neuron and then use calculus. Suppose that instead of having a hard limiting threshold function, we allow each neuron to offer a graded level of output $\in [0,1]$. The greater the weighted input activity to that neuron, the closer its output would be to 1.0. To make this more specific, suppose each neuron was governed by the sigmoid function,

$$f(\beta) = \frac{1}{1 + e^{-\beta}},$$

where β is the dot product of all the incoming activations and their associated weights (much like $\mathbf{w}^T\mathbf{x}$ before). Then, each neuron's output would appear as shown in figure 12.8. The output would saturate at both the extreme positive and negative ends of the weighted inputs. When the output function is smooth and continuous, as is the case when using a sigmoid function like this, there is a "back propagation" method for adjusting the weights to minimize the squared classification error of the network of neurons.

The essential idea is to start at the output of the neural network and adjust the weights that connect to the output node in order to reduce the error between

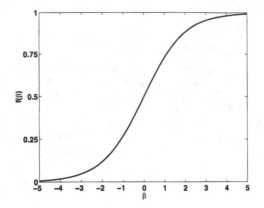

Fig. 12.8. A sigmoid function $1/(1+e^{-\beta})$ is often used as an approximate threshold function for neurons in a neural network. When the activation is strongly positive, the neuron responds with values close to 1.0. When the activation is strongly negative, the neuron responds with values close to 0.0. For activations in the range of $[-1,1]$ the sigmoid function is close to linear, but then saturates in both directions.

the network's actual and desired behavior. To make this more explicit, suppose that there are again n input patterns $\mathbf{x}_1, \ldots, \mathbf{x}_n$ and that there is a desired classification of each pattern, $d_j(1), \ldots, d_j(n)$, where $d_j(p)$ refers to the desired output for the j-th output node when presented with the p-th input pattern $\mathbf{x}_p$. (Note below that i will be used an index variable in different settings, so use care in interpreting the symbol in its proper context.) These "targets" can be real values (e.g., 0 or 1, but not limited to this Boolean choice). For a given set of weights, $\mathbf{w}$, the j-th output node of the network responds to each input pattern with a set of real-valued outputs $y_j(1), \ldots, y_j(n)$.

The goal is defined to adjust $\mathbf{w}$ so as to minimize the sum of the squared error associated with the network's outputs for each input pattern. Thus, we must first compute the sum of squared errors over all j outputs for any input $\mathbf{x}_p$:

$$E(p) = \tfrac{1}{2} \sum_j e_j^2(p),$$

where $e_j(p) = d_j(p) - y_j(p)$, and then compute the average of these squared errors over all patterns $\mathbf{x}_1, \ldots, \mathbf{x}_n$:

$$E = \tfrac{1}{n} \sum_{p=1}^{n} E(p).$$

Recall that each $y_j(p)$ is a function of the input vector $\mathbf{x}_p$ and the weight vector $\mathbf{w}$:

$$y_j(p) = g(\mathbf{x}_p, \mathbf{w}),$$

where g is a nonlinear function. Let's suppose that the neural network is constructed in l layers, where the inputs form layer 1, and the output nodes form layer l (see figure 12.9). Due to this hierarchic construction, it's possible first to focus attention on the weights that connect nodes from layer $l-1$ to each output node and adjust them to compensate for the error incurred in each network output taken over all input patterns.

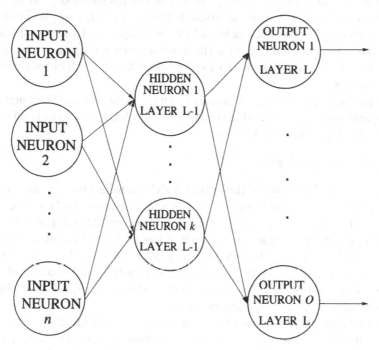

Fig. 12.9. A multiple-layer feed forward neural network. There are n input neurons which are connected to "hidden" nodes. The figure shows a single layer of hidden nodes, but there is no limit to the number of hidden layers that might be used. The final layer is the output layer, L, here consisting of O output neurons. A fully connected feed forward network has all neurons at each layer connecting to all of the neurons at the next layer. Note that the neural network can have more than one output (i.e., it can be multivalued).

One common method for adjusting parameters (weights) so as to minimize the sum squared error of a model (in this case, the neural network is a "model") is to alter them in proportion to the amount of error incurred. In this way, when the model is performing well and there is little error, only small adjustments are made to the parameters. Conversely, when the model is performing poorly, the error is large and the corresponding changes to the parameters might be made more efficacious by being larger. Another sometimes related method is to adjust the parameters in proportion to the local gradient of the error function. In addition, parameters are often changed in proportion to the input activations so that larger alterations are made when the inputs are larger and correspondingly smaller changes are made to parameters that are of smaller magnitude.

Returning to the neural network's output nodes, if we were to adjust the i weights that connect to each j-th node based on the error observed for the p-th input pattern using the above principles, we might perform the operation

$$\triangle w_{ji} = \eta \delta_j(p) y_i(p), \tag{12.1}$$

where p is the index of the input pattern, j is the output node, i is an index that runs from 1 to the number of nodes in the layer $l - 1$, $\triangle w_{ji}$ is the change to make to the weight connecting the i-th node in layer $l - 1$ to the j-th output node, η is a scaling constant, $y_i(p)$ is the input activation coming out of the i-th node in layer $l - 1$, and $\delta_j(p)$ is the local gradient for the j-th output node for input pattern $\mathbf{x}_p$.

The gradient $\delta_j(p)$ must be calculated to implement the weight change rule. This requires calculus. A suitable review of the derivation of $\delta_j(p)$ is offered in [210], p. 144. The main result is that

$$\delta_j(p) = e_j(p) f'(\beta_j(p)),$$

where $f'(\cdot)$ is the derivative of the output node's transfer function, and $\beta_j(p)$ is the dot product of the incoming i-th weight and activation to the output node when the p-th pattern, $\mathbf{x}_p$, is presented to the network. This happens to make the choice of a sigmoid function for each of the nodes fortuitous because the derivative of $f(\beta_j)$ is easy to compute as $y_j(p)(1 - y_j(p))$. Thus, the weights from all of the i nodes at layer $l - 1$ to each j-th output node can be adjusted by adding $\triangle w_{ji}$ to each w_{ji}. We simply need to choose η and use $y_i(p)$ and the product $y_j(p)(1 - y_j(p))$ to determine $\triangle w_{ji}$.

The problem, then, becomes one of adjusting the weights that connect the nodes in layer $l - 2$ with those in layer $l - 1$. In this case, we do not have direct information with regard to the "error" associated with each node; there is no given desired output value. Yet we can still determine the value of $\delta_j(p)$ for each "hidden node" as

$$\delta_j(p) = y_j(p)(1 - y_j(p)) \sum_k \delta_k(p) w_{kj}(p),$$

where k runs from 1 to the number of output nodes (or for multiple hidden layers, the corresponding number of nodes in the next higher layer). The summation term is essentially the weighted gradient associated with each node in the layer l. The update for weights connecting to the hidden layer $l - 1$ is again found using (12.1).

After updating all of the weights associated with the nodes in layer $l - 1$, we could proceed backward further, layer by layer, until all of the weights were updated. Then, we would need to process the next input pattern and recalculate the weight updates, and continue to iterate this process until the weights tended to converge. There are many variations of this procedure, but the essence is the same in each: a gradient-based minimization of the squared error provided by a nonlinear function (the neural network) in trying to match the function's output to a desired target for every presented input.

This algorithm is guaranteed to converge eventually, but only to a locally optimum vector $\mathbf{w}^*$. That is, for some ϵ-neighborhood around the discovered weight vector, $\mathbf{w}^*$, the error of the neural network will be at a minimum. Unfortunately, there is no guarantee that there won't be other vectors of weights that would yield even lower error than is obtained with $\mathbf{w}^*$ that reside outside the ϵ-neighborhood. This is a serious problem because the error function (the sum of the squared difference between the desired and actual network output taken over all input patterns) may be pocked with many local minima, and these may be quite distant from the true global minimum (or possibly minima). The back propagation algorithm can stagnate at suboptimal weights. Unless the neural network yields verifiably perfect performance (e.g., has zero classification error), the only real method for gaining confidence in having discovered a suitable set of weights is to continually iteratively restart the procedure using different initial weight vectors selected at random. If the procedure converges to essentially the same weights, repeatedly, this can serve to support a belief that the weights are indeed optimal, but we can never be certain.

The neural network pictured in figure 12.9 is called a multilayer perceptron, based on the single perceptron from Rosenblatt [396] with a threshold function. It is also described as a "feed forward" network because information flows in only one direction: from the input nodes to hidden nodes to the output node(s). There are other feed forward neural networks that utilize different transfer functions. Some use a Gaussian function instead of a sigmoid. Others use Gaussian density functions that operate on a distance metric of an incoming activation vector and a specified vector of coordinate means and standard deviations (described as a "radial basis function"). There are many other possibilities, each having the potential to match the structure of the classification problem at hand. Different input data and classifications may require alternative transfer functions in a neural network.

One important result that was offered in the late 1980s is that variations of multilayer feed forward neural networks can be designed as "universal function approximators" [235, 358]. In essence, this means that given a sufficient number of nodes in a hidden layer, by properly adjusting the weight vector, $\mathbf{w}$, the neural network can compute any measurable function to within a given degree of accuracy. This property holds both for neural networks with sigmoid or Gaussian functions; thus, these choices are common because there is a sense that the design is not inherently limited to being unable to compute some desired transform a priori. This choice, however, is often more of a "security blanket" than of any real practical substance, because these networks can only map arbitrary functions as the number of hidden nodes tends to infinity. In practice, the complexity of most networks is nowhere near what would be required to perform arbitrary mappings to the precision of the available computer.

Further, many neural network applications are designed to use only a single hidden layer because of the universality property that holds for an unlimited number of nodes. But it's possible that a network with only a single hidden layer may require a great many more nodes to compute what another network with several hidden layers can compute with only a few nodes in each layer. Thus the reliance on single hidden layer neural networks with sigmoid or Gaussian transfer functions is often as much a hindrance to finding the right mapping as it is an aid. Choosing the correct network structure (topology) still remains much of an art rather than a science.

12.3 Training and testing

Let's just presume at this point that you have a source of data and are posed the problem of designing a classifier (i.e., a procedure or device that is used to classify whether or not data belong to some particular class). For example, you might have image features taken from 100 mammograms and you must design a classifier to indicate whether or not the features in any chosen mammogram suggest that a malignancy is present. Let's also say that for each set of features from every mammogram you know the right answer. You know whether or not the patient in question had cancer. This assumption is not unrealistic. You might have cases where patients had suspicious features in their mammograms and the associated masses or microcalcifications were biopsied [159, 149]. Finally, let's presume that the data are not linearly separable and you've made a decision to design a neural network to serve as a classifier.

The next issue is to determine the size of the training set. In the case at hand there are 100 available input patterns. It might be tempting to use all of these data to train a multilayer perceptron using the back propagation algorithm. The problem with this approach is that the resulting neural network can be "brittle," essentially learning to fit all of the noise in the data with exacting precision. When new data are collected and presented to a neural network that was trained with all of the previously available data, the neural network often fails to perform to the same level as was indicated in training. It is necessary to hold out some of the available data for testing the neural classifier that is trained on the remaining subset of the data. The amount of data to separate into training and testing sets is problem dependent and requires statistical judgment. Often the available data are separated at random into equal sets for training and testing, but there is no such rule that will work in all cases. Further, you might anticipate that once you have completed training and have tested the designed neural network on new data, you might want to redesign or retrain the network and try again. Some practitioners therefore recommend holding out a third set of data for final testing. In this way, the iterative procedure of training and testing doesn't degenerate into essentially training on all the available data.

An alternative procedure for using the available data involves resampling. In a method called leave-one-out resampling, each set of data, in turn, is held out from training. After training, the held-out sample is classified by the best neural network and the error is recorded. This datum is then returned to the population of data and the next datum is removed. Training is then repeated on the entire set of remaining data, and the process is repeated. In this way, each data point is classified by a neural network which did not use that data in training. More information can be found on statistical resampling techniques in [110].

If several attempts to train the neural network do not result in an acceptable error, it may be useful to add more nodes to the structure of the network. A larger network can approximate more functions, but larger networks are also prone to overfitting the available data. This is the same problem that occurs when trying to fit polynomials to data. In figure 12.10, our intuition suggests that the data are well represented by a line, but a ninth-degree polynomial fits the available data with less error. The problem with the ninth-order model is that it does not provide a means for generalizing over values that have not yet been observed. Its error on these unseen points is likely to be high. Large neural networks face the same problem. Unfortunately, when using a gradient-based training algorithm such as back propagation it's often possible to become trapped in a locally optimal set of weights that do not yield the desired level of error. Sometimes the only solution to this problem is to accept an overly complex neural network or shift to an alternative training method.

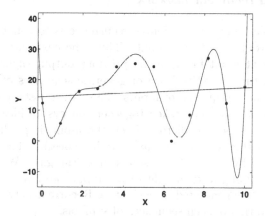

Fig. 12.10. Fitting a minimum squared error line and a ninth-order polynomial to data generated from a line with additive Gaussian noise. Even though the error is lower with the ninth-order polynomial fit, the generalizability of the polynomial is also lower. Using the ninth-order polynomial will portend large errors inbetween points, and even larger errors beyond the range of the available data, where the function tends to infinity.

One such alternative technique is simulated annealing (see chapter 5). Annealing is a stochastic search algorithm, and we have already seen how this procedure can be parallelized and modified into an evolutionary algorithm. It's not surprising, then, that evolutionary computation is potentially useful for designing and training neural networks. This is explored in greater detail at the end of this chapter.

12.4 Recurrent networks and extended architectures

Feed forward neural networks are essentially static pattern classification devices. They treat each input pattern independently of other input patterns and outputs. Sometimes this is appropriate, but suppose you wanted to model the dynamics of an automobile on a highway. The car's position, velocity, and acceleration at the next instant in time will be a function of its current position, velocity, and acceleration. That is, there is a recursive relationship between the input and output. It would be helpful to take advantage of this relationship by allowing feedback from the output of the network to the input. In other cases, it might be advantageous to use feedback loops from nodes in subsequent layers to nodes in prior layers to offer a form of memory to the neural network. Any neural network that incorporates such feedback is said to be *recurrent* rather than simply feed forward. Several recurrent neural networks have been proposed.

12.4.1 Standard recurrent network

Perhaps the simplest implementation of recurrence is to connect the outputs of the neural network back to the inputs. Thus, the next output is a function not only of the next input but also the previous output (figure 12.11). Such networks are sometimes used to learn or generate sequences of outputs based on a single input or a sequence of inputs. This is the essence of language: a sequence of symbols that when taken together conveys an idea. In this sense, the recurrent network is a predictor. Given the input, it predicts a sequence of outputs. We do this all the time in everyday speech. "The boy hit ..." is most often followed by "the ball" or sometimes "the girl." We anticipate the symbols that are coming next in light of the inputs we have received and the likely outputs that are generated at each step. Recurrent networks can offer a compact representation of such sequences of symbols.

12.4.2 Hopfield network

An entirely different form of recurrent neural network is the *Hopfield network* [232], which consists of a set of fully interconnected neurons (each communications with every other). Each neuron takes on a value (described as a state) of ± 1 depending on whether or not the weighted sum of the connected states offset by a fixed threshold is greater than 0. Symbolically,

$$y_j = \sum_{i=1}^{N} w_{ji} s_i - \theta_j,$$

where w_{ji} is the weighted connection from neuron i to j, s_i is the state of the i-th neuron, and θ_j is the offset of the j-th neuron. Then, the state of neuron j is

$$s_j' = \text{sgn}[y_j],$$

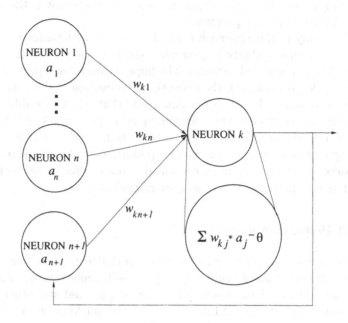

Fig. 12.11. A simple recurrent network. The output of the network is fed back as an input in neuron $n + 1$.

where sgn is the *signum function* which takes on a value of $+1$ if the argument is positive or -1 if the argument is negative. For the case where y_j is identically zero, the function's value can be chosen arbitrarily from ± 1.

Rather than adjust the weights of the network to achieve some desired output for each input, the network is used to store a number of N-dimensional vectors comprising symbols ± 1, represented by the weights that connect the neurons and their associated states. For each stored pattern $\mathbf{v}_i$, $i = 1, \ldots, p$, the weights of the network are determined by

$$w_{ji} = \begin{cases} \frac{1}{N} \sum_{k=1}^{p} v_{kj} v_{ki}, & \text{if } j \neq i \\ 0, & \text{if } j = i, \end{cases}$$

where v_{kj} is the j-th element of the pattern $\mathbf{v}_k$. Thus, the weights are a function of the outer product of the patterns to be stored. Once calculated they are fixed rather than being iteratively adjusted.

The Hopfield network is then used to remember stored patterns as generalizations over possible patterns that are presented to the network. For any pattern to be stored, $\mathbf{x}$, the network starts by setting each state $s_j = x_j$, for $j = 1, \ldots, N$. Then elements of s_j are updated at random according to the rule

$$s_j \leftarrow \text{sgn}[\sum_{i=1}^{N} w_{ji} s_j].$$

Once each state remains the same, the resulting vector of states, $\mathbf{s}$, is denoted as the *fixed point* or output of the network. In essence, it is the way that the network remembers the pattern $\mathbf{x}$.

The utility of this approach is not for storing the p patterns $\mathbf{v}_1, \ldots, \mathbf{v}_p$ but rather the resultant ability to generalize about a new pattern that is similar to one of the already stored patterns. The hope is that when a pattern that is similar to, say, $\mathbf{v}_1$, is presented, the network will transition to the same set of states, $\mathbf{s}$, that is obtained when $\mathbf{v}_1$ is presented, and that this will be different from the set of states encountered when offering up other patterns. The reliability of this process depends on the number of stored patterns, their inherent similarity, and the degree of noise imposed on the new pattern when it's offered to the network. Networks of this type are often described as *associative memories* because they associate new patterns with ones they have already seen.

12.4.3 Boltzmann machine

Another related form of neural network is the *Boltzmann machine* [225], which is a freely connected set of neurons (except no self-connections are allowed) where some are designated for input, others for output, and still others are treated as hidden nodes (figure 12.12). As with the Hopfield network, the processing units take on binary values (± 1) and all of the connections are symmetric. The updating for the state of each node proceeds one node at a time, with each randomly chosen. The state of the selected node is updated by flipping between ± 1 based on the Boltzmann distribution

$$\Pr(s_j \rightarrow -s_j) = \frac{1}{1 + e^{\frac{\Delta E_j}{T}}},$$

where s_j is the state of the j-th node, ΔE_j is the change in energy (the evaluation function), and T is the temperature associated with system. Thus, the states (nodes) of the network are fundamentally random, and can toggle endlessly depending on the effect in ΔE_j. Also, as the temperature T is decreased, the likelihood of a state flipping its sign decreases. The derivation of a gradient descent rule for updating the weights in the network is more complicated than the derivation of back propagation and Haykin ([210], pp. 321-330) provides the necessary details. The update rule for the weight w_{ji} connecting nodes i and j turns out to be a function of the correlation between s_i and s_j.

12.4.4 Network of multiple interacting programs

In each of the preceding designs, each node in the network is identical in the sense that it performs the same transfer function. What if we extended this protocol by allowing every neuron to actually become an arbitrary transfer function written in the form of a symbolic expression? The connections between neurons could be arbitrary and reflexive (i.e., connected to themselves), with

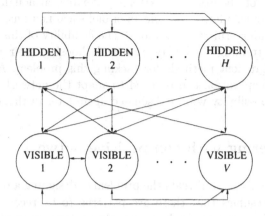

Fig. 12.12. The design of a Boltzmann machine (neural network). All neurons are fully connected. Some are "visible" and can be designated as input or output. The others are "hidden".

certain neurons defined for input and others for output. This architecture was offered by Angeline [9] to generate recurrent "programs." For example, figure 12.13 shows a possible neural network where the neurons T_0, T_1, and T_2 might be defined by the symbolic expressions

$$T_0 = (\text{In} + T_1)T_0$$
$$T_1 = (\text{In}/(T_2 + 0.97831))(T_0 - 0.11345)$$
$$T_2 = T_2 + 0.3745 \times \text{In} \cdot (T_0 + T_1).$$

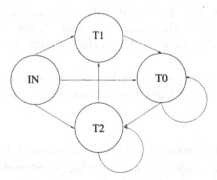

Fig. 12.13. The design of a network of multiple interacting programs (MIPs). The neuron "IN" is designated for input, whereas T_0, T_1, and T_2 are adaptable programs, and may be recurrent.

A prerequisite for constructing such a network is to define the expressions that can be used. Here those would include the operators $\{+, -, \times, /\}$ and terminals $\{\text{In}, \mathbb{R}, T_0, T_1, T_2\}$, where $\mathbb{R}$ is the set of real numbers. Clearly, such a network of "neurons" is extremely flexible, although it deviates from the initial idea of simulating the effects of biological neurons. Now we can have neurons performing arbitrary functions! The flexibility of this design also poses a challenge in training such networks, and choosing their topologies. There are no known gradient methods for tackling this problem. As an alternative, evolutionary computation can be used to adapt the network to meet the task at hand. This possibility will be discussed again later in this chapter.

12.5 Clustering with competitive networks

All of the above discussion treats the problem of designing a pattern recognition device. This presumes that there are patterns to be recognized. An entirely different situation often arises, however, when you are presented with data and need to determine if any patterns are even present. This is the problem of pattern *discovery*, not *recognition*. Such discovery is often described as *data mining*, the process of gleaming information from arbitrary data. In this situation, no a priori class labels are available for training. Instead, the task is to design a method for optimally clustering subsets of the available data according to useful rules. Neural networks provide a means for accomplishing this feat.

For simplicity, let's presume that the data consist of binary input strings from the alphabet $\{0, 1\}$, much like the Hopfield network operated on patterns as vectors of symbols from $\{-1, 1\}$. The goal is to find patterns in the collection of input strings by recognizing certain strings as being clustered. Consider the hierarchical arrangement of neurons as shown in figure 12.14. Each input neuron is connected to the neurons at the next level of the hierarchy, and so forth. The neurons in the higher levels are grouped in clusters and may also be linked with inhibitory connections (i.e., the output from any neuron in the cluster tends to diminish the output of all other neurons in that cluster). Only one neuron in each cluster will fire: the one with the greatest activity (assume the activity is scaled over $[0, 1]$).

Each neuron that is not an input has a fixed amount of weight (1.0) distributed among all of its incoming connections. That is, for the j-th neuron,

$$\sum_i w_{ji} = 1,$$

where w_{ji} is the weight connecting neuron i to neuron j. Initially, these values are chosen at random, subject to the preceding constraint. These weights are updated for the j-th neuron only if it is the one with the maximum output (the sum of the weighted inputs) among its cluster for the particular input that is offered to the network. The weight update is given by:

$$\Delta w_{ji} = \begin{cases} 0, & \text{if neuron } j \text{ does not have the maximum output} \\ & \text{of its cluster,} \\ g\frac{c_{ik}}{n_k} - gw_{ji}, & \text{if neuron } j \text{ has the maximum output of its cluster,} \end{cases}$$

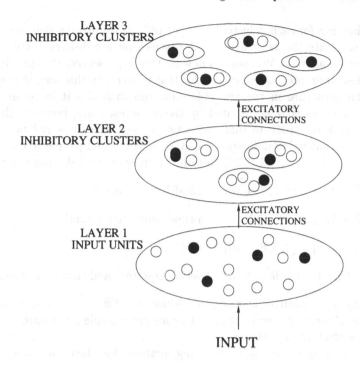

LAYER 3
INHIBITORY CLUSTERS

EXCITATORY
CONNECTIONS

LAYER 2
INHIBITORY CLUSTERS

EXCITATORY
CONNECTIONS

LAYER 1
INPUT UNITS

INPUT

Fig. 12.14. The design of a competitive learning network. Input excites certain neurons at the first layer. The outputs from these neurons are amplified and set to clusters of nodes in the second and higher layers in a hierarchy where only a single node in each cluster "fires." Further, nodes within the same cluster can be connected to other nodes in that cluster with inhibitory weights. Over successive presentations of input patterns, the network transitions to a "firing pattern" associated with each input. When new inputs are offered to the network, the hope is that the network will again exhibit the same firing patterns for inputs that are "close" to those that it has already observed.

where c_{ik} is 1 if the i-th element of the presented stimulus pattern, S_k, is 1, and c_{ik} is 0 otherwise, n_k is the number of active neurons in the input layer (i.e., the number of 1s in S_k), and g is a constant.

The update rule means that the change in the weight w_{ji} is $-gw_{ji}$ if the corresponding i-th element of the input pattern is 0. In this case, the weight gives up a proportion of its value to be redistributed to the weights that are associated with active connections. For those other active connections, the weight change is 0.0 if $c_{ik}/n_k = w_{ji}$, which means that the weight w_{ji} is distributed evenly across the active inputs. Otherwise, those weights that are larger than c_{ik}/n_k will be reduced while those that are smaller will be increased.

Over successive presentations of the input patterns, the weights will tend to distribute evenly across the active connections, but recall that the weight update occurs only for the neuron that has the maximum output. Initially, this is simply a random consequence of the initial weight values, but as the input patterns are presented iteratively, the network moves toward a stable set of neurons that fire for each "cluster" of input patterns. Patterns that are similar will tend to excite the same neurons. Patterns that are disparate will tend to excite different neurons. Whenever a new pattern is presented, the specific set of neurons that fire represent the cluster for that pattern. In this way, the network searches for structure in the available data and identifies it in terms of the pattern of neurons that are excited by the stimulus. Note, however, that the particular form of structure that the network may identify is not necessarily the structure that we would expect. That is, if you asked people to cluster the numbers $\{1,2,3,4,5,6,7,8,9,10\}$, you might observe the following responses:

- $\{1,2,3,4,5\}$; $\{6,7,8,9,10\}$ "small" and "large"

- $\{1,2,3,5,7\}$; $\{4,8,9,10\}$ "prime" and "not prime"

- $\{1,3,5,7,9\}$; $\{2,4,6,8,10\}$ "odd" and "even"

- $\{1,4,9\}$; $\{2,3,5,6,7,8,10\}$ "perfect squares" and "not perfect squares".

When using a competitive learning procedure, it's difficult to know what type of structure the network will discover. Uncovering the rule for determining that structure is often quite a challenge.

There are many other self-organizing methods for clustering using neural networks. For more information on these other competitive learning techniques, see [210], pp. 53, 397-434.

12.6 Using neural networks to solve the TSP

Even though it might not seem very natural at first, neural networks can be used to address combinatorial problems such as the TSP. Let's think about how we could do this. First, we have to decide between supervised and unsupervised learning approaches. In supervised learning, we need training data: input–output pairs where we know what the correct output is supposed to be and we can calculate the error that the network makes. What would the inputs to a supervised learning network be when tackling a TSP? The only data that we are given are the locations of the cities, as well as the distance metric (e.g., Euclidean). So it seems that the input would have to be a vector of city positions. What would be the output? It would have to be some ordering of the cities. For n cities, we'd need n output nodes. What then would be the desired output? That would be the order of the cities, perhaps each node sorted by its output strength in the order of appearance in the tour.

But there's a problem. How do we know what the order should be? After all, that's what we are trying to find out! We don't have any training data. Maybe if we had a variety of already-solved TSPs for some number of cities n, we could train the neural network to accept the city locations as input and give the corresponding order for the output, but this seems unlikely to work for large n because we would need many training examples and these are difficult to find when n gets big.

Instead of using supervised learning, we could try unsupervised learning. Again, the inputs are the city locations, but we don't have any specific preset training data. Instead, we want the network to self-organize into a configuration that tells us the order of the cities to visit. One way to do this is to construct a network with two input nodes, one for each dimension (x, y) of a city's location and connections to n output nodes. The nodes are initialized with random weights (w_x, w_y). City locations are presented to the network which then adjusts its weights in response. Let's say that the index for the city that is presented to the network is m. Then we find the *winning node* by determining which node minimizes

$$\min_i\{||\mathbf{x}_m - \mathbf{w}_i||\}.$$

The $|| \cdot ||$ norm here is the Euclidean distance. The node, say k, that wins the competition has its weights adjusted to move closer to the position of the city:

$$\mathbf{w}_k(new) = \mathbf{w}_k(old) + \alpha f(G, n)(\mathbf{x}_m - \mathbf{w}_k),$$

where G is a scaling function that decreases in time, and $f(G, n)$ is a function that amplifies the amount of correction to use in changing the weights of the k-th node, as is α. Things are a little more complicated than this because we can use $f(G, n)$ to adjust not only the node that wins the competition, but its neighbors as well, and use $f(G, n)$ as a function that puts more emphasis on the node that actually wins [61]. Furthermore, we can design $f(G, n)$ such that as G decreases, $f(G, n)$ focuses more change to the winning node and less to its neighbors. By updating the weights of the nodes in this fashion, the node that is closest to the current presented city is updated the most, but its neighbors are also updated and moved toward that same city.

As we cycle through the cities in the list, a different node will begin to win each competition. After many presentations of the cities, the nodes will have self-organized into a set where cities that are near each other in space will have nodes that are near each other in the ring. Traversing the ring of nodes then provides an approximate solution to the TSP.

There are many variations of this general procedure that can make it more effective. If you want to learn more about these methods, see [61, 60]. The basic procedure appears very similar to the learning rules used to discriminate between linearly separable data when using thresholds and is also similar to the competitive learning method described earlier. Neural networks were out of favor as models for intelligent systems in the 1970s and early 1980s until Hopfield and Tank [233], and others, demonstrated their apparent utility for pattern recognition and combinatorial optimization. Unfortunately, subsequent analysis has shown that many neural network methods for addressing the TSP are not very competitive with other heuristics [180]. Not every problem is amenable to neural networks, but sometimes it pays to consider alternative ways of solving problems that don't seem to have an immediate benefit.

12.7 Evolving neural networks

Training recurrent neural networks is often more difficult than finding the appropriate weights for a multilayer feedforward network. But even then, as we have seen, the standard gradient-based training methods (such as back propagation) can become stalled in locally optimal weight sets that do not offer the best performance that could be attained from the chosen neural architecture. This is a common problem in training neural networks because they are nonlinear functions. When the mapping from parameters (here, these are the weights of the neural network, and perhaps the topology) to output is a nonlinear function, and the output is then evaluated in light of another nonlinear function, the corresponding mapping from parameters to evaluation is often pocked with local optima. In these circumstances, gradient and other classic methods of optimizing the parameters often fail to yield the desired performance.

In contrast, we can turn to an evolutionary approach to training and designing neural networks. To start simply, if we are presented with a fixed architecture, then each individual is a set of weights that defines the input–output behavior of that architecture. The weights are often real valued, so they can be represented in a vector of reals and subjected to random variation and selection in light of an evaluation function. This function might still be the squared error function we were interested in before, but note that it no longer has to be a smooth function. We chose the squared error function because it allowed us to perform some calculus and derive a gradient-based update equation for training the weights. In forgoing the gradient method, we can also forgo the constraint that the evaluation function be differentiable. This opens numerous additional possibilities for designing neural networks in light of evaluation functions that more properly represent the value of the classifier in terms of its real-world cost and benefit.

For example, if we were classifying mammograms in terms of whether or not they indicated a malignancy, there are two ways to be correct: (1) correctly assert that there is a malignancy, and (2) correctly assert that there is no malignancy. Operationally, we would likely set the desired output of the network to be 1 for a malignancy and 0 for a benign case. Then we would measure the squared error from the target value associated with each mammogram. In doing so, however, we would give equally correct classifications equal credit. A neural network that correctly identifies a mammogram as indicating a malignancy (say the network outputs 1.0) has no error. Neither does a network that correctly identifies a benign case (say the network outputs 0.0). These two correct outcomes are, however, of different worth to the patient and the physician. It is typically much more important to correctly identify the malignancies than it is to correctly identify the benign cases.

In addition, the errors are not equally costly. There are two types of error: (1) the Type I error, a false alarm, where the neural network indicates a malignancy when none was present, and (2) the Type II error, or a miss, where the neural network indicates a benign condition when in fact a malignancy exists. In our usual squared error function, we would take the squared difference from the target as the error. This means that a neural network that outputs, say, a 0.9 when the target is 0.0 (benign) is just a bad as a network that generates a 0.1 when the target is 1.0 (malignant). Operationally, however, the miss is often much more costly than the false alarm. By using evolution to find the best set of weights for the neural network, we can tailor the evaluation function to truly represent the operational cost and benefit of the different types of correct and incorrect classifications.

There has already been a great deal of effort directed toward evolving neural networks. The earliest research in this direction was conducted in the 1960s, but it wasn't really until the early 1980s that the method became viable [149, 250]. The evolutionary process can not only search for the best set of weights for a fixed neural architecture, but also for the best architecture at the same time. All that is required is to encode the manner in which the connections between neurons are represented. The actual transfer function in each node could also be encoded and varied by evolution, as was shown above in Angeline's multiple interactive programs, or by choosing between more traditional transfer functions of sigmoid or Gaussian form, as offered in [422].

Just for illustration, let's consider a neural network with a maximum of ten neurons, two of which are designated as inputs (N1 and N2). One neuron (N10) will be the output. The connectivity of the network might be represented as a 10×10 matrix of 1s and 0s, where a 1 indicates that there is a connection coming from the neuron in the i-th row to the neuron in the j-th column. A 0 indicates no connection. These connections could be varied by randomly toggling a number of entries in the matrix. The more change that is imposed, the greater the effect on the resulting network. Each connection also has to have a weight associated with it. We could also encode these in a 10×10 matrix of real values. Each element in the matrix would represent the weight from neuron i to j if a connection exists; otherwise, the value for that entry would be ignored in evaluation. The weights could be varied by imposing zero mean Gaussian noise, dithering the values of all of the connections. Again, the more variation that is imposed, the greater the expected behavioral change of the network. Finally, we might also choose between sigmoid or Gaussian transfer functions for each node by having a binary vector of length 10 where a 1 indicates that the associated neuron is a sigmoid transfer function; otherwise, it has a Gaussian transfer function. These bits could be flipped at random, just like the elements of the connectivity matrix.

12.8 Summary

The dream of building artificial brains is as alive today as it has been since the beginning of computing. The dream is illusive, and will likely remain a dream for many decades (or even centuries) to come. In fact, some have suggested (e.g., Walter Freeman) that digital implementations of neural networks may fail to emulate the biological reality because real brains are chaotic analog devices. Digital computers can only approximate the behavior of analog devices to a given degree of precision. This precision may not be sufficient to capture the chaotic dynamics that exist in the real brain. Nevertheless, it's possible to make idealistic models of facets of biological brains and convert those models into computer algorithms that can help us solve problems. The advantages of parallel distributed information processing are multifold:

1. Neural networks offer flexible mapping functions that can be tailored to meet almost any real-world demands.

2. Rather than necessarily break down when a component of the neural network fails, it's possible to design these architectures to be fault tolerant because the information processing of the network is spread throughout the web of connections, rather than being stored in any single location.

3. There are training methods that can rapidly find locally optimal parameters for many neural architectures.

4. There are other (evolutionary) methods for training neural networks that can overcome multiple local optima and also adapt the topology and local processing functions of the network.

5. Many neural networks can be implemented in hardware (VLSI) with the result being very fast computation.

Some of these advantages are also disadvantages. For example, the flexibility that neural networks offer is both a blessing and curse. The natural desire is to make the networks larger and larger so that they can be more flexible and compute more complex functions. The larger the network is, however, the more brittle it often becomes as it fits the available data exactly and then fails to generalize over any noise in the data. Moreover, the larger and more interconnected the neural network becomes, the less explainable it becomes. This can be an important issue in medical and other critical applications where the user needs to know not only the right answer to their problem but the rationale that leads to that right answer. Many users have to be able to defend themselves!

The survey of techniques offered here is incomplete in some important respects. Several common neural architectures have been omitted, including radial basis functions, adaptive resonance theory, Hebbian learning, self-organizing feature maps, etc. A complete review of these designs requires the discourse that is possible in a textbook or handbook. Haykin [210] and Fiesler and Beale [135] are two good sources of information.

XIII. What Was the Length of the Rope?

Problems are very often expressed in fuzzy language and its takes some time to understand all the necessary specifications. The process of understanding the requirements — which is necessary for modeling the problem — isn't always straightforward. The following puzzle illustrates the case.

A rope ran over a pulley. At one end was a monkey. At the other end was a weight. The two remained in equilibrium. The weight of the rope was one-quarter of a pound per foot, and the ages of the monkey and the monkey's mother amounted to four years. The weight of the monkey and the weight of the rope were equal to one-and-a-half of the age of the monkey's mother.

The weight of the weight exceeded the weight of the rope by as many pounds as the monkey was years old when the monkey's mother was twice as old as the monkey's brother was when the monkey's mother was half as old as the monkey's brother will be when the monkey's brother is three times as old as the monkey's mother was when the monkey's mother was three times as old as the monkey was in the previous paragraph.

(You see what we mean about things not always being straightforward?)

The monkey's mother was twice as old as the monkey was when the monkey's mother was half as old as the monkey will be when the monkey is three times as old as the monkey's mother was when the monkey's mother was three times as old as the monkey was in the first paragraph.

The age of the monkey's mother exceeded the age of the monkey's brother by the same amount as the age of the monkey's brother exceeded the age of the monkey.

What was the length of the rope?[1]

If you find the problem confusing, try reading it again and make some notes for yourself along the way. This is the second-to-last puzzle in the book, so give it your best shot!

Actually, it's not that difficult to solve the problem once we understand it — when we know how to model it with appropriate equations. The second-hardest part here is simply understanding all of the information given in the description of the problem. The hardest part is undoubtedly having the patience

---

[1] It should be underlined here that the unknowns of this problem need not be integers, as was the case with the ages of the three sons in the first puzzle of this volume. Thus, the age of the monkey might be, for example, $1\frac{3}{7}$.

to understand the information. Patience is a prerequisite for effective problem solving.

Let's first introduce some variables:

x – age of monkey,
y – age of monkey's brother,
z – age of monkey's mother,
W – weight of the weight, and
r – the length of the rope.

Now we can translate each sentence of the puzzle into equations. It would be easier if we "read" the puzzle from the bottom up, i.e., if we start interpreting it from the last paragraph. The fourth (last) paragraph of the puzzle says simply that

$$z - y = y - x. \tag{12.2}$$

This is straightforward and doesn't require any further comments.

Now, let's try the sentences of the third paragraph. "The monkey's mother was twice as old as the monkey was..." can be written as

$$z - A = 2(x - A),$$

where A denotes some (not necessarily integer) number of years, as the sentence refers to the past. Similarly, we can interpret the remaining sentences of this paragraph. The third paragraph says that for some A, B, and C, the following equalities hold:

$$
\begin{aligned}
z - A &= 2(x - A), \\
z - A &= 0.5(x + B), \\
x + B &= 3(z - C), and \\
z - C &= 3x.
\end{aligned}
$$

This can be simplified into

$$4z = 13x. \tag{12.3}$$

The second paragraph is very much the same as the third and can be interpreted in the same way. (Note that the weight of the rope was $r/4$.) It says that for some A, B, C, and D, the following equations hold:

$$
\begin{aligned}
W - r/4 &= x - A \\
z - A &= 2(y - B) \\
z - B &= 0.5(y + C) \\
y + C &= 3(z - D), and \\
z - D &= 3x,
\end{aligned}
$$

so

$$W - r/4 = 11x - 2z. \tag{12.4}$$

The first paragraph provides more information. First, the ages of the monkey and the monkey's mother amounted to four years:

$$x + z = 4. \tag{12.5}$$

The weight of the monkey and the weight of the rope were equal to one-and-a-half of the age of monkey's mother:

$$W + r/4 = 3z/2. \tag{12.6}$$

The weight of the monkey is W, and the monkey and the weight remained in equilibrium. From equations (12.2), (12.3), and (12.5) we obtain

$x = 16/17$,
$y = 2$, and
$z = 52/17$.

Then equations (12.4) and (12.6) give

$W - r/4 = 72/17$, and $W + r/4 = 78/17$,

which yields $r = 12/17$. Thus the length of the rope was 12/17 feet.

No doubt you might have some interesting ideas of what to do to the authors of this puzzle using that rope! Fortunately, it's only about $8\frac{1}{2}$ inches long!!

Even when a problem is expressed very clearly and precisely, with all the necessary specifications, it's still possible to overlook some obvious things and make what amounts to a silly mistake in modeling the problem (see chapter II). The following puzzle illustrates the case.

There are three switches outside a room that control three electrical bulbs inside the room. They each hang on a long wire so you have to be careful not to bump into them when you enter the room! You know that there is a one-to-one connection between the switches and the bulbs. Each switch controls precisely one bulb. You know also that the switches are in the "off" position and the room is dark. Now we shut the door and you can't see anything inside the room.

Your task is to figure out which switch connects to which bulb, but you face a rather imposing constraint: you only get to make one trial. You can set the switches any way you like and then you have to enter the room. You can examine the room and based on this examination you have to be able to tell us which switch controls which bulb. Note that you're not allowed to go back out and touch any of the switches again after you've entered the room. You can do anything you like with the switches, but once you open the door to the room, that's it.

At first you might think that this is an "impossible" problem! After all, if you don't touch any of the switches, then it would be pointless to enter the room, since you already know it is dark in the room and there will be no way to figure out which switch goes with which bulb. So you have to do something with the switches. But what?

If you flip only one switch to the "on" position and then open the door, certainly one of the bulbs will be lit, but you won't be able to distinguish which of the two remaining switches goes with which of the two remaining bulbs.

Similarly, if you flip some pair of switches to the "on" position, you'll know from the light that remains unlit which switch it corresponds with, but you won't be able to differentiate between the two bulbs that are lit and the two switches that you flipped. What a dilemma.

The crux of the issue here is to start with the right model. The preceding thinking relies on a model that captures only one attribute of light bulbs, namely whether they are "on" or "off." But light bulbs have other attributes. Like what? How about temperature? There was nothing in the description above that restricted our attention just to the illumination of the bulb. In fact, that's really only part of your becoming illuminated about the problem.

Once we include the bulbs' temperatures in our model, the problem really is quite easy to solve. All you have to do is set two switches to the "on" position, wait a few minutes, then switch one of these to the "off" position and enter the room. When you do, one of the bulbs will be on, and the other two will be off. One of those two unlit bulbs, however, is going to be hot and that does make it very easy to give the proper answer. But watch your fingers and don't get burned!

13. Fuzzy Systems

Sometimes I think the surest sign
that intelligent life exists
elsewhere in the universe is that
none of it has tried to contact us.

Bill Watterson, *Calvin and Hobbes*

Prior to the advent of the digital computer, calculations were often performed on slide rules. By necessity, answers were almost always "good enough" rather than precise because our eyesight and manual dexterity weren't sufficient to manipulate a slide rule to yield an arbitrary number of significant digits. To multiply, say, π by e you would line up the slide rule at about 3.14, give or take, and then read off the answer next to 2.72 on the upper index. With a little practice, you could generate answers that were within the degree of precision offered in the original problem (here, two decimal places). No one could expect finer tuning than that. Perhaps this is where the common saying "close enough for government work" originated. Today, in contrast, the precision of the modern digital computer is a great asset.

When people communicate with others, however, they rarely resort to the level of exactness that might be possible. It simply isn't practical. For example, the adage when playing the stock market is *buy low and sell high*. It's good advice, but how low is "low" and how high is "high"? If the answer is, say, that a price of $10 per share for some particular stock is low, then isn't $10.01 still low? What about $10.02, or $11, or $20? It's unreasonable to expect someone to behave as if there were a firm clear dividing line that separates the prices that are low from those that are not. People understand and operate on terms based on their own individual understanding of the degree to which those terms represent some particular condition. Words like low, high, close, very old, red, early, and so forth, each have a general meaning that we understand. Those meanings are imprecise, but useful nevertheless. Trying to impose a precise meaning to each term isn't just impractical, it's truly impossible, because each term means something a little different to each person. The concept of "middle age" varies greatly from those who are under 30 years old to those who are in their 50s. Effective communication not only benefits from imprecise descriptions, it demands them.

It's natural to consider the possibility of programming algorithms that would treat concepts in the same manner as we do, essentially, computing with words instead of numbers. But all computation is performed ultimately on numbers. Thus, we must seek to describe a mathematical framework that encompasses our common usage of terms that are not black and white Boolean descriptions, but instead are indistinct, vague, and *fuzzy*.

13.1 Fuzzy sets

The basic foundations of operating on fuzzy descriptions of the real world were first offered by Lotfi Zadeh in 1965 [505]. The framework is really a straightforward extension of classic set theory. In a classic set, elements either are or are not members of that set. The set of even natural numbers that are less than 10 is a precise collection of four elements $\{2, 4, 6, 8\}$. Alternatively, this set could be described as a membership function, $m_A(x)$, defined over some "universe of discourse," where the function takes on the value 1 when the argument x is an element of the set A and is 0 otherwise. A fuzzy set offers the possibility for $m_A(x)$ to take on values other than $\{0, 1\}$, say in the range $[0, \alpha]$, where the value assumed indicates the degree of membership of x in A. For consistency, the possible range of values is often scaled to the real numbers between 0 and 1, inclusive. A membership $m_A(x) = 1$ indicates that the element x definitely is a member of A. A membership $m_A(x) = 0$ indicates that the element x is definitely not a member of A. Intermediate values indicate intermediate degrees of membership. It is this intermediate range that represents the extension of fuzzy sets beyond classic "crisp" sets.

It's often convenient to describe membership functions in terms of continuous-valued mappings, rather than $(x, m_A(x))$ pairs. For example, a membership function describing the fuzzy set of ages that are "old," might be given by

$$m_A(x) = 0, \qquad \text{if } x \leq 50$$
$$m_A(x) = (1 + \tfrac{25}{(x-50)^2})^{-1}, \quad \text{if } x > 50$$

(see figure 13.1) as offered by Zadeh [506]. Note that this isn't the only possible membership function for "old." In fact, there's no right answer to the question of which function to choose. The choice is subjective: each individual can postulate a membership function that captures their own perspective. This is an advantage for fuzzy systems because they can more readily reflect personal biases and can be tailored to fit each person's viewpoint. This, however, also poses a challenge to those who want to use membership functions for computation.

13.2 Fuzzy sets and probability measures

Since membership values are typically normalized over $[0, 1]$, and this is also the range of values for probabilities, there is often confusion regarding the relationship between fuzzy sets and probability measures. The first thought might be that they are identical, that fuzzy measures are really just another name for

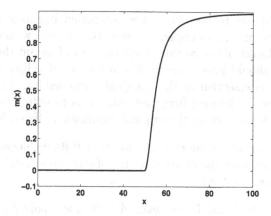

Fig. 13.1. A possible fuzzy membership function for ages that are "old." The function is 0.0 up until age 50, and climbs rapidly toward 1.0 after age 50.

probabilities. This is not correct: fuzziness and probability have two distinctly different interpretations.

Bezdek and Pal [44] provide an excellent example. Suppose you were thirsty and were offered a choice between two bottles of liquid. You're told that the first bottle has membership 0.91 in the fuzzy set of all potable (i.e., drinkable) liquids. In contrast, you're told that the probability that the second bottle contains potable liquid is 0.91. Which would you rather drink from? The fact that you must contemplate a decision immediately verifies that there is a difference in interpretation, otherwise you'd have no decision at all, but let's carry this further to see the implications.

A liquid that is 0.91 a member of the fuzzy set of potable liquids might not taste very good, but by any reasonable interpretation, we wouldn't expect it to be poisonous. In contrast, a liquid that has a 0.91 probability of being potable has a corresponding complementary 0.09 probability of not being potable — of being, say, arsenic. The anticipation is that drinking from the first bottle isn't deadly. The same cannot be said for drinking from the second bottle. Further, after observing the contents of both bottles, say by having a brave friend drink from each, the probabilistic interpretation of the second bottle must be resolved: either the liquid contained in that bottle is or is not potable. The probability that it's potable becomes either 1 or 0. The fuzzy description of the liquid in the first bottle can remain the same after sampling. Clearly then, probabilistic descriptions are not the same as fuzzy descriptions.

13.3 Fuzzy set operations

As with classic sets, we need to be able to operate on fuzzy sets and describe their intersection, union, complements, and so forth. This is particularly true because we would like to be able to use compound linguistic descriptions in a mathematical way. For example, a flower that is yellow and fragrant has membership in the set of yellow things and the set of fragrant things. A "yellow fragrant" flower should have membership in the set of yellow fragrant things that reflects the intersection of the marginal memberships. There are several alternative means for defining fuzzy set operations to reflect these properties. For clarity, we'll focus here on the original formulation offered in [505]:

- A fuzzy set is said to be *empty* if and only if its membership function is identically zero on the entire universe of discourse X (all possible arguments to the function).

- Two fuzzy sets A and B are *equal*, $A = B$, if and only if $m_A(x) = m_B(x)$ for all $x \in X$. For convenience, this can be written as $m_A = m_B$.

- The *complement* of a fuzzy set A is denoted A' and defined as $m_{A'} = 1 - m_A$.

- The fuzzy set A is *contained in* or is a *subset of* B if and only if $m_A \leq m_B$.

- The *union* of two fuzzy sets A and B is a fuzzy set $C = A \cup B$ with a membership function $m_C(x) = \max\{m_A(x), m_B(x)\}$, $x \in X$.

- The *intersection* of two fuzzy sets A and B is a fuzzy set $C = A \cap B$ with a membership function $m_C(x) = \min\{m_A(x), m_B(x)\}$, $x \in X$.

To return to our example of a yellow fragrant flower, suppose the flower in question has membership $m_{Yellow}(x) = 0.8$ and $m_{Fragrant}(x) = 0.9$. Then its membership in the fuzzy set of yellow fragrant flowers would be $\min\{0.8, 0.9\} = 0.8$. Its membership in the fuzzy set of yellow or fragrant flowers would instead be $\max\{0.8, 0.9\} = 0.9$. Its membership in the fuzzy set of flowers that are not yellow would be $1 - 0.8 = 0.2$. These are basic mechanisms for treating combinations of fuzzy descriptions.

Another common fuzzy descriptor is the word "very," which acts as an amplification of a condition (e.g., very old, very quiet). Intuitively, applying a condition of "very" to a fuzzy description should have a corresponding amplification effect on the membership function. In many cases, this can be designed explicitly. For example, let's reconsider the membership function for the fuzzy set A of ages that are old:

$$m_A(x) = 0, \qquad\qquad \text{if } x \leq 50$$
$$m_A(x) = (1 + \tfrac{25}{(x-50)^2})^{-1}, \quad \text{if } x > 50$$

If we say A^* is the fuzzy set that is "very A," this new set might be defined as

$$m_{A^*}(x) = (m_A(x))^2, \text{ for all } x \in X$$

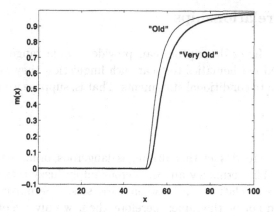

Fig. 13.2. The application of the hedge term "very" as a modifier to the previous fuzzy membership function for ages that are "old." The word "very" is taken to mean a corresponding squaring of the base fuzzy membership function. The memberships of ages over 50 in the fuzzy set of "very old" is less than their corresponding memberships in the fuzzy set of "old." Intuitively, if a person has a certain positive membership in the set of "old" people, then another person has to be much older than that to have the same membership in the set of "very old" people.

(see figure 13.2). Here, the effect of squaring the membership values is one of intensification. It requires an even more extreme value of x to generate the same membership in A^* as in A.

Other common linguistic terms can also be handled mathematically using fuzzy sets:

- If A is very inexact, then $A =$ very (not exact), which is an amplification of the complement of fuzzy set of "exact." Thus the membership function for A might be defined as $m_A = (1 - m_{Exact})^2$. Note that this is not the same fuzzy set as $A =$ not very exact, which is the complement of "very exact," or $m_A = 1 - m_{Exact}^2$.

- Zadeh [506] introduced additional notions of "plus" and "minus" which serve as milder degrees of "very":

$$m_{A+}(x) = m_A(x)^{1.25}, \text{ and}$$
$$m_{A-}(x) = m_A(x)^{0.75}.$$

He also defined the term "highly" to be roughly equivalent to "minus very very" and "plus plus very."

These modifiers are called *hedges*. They are commonly used in everyday language and can be systematically applied to modify fuzzy sets in a corresponding appropriate manner.

13.4 Fuzzy relationships

While the above fuzzy descriptors can provide a wide range of useful terms and an associated mathematics to treat such linguistics, they do not cover the natural extension to conditional statements. That is, suppose we wanted to treat the case where

If A then B.

For example, if the road is icy then driving is dangerous, or more succinctly, if *icy* then *dangerous*. The terms icy and dangerous imply fuzzy sets and the *if-then* condition implies a relationship between these sets. The universe of discourse for A and B need not be the same; therefore, the new universe of discourse that describes the relationship between A and B is a Cartesian product $A \times B$ of all pairs (a, b) where $a \in A$ and $b \in B$. More specifically, if R is a *fuzzy relation* from the fuzzy set A to the fuzzy set B, R is a fuzzy subset of $A \times B$ where every pair (a, b) has some associated membership value $m_R(a, b)$.

Suppose that $A =$ small numbers, over a universe of discourse $U = \{1, 2\}$, and $B =$ large numbers, over a universe of discourse $V = \{1, 2, 3\}$. The membership values for elements in A are given in ordered pairs $(a, m_A(a))$ as $(1, 1)$, $(2, 0.6)$, and similarly for B, $(1, 0.2)$, $(2, 0.5)$, $(3, 1)$. Then $R = A \times B$ is defined to take on a membership that equals the $\min\{m_A(a), m_B(b)\}$ for each (a, b) pair. This is described conveniently by a matrix:

$$R = \begin{array}{c|ccc} & 1 & 2 & 3 \\ \hline 1 & 0.2 & 0.5 & 1.0 \\ 2 & 0.2 & 0.5 & 0.6 \end{array}$$

R, then, is the fuzzy subset of $U \times V$ that describes the membership of each of its constituent elements (u, v) in terms of the relationship between fuzzy sets A and B.

Fuzzy relationships can be chained together as well. If R is a fuzzy relation from X to Y and S is a fuzzy relation from Y to Z, then the composition of fuzzy relations R and S is defined over the pairs $(x, z) \in X \times Z$ as

$$R \circ S = \max_y \left[\min\{m_R(x, y), m_S(y, z)\} \right].$$

If the domains of X, Y, and Z are finite, then $R \circ S$ is called the *max-min* product of the relation matrices R and S. In a *max-min* matrix product, the operations of addition and multiplication are replaced by *max* and *min*, respectively. With the forgoing algebraic framework, it's now possible to treat compound fuzzy conditional statements.

Consider the conditional "if A then B" to be a special case of "if A then B else C" where B and C are defined on the same universe of discourse V, and A is defined on a universe of discourse U. The statement "if A then B else C" is defined as

$$(A \times B) \cup (\sim A \times C)$$

and thus the simpler "if A then B" can be considered as covering an unspecified fuzzy set C. If this is interpreted as indicating that $(\sim A \times C)$ can be essentially any fuzzy subset of $(U \times V)$ then we can define "if A then B" as $(A \times B)$ because we can choose $(\sim A \times C)$ such that

$$(A \times B) \cup (\sim A \times C) = (A \times B).$$

This is a so-called *Mamdani method* of approximate reasoning. Note that this is not the only plausible definition because we could make other choices for the effect of $(\sim A \times C)$.

To explore one of these alternatives, suppose we took A and B as above. Then we already know that

$$(A \times B) = \begin{array}{c|ccc} & 1 & 2 & 3 \\ \hline 1 & 0.2 & 0.5 & 1.0 \\ 2 & 0.2 & 0.5 & 0.6 \end{array}$$

But what about $(\sim A \times C)$? The fuzzy set $\sim A$ is $\{(1,0),(2,0.4)\}$, and therefore if C is any fuzzy subset of V, say $\{(1,\alpha),(2,\beta),(3,\gamma)\}$,

$$(\sim A \times C) = \begin{array}{c|ccc} & 1 & 2 & 3 \\ \hline 1 & 0 & 0 & 0 \\ 2 & \min\{0.4,\alpha\} & \min\{0.4,\beta\} & \min\{0.4,\gamma\} \end{array}$$

Taking $(A \times B) \cup (\sim A \times C)$ yields

$$\begin{array}{c|ccc} & 1 & 2 & 3 \\ \hline 1 & 0.2 & 0.5 & 1.0 \\ 2 & \max[0.2, \min\{0.4,\alpha\}] & 0.5 & 0.6 \end{array}$$

We can determine without ambiguity that $(2,2)$ and $(2,3)$ have membership 0.5 and 0.6, respectively, because the $\min\{0.4,\beta\}$ and $\min\{0.4,\gamma\}$ will never be larger than 0.5 and 0.6. For the case of $(2,1)$, however, we cannot know if 0.2 will be larger than the minimum of 0.4 and α. This is a curious case of fuzzy computation where the answer to the question of what is "If A then B" depends in part on a fuzzy set C associated with a conditional dependence on the complement of A. Note that if the membership value for $(2,1)$ in $(A \times B)$ were 0.4 or larger, then the membership for $(2,1)$ in $(A \times B) \cup (\sim A \times C)$ would also be 0.4 without ambiguity. Under the Mamdani method for determining the *if-then* relationship, the consequence of $(\sim A \times C)$ is effectively removed from consideration.

An important extension of the principle of fuzzy relations comes when, under a particular specified fuzzy subset A, we wish to determine the resulting fuzzy subset B when $R = A \times B$. Here, $B = A \circ R$, the max-min product of A and R. For instance, if $A = \{(1,1.0),(2,0.6)\}$ and

$$R = \begin{array}{c|ccc} & 1 & 2 & 3 \\ \hline 1 & 0.2 & 0.5 & 1.0 \\ 2 & 0.2 & 0.5 & 0.6 \end{array}$$

from above, then $B = \{(1, 0.2), (2, 0.5), (3, 1.0)\}$.

To illustrate a compound fuzzy conditional statement, consider the following example from [506]. Let the universe of discourse be $U = \{1, 2, 3, 4, 5\}$, with the fuzzy sets

A = "small" $\equiv \{(1, 1.0), (2, 0.8), (3, 0.6), (4, 0.4), (5, 0.2)\}$,
B = "large" $\equiv \{(1, 0.2), (2, 0.4), (3, 0.6), (4, 0.8), (5, 1.0)\}$, and
C = "not very large" $\equiv \{(1, 0.96), (2, 0.84), (3, 0.64), (4, 0.36), (5, 0)\}$.

Note that while A and B are defined fuzzy sets, C is a computed fuzzy set, determined as the complement of the fuzzy set "very large," which has membership values

$$m_{very\ B}(x) = m_B^2(x).$$

Suppose that we want to find the relation R that describes the fuzzy conditional statement: "If x is small then y is large else y is not very large." In other words, "If A then B else C."

From above, we know R is defined by $(A \times B) \cup (\sim A \times C)$. First, we must solve for each part of this compound expression separately. $(A \times B)$ is the matrix

	1	2	3	4	5
1	0.2	0.4	0.6	0.8	1.0
2	0.2	0.4	0.6	0.8	0.8
3	0.2	0.4	0.6	0.6	0.6
4	0.2	0.4	0.4	0.4	0.4
5	0.2	0.2	0.2	0.2	0.2

$(\sim A \times C)$ is the matrix

	1	2	3	4	5
1	0	0	0	0	0
2	0.2	0.2	0.2	0.2	0
3	0.4	0.4	0.4	0.36	0
4	0.6	0.6	0.6	0.36	0
5	0.8	0.8	0.64	0.36	0

The union of these two matrices is the *max* value of each cell, thus,

$R =$

	1	2	3	4	5
1	0.2	0.4	0.6	0.8	1.0
2	0.2	0.4	0.6	0.8	0.8
3	0.4	0.4	0.6	0.6	0.6
4	0.6	0.6	0.6	0.4	0.4
5	0.8	0.8	0.64	0.36	0.2

Suppose that we now ask, what will the membership function for Y be if X is *very small*? (Recall that "very small" is the square of the membership values of the fuzzy set "small" above.) This can be determined by taking the *max-min* product of X and R:

$$X \circ R = [1.0\ 0.64\ 0.36\ 0.16\ 0.04] \circ R = [0.36\ 0.4\ 0.6\ 0.8\ 1.0].$$

Thus if X is *very small* then, according to the rule "If X is *small* then Y is *large* else Y is *not very large*," the membership function for the resultant Y is quite a bit closer to *large* than to *not very large*. This example leads naturally to an extension where fuzzy rules are applied to situations that are often found in control systems, where several fuzzy rules are used to match observed conditions and consequently to recommend alternative courses of action.

13.5 Designing a fuzzy controller

By aggregating multiple fuzzy rules of the form described above, it is possible to design complex algorithms that can treat real-world problems. One such class of problems is found in the area of control. These problems require allocating available resources so as to elicit a particular response from a system. Some examples include

- Maintain an airplane in straight-and-level flight.

- Hold the temperature of a room at 20 degrees centigrade.

- Keep a car engine at 800 revolutions per minute at idle, even when the air conditioner is turned on or off.

- Dispatch the elevators in a building to respond to all requests in minimum time.

Each of these systems — an airplane, a room, a car, a set of elevators — can be characterized by differential equations, finite state machines, or some other mathematical constructions. Collections of fuzzy rules can then be used to describe the state of the system and the best course of action to bring about the desired results.

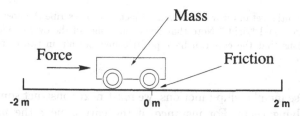

Fig. 13.3. A cart on a four-meter-long track. The goal is to return the cart to the center of the track with zero velocity. The available control is to push or pull on the cart, which has a given mass and coefficient of friction.

Consider the following typical example. There is a cart on a track that is four meters long (see figure 13.3). The cart has a certain mass and coefficient of friction. The control goal is to push or pull on the cart so as to return it to the center of the track, perhaps with the additional objective of accomplishing this feat in minimum time. The variables in question are the position x and velocity v of the cart, and the control force u. We might describe the cart's position simply in terms of the fuzzy sets *left, middle*, or *right*. Figure 13.4 provides possible fuzzy membership functions for these linguistic terms using trapezoidal and triangular shapes. Similarly, the cart's velocity might be described by *moving left, standing still*, and *moving right*. Figure 13.5 provides the analogous fuzzy membership functions for these fuzzy sets. Finally, the force to apply might be described by *pull, none*, and *push*, where *pull* indicates applying force to move the cart to the left, and similarly *push* indicates applying force to move the cart to the right. The fuzzy membership functions for the force applied to the cart might be as shown in figure 13.6. Finally, for the sake this example, let's assume that only three of the options for the force to apply are given fuzzy membership values: *push* or *pull* with one newton of force, or apply no force at all. We simply won't consider the intermediate values for calculating fuzzy control actions here, but we could still push or pull on the cart with any degree of force ranging between puls/minus one newton.

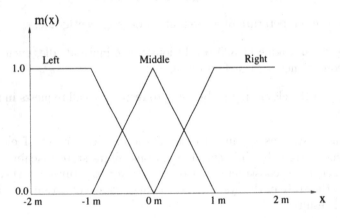

Fig. 13.4. A potential set of fuzzy membership functions to describe the position of the cart as "left," "middle," and "right." Note that from the design of the overlapping triangle and trapezoid functions that the cart can have positive membership in two of the three fuzzy functions at once

Given these membership functions, we must next construct some reasonable rules for applying them. For instance, if the cart is near the *middle* and is *standing still* then the force to apply is *none*. That's the first rule. Here are some other rules that we might consider:

- If *left* then *push*

- If *right* then *pull*

- If *middle* then *none*

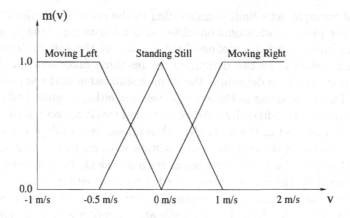

Fig. 13.5. A potential set of fuzzy membership functions to describe the velocity of the cart as "moving left," "standing still," and "moving right"

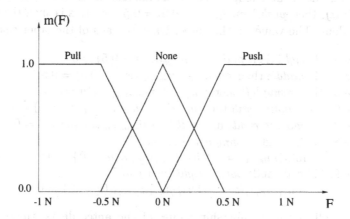

Fig. 13.6. A potential set of fuzzy membership functions to describe the action applied to the cart (force) in terms of "pull," "none," and "push".

- If *moving left* then *push*

- If *standing still* then *none*

- If *moving right* then *pull*

- If *left* and *moving left* then *push*

- If *right* and *moving right* then *pull*

It's more difficult to prescribe what to do in cases such as when the cart is *left* but is *moving right*. Should we push the cart? This might send it more rapidly toward the center, but since it's already moving right this extra push might send the cart past the center and on to the right-hand side of the track. For the sake of example, let's limit consideration to the above nine rules (the eight bulleted rules plus the additional one listed at the beginning of the paragraph).

For any input condition defined by the state of the cart in terms of its position and velocity, we can determine the resulting fuzzy outcome for each rule. The first step is to determine the membership value of the antecedent for each rule. The antecedents in the example here are either singular fuzzy sets, or compound fuzzy sets defined by the conjunction of two fuzzy sets. If the rule has only a single fuzzy set in the antecedent then all we have to do is calculate the membership value of the current system's state with respect to that fuzzy set. If the rule has more than one fuzzy set as the antecedent, then the membership value is defined by the minimum membership value in either fuzzy set.

Suppose that the cart is at the position $x = -0.5$ meters and is standing still, $v = 0$. Then, for Rule 1, "If *middle* and *standing still* then *none*," the membership of $x = -0.5$ in the fuzzy set middle is $m_{middle}(x = -0.5) = 0.5$ and the membership of $v = 0$ in the fuzzy set *standing still* is $m_{standing\ still}(v = 0) = 1$. Therefore, the membership value of the antecedent in this case is 0.5. For Rule 2, "If *left* then *push*," $m_{left}(x = -0.5) = 0.5$, and this is then the value of the antecedent. The values of the antecedents for each of the other rules are

Rule 3: "If *right* then *pull* " — $m_{right}(x = -0.5) = 0$
Rule 4: "If *middle* then *none*" — $m_{middle}(x = -0.5) = 0.5$
Rule 5: "If *moving left* then *push*" — $m_{moving\ left}(v = 0) = 0$
Rule 6: "If *standing still* then *none*" — $m_{standing\ still}(v = 0) = 1$
Rule 7: "If *moving right* then *pull* " — $m_{moving\ right}(v = 0) = 0$
Rule 8: "If *left* and *moving left* then *push*" —
$\qquad$ $\min\{m_{left}(x = -0.5), m_{moving\ left}(v = 0)\} = 0$
Rule 9: "If *right* and *moving right* then *pull* " —
$\qquad$ $\min\{m_{right}(x = -0.5), m_{moving\ right}(v = 0)\} = 0$

Next, given all of the membership values of the antecedents, these must be combined with the membership values of the consequents (i.e., the fuzzy sets that define the action to be taken on each rule). The combination is again a minimum between the membership associated with each possible action and the membership value of the antecedent, computed rule by rule. The result is a collection of fuzzy sets that describe the output based on each rule; these computed values are given in table 13.1.

The third step is to aggregate all of the results from each rule. This is accomplished by taking the maximum membership for each state in the consequent (here, -1, 0, or 1) across all of the results of each rule. This yields $(-1, 0)$, $(0, 1)$, $(1, 0.5)$. This is the fuzzy set that describes the membership of each possible action of pulling, none, or pushing on the cart given the input conditions and the implications of the rules that were assembled.

The final step is to "defuzzify" this fuzzy set to arrive a single output value. There are many ways to accomplish this. In fact, Mizumoto [325] outlined 11 different defuzzification procedures that have been offered in the literature. The most common is called the "center of area" method, which takes the ratio

Table 13.1. The result is a collection of fuzzy sets that describe the output based on each rule. Here we are restricting our computation to consider only three values of the force to apply: −1, 0, or +1 newton. The consequents indicate the degree of membership of each option in the corresponding fuzzy set.

Membership of Antecedent	*Consequent*
Rule 1: 0.5	*none* $\{(-1,0),(0,1),(1,0)\}$
Result: $\{(-1,0),(0,0.5),(1,0)\}$	
Rule 2: 0.5	*push* $\{(-1,0),(0,0),(1,1)\}$
Result: $\{(-1,0),(0,0),(1,0.5)\}$	
Rule 3: 0	*pull* $\{(-1,1),(0,0),(1,0)\}$
Result: $\{(-1,0),(0,0),(1,0)\}$	
Rule 4: 0.5	*none* $\{(-1,0),(0,1),(1,0)\}$
Result: $\{(-1,0),(0,0.5),(1,0)\}$	
Rule 5: 0	*push* $\{(-1,0),(0,0),(1,1)\}$
Result: $\{(-1,0),(0,0),(1,0)\}$	
Rule 6: 1	*none* $\{(-1,0),(0,1),(1,0)\}$
Result: $\{(-1,0),(0,1),(1,0)\}$	
Rule 7: 0	*pull* $\{(-1,1),(0,0),(1,0)\}$
Result: $\{(-1,0),(0,0),(1,0)\}$	
Rule 8: 0	*push* $\{(-1,0),(0,0),(1,1)\}$
Result: $\{(-1,0),(0,0),(1,0)\}$	
Rule 9: 0	*pull* $\{(-1,1),(0,0),(1,0)\}$
Result: $\{(-1,0),(0,0),(1,0)\}$	

$$\frac{\sum_{i=1}^{n} F(y_i) y_i}{\sum_{i=1}^{n} F(y_i)},$$

where y_i is the i-th possible element in the output set Y (here, −1, 0, or 1) and there are n such elements, and $F(y_i)$ is the membership value associated with each of these values. For the example, we have:

$$[(0)(-1) + (1)(0) + (0.5)(1)]/[0 + 1 + 0.5] = 0.5/1.5 = 1/3.$$

This means that the result of the fuzzy control rules suggests that we apply a force of 1/3 of a newton in pushing the cart. In light of the cart's position, this action appears reasonable.

The design of a fuzzy controller requires an iterative adjustment of the rules and the membership functions that comprise those rules. It's very unlikely that you'll be able to design a perfect fuzzy controller for a complex unstable system without a few rounds of trial and error, but it's quite likely that you'll be able to design a more compact controller using fuzzy rules than you would if you had to rely only on crisp rules. Fuzzy rules can often handle cases that are noisy or unforeseen more gracefully.

13.6 Fuzzy clustering

We've already seen how neural networks can be used for clustering data when there aren't any class labels by searching for structures in data. Fuzzy sets can also provide a means for clustering data. There are perhaps two main methods for performing fuzzy clustering (as suggested by Klir and Yuan [261]. The first is the fuzzy c-means method, where the number of clusters, c, is decided before the clustering is performed. The second is fuzzy equivalence relation-based hierarchical clustering, where the number of clusters is determined by the clustering algorithm rather than the user. This latter method requires a bit more fuzzy mathematics to understand than fuzzy c-means. Thus, for the sake of presentation, we will focus on fuzzy c-means here and refer to [261] for more information on using fuzzy equivalence relations for clustering (also see [43]).

Bezdek [42] offered the following concept for fuzzy clustering. Suppose that we are given observations $\mathbf{x}_1, \ldots, \mathbf{x}_n$ where each $\mathbf{x}_k$ is a p-dimensional real-valued vector for all $k \in \{1, \ldots, n\}$. The goal is to find a collection of c fuzzy sets A_i, $i = 1, \ldots, c$ that form a "pseudopartition" of the data. The objective is to assign memberships in each of the fuzzy sets so that the data are strongly associated within each cluster but only weakly associated between different clusters. Each cluster is in part defined by its centroid, $\mathbf{v}_i$, which can be calculated as a function of the fuzzy memberships of all the available data for that cluster:

$$\mathbf{v}_i = \frac{\sum_{k=1}^{n}[A_i(\mathbf{x}_k)]^m \mathbf{x}_k}{\sum_{k=1}^{n}[A_i(\mathbf{x}_k)]^m} \tag{13.1}$$

for all $i = 1, \ldots, c$. The parameter $m > 1$ controls the influence of the fuzzy membership of each datum.[1] The ratio above should look familiar because it is closely related to the defuzzification procedure given earlier. It's essentially a weighted average of the data in each fuzzy cluster, where the weights are determined by the membership values.

Given any initial collection of fuzzy sets, $\mathbf{A} = \{A_1, \ldots, A_c\}$, we can compute their associated centroids and determine the effectiveness of the clustering using the evaluation function,

$$J_m(\mathbf{A}) = \sum_{k=1}^{n} \sum_{i=1}^{c} [A_i(\mathbf{x}_k)]^m ||\mathbf{x}_k - \mathbf{v}_i||^2,$$

---

[1]Note that we have changed notation here to be consistent with Klir and Yuan [261].

where $|| \cdot ||$ is an inner product norm in $\mathbb{R}^p$ and $||\mathbf{x}_k - \mathbf{v}_i||^2$ is a "distance" between $\mathbf{x}_k$ and $\mathbf{v}_i$. The objective is to minimize $J_m(\mathbf{A})$ for some set of clusters $\mathbf{A}$ and chosen value of m. The choice of m is subjective, however, many applications are performed using $m = 2$. That is, the goal is to iteratively improve a sequence of sets of fuzzy clusters $\mathbf{A}(1)$, $\mathbf{A}(2)$, ... until a set $\mathbf{A}(t)$ is found such that no further improvement in $J_m(\mathbf{A})$ is possible.

The following algorithm [42] will converge to a local minimum or saddle point of $J_m(\mathbf{A})$, but is not guaranteed to converge to a global minimum.

1. Given a preselected number of clusters, c, and a chosen value of m, choose $\mathbf{A}(0)$ $(t = 0)$.

2. Determine the c cluster centers, $\mathbf{v}_1(t), \ldots, \mathbf{v}_c(t)$, using (13.1) above.

3. Update the fuzzy clusters $\mathbf{A}(t + 1)$ by the following procedure. For each $\mathbf{x}_k$:

   (a) if $||x_k - \mathbf{v}_i(t)||^2 > 0$, $i = 1, \ldots, c$, then update the membership of $\mathbf{x}_k$ in A_i at $t + 1$ by:

   $$A_i(t + 1)(\mathbf{x}_k) = \left(\sum_{j=1}^{c} \left(\frac{||\mathbf{x}_k - \mathbf{v}_i(t)||^2}{||\mathbf{x}_k - \mathbf{v}_j(t)||^2} \right)^{\frac{1}{m-1}} \right)^{-1}$$

   (b) if $||x_k - \mathbf{v}_i(t)||^2 = 0$ for some $i \in I \subseteq \{1, \ldots, c\}$, then for all $i \in I$, set $A_i(t + 1)(\mathbf{x}_k)$ to be between $[0,1]$ such that:

   $\sum_{i \in I} A_i(t + 1)(\mathbf{x}_k) = 1$, and
   set $A_i(t + 1)(\mathbf{x}_k) = 0$ for other $i \notin I$.

4. if $|\mathbf{A}(t + 1) - \mathbf{A}(t)| < \epsilon$, where ϵ is a small positive constant, then halt; otherwise, $t \leftarrow t + 1$ and go to step 2.

Klir and Yuan [261] suggest measuring the difference between $\mathbf{A}(t+1)$ and $\mathbf{A}(t)$ in step 4 by

$$\max_{i \in \{1, \ldots, c\}, \, k \in \{1, \ldots, n\}} |A_i(t + 1)(\mathbf{x}_k) - A_i(t)(\mathbf{x}_k)|.$$

Here is an example to indicate the results of this procedure. We sampled ten data points in two dimensions.[2] The first five points were drawn from a Gaussian density function with a mean position of (1,1) and a standard deviation in each dimension of 1.0. The other five points were drawn from a Gaussian density function with a mean position of (3,3) and standard deviation of 1.0. The data we generated were

---

[2]We thank Kumar Chellapilla for his assistance in programming the fuzzy c-means algorithm for this example.

$$
\begin{array}{rl}
1: & (1.7229,\ 0.2363) \\
2: & (1.0394,\ 3.1764) \\
3: & (2.5413,\ 1.4316) \\
4: & (0.7011,\ 0.5562) \\
5: & (0.0337,\ 1.0300) \\
6: & (2.6843,\ 2.7687) \\
7: & (3.9778,\ 2.8863) \\
8: & (3.0183,\ 3.1279) \\
9: & (3.8180,\ 2.2006) \\
10: & (3.7023,\ 2.7614)
\end{array}
$$

We chose $c = 2$ clusters (to match the way we generated the data) and selected $m = 2$. We randomly selected the initial memberships of each point in each cluster to be

					Data					
Cluster	1	2	3	4	5	6	7	8	9	10
$A_1(\mathbf{x})$	0.4717	0.4959	0.3763	0.6991	0.4651	0.4179	0.5109	0.4664	0.4448	0.6386
$A_2(\mathbf{x})$	0.5283	0.5041	0.6237	0.3009	0.5349	0.5821	0.4891	0.5336	0.5552	0.3614

Given this initialization, the final clusters generated were located at

   $(0.2748,\ 0.8913)$ and $(3.2105,\ 2.6175)$

with an associated $J_m(\mathbf{A}) = 9.1122$. Note that the cluster centers don't fall directly on the means of the two Gaussian distributions, but we shouldn't expect them to because they reflect the actual sampled data rather than the underlying distributions. The final fuzzy membership for each data point was

					Data					
Cluster	1	2	3	4	5	6	7	8	9	10
$A_1(\mathbf{x})$	0.7573	0.4639	0.2546	0.9484	0.9913	0.0311	0.0360	0.0232	0.0366	0.0169
$A_2(\mathbf{x})$	0.2427	0.5361	0.7454	0.0516	0.0087	0.9689	0.9640	0.9768	0.9634	0.9831

Figure 13.7 shows the distribution of sampled data and the centroids of the fuzzy clusters.

13.7 Fuzzy neural networks

If you are thinking ahead then you are already pondering about the possibilities for combining neural networks and fuzzy systems. In the previous chapter, we discussed the potential for neural networks to perform clustering and classification using crisp values. It's entirely reasonable to think about replacing those values with fuzzy numbers. Each node in the fuzzy neural network performs fuzzy arithmetic and instead of generating a single value, it generates a set of

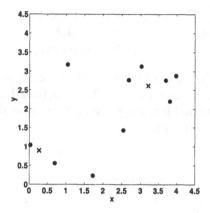

Fig. 13.7. The result of a fuzzy c-means clustering on 10 points generated from two Gaussian distributions. The points are shown as dark circles. The final positions of the clusters are shown by the two x's. The associated membership in each cluster for each point is indicated in the text.

fuzzy memberships in different possible values. The inputs to the fuzzy neural network are fuzzy numbers, the weights are fuzzy numbers, and the aggregation in each node that was a dot product when using crisp numbers is instead performed using fuzzy multiplication and addition. To give a better sense of what this involves, we must step back to the definition of fuzzy numbers and describe how to perform fuzzy arithmetic.

A fuzzy number is a fuzzy set A on R that possesses at least the following three properties:

1. The maximum membership of any element of A is 1.0 (i.e., A is a *normal* fuzzy set).

2. The support of A must be bounded.

3. $^{\alpha}A$ must be a closed interval for every $\alpha \in (0, 1]$.

This introduces some new concepts and notation that must be explained before we can proceed further.

- The support of a fuzzy set A on a universe of discourse U is the crisp set that contains all of the elements of U that have nonzero membership in A.

- The notation $^{\alpha}A$ corresponds to all values $\{u|A(u) \geq \alpha\}$. This is called the α-cut of A.

With these preliminary definitions, it's possible to prove that a fuzzy set A is a "fuzzy number" if and only if there exists a closed interval $[a, b] \neq \emptyset$ such that

$$A(x) = \begin{cases} l(x) & \text{for } x \in (-\infty, a) \\ 1 & \text{for } x \in [a, b] \\ r(x) & \text{for } x \in (b, +\infty) \end{cases}$$

where l is a monotonic increasing function from $(-\infty, a)$ to $[0, 1]$ and r is similarly monotonic decreasing over $(b, +\infty)$. The functions l and r must also be continuous from the right and left, respectively. Some typical examples of fuzzy numbers are shown in figure 13.8.

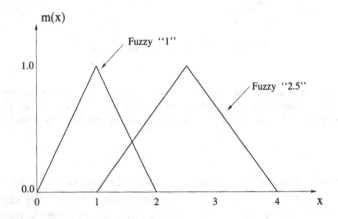

Fig. 13.8. Two fuzzy numbers: 1 and 2.5. Fuzzy numbers have certain mathematical properties as fuzzy sets. These are described in the text.

Having defined a fuzzy number, we can now proceed to define fuzzy arithmetic, particularly with regard to operations of addition, subtraction, multiplication, and division. Each α-cut of a fuzzy number is a closed interval of real numbers for all $\alpha \in (0, 1]$ and every fuzzy number can be uniquely represented by its α-cuts. Thus it is convenient to describe the fuzzy arithmetic operations in terms of operations on closed intervals associated with those α-cuts.

In particular, the four arithmetic operations above are defined as

$$[a, b] + [d, e] = [a + d, b + e]$$
$$[a, b] - [d, e] = [a - e, b - d]$$
$$[a, b] \cdot [d, e] = [\min(ad, ae, bd, be), \max(ad, ae, bd, be)]$$
$$[a, b]/[d, e] = [a, b] \cdot [1/e, 1/d], \text{ given } 0 \notin [d, e].$$

Since fuzzy numbers can be represented by their α-cuts, we can apply the above operations to the α-cuts of fuzzy numbers and in turn generate the corresponding resultant fuzzy number.

For example, suppose we have two fuzzy numbers denoted as

$$A(x) = \begin{cases} 0 & \text{for } x \leq 0 \text{ and } x > 2 \\ x & \text{for } 0 < x \leq 1 \\ -x + 2 & \text{for } 1 < x < 2 \end{cases} \qquad B(x) = \begin{cases} 0 & \text{for } x \leq 3 \text{ and } x > 5 \\ x - 3 & \text{for } 3 < x \leq 4 \\ -x + 5 & \text{for } 4 < x < 5 \end{cases}$$

$A(x)$ is a fuzzy "1" and $B(x)$ is a fuzzy "4." To determine the α-cuts of $A(x)$ and $B(x)$, simply set the formulas where $A(x)$ and $B(x)$ have support to be equal to α and solve for x. Therefore, for $A(x)$ we obtain ${}^{\alpha}A = [\alpha, 2 - \alpha]$ and for $B(x)$, ${}^{\alpha}B = [\alpha + 3, 5 - \alpha]$. The α-cut of a set $(A * B)$, where $*$ is one of the arithmetic operators above is defined as

$${}^{\alpha}(A * B) = {}^{\alpha}A * {}^{\alpha}B, \text{ for } \alpha \in (0, 1],$$

(if $* = /$ then an additional requirement is for $0 \notin {}^{\alpha}B$ for $\alpha \in (0, 1]$). If we wanted to add $A + B$, then ${}^{\alpha}(A + B) = [2\alpha + 3, 7 - 2\alpha]$. Similarly,

$${}^{\alpha}(A - B) = [2\alpha - 5, -1 - 2\alpha]$$
$${}^{\alpha}(A \cdot B) = [\alpha^2 + 3\alpha, \alpha^2 - 7\alpha + 10]$$
$${}^{\alpha}(A/B) = [\alpha/(5 - \alpha), (2 - \alpha)/(\alpha + 3)].$$

Focusing on the example of addition, the resulting fuzzy $(A + B)$ is then

$$(A + B)(x) = \begin{cases} 0 & \text{for } x \leq 3 \text{ and } x > 7 \\ (x - 3)/2 & \text{for } 3 < x \leq 5 \\ (7 - x)/2 & \text{for } 5 < x \leq 7 \end{cases}$$

and the result is a fuzzy number "5."

These sorts of fuzzy arithmetic operations are used to evaluate the output of each fuzzy neuron in a fuzzy neural network. More specifically, the output of the k-th node is defined as

$$Y_k = f(\textstyle\sum_{j=0}^{n} W_j X_{kj}),$$

where $f(\cdot)$ is the nonlinear transfer function (often a sigmoid), W_j is the incoming fuzzy weight from the j-th node, and X_{kj} is the j-th fuzzy activation coming into the k-th node. Thus the sum $\sum_{j=0}^{n} W_j X_{kj}$ is a fuzzy arithmetic sum on a series of fuzzy arithmetic products. The subsequent transform of this product by the function $f(\cdot)$ is accomplished using the "extension principle" for fuzzy sets (see [261] for algorithmic details). Although an example of fuzzy neural networks is beyond the scope of this book, you should be able to envision how such a construction would be assembled, and have an intuition about the types of operations that it would perform. Please see [43] for a more thorough treatment of various interpretations of fuzzy neural networks.

13.8 A fuzzy TSP

It's also possible to apply fuzzy logic to the traveling salesman problem (or other optimization problems). We could define fuzzy membership functions to describe the length of any given path between two cities. Even before determining the location of each city, we could construct some fuzzy membership functions that would represent the possible path lengths. The shortest length is arbitrarily close to zero, which occurs when two cities are placed essentially at the same location.

The longest possible length is the distance from the opposite diagonal vertices of the bounding rectangle in which all of the cities must be placed. Let's call this maximum length $MaxL$. Knowing, then, that the range of possible lengths between each pair of cities can vary from zero to $MaxL$, we could divide this range into four equal portions using $\frac{1}{4}MaxL$, $\frac{1}{2}MaxL$, and $\frac{3}{4}MaxL$, and then construct the following fuzzy membership functions to describe the length of a path segment in linguistic terms:

- A path has a membership of 1.0 in the set of *short paths* if its length is between zero and $\frac{1}{4}MaxL$. Beyond $\frac{1}{4}MaxL$, its membership in the set of short paths decreases linearly to zero at the value of $\frac{1}{2}MaxL$ and remains zero for all longer path segments.

- A path has a membership of 1.0 in the set of *medium-short paths* if its length is between $\frac{1}{4}MaxL$ and $\frac{1}{2}MaxL$. From a length of zero to $\frac{1}{4}MaxL$ its membership increases linearly from zero to 1.0. From a length of $\frac{1}{2}MaxL$ to $\frac{3}{4}MaxL$ its membership decreases linearly from 1.0 to zero, and it remains at zero for all longer path lengths.

- A path has a membership of 1.0 in the set of *medium-long paths* if its length is between $\frac{1}{2}MaxL$ and $\frac{3}{4}MaxL$. Its membership is zero for all lengths less than $\frac{1}{4}MaxL$ and its membership increases linearly from zero to 1.0 from $\frac{1}{4}MaxL$ to $\frac{1}{2}MaxL$. Its membership decreases linearly from 1.0 to zero from $\frac{3}{4}MaxL$ to $MaxL$.

- A path has a membership of 1.0 in the set of *long paths* if its length is between $\frac{3}{4}MaxL$ and $MaxL$. Its membership is zero for all lengths less than $\frac{1}{2}MaxL$ and its membership increases linearly from zero to 1.0 from $\frac{1}{2}MaxL$ to $\frac{3}{4}MaxL$.

Based on these fuzzy descriptors, we could create an evaluation function that favors those complete tours that have higher overall membership in the set of short paths, as opposed to the other fuzzy sets.

For example, we might choose the evaluation function

$$f(x) = \sum_{j=1}^{n} \sum_{i=1}^{4} \alpha_i M_{ij},$$

where i serves as the index of the i-th membership function, α_i is a scaling factor on the i-th membership function, and M_{ij} is the membership of the j-th path segment in the i-th membership function (i.e., M_{11} is the membership of the first segment of a given tour in the set of short paths). Each path has n segments and each segment will have a membership value in each linguistic class of path length. The goal is to minimize the value of $f(x)$ based on a prior choice of scaling factors α_i. It seems more reasonable to place increasingly more weight on the membership functions that are associated with longer path lengths because these are the things we want to minimize.

Would this be a good approach to solving the traveling salesman problem? Perhaps. But it likely would be insufficient to generate near-optimum solutions because the membership function of short paths is not sufficiently fine grained. It's probably not enough to simply have short paths. We might need to define more membership functions that describe *very short paths, extremely short paths*, and so forth. In the end, we would return to simply using the distance between each pair of cities, and the computational advantage of the fuzzy approach seems unclear. Nevertheless, it is certainly possible to think about addressing a fuzzy traveling salesman problem.

13.9 Evolving fuzzy systems

Every fuzzy system requires several design choices. How many fuzzy membership functions should be employed? What shape should they take? Triangular, trapezoidal, Gaussian, or some other possibility? What sort of relative scaling should they have? In a fuzzy control problem, what actions should be taken as a result of the fuzzy descriptors of the current state? If the cart is *left* should we *push*, or *push hard*, or perhaps *push lightly*? Each of these choices must be made before the complete fuzzy system can be evaluated. The process of tuning a fuzzy system to give the desired response is much like adjusting the topology of a neural network until the desired results are obtained. It's often a tedious matter of trial and error.

We have already seen how this sort of trial and error can be automated through evolutionary computation. To give a specific example, suppose we consider the possibility for adjusting the fuzzy control system we studied above for centering a cart. What were the choices available to us for designing this system? We had to first decide which parameters of the cart were important. Here, we focused on the position and velocity, but perhaps we could have also included the acceleration of the cart, in which case we would require a set of fuzzy descriptors about that as well. For each parameter in question, we needed to determine how many fuzzy descriptors would be employed and what shape those fuzzy descriptors would take on.

If we restrict our attention to triangles and trapezoids, just to keep things simple, then each fuzzy set is defined by the location of the vertices of their triangle or trapezoid. In figure 13.4, the fuzzy *middle* position for the cart is centered at zero meters and extends with support from minus one to one meter. Perhaps it would be possible to simply dictate here that the center of this fuzzy set should be at zero, since no other choice is rationale, but there is no way to know a priori that the membership function for *middle* should take on support over plus/minus one meter. Perhaps a better controller might be effected by using a support over ± 0.8945 meter, or even a range that isn't symmetric about the center! These are values that could be optimized by an evolutionary algorithm that would generate and test alternative controllers based on the fuzzy descriptors that they employ. Similarly, each of the fuzzy control rules could also be evolved, constructed by adapting the conditions that are required to satisfy each "if" and the actions implied as a consequence.

13.10 Summary

It's tempting to strive for precision and accuracy when using computers to solve real-world problems. The seemingly great advantage of the computer is the ability to calculate otherwise difficult mathematical expressions with exactness. Its capabilities in this regard far exceed those of the human brain and the sensors that brain relies on. But computers and humans were designed (evolved) for quite different environments. The computer's environment is a binary world of black and white. The human environment is a noisy, chaotic, time-varying place. Moreover, we perceive this world solely through imperfect sensors that provide only a fuzzy image of what surrounds us. By necessity, we have evolved a language of words to handle our incomplete, imperfect knowledge of the real world, and have crafted an altogether different language of mathematics to describe an idealized world that operates with precision, in a vacuum, where the mass of physical objects is concentrated at a point, and all lines have no width.

Despite the vague quality of our natural language, we can still treat it mathematically and, essentially, compute with words. The advantage of being able to treat classes of situations with ambiguous terms often carries over to the design of practical controllers and pattern recognition algorithms. We have seen some simple examples of this in the current chapter, and have also reviewed the basics of how to perform fuzzy arithmetic and describe the relationship between two more fuzzy sets. To learn more about the potential of fuzzy logic see [44, 503, 261, 403].

XIV. Everything Depends on Something Else

There's an old story about a boy who saw a thunderstorm and asked his father what caused the storm. The father looked down at his son and said "A butterfly, on the far side of the world." With his son intrigued, the father continued "it flapped its wings, and caused the air to move, spinning away from the butterfly. The spinning air swirled around with other air currents, growing bigger and bigger, cascading around the planet, until it formed this thunderstorm here." The so-called "butterfly effect" comes in many such stories, but they all have one thing in common: the fact that a single action can cause a sequence of other actions of growing size and unintended consequences, and that everything in the world depends on something else.

You might remember a problem from section 1.3 about finding the best place for a new store. You might gather demographic information about your customers, make some assumptions about how far they would travel to get to your store, and use that to locate the best spot. If you do that, however, you'll invariably pick a poor location, because this analysis leaves out an important ingredient: your competitors! Just as you're choosing where to put your store, so are they. If they take into account where you're likely to put your store, and you don't take into account where they are likely to put their stores, you lose.

Many real-world problems are of a similar nature. To solve these problems, you have to take into account other parts of the problem that are changing at the same time you are changing. They may be competitors trying to beat you, or friends trying to help you, or bystanders who are just getting in the way, but you always have to consider how parts of a problem interact. So here are some examples to get you warmed up.

Let's begin with a simple case of a dog chasing a cat. Take a look at figure XIV.1 and you'll the see starting point of the dog D and the cat C. The dog sees the cat and barks. The cat hears the dog and starts running in the direction across the page from left to right. Suppose that the dog and the cat both run at the same speed v and the dog chases the cat by heading straight for it, aiming constantly at the cat, as it runs away. Here's the first question, which is an easy one: Will the dog ever catch the cat? Right, it's easy, the answer is no because they are both running at the same speed and the dog has to cover more ground than the cat if it is ever going to catch its feline foe. So here's the next question, which is a bit more challenging. After a sufficiently long period of time, the dog's track will become almost identical to the cat's track. Once the dog is effectively directly behind the cat, after this long time, how far apart will they be, supposing that the original distance DC was 20 meters?

Fig. XIV.1. Initial positioning at the start of the chase

If you can't think of an immediate path to the solution, perhaps at least having an upper bound to the solution might help. Do you think the distance can be greater than 20 meters? Clearly not, because this is the starting distance, and if the dog ran straight to the cat's initial position C and then followed the cat, it'd be 20 meters behind. Instead, the dog is asymptotically closing on the cat, like a heat-seeking missile that flies at the same speed as its target. So we know the answer has to be less than 20 meters. How much less?

Some trigonometry will help here. Let's use α to describe the temporary angle between the line segment DC from the dog's current position D_t and the cat's current position C_t as they both run at a speed of v. Figure XIV.2 presents the picture.

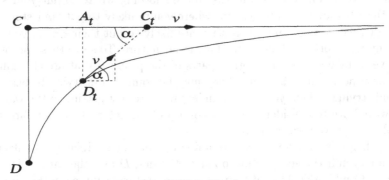

Fig. XIV.2. The chase

The rate at which the dog approaches the cat depends on two factors: (1) the speed v that they are both running, and (2) the component $v \cos \alpha$ that is oriented in the direction that the cat is fleeing. The point A_t in the figure shows the projection of the dog's position onto the cat's path. You can imagine the problem now as figuring out how quickly A_t approaches C_t as t increases. The cat is running at v but the dog's projection in the cat's path A_t is following at

$v \cos \alpha$. The difference between these speeds is $v(1 - \cos \alpha)$ and that's the rate at which the dog closes in on the cat.

There is a remarkable relationship between the distance $D_t C_t$ and the distance $A_t C_t$. As $D_t C_t$ decreases, $A_t C_t$ increases with the same speed. Thus the sum of distances $D_t C_t$ and $A_t C_t$ is always a constant, and in this case it's always 20 meters, just as it is when the dog first barks. As t grows and grows, D_t will eventually coincide with A_t so that the distances $D_t C_t$ and $A_t C_t$ will be equal. And that yields the answer to the question: Since the sum of $D_t C_t$ and $A_t C_t$ is always 20 meters, and in the end they are both equal, the final distance between the dog and cat will be 10 meters! Ruff!!

Now let's jump to the plains of the Masai Mara in Kenya, where four lions have stalked out a wildebeest. The lions find themselves in the improbable predicament of being at the four corners of a square with the wildebeest in the center. None of these lions is dominant over another, so they each follow the other in a circling spiraling path that closes in on their prey. Like the dog chasing the cat, they aim directly for their closest neighbor, as shown in figure XIV.3.

Fig. XIV.3. Four lions on the African plains

Each lion runs directly after its neighbor. The lion in the top-left corner runs after the lion in the top-right corner, the lion in the top-right corner runs after the lion in the bottom-right corner, and so on. At the start, the lions find themselves separated by 1000 meters, and they run at 20 meters per second. Your problem is to determine how long it will take for the lions to meet. Oh, and will their tracks ever cross? And if they do, where? Oh yes, and how long will their tracks be? Okay, it's a multipart problem!

With four lions approaching each other, the problem seems at once related to the dog chasing the cat, except that there are four moving parts instead of just two. Actually, the solution is a bit simpler.

Every lion runs at a right angle to its neighbor, since they are all linked to each other, and therefore every chasing lion approaches its neighboring lion at a rate of 20 meters per second. The lions will therefore all meet in 50 seconds because they were all 1000 meters apart at the start. Since each lion will meet in 50 seconds, each also covers 1000 meters in the process. Also, because the lions are always at right angles, they are always on the vertices of a square, which is rotating in time. The area of the square decreases as its sides shrink at 20 meters per second. After 50 seconds, the area of the square will be zero and all the lions will be standing on top of each other (and the wildebeest!) at the center of the original square. The lions will mark out logarithmic spirals that don't intersect, because if they did intersect that would mean another lion had earlier been in the same spot as a lion is presently and that's impossible because the four distances from the lions to the center of the square are always equal and diminishing steadily. The reason this puzzle is a bit easier than the dog and cat is that the fourth lion is linked back to the first.

Instead of nature red in tooth and claw, here's a puzzle where the participants get to cooperate, which was described in [346] but we've updated it for the twenty-first century.

Four people on a reality TV show have to trek into the Sahara desert. Each person can carry enough water to last 10 days and can walk 24 kilometers each day. (So each person starts with 240 units of water, where 1 unit is the amount that an individual uses when walking 1 kilometer.) Their collective objective is to have at least one member of their team get as deep into the Sahara as possible — and return! The team has studied the desert and knows that it's uninhabited, so it's safe to leave water behind for the return trip. The rules of the TV show mandate that no team member can return to civilization to replenish his or her water supply and then go back into the desert. The rules of the TV show also mandate that everyone has to stay alive. The TV producer's lawyers didn't want any lawsuits and the show's advertisers didn't want to see anyone walk into the desert, leave their remaining water for their companions, and then crawl off to die of thirst! So how far can they get?

If they all stay together they can manage a trip of only five days into the desert, or 120 kilometers, leaving enough water to return. If they cooperate, however, they can get much further. Suppose we model the problem as shown in figure XIV.4. The team starts at S. All four people go to the point A, where one returns to S with enough water to make the journey, and perhaps leaves some water at A for the remaining teammates on their return leg. The remaining teammates head to B, and then one returns to S as before. Then it's on to C, with the same strategy, and finally the last person treks to D and returns. That's a good framework for solving the problem. The question now is: How far away from S is D?

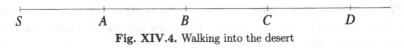

S A B C D

Fig. XIV.4. Walking into the desert

Each team member must cross the segment SA twice, once in each direction, so SA is traveled eight times. Similarly, AB is traveled six times, BC is traveled four times, and CD is traveled twice. Let's say that it takes a units of water for a person to travel SA. Then at A, $3a$ units of water must be left behind for the return trip AS for each of the three teammates who travel past A. With similar notation, $2b$ units must be left at B, and c units left at C.

The goal is to maximize the distance SD. Each traveler has 240 units of water when he or she starts. It takes a units of water to get to point A, so each traveler is left with $(240 - a)$ units at that point. The first person to head back to the start will need an additional a units to make the return trip and will need to leave $3a$ units for the other travelers so that each can make it back on the last leg. That means the first person will need to start with 240 units and then consume a to get to A, leave $3a$ at A for the others, and consume a more on the way back. This leaves him or her with $(240 - 5a)$ units of water. Each of the other travelers will have expended a units when getting to A, so the first traveler can replenish all of the remaining teammates by giving each a units of water, and optimally the first traveler would have no more water when getting back to A, so $(240 - 8a) = 0$, or $a = 30$.

After addressing the first segment of the journey into the Sahara, the remaining three legs can be addressed by the same logic, which leads to $(240 - 6b) = 0$ or $b = 40$, $(240 - 4c) = 0$ or $c = 60$, and similarly $(240 - 2d) = 0$ or $d = 120$. Since each person uses 1 unit of water per kilometer, the positions for A, B, C, and D are at 30, 70, 130, and 250 kilometers away from S. So the answer is 250 kilometers.

By the way, the depth of the penetration into the Sahara can be expressed mathematically as

$$120(1 + \tfrac{1}{2} + \tfrac{1}{3} + \tfrac{1}{4}) = 250,$$

which is handy in generalizing the problem to the case where n teammates show up for the TV reality program. In that case the maximum distance they can travel is

$$120 \sum_{i=1}^{n} 1/i,$$

which, of course, can be made as large as necessary because the sequence diverges to infinity.

A fabulous puzzle on unusual "competition and cooperation" can be found in [456]. The puzzle comes in two versions: small and large. Let's start small.

Ten pirates have plundered a ship and discovered 100 gold pieces. They need to divide the loot among themselves. They want to be fair and abide by the law of the sea: to the strongest go the spoils. So they have an arm-wrestling match to determine how strong each pirate is and then they sort themselves from strongest to weakest. No two pirates are equally strong so there's no doubt about the order, which they all know well. We can label the pirates from weakest to strongest as $P1$, $P2$, and so forth, up to $P10$ in this small version of the problem. The pirates also believe in democracy (who says there is no honor among thieves?!) and so they allow the strongest pirate to make a proposal about the division, and everyone votes on it, including the proposer. If 50 percent or more of the pirates vote in favor, then the proposal is accepted and implemented. Otherwise, the proposer is thrown overboard, into shark-infested waters, and the procedure is repeated with the next strongest pirate.

All pirates like gold, but these pirates hate swimming with sharks even more than they like gold. So any one of them would rather stay onboard the ship and get no gold at all than be thrown overboard to fend off the sharks. All the pirates are rational, and they know that if they damage any of the gold pieces, e.g., by trying to divide them into smaller pieces, then the bullion will lose almost all of its value. Finally, the pirates can't agree to share pieces as no pirate would trust his fellows to abide by such an agreement.

At last, the question: What proposal should the strongest pirate make to get the most gold?

Sometimes, the best way to analyze such possible strategies is to work backward from the end, as at that stage it may be obvious as to which strategy is good and which is bad. Then we can transfer that knowledge to the next-to-last decision, and we can continue in succession because each of these strategic decisions are all centered on the same question: What will the next pirate do if I do this? For any n pirates, the question is based on the decision for $n - 1$ pirates.

So, let's start at the end, when there are just two pirates, $P1$ and $P2$. In that case the strategy of the strongest pirate, $P2$, is obvious: propose 100 gold pieces for himself, and none for $P1$. His vote would carry 50 percent of the vote necessary for the acceptance of his proposal and he would be one rich pirate!

Now, consider the case with three pirates. Note that pirate $P1$ knows (and $P3$ knows that $P1$ knows!) that if $P3$'s proposal is turned down, the procedure would proceed to the two-pirate stage where $P1$ gets nothing. So $P1$ would vote for absolutely any proposal from $P3$ that gets him *something*. Knowing then that the optimal strategy for $P3$ is to use a minimal amount of gold to bribe $P1$ to secure his vote, $P3$ should propose 99 gold pieces for himself, 0 for $P2$, and 1 gold piece for $P1$.

The strategy of $P4$ in the scenario with four pirates is similar. As he needs 50 percent of the vote, he needs a vote of one additional pirate. Again, he should use a minimum amount of gold to secure this vote, so his proposal is 99 gold pieces for himself, 0 for $P3$, 1 gold piece for $P2$, and 0 for $P1$. Of course, $P2$ would be happy to vote for this proposal; otherwise $P4$ is thrown overboard, the procedure reduces to three pirates, and $P2$ gets nothing.

Now, the strategy of $P5$ in the scenario of five pirates is just slightly different. He needs two additional votes from his fellows. Thus he proposes 98 gold pieces for himself, 0 for $P4$, 1 gold piece for $P3$, 0 for $P2$, and 1 gold piece for $P1$. Clearly, the votes of $P3$ and $P1$ are secure, because in the four-pirate scenario they would get nothing.

It's straightforward now to design a proposal for $P6$ in a six-pirate scenario, for $P7$ in a seven-pirate scenario, etc. In particular, the proposal for $P10$ is: 96 gold pieces for himself, 1 gold piece for each of the pirates $P8$, $P6$, $P4$, and $P2$, and none for the rest. This solves the small version of the puzzle. It's good to be the strongest pirate, at least when there's a small number of pirates and a whole lot of gold.

Now, let's move to the large version of the puzzle. Let's leave all the assumptions as they were before but increase the number of pirates to 500. The same pattern emerges, but there's a catch, because it works only up to the 200th pirate. The pirate $P200$ will offer 1 gold piece for himself, 1 gold piece for each even-numbered pirate, and none for the rest. And it's now when the fun starts in this version of the problem.

The pirate $P201$ still can follow the previous strategy except that he runs out of bribes and he proposes nothing for himself. So he proposes 1 gold piece for each odd-numbered pirate from $P199$ to $P1$. In that way he gets nothing but at least he stays on board and avoids being eaten alive by sharks.

Pirate $P202$ also gets nothing. He has to use all 100 gold pieces to bribe 100 pirates and stay dry. The selection of these pirates is not unique, as there are 101 pirates who are willing to accept the bribe (pirates who don't get anything in the 201-pirate scenario), so there are 101 ways to distribute these bribes.

What about the 203-pirate scenario? This pirate must get 102 votes for his proposal including his own vote and he does not have enough gold pieces to bribe 101 of his fellow pirates. So $P203$ will go overboard regardless of what he proposes! Too bad for him.

This is important for $P204$ though, as he knows that $P203$ would vote for anything to save his life! So $P204$ can count on $P203$ no matter what he proposes. That makes his task easy, as he can count on $P203$, himself, and 100 bribed fellows, so he can secure 102 votes. Again, the recipients of bribes should be among the 101 pirates who would receive nothing under $P202$'s proposal.

The pirate $P205$ in the 205-pirate scenario faces an impossible task. He can't count on $P203$ or $P204$ for support: each will vote against him, and will save himself. So $P205$ will be thrown overboard no matter what he proposes. The moral is: don't be the strongest in a group of 205 democratic pirates.

The same fate is awaiting $P206$: he can be sure of $P205$'s vote, but that's all he can count on, so overboard he goes.

Similarly, $P207$ faces a soggy end to his existence, as he needs 104 votes: his own, 100 from bribes, and 3 additional followers. He can get votes from $P205$ and $P206$, but these are not enough... so overboard he goes.

The fate of pirate $P208$ is different, as he also needs 104 votes, but $P205$, $P206$, and $P207$ will vote for him to save their lives! With his own vote and 100 bribed votes, his proposal will be accepted and he will survive. Of course, the recipients of his bribes must be among those who would get nothing under $P204$'s proposal: the even-numbered pirates $P2$ through $P200$, and then $P201$, $P203$, and $P204$.

Now, we can see the pattern, which continues indefinitely. Pirates who are capable of making successful proposals (even though they get no gold from their proposals, but at least they get to stay on the ship) are separated from one another by ever longer sequences of pirates who would be thrown overboard no matter what proposal they made, because their votes are ensured for a stronger pirate's proposal! So the pirates who can make a successful proposal are $P201$, $P202$, $P204$, $P208$, $P216$, $P232$, $P264$, $P328$, $P456$, and so on, i.e., pirates whose number equals 200 plus a power of 2.

It's also easy to see which pirates are the lucky recipients of bribes. As we saw before, the solution here is not unique, but one way to do this is for $P201$ to offer bribes to the odd-numbered pirates $P1$ through $P199$, for $P202$ to offer bribes to the even-numbered pirates $P2$ through $P200$, for $P204$ to the odd numbers, for $P208$ to the even numbers, and so on, alternating between odd and even.

So, as the puzzle clearly illustrates, being the strongest and having a chance to put forward the first proposal is not always the best unless, of course, the number of pirates is quite small. Sometimes it's good to be a big fish in a small pond.

Finally, here's a great puzzle derived from a Sherlock Holmes mystery written by the venerable Arthur Conan Doyle [107]. The puzzle (found in [265]) poses an incredulous conundrum of interconnected components. Here we go.

Sherlock Holmes was asked to investigate a great mystery: the disappearance of ten-year-old Lord Saltire, the only son and heir of the Duke of Holdernesse. Lord Saltire disappeared from the Priory School one night. Dr. Huxtable, the founder and principal of this school, asked Holmes for help. During his investigation, Holmes discovered the body of murdered German teacher, Mr. Heidegger. He also noticed a 25-foot wide mud patch with recent front and real-wheel bicycle tracks, as illustrated in figure XIV.5. Holmes knew that deducing the direction of the tracks was crucial: Were they heading toward the school or away from it? Holmes wasted no time in coming up with the right answer.

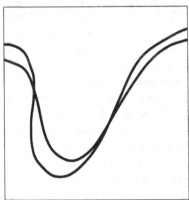

Fig. XIV.5. Bicycle tracks in the mud

Now here's "the rub." In Conan Doyle's version [107], Holmes' explanation of the problem to Watson is actually no explanation at all! Holmes "deduced" the direction of the cyclist from the degree of depression of the tracks, saying that the rider's weight made the rear track deeper. But this can't explain the direction of the cyclist! Can you best Sherlock? Was it from left to right or from right to left, and why? Justice hangs in the balance! Do your best!

14. Coevolutionary Systems

'Well, in *our* country', said Alice, still panting a little,
'you'd generally get to somewhere else — if you
ran very fast for a long time, as we've been doing.'
'A slow sort of country', said the Queen. 'Now,
here you see, it takes all the running you can do,
to keep in the same place. If you want to get somewhere else,
you must run at least twice as fast as that!'

Lewis Carroll, *Through the Looking-Glass*

The typical approach to problem solving starts with understanding the problem in quantifiable terms. You craft an evaluation function that captures the trade-off between the pertinent parameters and consider which method or collection of methods you will use to find the settings of the parameters that optimize the function. We've seen this in the traveling salesman problem, the satisfiability problem, and the general nonlinear programming problem. In each case, we establish an evaluation function that considers distances, logical truth and falsehood, or mathematical relationships, respectively, as well as perhaps some constraints on the variables involved. A thorough understanding of the evaluation function allows us to create the correct combination of representation, search operators, and selection criteria to find the extrema of the evaluation function, and hopefully a good solution or set of solutions.

There are, however, cases where we really have no idea of how to begin creating a good evaluation function. Some of these involve discovering optimal strategies in games. The trouble with games is that there's often no clear idea of how to measure performance beyond saying simply that you won, lost, or perhaps played to a draw. Suppose a friend challenges you to play a game, and you've never played it before. Maybe it's a new game, and your friend hasn't played it either. What would you do? You'd read the rules to understand what moves you can and cannot make, and you'd look for the objective: the criterion for winning and losing. Beyond that, you're on your own, and you have to devise your strategy and tactics from scratch, your possible moves, your opponent's countermoves, and the decision criteria you'll use to choose one move over another. This is quite possibly the most difficult problem to solve, one where you don't have any guidance to rely on, no ideas about what to do, and you have to learn simply through trial and error.

This situation is very similar to that faced by living creatures in the natural environment. Animals (and many plants too) are consummate problem solvers, constantly facing the most critical problem of avoiding being someone else's lunch. Many of the defensive and offensive survival strategies that animals employ are genetically hardwired instinctual behaviors. But how did these behaviors begin? We can trace similarities in survival strategies across many species. For example, many animals use cryptic coloration to blend into their background. These animals might be only very distantly related, such as the leafy sea dragon and the chameleon, and yet their strategy is the same: don't be noticed. Other animals have learned that there is "safety in numbers," including schooling fish and herd animals such as antelope. Furthermore, herding animals of many different species have learned to seek out high elevations and position themselves in a ring looking outwards, this providing the earliest possible sighting of a potential predator. These complex behaviors were learned over many generations of trial and error, and no doubt a great deal of life and death in the process.

This is a process of coevolution. It is not simply one individual or even one species against its environment, but rather individuals against other individuals, each competing for resources in an environment, which itself poses its own hostile conditions but doesn't care which individuals win or lose in the struggle for existence. Competing individuals use random variation and selection to seek out superior survival strategies that will give them an edge over their opposition. Monarch butterflies learned to be toxic, leaving a bad taste in the mouths of their avian predators. Viceroy butterflies learned that they didn't have to be toxic; simply appearing like a monarch butterfly was enough to fool the birds, and easier to evolve. Antelope learned to form a ring to spot predators more quickly; predators, such as lions, learned to hunt in teams, and use the tall light grasses of the African savanna to mask their approach as they stalk. Each innovation from one side may lead to an innovation from another, an "arms race" of inventions for survival that involves individuals evolving to overcome the challenges posed by other individuals, which are in turn evolving to overcome those new challenges, and so forth.

The language we are using here to describe the learning involved in this coevolutionary process is chosen carefully. Individual antelopes did not gather together in a convention to discuss new ideas about "Surviving in the Age of Lions" and come up with the strategy of herding or defensive rings on high ground. Nevertheless, the strategies are unmistakable and clearly reflective of a process of learning. Where the learned behaviors are instinctual, they have been accumulated through random variation and selection within the species. The entire genome of the species in question represents the learning unit in these cases, with individuals representing potential genetic variations on a general behavioral theme. No one told the leafy sea dragon to look like its surrounding vegetation. No one told the monarch butterfly to be toxic. Yet these and many other even more astounding inventions were discovered by a coevolutionary process without using any explicit evaluation function other than life and death, and mostly death.

This might seem like a wasteful process: "learning by death" as it was reportedly described by Marvin Minsky to one of the pioneers of evolutionary computation, Hans Bremermann, in the early 1960s. But in the absence of a better approach, natural evolution provides a basis for solving problems for which only the most basic information might be available, such as the available moves and the rules of the game. As will be shown later in this chapter, this concept of coevolution can also be applied to cases where there is explicit information in the form of an evaluation function. But to introduce the concept, let's start from scratch.

14.1 Playing a game

Above, we mentioned the idea of playing a game for the first time. Before we make that more specific, let's review the fundamentals of games. Games are characterized by rules that describe the allowable moves each player can take. These moves constitute the behavior of each player, the manner in which each allocates his or her resources. More formally, a player's behavior is a series of stimulus–response pairs, in which resources are allocated in response to particular situations. When a player makes a move, he or she receives a payoff, and usually the player tries to maximize the cumulative payoff over some period of time. In some games, such as checkers or chess, the payoff comes at the end of the game, but it's easy to imagine a surrogate payoff that correlates with a player's chances of winning at any given time based on the history of moves in the game.

Some games are competitive, but others are cooperative, and still others are mixed. We'll see different examples in this chapter. The payoffs determine the form of the game. If the payoffs suggest that when one player is winning and gaining in payoff, the other player is losing and his or her payoff is reduced, that's clearly a competitive game. But some games are constructed where two or more sides can gain in payoff at the same time. Think about treaties that are signed between countries. Each signatory to the treaty may view the "game" it's playing as a cooperative venture where the outcome will be best for both countries, presuming of course that each intends to honor the treaty!

Over 70 years ago, von Neumann and Morgenstern [481] laid out the foundations for classes of games where two players engage in a series of moves, both players know all the possible moves that either could make, and also know the exact payoffs that each will receive. Think about real-world games — football, a corporate takeover, marriage — and you'll quickly remark about how simplified these conditions really are, and yet von Neumann and Morgenstern's game theory has been adopted widely in economics and other venues.

In *zero-sum* games, where whatever payoff one player earns is taken from the other player (such as trading cattle futures), a fundamental strategy is to minimize the maximum damage that an opponent can do on any move. This is called a minimax strategy, where you examine the possible responses to your actions and then choose an action where your opponent's reaction does the least possible harm to you. If both players in a zero-sum game adopt a minimax strategy, then the payoff that each earns is said to be the *value* of a play, or the value of the game. If the value is positive, that means that one player can always have the upper hand over the other, whose corresponding payoff will be negative.

Minimizing the maximum damage that your opponent can do sounds attractive, even in nonzero-sum games, but it is a very conservative way to play. Suppose you're in the middle of a game and you and your opponent have two moves, A and B. If you both play A, you'll both earn $1. If you play A and your opponent plays B, you'll earn $2. If you play B and your opponent plays A, you'll earn $100, but if you both play B then you'll get no payoff at all. The minimax player looks at this situation and says "If I play A, I know I'm going to get at least $1. If I play B, I might not get anything." That's true, but the difference between $1 and nothing is so small, as are both payoffs, that it hardly makes sense to lock in a $1 payoff and forgo the chance that by playing B, your opponent might make a mistake and play A, giving you the big $100 payoff. Minimax is risk averse, and in this case it seems to work against common sense.

So, back to the challenge we want to take on in this chapter. Suppose you are going to play a game. You don't know anything except for the rules of the game and the way the resources are allocated — for example, the way pieces are laid out on a board. How can you discover a winning strategy, which might or might not be a minimax strategy, even when you don't know the payoff function, and neither does your opponent? At first, this situation might seem overly harsh, since no one would engage in such a game for fun (it sounds more like punishment), but recall that this situation is what nature has been posing to living organisms for billions of years. Its discoveries are ingenious, and so too can be the products of simulated evolution.

14.2 Learning to play checkers

Suppose you were faced with the challenge of learning to play checkers (also known as draughts) but you had to do it essentially from first principles. You're given the rules of the game, which follow below, and you can see the pieces on the board, and know how many you have and how many your opponent has. Beyond that, you're on your own. In fact, let's make it even more challenging by saying that after you complete a game, you won't get to know if you won, lost, or played to a draw. Only after completing some random number of games would you be told that you earned some number of points, which represents your overall performance. So not only do you not know which moves you made in any game were good or bad, you don't even know which games were won, lost, or drawn. Quite a challenge! In fact, a simpler challenge was offered by Allen Newell, a father of artificial intelligence, in the early 1960s when he stated that there wasn't enough information in the outcome of a game such as checkers (or chess) to allow a machine to learn how to play the game (in [324]). The challenge here raises the bar even further, providing no feedback about which games were won, lost, or drawn, but this is still a challenge that can be addressed using coevolution.

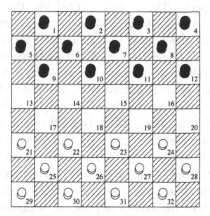

Fig. 14.1. The starting position in the game of checkers. Both players have 12 pieces. Different colors can be used, such as red and white or black and white. The white player moves second.

First, let's review the rules. Checkers is played traditionally on an eight-by-eight board with squares of alternative colors of red and black (figure 14.1). There are two players, denoted as "red" and "white." Each side has 12 pieces (checkers), which begin in the 12 alternating squares of the same color that are closest to that player's side, with the rightmost square on the closest row to the player being left open. The red player moves first and play then alternates between sides. Checkers are allowed to move forward diagonally one square at a time, or, when next to an opposing checker and there is a space available directly behind that opposing checker, by jumping diagonally over an opposing checker. In the latter case, the opposing checker is removed from play. If a jump would in turn place the jumping checker in position for another jump, that jump must also be played, and so forth, until no further jumps are available for that piece. Whenever a jump is available, it must be played in preference to a move that does not jump; however, when multiple jump moves are available, the player has the choice of which jump to conduct, even when one jump offers the removal of more opponent's pieces. When a checker advances to the last row of the board, it becomes a king, and can thereafter move diagonally one square in any direction, forward or backward.[1] The game ends when a player has no more available moves, which usually occurs by having his or her last piece removed from the board, but it can also occur when all existing pieces are trapped, resulting in a loss for that player with no remaining moves and a win for the opponent (the object of the game). The game can also end when one side offers a draw and the other accepts.

---

[1] There are other variations of the game where the king can move more than one square; however, the official rules of the European Draughts Association and the United States Checkers Federation both impose the limitation of moving only one square at a time.

In the late 1990s and into 2000, Chellapilla and Fogel [150] implemented a coevolutionary system that taught itself to play checkers at a ievel that's on a par with human experts. The system worked like this. Each checkerboard was represented as a vector of length 32, with components corresponding to an available position on the board. Components in the vector could take on elements from $\{-K, -1, 0, +1, K\}$, where K was an evolvable real value assigned for a king, 1 was the value for a regular checker, and 0 represented an empty square. The sign of the value indicated whether or not the piece in question belonged to the player (positive) or the opponent (negative). A player's move was determined by evaluating the presumed quality of potential future positions. This evaluation function was structured as a feed forward neural network with an input layer, multiple hidden layers, and an output node (figure 14.2). The output node was constrained to the range $[-1, 1]$. The more positive the output value, the more the neural network "liked" the position; the more negative, the more the neural network "disliked" the position. Minimax was used to select the best move at each play based on the assessments from the neural networks.

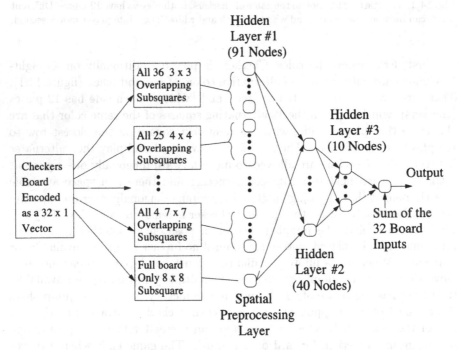

Fig. 14.2. The neural network architecture used by Chellapilla and Fogel. The input layer corresponded to the positions on the checkerboard. The first hidden layer comprised neurons for each square subsection of the checkerboard. The remaining hidden layers provided non-linear functions leading to an overall output. When the output was close to $+1$ this meant that the neural network favored the position on the board. In contrast, positions that yielded values close to -1 were not favored.

The coevolutionary system started with a population of 15 neural networks, each one having its weighted connections and K value set at random in the range $[1, 3]$. As a result, the neural networks initially made random evaluations, leading to random decisions, and mostly lousy random decisions, but some were more lousy than others. Each of the 15 parent networks created an offspring through mutation of each weight and K value, and then the 30 neural networks competed in games of checkers, with each playing as the red player in five games. Points were awarded to competing neural networks for winning $(+1)$, losing (-2), or playing to a draw (0). A selection routine then kept the 15 highest-scoring neural networks to be parents for the next generation, with this process of coevolutionary self-play iterating for hundreds of generations. Note that the neural networks did not receive feedback about the outcome of specific games, nor external judgments on the quality of moves. The only feedback was an aggregate score earned over a series of games.

The best-evolved neural network (at generation 840) was tested by hand against real people playing over the Internet in a free checkers game website (www.zone.com) using the screen name "Blondie24" to attract players. After playing 165 games, Blondie24 was rated in the top 500 of 120,000 registered checkers players on this site. All of the details regarding this research are found in [67, 68, 69, 150].

There are 10^{20} possible positions in checkers, a number far too large to enumerate. At the time of this writing, checkers remains an unsolved game, meaning that nobody knows with certainty if the game can be won by red or white, or whether the game should always end in a draw. But if a game of 10^{20} possibilities isn't enough for you, perhaps chess, at 10^{64}, might be. Fogel and Hays [158] have combined neural networks and coevolution to create a master-level chess playing program, again without providing the simulated players with any specific feedback about which games were won or lost during evolution. Simply by playing games against itself, the coevolution system was able to adjust the parameters of the strategies, which incorporated different neural networks, and rise to a very high level of play. Coevolution can be a versatile method for optimizing solutions to complex games, and a reasonable choice for exploring for useful strategies when there's very little available information about the domain.[2]

14.3 Optimization through competing populations

The Blondie24 experiment relied on the principle of coevolution within a single population. In nature, however, it's not only individuals within a species

---

[2]If such human expertise is available, and the goal is to develop a system with superior performance, that expertise should be used, and possibly hybridized with additional expertise garnered by a computer through a method such as coevolution. The objective of the experiments described here was to investigate how successful an evolutionary approach could be without such human expertise.

but, of course, individuals of different species that compete for survival. Sometimes these relationships are typified by predator-prey, and other times by host–parasite. Either way, coevolution can be implemented in simulation using multiple populations playing opposing sides and used to advantage. Let's examine a couple of examples.

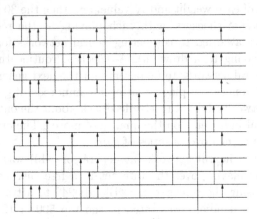

Fig. 14.3. A sorting network for ordering 16 elements that uses 63 comparisons, drawn following [218].

Hillis [218] offered a now-famous example of using a simulation of so-called hosts and parasites to improve an evolutionary optimization. He studied the problem of evolving a sorting network, which is a sequence of operations used to sort a fixed-length set of numbers. A sorting network can be represented graphically, as shown in figure 14.3, which is a network first offered in [34] and [138] for sorting 16 elements. The arrows indicate the sequence of operations. Each number in the list to be sorted is associated initially with one horizontal line. Comparisons are conducted between lines connected by arrows, progressing through the network from left to right. Two elements are exchanged in the list if the element at the head of an arrow is less than the element at the tail. After conducting all the comparisons indicated in the sorting diagram, the final list of numbers is sorted in descending order. This network requires 63 comparisons. The network in figure 14.4, however, invented by Green (in [264]), needs only 60 comparisons, and to our knowledge still stands as the best-known sorting network for 16 elements.

Hillis conducted an evolutionary search using a data structure to represent sorting networks, with the aim of discovering the minimal sorting network for 16 elements. Networks were scored based on the percentage of input cases (strings of unsorted numbers) that were sorted correctly. There were many ad hoc nuances to his representation and to the manner in which he implemented his evolutionary algorithm, including variation operators, selection, an imposed spatial nearness between solutions, and other facets, but jumping to the end result, after 5000 generations of evolution with a population of 65,536 candi-

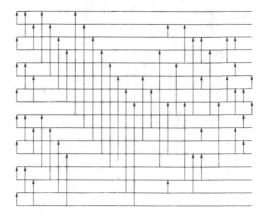

Fig. 14.4. A more parsimonious sorting network that requires only 60 comparisons, drawn following [218].

date solutions, the best-evolved sorting network required 65 comparisons. The representation that Hillis chose was restricted to a minimum of 60 comparisons, so there was no possibility of finding a truly superior design (one that would have fewer than the best-known 60 comparisons), but 65 comparisons served as a benchmark for further comparison using a coevolutionary approach based on hosts and parasites.

Hillis noted that many of the sorting tests during the initial evolution were too easy (it's likely that they were already mostly sorted to begin with) and served to waste computation time. He devised a method in which two populations evolved against each other to overcome this limitation. The first population comprised the sorting networks while the second population comprised sets of test cases. The sorting networks were scored according to the test cases that were available in a simulated local neighborhood for each network (assigning a nearness metric between test cases and networks). The sets of test cases (comprising 10 to 20 specific cases) were scored based on how well they found problems in the sorting networks. Variation and selection were applied to each population, resulting in the simultaneous optimization of the sorting networks and increasingly challenging test cases. Hillis reported that the coevolutionary approach did not exhibit tendencies to stall at local optima, even after tens of thousands of generations (the actual number was not specified), and that this procedure generated a sorting network comprising only 61 comparisons, which is shown in figure 14.5. Perhaps an even better network could be discovered.

Sorting networks is just a beginning when thinking about the use of coevolution in optimization. For example, Sebald and Schlenzig [421] studied the problem of designing drug controllers for patients during surgery, using complex models for how patients would respond to blood pressure medication. The objective was to discover a class of control strategies that would be able to treat any patient and hold the patient's blood pressure at a target level for a given

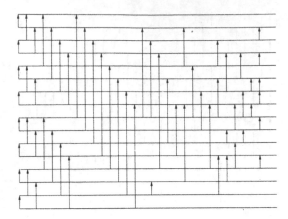

Fig. 14.5. The evolved sorting network that required 61 comparisons, drawn following [218].

duration. Based on real-world data, physical considerations, and prior research, they chose to model the patient's response with models of the form

$$X[t+1] = \alpha X[t] + \beta u[t-d] + D_0 + D_1 t + D_2 t^2 + D_3 t^3,$$

where $X[t]$ is the mean arterial pressure (MAP) in millimeters of mercury (mmHg) at time step t, d is the transport delay, and u is the drug infusion rate. The parameters D_0 through D_3 modeled the effects of external factors, such as the introduction of additional medications. At first glance, the patient's response to the drug is a simple linear function of the previous MAP and the amount of drug introduced. The complexity arises, however, when considering that the transport delay, d, is unknown to the controller. This is the amount of time it takes between the first injection of the drug into an intravenous (IV) unit and the onset of its effects on the patient. If the delay is long, the effects of any control will not be seen for a corresponding long period of time, and can lead to a controller attempting to "over control" the situation. Couple this with the disturbance effect from an unknown cubic equation, governed by $D_0 - D_3$, and the control task is very challenging. Figure 14.6 shows the uncontrolled trajectories of several different patients with varying parameter values. These trajectories represent the blood pressure of the patient if no drug were administered. There is a wide range of behaviors in this sample, and none show the additional effect of the transport delay.

Sebald and Schlenzig used coevolution to optimize a CMAC controller for these patients. The details of CMAC controllers are beyond our scope here, but broadly these are adaptive reinforcement learning controllers that have initial parameters that are used to adapt their behavior based on errors observed in their controls. Sebald and Schlenzig constructed a two-population scenario where a population of patients was evolved, searching for the most difficult patients for the CMAC controllers, while a population of CMAC controllers was evolved simultaneously to search for the most robust parameters that could

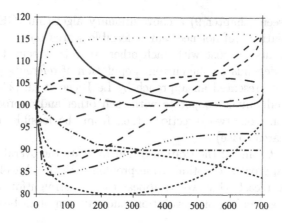

Fig. 14.6. The uncontrolled trajectories of the blood pressure of several patients based on various parameter values, following [421]; MAPs in mmHg as functions of time steps (2 seconds per step).

treat the widest range of patients. After only 50 generations of evolution, the best CMAC controllers could control almost all of the prospective patients to within 5 mmHg of a specified target value.

Other researchers have studied games such as pursuit-evasion, where an interceptor attempts to catch an evader, again using coevolution. For example, Reynolds [387] implemented a simulation of the child's game of "tag," where one vehicle would chase another until arriving successfully within a preset distance of the vehicle being chased, whereupon the chaser would become the "chasee" and the simulation would continue. The chaser was given the advantage being able to move faster. Coevolution optimized controllers for each vehicle, playing either side of the game as appropriate. A similar effort was conducted by Cliff and Miller in the mid-1990s, and you can find some excellent graphic examples of their work at http://www.cogs.susx.ac.uk/users/davec/pe.html. They have movies that depict the initial random behavior of a pursuer and evader, and the coevolved learned behavior at different generations. While the evader from the 900th generation can barely outmaneuver the pursuer from the same generation, the evader from the 200th generation cannot. And coevolution does not need to be limited to simulations. Floreano and Nolfi [136] have conducted coevolution pursuit-evasion experiments with real Khepera robots.

Also, as we discussed previously in section 9.1.12, an interesting approach to constraint satisfaction problems was reported by Paredis [350]. The method was based on a coevolutionary model in which a population of potential solutions coevolved with a population of constraints. Fitter solutions satisfied more constraints, while fitter constraints were violated by more solutions. There was no consideration of feasible or infeasible solutions in this case. The approach was tested on the N-queens problem and was compared with other single-population approaches [351, 350].

We've also seen (chapter 9) a coevolutionary algorithm (GENOCOP III) where two populations (of not necessarily feasible search points and fully feasible reference points) coexist with each other. In this system, the evaluation of search points depends on the current population of reference points. In the same chapter, we presented an approach by Le Riche et al. [280], where two populations of individuals cooperate with each other and approach a feasible optimum solution from two directions (i.e., from the feasible and infeasible regions of the search space).

These are just some of the examples that indicate the overall utility of the coevolutionary approach to optimization problems. With a bit of thought, it's often possible to recast what would appear to be a straightforward optimization problem as a problem of coevolution, and this can often be done to some advantage.

14.4 Another example with a game

Here's another example with a game, this time a military type of game, where a computer used coevolution to teach itself how to play. Recently, Johnson et al. [246] built a coevolutionary system that taught itself to play a version of the "TEMPO Military Planning Game." This is a two-sided competitive game used in courses by the Defense Resources Management Institute to introduce the basics of modern defense management. Teams of players compete in building force structures by dividing limited budgets over a succession of budgeting periods ("years") between categories such as "acquisition" and "operation" of "offensive units" and "defensive units." The rules are no more complex than the rules of Monopoly. However, players learn that the rules' apparent simplicity is deceptive: the rules pose challenging and difficult decision problems. No feasible algorithm for optimal play is known.

Melich and Johnson, at the U.S. Naval Postgraduate School, were studying force-structure planning and seeking approaches to answering the question: What investments should be made, and when, to develop the ability to win against an unknown opponent? Impressed by the success of evolutionary computation techniques, as demonstrated by the checkers program [150], they conceived the idea of evolving players for the TEMPO game as a way to study planning and budgeting behavior. In 2000, Johnson applied evolutionary computation to a highly simplified version of the TEMPO game.

The actual full set of investment categories for the TEMPO game comprises: (1) operation of existing forces, (2) acquisition of additional or modified forces, (3) research and development ("R&D"), (4) intelligence, and (5) counterintelligence. There are four types of forces: two offensive ("Offensive A and B") and two defensive ("Defensive A and B"). Each type comprises several weapon systems with varying acquisition and operation costs (measured in "dollars") and measures of effectiveness (in "utils"). "R&D" represents current investment that buys the possibility in a future year of acquiring new weapon systems, possibly with better price/performance ratios than those now available. Investment in intelligence buys information about the opponent's operating forces and investment activities. Investment in counter-intelligence degrades the information the opponent obtains through intelligence.

In Johnson's initial experiments, there was only one offensive and one defensive weapon system; R&D, intelligence, and counter-intelligence were omitted. The experiments were conducted using an evolutionary system working on Lisp code. Individuals (candidate algorithms) were represented as computer programs in a simple Lisp-like language, written in terms of variables that described the current state, and operations that attempted to allocate funds to various categories for the coming budgeting period. State variables included the current total available budget, probability of war, and current inventories, acquisition limits, prices, and operating costs for the offensive and defensive units. The budget categories were the acquisition and operation of offensive and defensive units.

Investment algorithms were evaluated for fitness by pitting each in games against a selection of others from the same population and recording wins and losses, in a manner similar to the experiments in [150]. The question was: Would anything reasonable emerge in such a simple framework? And indeed it did. Starting from an initial generation of completely random programs, an algorithm was evolved that allocated funds according to rudimentary sensible rules, characterized by Johnson as "dumb, but not crazy": it would not attempt to acquire units beyond the appropriate acquisition limits or to operate units beyond the number in inventory, and it incorporated a check to assure that an initial allocation to the operation of offensive units had not exhausted available funds before further allocations were attempted.

Encouraged by these results, Johnson et al. [246] developed a coevolutionary program that evolved investment algorithms for a version of the TEMPO game with significantly more features of the full game: multiple weapon types, intelligence and counter-intelligence, though not yet R&D. In the coevolutionary system, the two sides were modeled by separate populations. There were modules to control the execution of the system and to define the scope of experiments. Provision was made for varying the weapon parameters (e.g., costs, utils, year of first availability) from game to game during fitness evaluation so that the evolving players did not simply "learn" what weapons to expect to be available.[3]

The coevolutionary system contained populations, environments, and evolution parameters for each of two players, X and Y. The two populations were constructed initially as sets of randomly generated individuals. Each individual was represented by a fuzzy-logic rule base with a varying number of rules as phenotypes, which were encoded in a list. The two sets of environments were read initially from two input files. Each environment maintained a list of

---

[3]It is even possible to introduce new weapon classes ("Offensive and Defensive C, D, etc.") and (unlike the "official" TEMPO) to model asymmetric situations in which the two sides have different budgets and different weapon parameters.

weapons and weapon parameters, intelligence, counter-intelligence, war probability, and budget. During play, the individual obtained this information from the environment, evaluated the weapons, and distributed budgets accordingly. The decisions (how much to buy/operate) were reported back to the environments. The two sets of evolution parameters were also read from two files. These files controlled the coevolutionary system. For each generation, each individual of the first population played a certain number of games against individuals selected randomly from the second population. Likewise each individual of the second population played a certain number of games against individuals selected randomly from the first. Then each individual was assigned a fitness based on all the games it played. The individuals in a population were sorted by fitness, a certain percentage of the best individuals were kept in the population, and these were varied by mutation and crossover. The best individual of each population was recorded (i.e., saved) and played against a fixed "greedy expert" as a benchmark.

The fuzzy-logic rule base that represented an individual consisted of a variable number of rules with an "if–then" structure. The "if" part had a variable number of terms referring to war probability, budget, weapon parameters, intelligence results, and the like. These determined the degree to which the rule applies. The "then" part assigned a "goodness" value to each investment item to which it applies — weapon system, intelligence category, or counter-intelligence. The rules were applied to each separate possible investment item to assign a composite "goodness" to the item. The budget allocation was computed by linearly scaling the "goodness" values, followed by normalization.

The coevolutionary system evolved the complete structure and the values of the rules, i.e., the number of rules, the number and type of terms in the "if" clauses, and the "goodness" values in the "then" parts. The rulebase used fuzzy logic because this can capture important nonlinearities in a very compact way and presented in a form that can be understood easily by humans. Here's an example:

> **if** [PWar IS Very Low – Low][CATEGORY IS DEFENSIVE]
> [SUBTYPE IS 1 OR 2][Inventory IS Low]
> [MaxAcquisitonUnits IS Low – Medium]
> [AcquisitionCost IS Very Low]
> [UtilsPerAcquisitionCost IS Very Low – Low]
> **then** [Evaluation IS Low]

The terms of the "if" part referred to 7 of 16 variables that were available for constructing a weapon rule: PWar is the probability of war; CATEGORY is either "OFFENSIVE" or "DEFENSIVE"; "TYPE" (not shown) is A or B, as in "Offensive A" or "Defensive A"; "SUBTYPE" is a number that distinguishes between weapons of the same category and type, as in "Offensive A 1" or "Offensive A 2." "Inventory" is the number of units of a weapon on hand — the largest number that can be operated (if the operating cost is paid); "MaxAcquisitionUnits" is the largest number of units of the particular weapon that can be purchased in one year; "AcquisitionCost" is the cost in dollars per unit, and "UtilsPerAcquisitionCost" is the ratio of util value to acquisition cost.

A term like "AcquisitionCost IS Very Low" refers to the degree of membership of the acquisition cost in a certain fuzzy set represented internally by a bell-shaped (Gaussian) membership function with a given center c and "width" (standard deviation) σ. The program used the actual numeric values of c and σ internally, and these were the quantities that mutation and crossover operated on. But for the human reader, expressions like "Very Low" were presented, as these were presumably more palatable than a pair of numbers like $c = 48.7682$, $\sigma = 17.1056$. The range of meaningful acquisition costs was divided into five subranges running from "Very Low" to "Very High." Here, the interval $c \pm \sigma$ was within the "Very Low" subrange for acquisition costs. The "Evaluation IS Low" in the "then" part of the rule referred to the "goodness" value. Again, the program used a specific number. The human reader was told that the number was in the low subrange of possible "goodness" values. Thus, to the extent that the rule applied to a weapon, it was a reason *not* to buy it.

In addition to the coevolutionary system, there was a game system that allowed a human player to play against a saved computer player. The computer player distributed its budget according to its rule base, while the human player interacted with the game system through a spreadsheet interface. One evolved player beat Johnson twice, running "straight out of the box," to his combined chagrin and delight — one naturally has mixed feelings when surpassed by one's progeny.

The game system has been used in an economics course at the U.S. Naval Postgraduate School. Many of the students needed three or four tries in competition against the evolved program before achieving an outcome that they were willing to submit for a grade. The professor in the course (who shall remain nameless) was able through prolonged and concerted effort to beat the machine by a small margin on a first try. During play, he was ascribing all manner of sophisticated motivations to the machine for its moves. He was dismayed to learn afterward that he had been competing against a set of precisely three rules of the form illustrated above. In short, the evolved players very definitely showed human-competitive performance!

14.5 Artificial life

Instead of addressing optimization problems, coevolution can also be used to study models of natural phenomena, including aspects of life itself. This idea of studying "life as it could be, rather than life as it is" received great attention in the late 1980s and early 1990s under the rubric of a field called *artificial life*. As with just about everything in evolutionary algorithms, however, the roots of this field go back many decades, even to the earliest efforts by Nils Barricelli in the 1950s. In 1954, Barricelli published a paper [32] that described an experiment of numbers moving about in a grid. The numbers represented organisms migrating, and when different organisms entered the same grid location, rules were applied to determine whether or not either number might reproduce, mutate, or move. This simple setting generated interesting patterns of numbers over time in the grid, and was evidence of what would later be described as *emergent properties*, unpredictable behaviors that emerge from the interaction of local rules.

In 1969, Conrad put forth a far more advanced effort that simulated an ecosystem [76, 77]. The model started with a *world*, which was a one-dimensional circular string of *places*. Resources called *chips* were distributed at these places, and 200 to 400 organisms were initialized to "live" in this world at particular places. The coding for the organisms was quite intricate, incorporating essentially a computer program for executing a series of instructions that constituted each organisms' behavior. Organisms used chips in the environment as energy, to improve their internal condition, repair themselves, and to move from one place to another. This was perhaps the first evolutionary program with a hierarchic relationship between the organisms' simulated behaviors and their underlying genetic program in which no explicit criterion for measuring performance was given. Instead, the appropriateness of alternative behaviors emerged from the organisms' interactions in the places in the world model.

Many years later, Ray created a simulation called *Tierra* [382], in which programs interacted in a two-dimensional grid. The programs, written in assembly code, had the function of replicating themselves, with shorter programs replicating more quickly than longer programs. Thus there was an intrinsic selection pressure for shorter programs as they competed to fill the available grid. Ray noticed, however, that in addition to this selective feature, new programs were created that served as parasites to their hosts. The programs were written originally by Ray in three modules: one to determine the length of the program, another to monitor a reproduction loop, and the last to actually copy the program to a new location in the grid. The newly evolved parasite programs took the first and second modules but eschewed the third. Instead, they used a pointer feature to borrow the copy routines of older "host" programs to make copies of themselves. Thus, as long as there were host programs in the grid, these much shorter programs could outcompete the hosts because they replicated faster. This and other features of the Tierra experiment were again emergent unpredicted properties of the simulation.

If you find this sort of experimentation fascinating, as many do, you should examine [3, 450] for more details. Musing about life as it could be is entertaining and possibly insightful, but this book is about problem solving so let's return to games and examine another setting that is related to artificial life where there is a real problem to solve.

14.6 Games of cooperation and competition

Suppose you're playing a game and your problem is to generate the maximum payoff. As indicated earlier, not all games are competitive. In some cases, you can realize a better payoff if you cooperate with another player, rather than attempt to compete with him or her. In other cases, the situation is somewhat more complicated. For example, one simple game that we've already seen in chapter 11 that embodies these complex dynamics is the iterated prisoner's dilemma (IPD). Just to refresh your memory, in the basic IPD, there are two players, each with two options for each play: cooperate or defect. Cooperating implies attempting to increase the payoff for both players. Defecting implies attempting to increase one's own payoff at the expense of the other player. The game is "nonzero sum," meaning that payoffs that accrue to one player are not necessarily taken from the other player, and it is "noncooperative," meaning that players do not get to communicate about what play they will make ahead of time.

As we saw before, the payoffs can be written in a 2×2 matrix, as shown in table 14.1. To be a prisoner's dilemma, two constraints must apply to the payoffs R, S, T, and P:

(1) $R > (S + T)/2,$
(2) $T > R > P > S.$

The first constraint removes any obvious desire for mutual cooperation and also ensures that the payoffs for a series of mutually cooperative moves are greater than a sequence of alternating plays of cooperate–defect against defect–cooperate (which might represent a more sophisticated form of cooperation [5]). The second constraint ensures that defection is a dominant action, and also ensures that the payoffs accruing to mutual cooperators are greater than those accruing to mutual defectors.

Table 14.1. The general payoff matrix for a prisoner's dilemma. The entries in the form (A,B) are the payoff to A and B, respectively. The letters R, S, T, and P are variables representing the payoff for alternative combinations of moves. R is the "reward" both players receive when mutually cooperating. T is the "temptation" payoff for defecting against a player who cooperates. S is the "sucker" payoff for cooperating when the other player defects. P is the "penalty" payoff awarded to each player when both defect.

		Player B	
		Cooperate	Defect
Player A	Cooperate	(R, R)	(S, T)
	Defect	(T, S)	(P, P)

If you're only going to play this game for a single move, defecting is the only rational thing to do. If the game is iterated over a series of moves, however, defection is not so compelling, as can be seen when using a coevolutionary simulation.

In 1987, Axelrod [15] conducted a simulation in which strategies in the IPD were represented as look-up tables based on the three previous moves made in the game using the payoff matrix shown in table 14.2. A strategy in the

Table 14.2. A specific potential payoff matrix for the prisoner's dilemma

		Player B	
		Cooperate	Defect
Player A	Cooperate	(3, 3)	(0, 5)
	Defect	(5, 0)	(1, 1)

game was a series of 64 instances of all combinations of previous moves by both players on the past three moves, as well as the cases where players were just starting and didn't have a three-move history. For example, one element in a player's strategy might be [SELF(C,C,C) and OPPONENT(C,C,C) yields C] where SELF refers to the three moves made by the strategy in question, OPPONENT refers to the other player's moves, and C stands for cooperate.

Axelrod set up an evolutionary simulation in which strategies competed against each other in a round-robin format (everyone plays against every possible opponent) for 151 moves in each encounter. The strategies earned points for each encounter and after all pairings were complete, those with higher scores were favored for survival using proportional selection. New variations of the strategies were created by mutation (changing the outcome associated with a particular combination of moves) as well as one-point crossover. In 10 trials, Axelrod made two interesting observations. The first was that the mean score of the surviving parents in the early generations of evolution decreased, which meant that most of the winners were defecting. Shortly thereafter, however, the mean score of the surviving parents increased and headed close to 3, indicating that the population had learned to cooperate. Nobody instructed the solutions to cooperate; this property simply emerged from the simulation and showed that mutual cooperation can arise between players even when defecting is the rationale play in a "one-shot" prisoner's dilemma.

The second observation concerned the form of the strategies that were evolved. Many resembled the strategy called "tit-for-tat," which is very simple but quite effective. It cooperates on the first move, and then mirrors whatever the opponent did on the last move. If the opponent keeps cooperating, so does tit-for-tat. If the opponent takes advantage of the initial cooperation by defecting, then tit-for-tat will retaliate with a defection on the next play, and will continue to defect until the opponent cooperates again. Therefore, tit-for-tat is never taken advantage of too much by other strategies. Axelrod had earlier conducted two competitions with the IPD involving game theorists from around the globe, and the winning strategy in each of those instances was tit-for-tat. It was thus very interesting and even reassuring to see forms of this strategy emerge from his simple simulation.

It is also of interest to determine the effects of other representations for strategies. For example, in 1993, Fogel studied the IPD by replacing Axelrod's look-up tables with finite state machines [146, 148]. Figure 14.7 shows an exam-

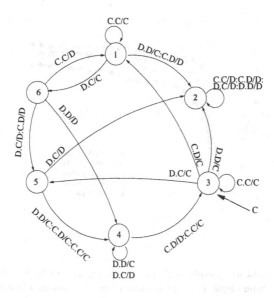

Fig. 14.7. A potential finite state machine for playing the iterated prisoner's dilemma, from [146]. C stands for cooperate, D stands for defect, and the bold arrow indicates the start state.

ple of a finite state machine used as a strategy in the IPD. The same procedure for playing games between strategies was followed, but new strategies were created by applying mutation to surviving parents, either by adding or deleting states at random, or by changing next-state transitions, output symbols, the start state, or the starting move. Figure 14.8 shows the average results for population sizes ranging from 50 to 1000 parents. The results are essentially the same, and are very close to what Axelrod observed: an initial tendency to mutual defection followed by a rise to mutual cooperation. The result was robust to this change in representation.

This might give you some hope that even when people face decisions where they could take advantage of each other and perhaps experience some gain for that behavior, they might still end up cooperating. From personal experience, however, you might be skeptical. There are always many cases in the news where people are caught "defecting" in one form or another for some tempting payoff, when the equivalent of mutual cooperation could have yielded a satisfying gain.

One possible explanation for this result is that the standard IPD offers only two plays: cooperate or defect. In the real world, however, there are degrees of cooperating and defecting. The range of possible actions is continuous, not bipolar. Harrald and Fogel [208] studied a version of the IPD where players could offer plays based on a numeric scale of -1 and 1, where -1 was complete defection and 1 was total cooperation. These values were fed into artificial neural networks, used to represent strategies. Each neural network had six input nodes, one for each of the previous three moves by each player (see figure 14.9), and a hidden layer of neurons that varied in size. The neural networks generated

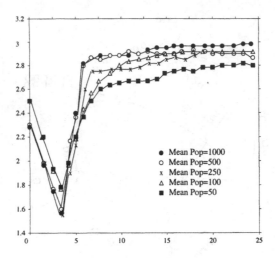

Fig. 14.8. Fogel's results when evolving finite state machines to play the iterated prisoner's dilemma [146]. The graph displays the mean of all parents' scores as a function of the generation number.

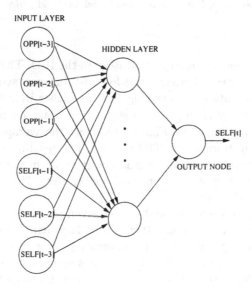

Fig. 14.9. The neural network architecture used to play a continuous version of the iterated prisoner's dilemma.

numbers between -1 and 1, with payoffs accruing based on a linear approximation to the payoff function from table 14.2, which is shown in figure 14.10. The surviving neural networks at each generation created offspring through random Gaussian mutation of every weight and bias term (by now, you should be quite familiar with these sorts of operations) and each game was played again for 151 moves.

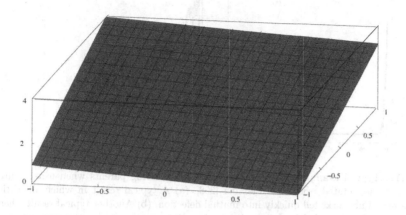

Fig. 14.10. A planar approximation to Axelrod's payoff function for the prisoner's dilemma [208]; the vertical axis indicates the payoff for plays made along the horizontal axes.

The results in this new framework were not at all like what was observed before. Figure 14.11 shows some typical examples. Several look like the patterns we saw in the stock market after March 2000 (steep declines). Note how different these patterns are from those in figure 14.8. All of the graphs show results when using 20 hidden nodes in the neural networks, so these strategies were potentially very complicated. The vast majority of trials showed results tending toward decreasing payoffs, not increasing payoffs. The obvious question is: What accounts for this change?

Darwen and Yao [82] studied a variation of the IPD with eight options, rather than two (i.e., cooperate or defect). Their results were very similar to those of [208] and they came to two conclusions. The first is that as the number of options to play increases, the fraction of the total game matrix that is explored decreases. With only two options, the entire set of four possible outcomes is generally explored by an evolving population of strategies. In the eight-option case that they studied, many of the 64 possibilities were not explored. Darwen and Yao showed that each 2×2 subset of the 8×8 payoff matrix was itself a prisoner's dilemma, and thus it was possible that the evolving solutions were settling into series of plays that were somewhat like local optima in the larger 8×8 matrix. Their second conclusion, perhaps better described as an observation, was that when the IPD had more choices, strategies evolved into two types, where both types depended on each other for high payoffs and did not necessarily receive high payoffs when playing against members of their own type. This mutualistic dependence was surprisingly resistant to invasion by other strategies that might have increased the eventual payoff within the population.

Darwen and Yao's observations are very interesting, yet they perhaps do not fully explain the consistent degradation in payoff that is seen when a continuous

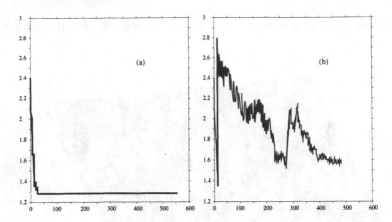

Fig. 14.11. Typical examples of the mean payoff to surviving parents when using neural networks in the iterated prisoner's dilemma [208]. (a) A typical result in which all of the evolving neural networks fell quickly into mutual defection. (b) Another typical result where initial cooperation degraded into mutual defection. Both graphs display the mean score of parents as a function of generation number.

range of options is employed. Nevertheless, these few examples illustrate a much richer opportunity for using coevolution to study the emergence of strategies in more complex games. There are hundreds of papers written about the prisoner's dilemma each year, and very many of the historical contributions to this litera-ture have involved evolutionary algorithms in different forms. These and many other studies indicate the potential for applying coevolutionary simulations to simple and complex nonzero-sum noncooperative games.

14.7 Modeling intelligent agents

Each player in an IPD is an "agent," presumed to allocate resources to achieve its objectives based on its current environment. What if we viewed each of these agents as having its own purpose, and its own evolutionary mechanism for improving its behavior to achieve its purpose? Instead of one or two populations of coevolving agents, we might have hundreds. In fact, that is exactly what was done in one study [154], within a setting called the *El Farol Problem*.

The problem was introduced by Arthur [13], based on a bar, the El Farol, in Santa Fe, New Mexico, which offers Irish music on Thursday nights. Sup-pose there are 100 Irish music aficionados and that each chooses independently whether or not to go to the El Farol on a certain Thursday night. Further, sup-pose that each attendee will enjoy the music if there are fewer than 60 people in the bar; otherwise, it's too crowded and nobody will have a good time. The problem then, for each music lover, is to predict how many people will be at the bar, and go only if the prediction indicates fewer than 60 expected people.

Here's where it gets a little tricky. If the majority of people predict that the bar will be empty, they will go and instead it will be full, leading to a miserable time for all. If instead most people predict that the bar will be full, then they will stay home, and miss out on an otherwise fine night of Irish music. To help matters, the bar publishes the weekly attendance for everyone to use as data in their models. So the challenge for each agent is to use those data and anticipate the attendance for the next week, and do so while the 99 other music lovers are doing the same thing.

Arthur's idea was to model each person as a computational agent, each of which had some number of alternative predictive models — such as anticipate the same attendance as last week, or use the average of the last four weeks. Across all the alternatives that each agent would have, some models would naturally do better than others. Each agent could use its best model given the prior data to make its prediction, and then either go to the bar or stay home. None of these models would change for any agent, but the model that was best at any time might, because of the fluctuations in the weekly attendance as other agents use their models. Interestingly, in a simulation, Arthur found that this protocol yielded an average of about 60 agents going to the bar, with a fairly stable fluctuation around that average.

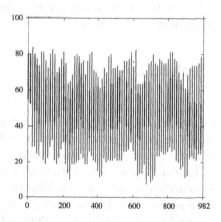

Fig. 14.12. A typical example of the attendance (attendance as a function of week) at the bar when each agent could evolve its own models for predicting future attendance [154].

Perhaps even more interesting, however, is what happened when each agent was allowed to evolve its own models during each simulated week, as was studied in [154]. Each agent was allowed 10 parent models, all in the form of autoregressive moving averages, and 10 offspring models and used 10 generations of variation and selection during each week to optimize the models before predicting the attendance at the bar and going or staying home. Instead of stable fluctuation of the attendance around 60 bargoers, the attendance swung wildly, as shown in figure 14.12. Arthur's result of generally stable attendance when each agent used fixed sets of models turned into a highly variable result when the agents evolved their models as an adaptive process. Furthermore, [154] showed that the overall attendance was modeled well as a Markov chain, where no stable state would ever be expected. Adding evolution, and in particular coevolution, to a system can generate many unintended consequences.

The broader area of developing models of intelligent agents has potential impact in the way we design business environments and even education systems. Blair [47] offered an analysis of the prospects of student–teacher mutual evaluations, where teachers are evaluated by students and therefore have an incentive to lower standards so that students will give inflated reports, and the teachers can reciprocate by evaluating the students as having performed "better" (better is in quotes here because it is better only on paper in terms of a grade, and not better in terms of knowledge or the ability to apply knowledge). Blair showed that with some simple assumptions about the costs of the effort to prepare lectures, and the individual and collective benefits of teaching, the disappointing conclusion was that the best thing a teacher could do would be to lower the standards routinely, thus buying better student evaluations at the expense of ill-prepared students, with a graveyard spiral of descending performance on both sides likely to emerge. Coevolutionary models of the payoffs to students and teachers involved in this sort of mutual evaluation system can suggest those that lead to disaster and those that may in fact serve to the betterment of the overall product for both students and teachers.

14.8 Issues in coevolution

Coevolution can be a very valuable optimization tool. As with all tools, it is important to understand when to use the tool, when not to use the tool, and what the obstacles to its successful application might be. One possible impediment to successful coevolution occurs if the process stagnates at solutions of less than desirable quality. This can happen for a variety of reasons, some of which are related to issues of premature convergence in function optimization, which were discussed earlier (see section 11.2.1). For example, relying too heavily on recombination or other variation operators that are limited by the amount of diversity present in the population can result in a restricted search of the available problem domain. In competitive coevolution, particularly in a two-population framework — such as a predator/prey process — this can result in stagnation. The pressure for one population to improve comes when the competing population is racing ahead of it. If one population stagnates, so does the selection pressure that can be applied to the opposing population, and so the progress that the opposing population makes can slow down or halt. By consequence, the pressure applied back to the first population also relaxes, and the result is no longer an arms race but rather something resembling mutual indifference, where neither side really appears to care much about beating the opposition.

Within the past decade there's been conjecture that stochastic environments are a key ingredient to successful coevolution [48]. One example, which was at least moderately successful, was an effort to evolve strategies for backgammon [356, 355]. Backgammon is an interesting game in part because the options available to the players are determined by rolls of the dice, and weaker players can sometimes escape defeat against otherwise superior competition simply by obtaining a lucky series of dice rolls. The stochastic character of the game, and its instability with respect to predicting a winner, provides an intrinsic element of exploration in the course of play. It is therefore not possible for strategies to collude by fixing on specific deterministic move sequences that may not be particularly effective in general, but suffice for the available competition. More recent results with checkers and chess [67, 68, 69, 150, 158, 258] indicate that stochastic games are not required for successful evolution, even when the games are complex and pose many alternative states. Stochastic environments may yet be shown to aid in coevolutionary advancements, but they are not necessary for successful coevolution — neither are environments that are ergodic (where every possible state can be reached eventually from every other) (cf. [48]).

There are other causes of coevolutionary stagnation. Probabilistically, an evolutionary algorithm will discover the easiest workable strategies. More specifically, if there is a simple strategy that can be discovered quickly, and a more complex, elaborate strategy that is more difficult to assemble but equally good or perhaps even slightly more effective, it is likely that evolution will locate and exploit the simple strategy first. From the vantage of competitive coevolution, this presents another difficulty, because it may be that populations full of simple strategies may be unable to use random variation to catapult themselves into regions of the solution space that involve more complex but better solutions.

In the IPD, suppose that one strategy employed in a population is tit-for-tat. We have seen this is a very simple strategy that can be captured in just two Spartan rules. Thus, it is improbable by most typical variation operators, regardless of the representation that you might employ, that you could vary tit-for-tat into a very complex yet effective strategy that relied on, say, a function of the last 12 moves. The most likely pathway to this complex strategy would come via a series of intermediate steps, each representing solutions of increasing complexity. But with increasing complexity often comes a decreasing likelihood of discovering something effective, or more properly, something more effective than the simple parents it was created from.

The situation is something of a so-called "catch-22": simple but perhaps weakly effective solutions may be easy to find, but then once found, these solutions may not provide the basis for leaping to other more complex and highly effective solutions. Yet, we cannot hope to create complex highly effective solutions out of nothing, and thus the catch is that we are stuck with having to start somewhere and face the prospect of stalling out at solutions of lower complexity and perhaps lower effectiveness. It may also be true that simple solutions use all of the available data structure to represent a solution [451] (e.g., in a simple neural network, all of the weights and bias terms may contribute to the output, rather than having some connections that are superfluous). In this case, it becomes very difficult to successfully vary a good strategy, because it means changing parameters that are actively involved in generating the behavior(s) that has been successful to that point.

Many ideas have been developed to help overcome this situation. One such method provides incentives in individual payoff functions for complexity. Instead of using the maxim of parsimony that favors the simplest possible solution, more complex solutions are favored, with the hope that the additional complexity will offer a source of potential diversity in the strategies that can be exploited. Alternatively, the coevolutionary process can be started with only simple solutions, but random variation can be applied to increase the coding length of the solution, e.g., by adding additional rules to a rule base, or weighted connections to a neural network, or states to a finite state machine. In so doing, the core of the encoding remains the same, allowing the solution to retain its prior behaviors that were proven effective, while also providing new genetic material for variation that is not in conflict with what has already succeeded. In essence, the process is allowed to grow solutions that do not inherently forget what they have already learned during the generations of variation and selection.

Another idea to prevent evolutionary stagnation is simple: record the best solutions at each generation, or at some intervals, and then use those solutions as a "hall of fame" [398]. New evolving solutions can then be pitted not only against other evolving solutions but also against the best representatives from prior generations in the hall of fame (or a so-called "master tournament" [136]). This forces solutions to remember behaviors that were effective against earlier competition. Otherwise, it is possible that those behaviors will be forgotten, and a best representative from an early generation might defeat one from a later generation, even by some seemingly simply tactical trick that was overcome earlier by intermediate populations along the way [355, 136]. The idea can be extended to something more strict in a "dominance tournament" [451] where the best solutions are examined in each generation and a dominant solution is one that can defeat every other dominant solution from an earlier generation. Finding new dominant solutions provides real evidence of evolutionary progress in a coevolutionary framework.

Still another idea for encouraging progress in coevolution is to apply finite life spans to the surviving parents at each generation. This prevents the situation in which a parent can locate itself in a part of the solution space where it creates poor offspring and by which the parent benefits directly by defeating these poor offspring, which was observed in [146, 148]. By ensuring that parents die off, at least occasionally, the momentary loss of possibly superior solutions may be overcome by subsequent advances generated from progeny of the offspring that would otherwise die. Situations such as this are all problem dependent, so experimentation and analysis is required to identify when such measures may be appropriate.

One notable procedure for improving the optimization properties of coevolutionary simulations that involve multiple populations of competing solutions (as in a host–parasite setting) concerns carefully selecting opponents worth competing against [397]. Think back to the example of coevolving models of patients

on the operating table with models of control strategies for administering drugs
to control the patient's blood pressure. The control strategies were scored based
on how well they maintained the blood pressure at a target level for each patient tested. In contrast, the patients were scored based on how quickly they
could cause the controller to fail, because in this case we were interested in finding those patients that are most challenging. Suppose that there are multiple
control strategies in their respective population that can each handle the vast
majority of existing patients. Although this sounds good, from the perspective
of advancing to improved solutions, it implies that there is very little information in these patients that can be applied usefully. If all of the controllers can
handle the patients, there is no selection pressure to use in discriminating between one controller and another. A similar argument can be made for patients
that defeat large numbers of controllers.

One idea for improving the chances of further coevolutionary advancement
in situations such as this reduces the score for members of each population
based on the number of other members in that population that can defeat a
particular individual in the opposing population. An example will help. Suppose
that a particular patient in a population can be controlled successfully by 97
of 100 controllers in a competing population. Instead of assigning a success,
perhaps one point, to each of these 97 controllers, each would receive 1/97 of a
point. In general, if there are N individuals that can address an opponent with
success (whatever the definition of success for the application), then the score
received by each individual is $1/N$. In this manner, when individuals can treat
opponents successfully that are by and large difficult for most others to treat,
those individuals earn more for their efforts. Similarly, opponents that are easy
to defeat yield only small incremental point gains.

This idea is somewhat similar to what happens in pari-mutuel wagering
on horse races. The payouts are determined by the amount of money bet on
each horse. If a horse is an obviously superior competitor, its odds are lower
— not because someone thinks that the real chances of the horse winning are
high but because lots of people are betting lots of money on the horse. When
that horse wins, the payout is low. When everyone can figure out which horse
will win, it hardly becomes worth the time and effort to bet. In contrast, the
reason longshots pay out with big prizes is because relatively fewer dollars are
bet on these horses. If you can successfully pick the longshot, you are like one
of the controllers who can successfully manage a patient when all of your fellow
controllers are failing, and you deserve some extra credit.

An analogous idea is to focus the sampling of opponents on those that might
provide the biggest challenge to the widest range of competitors. This can save
time in the competition by reducing the contests between solutions where one
is known to be weak against all competitors. One method for implementing
this idea is to choose opponents at random in proportion to their fitness, which
would be computed using the method explained directly above. No doubt you
can craft other reasonable methods as well, or extend them to treat other as-

pects of cooperative problem solving, such as having individual neurons learn to cooperate in pattern recognition problems [290, 288, 289].

14.9 An example of cooperative coevolution on the TSP

Another approach to coevolution seeks to develop solutions that cooperate to solve parts of problems. This is called cooperative coevolution. It has roots going back at least to [360], with many other more recent examples, including [126, 127, 361]. The central idea is to decompose a large problem into smaller subsections and evolve solutions to the subsections, but the trick is that the score a solution receives depends on the other solutions that it is paired up with in solving the overall problem. Thus, no attempt is made to directly partition credit to any particular solution to part of the problem; instead, each solution receives a payoff based on how well the team it belongs to addresses the entire challenge.

Here's an example with the traveling salesman problem. Suppose there are 100 cities in the problem and, for the sake of the example we distribute 25 cities in each of the four quadrants, uniformly at random. Figure 14.13 shows the distribution of the cities.

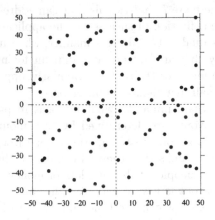

Fig. 14.13. The distribution of 100 cities in the TSP example. Each quadrant has 25 cities

We could address the problem of finding the minimum length tour of these cities with a straightforward evolutionary approach. As a baseline for measuring performance in this example, we used a permutation representation, with 50 parents initialized at random. Each parent generated one offspring via an inversion operator, which selected two points in the permutation at random and reversed the order of the cities between the points. All 100 solutions in the population at each generation were scored with respect to their distance in covering all the cities and returning to the first city in their list. The 50 lowest–scoring (best) solutions were retained as parents for the next generation and the process was iterated over 10,000 generations. Figure 14.14 shows the rate of optimization of the best solution in the population as a function of the number of generations, and figure 14.15 shows the best-evolved final solution. It has a total tour length of 833.813 units.

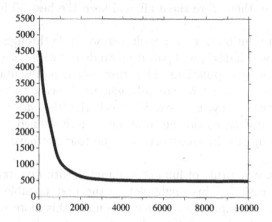

Fig. 14.14. The rate of optimization of the best solution to the TSP when treating it as one holistic problem; the graph displays cost (distance) as a function of generation number

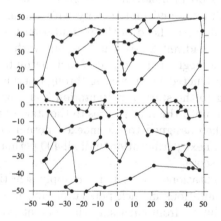

Fig. 14.15. The best-evolved solution after 10,000 generations

Now, instead, let's explore how we can treat the problem using cooperative coevolution. Let's divide the problem into four subproblems, each with its own population, each associated with a quadrant of the space. There is one population for quadrant 1, one for quadrant 2, and so forth. Each population has 50 parents and will have 50 offspring, one per parent, and we're still using the permutation representation, but now each permutation is a list of cities to visit only in the specific quadrant that corresponds to the particular population.

Each member of every population is created at random, with a jumbled ordering of cities numbered 1 to 25, for each city in the quadrant. Tours are created by selecting one permutation from each of the four populations and combining them to define a complete circuit of the cities. Initially, we create 100 solutions in each population, then score them all, and keep the best 50 from that point forward.

To score each solution in each population, we cycle through each one (designated by the "cycle index") and pair it up randomly with one solution chosen from each of the other populations. Thus there is one representative from each population, but at this point we are only assigning a score to the permutation selected by the current cycle index. We'll call this the *indexed permutation*. The score is assigned based on the total tour length defined by the four permutations. To complete the construction of the tour, we have to link the four permutations.

Suppose we use the rule of linking quadrant 1 with quadrant 2, 2 with 3, 3 with 4, and 4 with 1. This might not be the best possible approach, but for the sake of the example it's sufficient. To make this more specific, just for initialization (later in the evolution, the links between quadrants will be free to evolve), let's start with quadrant 1 and find the city with the lowest value in the y dimension, effectively the city that is closest to quadrant 2. We can move that to the last position in the permutation taken from population 1, representing the last stop in quadrant 1 before heading to quadrant 2. Then we can move to quadrant 2 and find the city with the highest value in the y dimension and move it to the head of the permutation taken from population 2 as the first stop in quadrant 2. We can continue with this process through the quadrants until the beginning and end of each permutation is defined. The total tour is then constructed by transiting through permutation 1 from start to finish, then linking to the start of permutation 2 and proceeding from start to finish, and so forth, until the tour returns to the start of permutation 1. The total tour length is then assigned to the indexed permutation. The process is then repeated for each next permutation until all 400 permutations (100 in each population) receive a score.

Once all the permutations are scored, we apply selection and remove the 50 worst-scoring permutations in each population and use variation operators to create one new solution from each surviving parent. We can use the same inversion operator as indicated above, but we can also combine it with another operator that will allow the cooperative coevolutionary procedure a chance to learn which cities should be the connection points between the quadrants, rather than forcing those connections to occur with greedy heuristics, as we did above.

In our example here, each parent always uses the inversion operator, but also as a 0.5 probability of using a shift operator that can translate the permutation of cities forward or backward up to 13 (essentially half of the length of the permutation of 25 cities) steps. In order to facilitate a nearness between parents and offspring, the number of steps to shift in either direction is made to scale down linearly in probability. That is, the probability of choosing to shift by 1 unit is 13/91, by 2 units is 12/91, and so forth, down to the probability of shifting by 13 units, which is 1/91. The shift operator will change the first and last cities in a permutation while keeping the general path through the cities unchanged. The first city in a permutation is designated as the city to transition to first when coming in from another quadrant, while the last city in a permutation is the exit city that connects to the first in the next permutation. In this manner, the cooperative coevolutionary process can learn which cities to use as entry and exit points, and not rely on the rule of using the city that is closest to the neighboring quadrant. Note that we're not claiming any optimality to this scheme. It is simply meant to illustrate the potential for the overall method.

Once all the offspring permutations are created in each population, scoring must again be applied to assess the worth of each parent and offspring. Rather than combine solutions from populations at random, however, there is now a basis for favoring the selection of those solutions that have better scores from the previous generation. At the first generation, we have no such information, but at the second generation we do. One approach is to pair each indexed permutation with the best-scoring permutations from each of the other populations, and that's the approach we've taken for the example here. Each indexed permutation receives a score based on the total tour length it generates when combined with the three best-scoring representatives of the other populations. After all the scores are computed, the selection and variation process continues until we reach 10,000 generations.

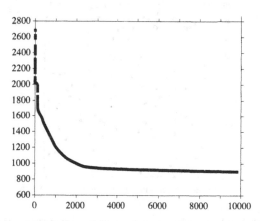

Fig. 14.16. The rate of optimization of the best tour length in the cooperative coevolution experiment; the graph displays cost (distance) as a function of generation number

Figure 14.16 shows the rate of optimization of the overall best tour length in the cooperative coevolution experiment. Figure 14.17 overlays the rates between the coevolutionary method and the standard method for easy comparison. Note that the cooperative coevolution method improves faster, and takes advantage of the head start that the heuristic of connecting neighboring quadrants provides. Figure 14.18 shows the best-evolved tour, which has a total length of 814.565, which is about 2.3 percent shorter than the tour evolved using the standard approach. Note that the coevolutionary method learned to transition between quadrants using cities that aren't the closest neighboring cities. It was able to take the initial hints at which cities to use and improve on those ideas.

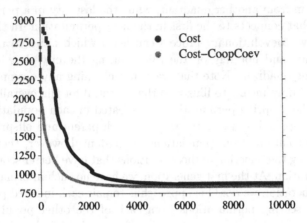

Fig. 14.17. Comparing the rates of optimization between cooperative coevolution and the standard approach; the graph displays cost (distance) as a function of generation number

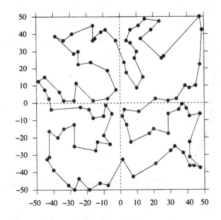

Fig. 14.18. The best-evolved tour when using cooperative coevolution

It's encouraging that the cooperative coevolution method generated a solution that is better than the standard method, in part because we constrained where it could cross the coordinate axes. The standard method was free to cross the axes wherever evolution determined best and could make multiple crossings if seemingly appropriate. For comparison, we reran the standard method using a penalty of 10,000 units for each crossing of a coordinate axis. This forced the evolutionary algorithm to cross the axes a minimum number of times (four). Figure 14.19 shows the rate of optimization, with attention focused on the range of tour lengths available after the tours had evolved to have only four coordinate crossings. Figure 14.20 shows the best-evolved solution after 10,000 generations, and it has a total tour length of 834.035, which is almost the same as the original best-evolved tour with the standard method.

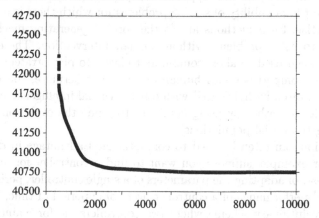

Fig. 14.19. The rate of optimization when enforcing a standard evolutionary approach to cross the coordinate axes a minimum number of times; the graph displays cost (distance plus a penalty of 10,000 per crossing each axis) as a function of generation number

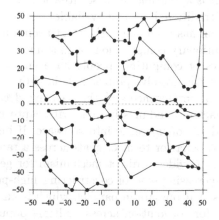

Fig. 14.20. The best-evolved tour when forcing a standard evolutionary algorithm to cross the coordinate axes a minimum number of times

Although these results represent only single trials of the methods and the comparisons do not take into account computational requirements or programming overhead, they do indicate the potential for an evolutionary method to optimize parts of a larger problem by combining the solutions to the subproblems and scoring them based on how well they all contribute cooperatively to the overall challenge. Not all coevolutionary methods have to be competitive to be successful.

14.10 Summary

The prospect of coevolving solutions simply by playing them off against each other, or seeking solutions that appear to cooperate well in addressing a problem, opens up the possibility of solving problems for which there is no available human expertise. Such methods are at the core of generating machines that will be able to solve problems without human intervention. These methods have already been used to allow computers to learn to play strategy games on a level that is competitive with human experts. Furthermore, coevolutionary techniques can often by hybridized with other artificial intelligence methods or human knowledge, either by using heuristics to guide the evolutionary search or by seeding the initial populations.

Coevolution can often be used to computational advantage in discovering solutions. For example, suppose you want to find a controller for an unstable system. Instead of adapting the parameters of a single controller, perhaps based on the overall performance of a controller over some period of time, you might set up a coevolutionary scheme where two controllers vie for maintaining the system in a stable range, and the first to fail is eliminated. At that point, it's not necessary to keep testing the winner. This can save considerable computing cycles. And although it's true that coevolution may sometimes "waste" computer cycles in arriving at a solution (or a set of solutions) by its process of "learning by death," it is applicable in those situations where there may be no other evolutionary or traditional method available.

To date, coevolution has been one of the most promising and yet underutilized methods of evolutionary computation, or machine learning more broadly. Since the earliest days of computing, people have envisioned machines that could teach themselves how to solve problems. It is sadly ironic then that most of the efforts in artificial intelligence over the past five decades have had little to do with learning machines and mostly to do with capturing what we already know as facts or methods of inference into software. If intelligence is viewed as the ability to adapt behavior to meet goals across a range of environments [148] then coevolution provides a direct mechanism for generating intelligent machines. With the increasing ease of constructing high-performance clusters of personal computers that operate in parallel, which is an architecture that is extremely well suited for coevolution across multiple populations, we can expect significant advances in machine learning and problem solving to come from applications of coevolution. The examples in this chapter provide a glimpse into the future.

XV. Who's Taller?

If you've ever been in a sports bar, you've probably heard people comparing athletes from different historical times. "Babe Ruth was a better home run hitter than Barry Bonds," someone might say. Or perhaps you might have heard someone say "the summer of 1969 was the best summer of my life" — or maybe "the students in my class from 2003 were smarter than the students in 2002."

We usually know what the speaker means, intuitively, but on closer inspection we might find our intuition giving way to some lingering doubts about how exactly we should interpret these sorts of claims. For example, suppose your best friend says that he got a better deal buying a Ferrari than he would have received if he'd purchased a Porsche. What does this mean? Does it mean that he got a better price? Maybe. Does it mean that he got better financing? Maybe. What about better performance, acceleration, or cornering? Maybe those things too.

Trading off multiple criteria is a challenging aspect of decision-making and problem solving. The following story (adapted from [455]) illustrates the point.

Two groups of students are attending a small Ivy League college. The students in one group, which we'll call group A, boast that they are taller than the students in the other group, which we'll call group B. At the same time, the students in group B enjoy the opinion that they are better logicians than the students in group A.

One afternoon, while walking through the center of campus by some benches, one of the students from group A spotted another from group B and approached him, saying: "We are taller than you!" The student from group B, who found himself straining his neck looking skyward at his classmate from group A, replied: "What do you mean that you are taller than we are? Look, have a seat." Caught off guard by his diminutive classmate, the big man sat down.

"Now, let's say that a student from your group A is represented by the symbol a, and a student from my group B is represented by b. You follow me?" asked the short student, sitting down next to his larger classmate, who nodded. "Okay, so you say 'we are taller than you!' — big man — all right, but what does it mean? Does it mean that..." The short student took a breath and adjusted his wire-frame glasses. "...that:

1. each a is taller than each b?

2. the tallest a is taller than the tallest b?

3. each a is taller than some b?

4. each b is smaller that some a?

5. each a has a corresponding b (and each of them a different one) whom he surpasses in height?

6. each b has a corresponding a (and each of them a different one) by whom he is surpassed?

7. the shortest b is shorter than the shortest a?

8. the shortest a exceeds more b's than the tallest b exceeds a's?

9. the sum of heights of the a's is greater than the sum of heights of the b's?

10. the average height of the a's is greater than the average height of the b's?

11. there are more a's who exceed some b than there are b's who exceed some a?

12. there are more a's with height greater than the average height of the b's than there are b's with height greater than the average height of the a's?

13. or that the median height of the a's is greater than that of the b's?"

"Hmmm," pondered the giant student from group A. "I'm not sure. I'll need to think about that."

"You do that. I've got a logic class now, but I'll be back here in an hour. Tell you what. You know we students in group B think we are better logicians than you in group A. But if you can tell me which of my 13 questions are independent of each other, and which are mutually dependent, maybe I'll change my mind." And with that, the group B student picked up his logic book and headed to class.

So, can you help the student from group A? Can you find all of the pairs of questions where an answer of "yes" to the first question implies the answer "yes" to the second question? Are there any questions that are equivalent, where the answer to both questions must be the same? Are there any pairs that are dependent but not equivalent?

While you're thinking about that, on the last page of the last chapter (chapter 17) we mention Fermat's Last Theorem, which states that if x, y, z, and n are natural numbers and $n \geq 3$, then the relation $x^n + y^n = z^n$ doesn't hold. Here's a little something to think about. Suppose that we make the assumption that $n \geq z$. At first glance, this assumption seems pretty innocent, as we're interested mainly in the cases where n is large. The funny thing is that if you make this assumption, it becomes very easy to prove Fermat's theorem.

Suppose that there exist natural numbers x, y, z, and n such that $n \geq z$ and $x^n + y^n = z^n$. It's clear that $x < z$, $y < z$, and $x \neq y$. Also, because of symmetry, we can assume that $x < y$. Then

$$z^n - y^n = (z - y)(z^{n-1} + yz^{n-2} + \ldots + y^{n-1}) \geq 1 \cdot nx^{n-1} > x^n,$$

contrary to the assumption! As you see, a little "innocent" assumption changes the problem is a significant way.

By the way, if you're able to find a short, elegant proof of this theorem for the case where $n < z$, please let us know immediately!

So, what about our friend that we left on the campus bench? "I see you're still here," said the short student from group B, returning from his logic class. "Did you get the answers?" The big man just shook his head. "Maybe we in group B are better logicians after all," said the short student, reaching into his pocket and handing his classmate a slip of paper. "Okay, here are the answers. You can verify them for yourself. Make sure I didn't make any mistakes!" The slip of paper read: "Let us denote the items of the problem by the numbers they are designated with, and let the symbol $p \rightarrow q$ denote the fact that the answer *yes* to question p implies the answer *yes* to question q.

Your homework assignment is to check that

$1 \rightarrow 2$, $1 \rightarrow 3$, $1 \rightarrow 4$, $1 \rightarrow 7$, $1 \rightarrow 8$, $1 \rightarrow 10$, $1 \rightarrow 11$, $1 \rightarrow 12$, $1 \rightarrow 13$,
$2 \rightarrow 4$,
$3 \rightarrow 7$,
$4 \rightarrow 2$,
$5 \rightarrow 7$,
$6 \rightarrow 2$, $6 \rightarrow 4$, $6 \rightarrow 9$,
$7 \rightarrow 3$,
$8 \rightarrow 3$, $8 \rightarrow 7$.

Thus the answers to questions 2 and 4 are identical and the same is true for the answers to 3 and 7."

So, did the short guy get them all right?

15. Multicriteria Decision-Making

> It is not greedy to enjoy a good dinner,
> any more than it is greedy to enjoy a good concert.
> But I do think that there is something greedy
> about trying to enjoy the dinner and the concert
> at the same time.
>
> Gilbert K. Chesterton, *Generally Speaking*

Real-world problems usually pose multiple considerations. It's rare that you might face only one parameter of concern. Think about buying a car. You might be the person who just simply has to have the Porsche 911 Carrera, with the "speed yellow" coloring, no matter what the price. If you are this person, then your choice of cars is essentially a binary decision problem: either you have the car or you don't! Maybe you could sacrifice a little and say that although you like yellow, you'd be happy with red or black, but you definitely don't want blue or white. Then the problem becomes one of optimizing along one dimension: color. But if you're like most people, the prospect of buying a car offers many things to think about.

What most people — but not everyone! — starts with when they consider purchasing a car is the price: Can you afford it? So price is clearly a factor. Then there's aesthetics. Do you like the way it looks? Then comes functionality: Will it carry what you need? Your children? Your surfboard? And there are other concerns, such as the CD stereo system, the crash safety record, the acceleration performance, the prestige of ownership, and so forth. When you trade your money for a car, you are buying all of these parameters.

Buying a car is simple compared to running a large corporation. If you're a shareholder of a company, you want the stock price to go up so you can hold a more valuable investment. If you're environmentally conscious, you might also want to know that the company is careful with its manufacturing efforts, being sure to recycle its products and avoid polluting the air and water. If you're running a company, however, you might have many more concerns [217].

You're certainly concerned about maximizing profit, but over which time period? The short-term time period is usually the next three months, before the next quarterly report is due to the shareholders. The mid-term time period is perhaps a year, because shareholders and corporate analysts tend to look to year-to-year comparisons. The long-term time period is perhaps two to five years. Frankly, most people don't look this far into the future when running a company. But note that it's not enough to say you want to maximize profits, you have to say when you want those profits, because things you do to maximize short-term profits might have adverse effects on long-term profits, and vice versa. You also have to consider the stability of those profits. Do you mind seeing them go up and down over time, or are you content with fluctuation as long as they trend upward?

Profits are nice, but not everything translates directly to the bottom line. What about market share? You might want to become like Microsoft and dominate the office software market. But to increase your market share, you might need to sacrifice profits. What about product diversification? Do you want to be known for one product or many? Being known for many products provides security but it takes time and money to advance multiple products in the business arena. What about worker morale? You probably want happy employees. What does it take to make them happy? Some people are happy when they work on interesting challenging problems. As an employer, that's great. Others are happy when they are receiving health care benefits, vacations, sick leave, maternity leave, education reimbursements, life insurance benefits, new equipment, leased cars, memberships to country clubs, and spacious offices overlooking beautiful vistas. As an employer, that's not so great.

What about control of the company? If you're the sole owner, you're the dictator. What you say goes, as long as it doesn't violate any laws. If you share ownership with others, then as long as you have more than 50 percent of the shares, what you say still probably goes. But if you have fewer than half the shares, then what you say doesn't necessarily matter at all! Maybe you'd be willing to sacrifice some profits to keep control over your business, especially if it's a family business that has been passed down from generation to generation.

Effective problem solving requires considering the trade-offs that are inherent in how you allocate resources to achieve your objectives. Addressing multiple objectives may require techniques that are quite different from standard single-objective optimization methods. As indicated above, if there are even two objectives to be optimized, it might be possible to find a solution that is best with respect to the first objective but not the second, and another different solution that is best with respect to the second objective but not the first. Multiple objective problems pose the challenge of defining the quality of a solution in terms of several, possibly conflicting, parameters.[1]

---

[1]One of the authors (Z.M.) had a head-on collision a few years ago while driving his convertible Porsche 911. The accident was "optimal" on many dimensions: (1) the object he hit was optimal — it was a tree, thus no one was hurt despite there being other cars on the road, and just a few meters behind the tree was a wall of a house so the tree was extremely well positioned, (2) the speed of the impact was optimal — it took four weeks for the insurance company to decide to total the car because it was a borderline decision — if the speed at impact had been just a bit lower, the car would have been repaired and who wants a repaired 911 when you can have a new one? — and if the speed at impact had been just a bit higher, the driver might not have been quite so fortunate, and (3) the timing of the accident was optimal — it was November in North Carolina, and who needs a convertible during the winter, anyway? Of course, these dimensions represented non-conflicting objectives!

Whenever you face multiple conflicting objectives you are involved in what's known as multicriteria decision-making. In such cases, you can seek to find a way to aggregate the contribution of each objective to a broader overall objective. Then you can find the optimum solution for that overall objective, or you can seek out solutions that are optimal in a different sense that treats objectives somewhat independently, where a solution is optimal if there is no other solution that is better across all concerns. Let's take a look at this concept with a simple example.

Suppose someone puts you in charge of a U.S. toy company. You want to minimize the cost of producing your toys and you also want to minimize the time it takes to make the toys. These two objectives are in conflict. You could lower the cost by exporting much of your assembly operations to a foreign country in Southeast Asia where the labor rates are cheap. This, however, would increase the time involved because it would incur extra weeks of shipping goods back and forth across the ocean. Instead, you could assemble everything in the United States and save the time, but then you'd have to pay higher labor rates and incur other costs involved with payroll taxes, worker's compensation insurance, and so forth. Figure 15.1 illustrates the point. There are two different solutions, **x** and **y**, such that the first is better with respect to *time* and the second is better with respect to *cost*. Neither is better than the other with respect to both objectives.

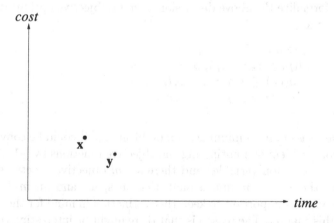

Fig. 15.1. Two "optimal" solutions: **x** and **y**

Let's explore the concept of optimality using this example. We can see that the solutions **x** and **y** are not comparable directly. Without creating some way of translating between *cost* and *time*, it's impossible to state which of these solutions is better. But is one or the other of the solutions **x** and **y** "optimal" in some sense?

It's intuitive that a particular solution **x** "dominates" an area of the search space in the sense that any solution **z** from this dominated area is inferior to **x** with respect to both objectives (see figure 15.2). Clearly, **z** can't be "optimal" in any reasonable sense of this term when compared to **x**. On the other hand, if for a solution **x** we find that there's no other solution that is better with respect to all of the objectives of the problem, then solution **x** is "optimal" in some sense.

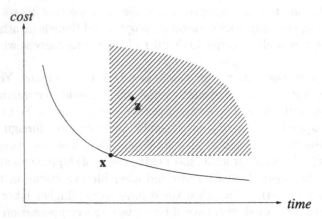

Fig. 15.2. A solution **x** generates a dominated (shaded) area. Any solution in this area, such as **z**, is a dominated solution.

We can formalize the above discussion. A multiobjective optimization problem is defined as

minimize $f_i(\mathbf{x})$, $i = 1, \ldots, m$,
subject to $l(p) \le x_p \le u(p)$, $p = 1, \ldots, n$,
$\quad\quad\quad\quad g_k(\mathbf{x}) \le 0$, $k = 1, \ldots, l$,
$\quad\quad\quad\quad h_j(\mathbf{x}) = 0$, $j = 1, \ldots, q$.

Here, we're assuming a minimization problem, and it could be converted to a maximization problem by multiplying the objective functions by -1. A solution vector **x** has n decision variables and there are m objective functions, from f_1 to f_m. Thus there is an n-dimensional decision space and an m-dimensional objective space. It's important to keep this distinction in mind for the remaining sections of this chapter. The reason is that there might be interesting (nonlinear) mappings between these two spaces. For example, when we discuss some aspects (e.g., diversity) of a multiobjective optimization method, it will be important to discern whether we are speaking about the diversity of points in the decision space or in the objective space.

Now, let's say that we a have a set of potential solutions to a multiobjective optimization problem. As discussed above, it might be convenient to classify these solutions into *dominated* solutions and *nondominated* solutions. A solution **z** is dominated if there exists a feasible solution **x** that is (1) at least as good as **z** with respect to every dimension, i.e., for all objectives f_i ($i = 1, \ldots, m$)

$$f_i(\mathbf{x}) \le f_i(\mathbf{z}) \text{ for all } 1 \le i \le m$$

and (2) strictly better than $\mathbf{z}$ on at least one objective i, i.e.,

$$f_i(\mathbf{x}) < f_i(\mathbf{z}) \text{ for some } 1 \leq i \leq m.$$

Any solution that's not dominated by any other feasible solution is called a *nondominated* solution. Again, figure 15.2 illustrates the case, where solution $\mathbf{z}$ is dominated.

Now we're able to discuss the issue of optimality. The nondominated solution set of the entire feasible search space is called the *Pareto-optimal* set, named after Italian economist Vilfredo Pareto (1848–1923). As solutions in this set are not dominated by any other feasible solution in the whole search space, these solutions are in this sense optimal for the multiobjective optimization problem.[2] A Pareto-optimal set must be a nondominated set, but the opposite need not be true. A nondominated set may contain some Pareto-optimal solutions and some non-Pareto-optimal solutions. The task of any algorithm for finding a Pareto-optimal set is to return a set of nondominated solutions (there are some efficient procedures that determine the nondominance property of any set of solutions [98]) that approaches the Pareto-optimal set as closely as possible. All Pareto-optimal solutions might be of some interest and, ideally, the system should report back the set of all Pareto-optimal points.

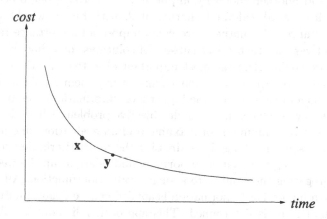

Fig. 15.3. The trade–off between *cost* and *time*

Following our earlier example, figure 15.3 illustrates the trade-off between *cost* and *time*. As before, we seek to minimize two objectives at once: *cost* (f_1) and *time* (f_2). The continuous line in the figure represents the Pareto-optimal front, which displays the trade-off between these objectives. For any solution on the front, an improvement with respect to *cost* will result in increased *time*, and

---

[2]For deeper studies on multiobjective optimization it's worthwhile to also define a local Pareto-optimal set where for each $\mathbf{x}$ in the set there is no dominating solution in the neighborhood — as opposed to the whole search space — of $\mathbf{x}$, as well as a concept of *strong* dominance, where a solution is strictly better on all objectives, but these are outside the scope of this chapter.

vice versa. Solutions **x** and **y** on this line represent two of many nondominated (Pareto-optimal) solutions: **x** does a better job of optimizing *time*, whereas **y** is more effective at minimizing *cost*.

Now it comes to making a decision. We know that there's a set of nondominated solutions (the complete Pareto-optimal set). Each of the solutions in this set might be of interest. The problem is that in most cases we're only able to implement one solution. Think back to the example where you are running a toy manufacturing company. You might have four potential solutions that appear to be Pareto-optimal: (1) Send 100% of your inventory overseas, (2) send 77% of your inventory overseas, (3) send 22% of your inventory overseas, and (4) keep everything in the United States. Well, you certainly can't send everything overseas and also keep everything in the United States. In fact, you can only choose one of these solutions. It's all very well to say they are all "optimal" in some sense, but you need a means for focusing on one of these solutions.

To find this means, we have to look beyond the formulation of dominated and nondominated solutions to include additional higher-level information. For example, we may assert relative importance to each objective, perhaps by assigning numeric weights or fuzzy linguistic terms. One objective may be "very important" and another "not very important," or perhaps one objective may be rated at 9 on a 10-scale and another at 2, with higher weight indicating greater importance. Alternatively, we could impose an importance ranking for all the objectives and then select subsets of solutions following this ordering. Or perhaps we could select the most important objective and then convert all the other $m - 1$ objectives into constraints, using them as thresholds to be satisfied. In each of these cases, the higher-level information is used to convert the multiobjective problem into a single objective problem, where the goal then becomes to find the minimum (or maximum) of an evaluation function. In section 15.1, we discuss methods for addressing the multicriteria decision-making problem by assessing the relative importance and relationships between the objectives, aggregating the result into a single evaluation function. Alternatively, the time when higher-level information is applied can be delayed until after the Pareto-optimal front is determined. This approach relies on first finding the Pareto-optimal front and is discussed in section 15.2.

15.1 Single numeric approach

There are classic methods for converting multiobjective optimization problems into a single objective formulation. Undoubtedly the best-known of these is the simple weighted-sum method, where evaluation functions f_i are developed for each objective and a numeric value is assigned to each individual evaluation function. The functions are then combined into one overall evaluation function F:

$$F(\mathbf{x}) = \sum_{i=1}^{m} w_i f_i(\mathbf{x}).$$

Typically, the weights $w_i \in [0..1]$ and $\sum_{i=1}^{m} w_i = 1$. In this linear case, any values of the weights (assumed positive) can be normalized without changing the solution $\mathbf{x}$ that maximizes or minimizes $F(\mathbf{x})$.

So where do the weights come from? This is a key question! The answer is that the weights are a subjective reflection of the relative importance of the objectives as determined by the problem solver. If you ask different people what weights should be used, you'll likely get different answers. That's the way it should be, because different things are important to different people. But it's important to keep sight of the numeric values involved. Some objectives may be measured in terms of dollars, others in terms of time, and still others in terms of distance or otherwise disparate measures. Later we'll see a method that provides a structure for handling these disparate items, but for now just know that you will likely have to massage your evaluation functions f_i in order to make them commensurate with each other in light of the assigned weights.

The linear sum (also known as the arithmetic mean) method for aggregating multiple objectives is attractive primarily because it is simple. It's also attractive because the contributions from the different functions f_i are linear. An increase in the value of one function f_i leads to a proportional increase in the overall evaluation. Moreover, connecting back to Pareto optimality, the optimum feasible solution for F is also Pareto-optimal if all the weights are positive. What's more, if $\mathbf{x}$ is a Pareto-optimal solution for a *convex* multiobjective optimization problem, then there exists a nonzero positive weight vector $\mathbf{w}$ such that $\mathbf{x}$ is a solution for F [321].

But if you are interested in the Pareto-optimal solutions, then there are some disadvantages. It might be difficult to set the weights in a way to produce a Pareto-optimal solution in a desired area of the objective space. For a specific weight vector there might be several minimum solutions. Each of these may represent a different Pareto-optimal solution and some of these solutions might be dominated by others (thus wasting search effort). Also, for nonconvex objective spaces (see figure 15.4) there might be segments of Pareto-optimal points that can't be generated by weighted linear sums.

Another method — the method of distance functions — combines multiple evaluation functions into one on the basis of a demand-level vector $\mathbf{y}$:

$$F(\mathbf{x}) = (\sum_{i=1}^{m} |f_i(\mathbf{x}) - y_i|^r)^{\frac{1}{r}},$$

where (usually) $r = 2$ (Euclidean metric). Sometimes the method includes weights as well:

$$F(\mathbf{x}) = (\sum_{i=1}^{m} w_i |f_i(\mathbf{x}) - y_i|^r)^{\frac{1}{r}},$$

being called the weighted metric method, or target vector approach [170]. When a large r is used, the problem reduces to the problem of minimizing the largest deviation $|f_i(\mathbf{x}) - y_i|$ and is called the weighted Tchebycheff problem. (Minimizing the maximum of n objectives is also sometimes termed the "minimax

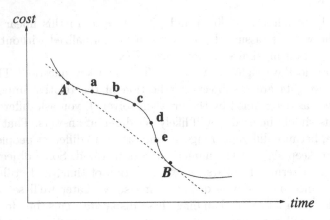

Fig. 15.4. Some Pareto-optimal solutions (**a**, **b**, **c**, **d**, and **e**) are "hidden" in a nonconvex objective space.

approach" [170]). Under this formulation, the optimum solution is the one that minimizes the distance (or a surrogate function) between $F(\mathbf{x})$ and $\mathbf{y}$.

Yet another possible approach — called the ϵ-constraint method — is based on the idea of keeping just one of the objectives (possibly the most important one), say, f_r, and converting the other objectives to constraints:

$$f_i(\mathbf{x}) \leq \epsilon_i, \text{ for } 1 \leq i \leq m \text{ and } i \neq r.$$

The parameters ϵ_i represent the upper bounds for the values f_i; intuitively they represent requirements for each objective. The main advantage of this approach is that it can be applied for both convex and nonconvex problems. Also, the solution of the ϵ-constraint problem is Pareto-optimal for any given upper bound vector. The main disadvantage is, as usual, in complexities connected with selecting the ϵ_i's.[3]

It's also possible to use goal programming methods [393] for multiobjective optimization problems. As with some of the methods identified above, these seek solutions that aim at predefined targets (these might be of different types: inequalities, equalities, or ranges) for some evaluation functions. If no solution hits the targets on all evaluation functions, the problem becomes one of finding solutions that minimize deviations from the targets:[4]

minimize $\sum_{i=1}^{m} \alpha_i p_i,$

where the parameters α_i are weighting factors for (positive) deviations p_i on the i-th objective, subject to

---

[3]A loosely related idea is called a "lexicographic" approach, which uses priorities on each objective and uses the objective with the highest priority first, with ties being decided by the second-highest priority objective, and so forth in succession [170].

[4]For "less-than-equal-to" targets.

$$f_i(\mathbf{x}) - p_i \le t_i,$$

and t_i's represent targets.

Frankly, each of the above methods is overly simplistic and does not fit the framework of multicriteria decision-making in real-world circumstances very well. Each of the methods requires considerable adjustments to "fit" a given situation, which is not unlike the remarks we've made about fitting linear methods to nonlinear problems.

People have studied multicriteria decision-making for many decades, going back at least to the seminal work of Keeney and Raiffa [255], and even von Neumann and Morgenstern [481]. Keeney and Raiffa [255] presented the concept of structuring a multicriteria decision problem from a top-down perspective of viewing the objectives first at the highest level, and then at successively lower sublevels, building a hierarchical structure. Fogel [162] described a similar approach, termed the "Valuated State Space" approach, that can provide a very flexible method for addressing multicriteria decision-making without regard to Pareto-optimality considerations.

15.1.1 The Valuated State Space approach

The Valuated State Space approach begins by asking people with authority over the domain in question to identify the parameters of concern. For example, at the beginning of this chapter, we discussed the decision of buying a car, which included parameters such as cost, aesthetics, comfort, utility, and safety. Let's limit it there. Following Miller [322] there are usually 7 ± 2 different parameters of concern for any real-world problem, at least as people perceive the problem. Here, we have five parameters and these form the first level of the Valuated State Space.

Before considering any sort of relative importance of the parameters, they must each be explicated into subparameters, and possibly subsubparameters, and so forth, until a specific degree of achievement can be made measurable. For example, with respect to cost, what are the elements of cost? There's the sticker price, but that's just window dressing on the car really, so we can ignore that. There's the negotiated price — now that's a real cost. What else? What about financing? Most people don't pay cash for their car but instead take out a loan for a period of time. So cost involves monthly payments and a total duration of making those payments. What else? How about insurance? The owner of the Porsche 911 Carrera might have to spend quite on bit on insurance. Then there are license fees to the state government and perhaps also luxury taxes that we can group together under the heading of "extras." And then there's depreciation, which describes the amount of money you'll likely get for your car when you sell it, and you'll need to have that in mind even before you buy the car in the first place. There are other things such as operating costs for fuel and servicing, and maybe even parking fees and other costs, but for the sake of our example, let's just leave it there.

Can we measure each of these subparameters? Let's see. The negotiated price is measurable. The financing isn't measurable, but we can make that measurable by further dividing it into subsubparameters of total costs of interest payments and the term period of the loan. The price of insurance is measurable. The so-called extras are measurable. Depreciation isn't measurable directly, but it can be estimated directly. So from a hierarchic view, we have the following:

1.0 Cost

    1.1 Negotiated Price

    1.2 Financing

        1.2.1 Interest Rate

        1.2.2 Term of Loan

    1.3 Insurance

    1.4 Extras

    1.5 Depreciation

Since each of the lowest-level parameters is measurable, the next step is to define the degrees of achievement on each parameter that make a difference to you. If you are a mathematician, you might be tempted to look for a continuum of degrees of achievement. For example, with respect to the negotiated price, it would range from "free" to some amount greater than, say, $80,000. But the worth of being somewhere in this range does not scale linearly with the price. Psychologically, people tend to have thresholds that set up class intervals — sometimes these are called "pain thresholds" — that form discrete segments that define alternative states in the Valuated State Space. For instance, with regard to negotiated price, we can imagine someone's class intervals might be "less than $5,000," "between $5,000 and $10,000," "between $10,000 and $20,000," "between $20,000 and $30,000," "between $30,000 and $40,000," "between $40,000 and $80,000," and "greater than $80,000." These intervals reflect concepts that are significant to the buyer, and the fact that $35,000 is $4,000 more than $31,000 doesn't make any substantive difference to the buyer. You may find this incredible, but it's human psychology. With the class intervals defined, it's possible to assign degrees of achievement values to each interval, usually on a 10-scale, where 10 is the best score and 0 is the worst score. Here, we can imagine the degrees of achievement being assigned as

10	"Less than $5,000,"
9	"Between $5,000 and $10,000,"
7	"Between $10,000 and $20,000,"
5	"Between $20,000 and $30,000,"
3	"Between $30,000 and $40,000,"
2	"Between $40,000 and $80,000,"
0	"Greater than $80,000."

The assignment of zero points to the last category would be symptomatic of someone who could never afford an $80,000 car.

With this method of assigning values to class intervals, there is no problem of combining some objectives that are best to minimize with others that are best to maximize. Each can be written such that the optimum class interval receives the highest score. There is also no problem of scaling evaluation functions or normalizing across functions that involve different types of units, such as distance and time. What remains is to assign relative importance weights to the parameters at each level that reflect an individual's subjective beliefs about each objective.

A complete Valuated State Space with respect to cost for a person buying a car might look like this:

10	1.0	Cost	
	10	1.1	Negotiated Price
		10	Less than $5,000
		9	Between $5,000 and $10,000
		7	Between $10,000 and $20,000
		5	Between $20,000 and $30,000
		3	Between $30,000 and $40,000
		2	Between $40,000 and $80,000
		0	Greater than $80,000
	4	1.2	Financing
		8	1.2.1 Interest Rate
		10	0%
		9	Between 0% and 1%
		7	Between 1% and 3%
		4	Between 3% and 5%
		1	Greater than 5%
		7	1.2.2 Term of Loan
		10	No term
		9	Less than 1 year
		8	1, 2, or 3 years
		7	4 or 5 years
		4	More than 5 years
	3	1.3	Insurance
		10	Under $100 per year
		9	Between $100 and $300 per year
		6	Between $300 and $500 per year
		4	Between $500 and $1,000 per year
		1	More than $1,000 per year

1	1.4	Extras
3		1.4.1 License Fee
10		Less than 1% of the negotiated price
8		Between 1% and 3% of the negotiated price
4		More than 3% of the negotiated price
10		1.4.2 Luxury Tax
10		No tax
1		Yes, the car qualifies for luxury tax
8	1.5	Depreciation (Predicted Retained Price at Sale)
10		More than 90% of its negotiated price
8		Between 70% and 90% of its negotiated price
6		Between 50% and 70% of its negotiated price
4		Between 30% and 50% of its negotiated price
2		Between 10% and 30% of its negotiated price
0		Less than 10% of its negotiated price

To construct the entire Valuated State Space, each of the other main parameters of concern (aesthetics, comfort, utility, and safety) would also need to be explicated in a similar fashion.

Once all of the degrees of achievement, class intervals, and relative importance weights are in place, the next step is to consider whether or not any parameter is *critical*. This has a special meaning: a parameter is said to be critical when a failure to gain any degree of achievement with respect to that parameter negates the contributions of all other parameters. A necessary but not sufficient condition for a parameter to be critical is that the degree of achievement value for the worst class interval must be zero. An obvious example of a critical parameter for a car would be the negotiated price: if you can't pay the seller's price, it doesn't matter how good the car's performance is, what color it is, or how many surfboards it can carry. You aren't buying this car!

If there is only one critical parameter, this fact can simply be noted with the understanding that if the worst-case outcome is realized, the overall worth for the situation, no matter what the other class intervals indicate, will be zero. If more than one parameter is critical, this has implications that will be discussed shortly.

The challenge of making the optimum decision with respect to all of the objectives when the problem is formulated as a Valuated State Space is to determine which decision — which allocation of resources — places you in the best set of states (class intervals) in the Valuated State Space. Determining this requires a normalization function to aggregate the impact of each measurable parameter, starting at the lowest levels. This normalization could be a simple arithmetic mean, or in the case of critical parameters, it could be a geometric mean (a weighted product function). Let's calculate one example just with respect to cost, from above.

Suppose that you went to the car dealer and found a nice new four-door sedan. This car might be described as indicated by the "X" marks:

```
10   1.0   Cost
     10    1.1   Negotiated Price
                 10    Less than $5,000
                 9     Between $5,000 and $10,000
           X     7     Between $10,000 and $20,000
                 5     Between $20,000 and $30,000
                 3     Between $30,000 and $40,000
                 2     Between $40,000 and $80,000
                 0     Greater than $80,000

     4     1.2   Financing
                 8     1.2.1  Interest Rate
                       10    0%
                 X     9     Between 0% and 1%
                       7     Between 1% and 3%
                       4     Between 3% and 5%
                       1     Greater than 5%

                 7     1.2.2  Term of Loan
                       10    No term
                       9     Less than 1 year
                 X     8     1, 2, or 3 years
                       7     4 or 5 years
                       4     More than 5 years

     3     1.3   Insurance
                 10    Under $100 per year
                 9     Between $100 and $300 per year
           X     6     Between $300 and $500 per year
                 4     Between $500 and $1,000 per year
                 1     More than $1,000 per year

     1     1.4   Extras
                 3     1.4.1  License Fee
                       10    Less than 1% of the negotiated price
                 X     8     Between 1% and 3% of the negotiated price
                       4     More than 3% of the negotiated price

                 10    1.4.2  Luxury Tax
                 X     10    No tax
                       1     Yes, the car qualifies for luxury tax
```

8	1.5	Depreciation (Predicted Retained Price at Sale)	
	10	More than 90% of its negotiated price	
	8	Between 70% and 90% of its negotiated price	
X	6	Between 50% and 70% of its negotiated price	
	4	Between 30% and 50% of its negotiated price	
	2	Between 10% and 30% of its negotiated price	
	0	Less than 10% of its negotiated price	

If none of the parameters is critical, an arithmetic mean can be used. The calculation must proceed from the lowest level upward. So first we focus on 1.2 and see that its constituent parameters at levels 1.2.1 and 1.2.2 have degrees of achievement of 9 and 8 respectively, with relative importance weights of 8 and 7, respectively. The contribution then to 1.2 is determined by taking the ratio of the degree of achievement attained to the maximum possible degree of achievement in each parameter and weighting that by the relative importance weights. Here, the calculation is

$$(9/10)(8/15) + (8/10)(7/15) = 0.480 + 0.373 = 0.853,$$

which is then rescaled to the 10-scale as 8.53. Note the value of 15 is the sum of 8 and 7, the respective relative importance weights. Thus 8.53 is the contribution from 1.2. Similarly, we compute the contribution from 1.4, based on 1.4.1 and 1.4.2, as

$$(8/10)(3/13) + (10/10)(10/13) = 0.185 + 0.769 = 0.954,$$

or 9.54 on the 10-scale. These can then be aggregated at the next level in the hierarchy:

$$(7/10)(10/26) + (8.53/10)(4/26) + (6/10)(3/26) + (9.54/10)(1/26) + (6/10)(8/26) = 0.691,$$

or 6.91 on a 10-scale. Note that this is the contribution to the parameter "cost," and the overall value of choosing this car would be computed by aggregating the contributions from all other parameters as well. Two alternative cars could then be compared in terms of how well they satisfy the overall multicriteria decision by comparing their single overall evaluations.

If all the parameters were critical at the first level of the hierarchic Valuated State Space,[5] rather than using the weighted arithmetic mean, the weighted geometric mean could be used. This uses the product of each contribution, raised to the scaled relative importance weight:

$$\prod_{i=1}^{m} \text{Contribution}_i^{w_i},$$

---

[5]Note that a critical parameter does not necessarily have to have critical subparameters. Criticality is a feature of a parameter at a particular level in the hierarchy. Also note that the example here is purely pedagogical, as we are hypothesizing criticality to parameters where not all have the potential to return a zero payoff, which is a necessary condition.

where there are m parameters and w_i is the relative importance, scaled between zero and one and summing to 1.0 over all weights. In this way, if the contribution from any parameter is zero, the product function makes the entire score go to zero. Under the geometric mean, the sedan above would score

$$(0.7)^{(10/26)} \times (0.853)^{(4/26)} \times (0.6)^{(3/26)} \times (0.954)^{(1/26)} \times (0.6)^{(8/26)} = 0.684.$$

The Valuated State Space provides a convenient structure for assessing parameters in terms that people are familiar with, allows individuals to apply subjective relative importance weights, and provides a mechanism for dealing with degrees of criticality of parameters. It can also be extended to handle purely linguistic descriptions of class intervals and fuzzy arithmetic for aggregating the contribution of each parameter to an overall assessment. Furthermore, it can be extended to describe the relationship between two or more players in a game, where the overall degree of success depends on the individual successes of the players. The Valuated State Space approach provides a rank ordering of all possible outcomes and rapid comparison of two or more potential decisions to determine which is better.

15.1.2 Fuzzy multicriteria decision-making

In chapter 13 we saw that precise or so-called *crisp* descriptions of parameters can be viewed as one extreme of a fuzzy description. Fuzzy logic can be applied in multiple ways to the problems of multicriteria decision-making. If you say that the relative importance of a particular parameter is, say, 9 on a 10-scale, you could alternatively say that the parameter in question is "very important," and you could construct a fuzzy membership function that describes the set of very important things, and less important things too. What's more, you could think about a Valuated State Space and the class intervals on a parameter. Those class intervals don't have to be separated by crisp lines of distinction. Instead, the intervals could be fuzzy. If you're buying a car, and price is a consideration, instead of using discrete intervals of, say, less than $10,000, between $10,000 and $30,000, and more than $30,000, you could have intervals of "cheap," "moderately priced," and "expensive." Any particular price might then have positive membership in more than one class. A car priced at $27,500 might have membership of 0.8 in the set of "moderately priced" cars, and 0.3 in the set of "expensive" cars. The methods of defuzzifying that we discussed in chapter 13 could then be applied to arrive at a single overall descriptor.

Fuzzy logic can also be applied to describe the aggregation of various parameters. The most typical aggregation function is the simple arithmetic mean, but we also saw that a geometric mean can be appropriate when treating parameters that are *critical*. An alternative aggregation function can be derived from a fuzzy technique called *ordered weighted averages* or OWA, which was offered in Yager [502] and has received considerable attention (see [177, 214, 139] and many others). An OWA operator is a mapping $F : R^n \to R$ that has an associated vector of length n, typically denoted by $\mathbf{w}$, where each w_i, $i = 1, \ldots, n$, is in the range of $[0, 1]$, with the sum of all $w_i = 1$, and

$$F(a_1, \ldots, a_n) = \sum_{j=1}^{n} w_j b_j,$$

where b_j is the j-th largest element from $\mathbf{a}$.

The OWA operator does not apply a fixed weight to any particular entry in $\mathbf{a}$ before we know the order of the elements in $\mathbf{a}$. For example, if $\mathbf{w} = (0.6, 0.2, 0.1, 0.1)^T$ and $\mathbf{a} = (0.9, 0.2, 0.8, 0.5)$, where $\mathbf{a}$ might represent the contribution from four different parameters, then

$$F(0.9, 0.2, 0.8, 0.5) = 0.6 \times 0.9 + 0.2 \times 0.8 + 0.1 \times 0.5 + 0.1 \times 0.2 = 0.77.$$

Note the reordering of the elements in $\mathbf{a}$. The OWA operator provides the most weight to the greatest element of $\mathbf{a}$, the next-most weight to the next-greatest element, and so forth. If $\mathbf{a}$ is ordered from greatest to least, then it is simple to see that the OWA operator can be manipulated to provide fuzzy max, min, and simple average operators, by setting $\mathbf{w}$ to $(1, 0, \ldots, 0)^T$, $(0, 0, \ldots, 1)^T$, or $(1/n, \ldots, 1/n)^T$, respectively. Fuller [177] provides an interesting example of using expert interviews to assess the fuzzy degree of relative importance of parameters, and develop fuzzy quantifiers to combine the expert opinions, and you can find in-depth reviews of applications of fuzzy logic to multicriteria problems in [71, 140, 293, 405]. The concepts of fuzzy logic can also be applied to case-based reasoning [392] by matching known cases using fuzzy descriptors such as "slightly similar," "definitely similar," and so forth, and describing the utility of the cases based on their degree of dominance with respect to one or more criteria with fuzzy terms such as "slightly useful" or "very useful." San Pedro and Burstein [406] provide an application of fuzzy case-based reasoning for multicriteria decision support in weather prediction.

15.1.3 Summary

We've seen many advantages and also disadvantages to aggregating many objectives into a single scalar value. The advantages include ease of comparisons, systematic structuring of objectives, and the ability to reflect in a Valuated State Space individual subjective opinions regardless of relative importance, degrees of criticality, and class intervals that make a difference. Some of these advantages also pose other challenges. For example, when two or more people are involved in decision-making, they may not agree on the relative importance parameters. And if it's desired to determine which solutions are Pareto-optimal, many of the single aggregate methods will not be able to offer an answer. Instead, it may be desirable to discover an assortment of potential solutions from the Pareto-optimal set and then consider each to determine which to implement. This shifts the focus from finding extrema of aggregate functions to finding collections of diverse solutions that all share a place on the Pareto front.

15.2 Evolutionary multiobjective optimization approach

When you think about finding a collection of points on a front, you should immediately think about using evolutionary algorithms. It's natural, because evolutionary algorithms process collections of points inherently. The trick is to craft an evolutionary algorithm that will spread out its candidate solutions along the Pareto front, rather than seek out a single "best" solution. All Pareto-optimal solutions are equally important at this stage in the decision-making, as it is presumed that there is no higher-level information that's available for assessing the relative trade-offs of the objectives. The criteria of concern in this approach are finding solutions as close to the Pareto-optimal front as possible while also finding as diverse a collection of solutions as possible.

Figure 15.5 illustrates the point. There are two sets, (a) and (b), of Pareto-optimal front solutions but each has very different diversity. We would want to favor (a) over (b) because it represents a better "cognizance" over the Pareto front. It's important to remember, however, that the set (a) of Pareto-optimal front solutions of figure 15.5 is diverse in the objective space only. It's often the case that diversity in the objective space implies diversity in the decision space, but for complex and highly nonlinear problems it need not be the case. So it may happen that the diversity of the vectors **x** of decision variables is very different than the diversity of corresponding solutions in the objective space.

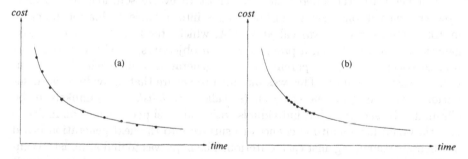

Fig. 15.5. Two sets of Pareto-optimal front solutions. The diversity of (a) is "greater" than the diversity of (b).

Evolutionary algorithms have become established as *the method* for exploring the Pareto-optimal front of solutions for challenging multiobjective optimization problems. Of course, you can use evolutionary algorithms for all of the optimization problems posed in the previous section too. But in the case of searching for a population of diverse points with desired properties (nondominated diverse solutions), evolutionary algorithms are the natural choice. Instead of requiring multiple runs from different initial starting positions in the hopes of landing at diverse points on the front, evolutionary algorithms can be crafted to accomplish this task in a single run. Let's look a bit at the history behind applying evolutionary computing to Pareto optimization.

15.2.1 Brief history

It's difficult to precisely categorize all the different evolutionary approaches to multiobjective optimization. A wide variety of approaches has emerged over the last 20 years. But broadly, evolutionary algorithms can be applied both in the case where a multiobjective problem is converted into a single–objective formulation and also to the case where a search is made with respect to Pareto optimality directly, without using any form of *aggregation selection* to combine the assessments with respect to individual objectives.

When using the aggregate method, parameters can be varied systematically to find a set of Pareto-optimal solutions. But an alternative is to use evolutionary algorithms to search for the Pareto-optimal solutions directly, within a single population and a single run of the algorithm. These two alternatives can be described as "non-Pareto" and "Pareto-based" approaches, respectively. In "non-Pareto" optimization, which is also termed *criterion selection* because the algorithm will switch between objectives during selection, the individual objective values of a solution (i.e., $f_1(\mathbf{x}), f_2(\mathbf{x}), \ldots, f_m(\mathbf{x})$) are used to determine which parents to select and how to replace parents with offspring. In "Pareto-based" methods, which are also called *Pareto selection*, the individuals in the evolving population are ranked based on their qualities of dominance and Pareto optimality.

Schaffer [410, 411] offered the first effort to evolve solutions across multiple criteria without aggregating objectives into a single evaluation function in 1984. This approach was called VEGA, which stood for "vector evaluated genetic algorithm." Given a problem with m objectives, VEGA's central idea was to select $1/m$ of the parents at each generation using each objective in turn, in each generation. This was intended to ensure that individuals with superior performance on one criterion (so-called *specialists*) were unlikely to be eliminated by selection, but individuals with mid-level performance on multiple criteria would have multiple chances to survive into the next generation based on their assessment against each criterion. The hope was to find a robust "Wunderkind" solution that was good with respect to every objective but wouldn't lose specialists along the way.

Schaffer [410] studied the problem of evolving a classifier system for a multiclass pattern discrimination task.[6] Prior efforts with a traditional evolutionary algorithm generated cases where some good discrimination rules for each parent class would be created in different individuals, even in early generations. Unfortunately, these rules could never be combined successfully by evolution into a Wunderkind. Why? Because the selection criterion favored classifiers based on the total number of correctly classified cases. If one individual inherited useful

---

[6]It was a simple problem that involved EMG signals reduced to 12-bit patterns. The interpretation was that each of six muscles in the human leg was either active or inactive (6 bits) during the stance and swing phases (times 2) of the gait cycle. The classes were "normal" and four classes of abnormal gait. Schaffer chose that problem because (1) it was small and (2) the inputs were bits (natural for classifier systems).

rules for, say, classes 1 and 2, it could displace another individual who had useful rules for, say, class 3, but not for classes 1 and 2. The former classifier could appear twice as fit as the latter, but the population would then select against the specialist before any successful recombination could take place. VEGA's selection mechanism was intended to protect specialists while providing more offspring to individuals that were competent across multiple objectives.

Schaffer realized that his selection scheme opened the door to multiobjective optimization with evolutionary algorithms. He created some simple multiobjective problems (two of them are presented further in section 15.2.2) to study VEGA's dynamics and determine its utility. This effort led to some heuristics to guide a design. For example, one heuristic was that the population size should increase with the number of objectives. The idea here was that sufficient exploration of the objectives would be supported by increased variation, and that would stem from a larger population. In other experiments that studied selection preferences for nondominated solutions, this was found to lead to rapid convergence to suboptimal individuals. VEGA was also observed to have a problem with *speciation*, where individuals in a population tend to excel at different criteria, and this may prevent finding more robust solutions. One possibility for overcoming this speciation was to crossbreed between individuals of these different species, as opposed to recombining randomly.

For historical accuracy we should also mention the work of Fourman [173], who around the same time as Schaffer investigated a rather similar idea. Fourman first tried lexical selection using a prioritized list of criteria, but this worked poorly for his circuit design problem. Finally he hit on the idea of randomly choosing a criterion whenever a selection had to be made. This scheme worked much better than a prioritized list of criteria.

A few years later, Goldberg [187] suggested an evolutionary algorithm that scored solutions based on dominance rather than with respect to values assessed with respect to individual objectives. The proposed method would operate in an iterative fashion by first identifying all solutions that were nondominated. These solutions would then be all assigned the same score and removed from consideration. Then the process would again determine, from the solutions remaining, which solutions were nondominated. These would all be assigned the same score, which would be worse than that assigned to the solutions removed in the prior iteration, and then removed from consideration. This repetitive process would be executed until every solution received a score and then selection would be applied based on these dominance scores. We can visualize this process as peeling an onion: each layer corresponds to the current nondominated set of solutions. Figure 15.6 illustrates the idea.

Goldberg [187] indicated the importance of preserving the diversity of solutions, either in the solution space or the objective space, and suggested a sharing technique (see section 11.2.1) to prevent the evolutionary algorithm from converging to a single solution at the Pareto-optimal front and to maintain diverse solutions across the nondominated frontier. For example, in figure 15.6, solutions x_1, x_2, x_3, and x_4 in the first nondominated front are not distributed

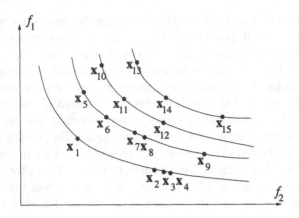

Fig. 15.6. Several solutions are classified into different nondominated fronts.

uniformly: solution x_1 is isolated whereas solutions x_2, x_3, and x_4 are clustered. Thus the role of fitness sharing would be to emphasize solution x_1, to prevent its disappearance in subsequent generations.

These two distinct approaches generated interest in the evolutionary computation community. From the early 1990s, many researchers started reporting on new approaches, first results, and new test problems. There were also more and more papers on applications of evolutionary techniques for solving real-world multiobjective optimization problems. This was hardly surprising as evolutionary computation techniques are not only appropriate for searching intractably large spaces (as demonstrated in earlier chapters of this text) but also capable of exploiting similarities between solutions, and can also approximate the whole Pareto-optimal front in a single optimization run.

Before this increased interest, the earliest efforts in evolutionary multiobjective optimization focused on the aggregate selection approach. These date back to the 1957 work of Box [53] who suggested an ϵ-constraint-based approach to his evolutionary operation method (before this approach was proposed formally years later) and who discussed how to determine which of the multiple objectives to select as the main criterion and which to reposition as constraints. In the early 1960s, L. Fogel [161, 167] studied a multiobjective problem, in which finite state machines were evolved to predict sequences of symbols. The multiobjective problem was to find finite state machines with superior predictive capability while also having minimum complexity. The aggregate function comprised a linear weighting of (1) the cumulative payoff for the predicted symbols in light of the symbols that were actually observed and (2) the number of states in the finite state machine.

In 1990, Fogel [143] extended this early work by applying it to a problem of system identification. The challenge was to find optimal autoregressive moving-average models (both with and without exogenous inputs, thereby denoted as ARMA or ARMAX models) based on observed data. Using information-theoretic criteria, Fogel [143] evolved models that were scored in terms of their goodness of fit to the observed data and their complexity in terms of degrees of freedom. Two information criteria were used: Akaike's information criterion (AIC) and the minimum description length (MDL) principle. These are common principles in model building that trade off the goodness of fit for model complexity, albeit with different mathematical forms for measuring complexity. Fogel extended this work in 1991 [144] to evolve neural networks in light of the aggregate selection of minimizing both model error and complexity.

The early 1990s saw many different applications of evolutionary multiobjective optimization with aggregate selection. For example, Syswerda and Palmucci [462] used it for evaluating schedules, and Jacob et al. [236] used it for task planning. Wienke et al. [497] applied an evolutionary approach to a problem of optimizing intensities of atomic emission lines of trace elements, using a goal programming method (also an aggregation selection approach). Ranjithan et al. [377] used the ϵ-constrained method for solving a groundwater pollution containment problem. Kursawe [276] built a general multiobjective evolution strategy system based on lexicographical ordering, where at each step one of the objectives was selected randomly according to some probability vector, this being similar to the criterion selection experiments of Schaffer and Fourman mentioned earlier. Hajela and Lin [206] optimized a 10-bar plane truss in which weight and displacement were to be minimized using an approach based on VEGA; however, the weights of each objective were included in the data structure and diversity was maintained through fitness sharing and mating restrictions. In contrast to the aggregate selection method, Hilliard et al. [216] used a Pareto-optimality ranking method for a two-objective scheduling problem incorporating time and delay. And Liepins et al. [283] found that a Pareto-based approach gave better results than a non-Pareto-based approach for a few set covering problems.

Each of the efforts in the above paragraph were published in 1991 or 1992. Following that time, more efforts were directed toward the Pareto-based approach. For example, in 1993, three independent studies, each based on the Pareto approach, were reported.

Srinivas and Deb [448] proposed a technique called a nondominated sorting genetic algorithm (NSGA). The population is ranked on the basis of nondomination, and the solutions in the best set — those in the first front — are assigned a fitness that's equal to the number of individuals in the population. Sharing functions are used to maintain population diversity, where each front has its own sharing function. For each solution $\mathbf{x}$ in a front, the normalized Euclidean distances from all other solutions $\mathbf{y}$ in the *same* front are calculated. Also, the fitness value of the $\mathbf{x}$ solution is modified by the so-called niche count $nc_{\mathbf{x}}$ of $\mathbf{x}$,

$$nc_{\mathbf{x}} = \sum_{\mathbf{y}} sh(d(\mathbf{x}, \mathbf{y}))$$

(see section 11.2.1 for the definition of the sharing function sh). The original fitness of an individual is divided by its niche count.

The assignment of original fitness values (i.e., fitness values before modification) for the solutions in the second front is done only after the above modification of fitness is calculated for the first-front solutions; these original fitness values for the second-front individuals are lower than the lowest modified value of the first-front individual. Note also that the sharing distance is calculated in the decision space (as opposed to other methods that consider sharing functions in the objective space).

Clearly, as the solutions in the first nondominated front have the largest fitness values, the chances for their selection (a stochastic remainder proportionate selection was used) for reproduction are improved relative to other solutions. This allows the system to explore nondominated areas of the search space and to converge toward such an area. For the full discussion on the system and first experimental results the reader is referred to [448].

Horn and Nafpliotis [234] presented a variation called a niched Pareto genetic algorithm (NPGA), which uses a binary tournament selection based on Pareto dominance. During such a tournament selection, two randomly chosen parents are compared against a selected subpopulation of size s, i.e., each of the two selected parents is compared with every one of s solutions from the subpopulation. The result is that either one of the parents is a clear winner (i.e., it dominates all s solutions in the selected subpopulation whereas the other parent does not), or there is a draw (e.g., both parents dominate all s solutions, or both are dominated by at least one of s solutions from the subpopulation). In the latter case, both parents are placed into offspring population (which is filled during this process) and their niche counts are calculated: the parent with the smaller niche count is a winner.[7] The same procedure is followed for selecting the second parent and then the two selected parents reproduce to create an offspring. At this stage, the population of offspring is not empty and from this point forward, the niche count is used as a tie-breaker during the tournament.

As with NSGA, nondominated solutions are favored because of the dominance check with s solutions selected from the population. Moreover, diversity is encouraged as parents from less-crowded areas are preferred. For the full discussion on the system and first experimental results see [234].

Fonseca and Fleming [168] introduced MOGA, an acronym for a multiobjective genetic algorithm. This was probably the first approach (to our knowledge) that emphasized nondominated solutions together with procedures to promote diversity. Fitness was assigned to solutions as follows.

First, for each solution $\mathbf{x}$, the algorithm assigns a score that is equal to one plus the number of solutions in the population that dominate $\mathbf{x}$. Any nondominated solution is assigned the score 1, because there are no solutions in the population that dominate it. Note that in every population, there is at least one solution with score 1. Also, the maximum value of any score is equal to population size, *pop_size*. The algorithm then assigns the raw fitness values, explained shortly, using a mapping function based on the scoring function indicated above. Solutions with a score of 1 receive the highest values of raw fitness: if there are k_1 solutions with a score of 1, they receive raw fitnesses of

---

[7] At the beginning of the process, when the offspring population is empty, one of the parents is selected randomly.

$pop\_size, pop\_size - 1, \ldots, pop\_size - k_1 + 1$ (in arbitrary order). Then solutions
with a score of 2 receive the next group of high numbers as their raw fitness:
$pop\_size - k_1, pop\_size - k_1 - 1, \ldots, pop\_size - k_1 - k_2 + 1$ (if there are no solutions with score 2, the algorithm takes a group of solutions with the lowest
score). This procedure continues until all individuals are assigned raw fitnesses.
The next step of the algorithm changes raw fitnesses into average fitnesses of
all individuals: each individual with the same score is assigned the same average fitness, which is the average of all raw fitnesses of individuals with the
same score. For example, all individuals with the score of 1 will be assigned the
average fitness of

$$1/k_1 \sum_{i=1}^{k_1} (pop\_size - i + 1).$$

Thereafter, the shared fitness values are assigned by dividing the average fitness
of an individual by its niche count. Finally, these shared fitness values are scaled
such that the average fitness value of all individuals with the same original scores
remains the same as they were earlier, just after averaging their raw fitness.

As with the two other approaches described above, nondominated solutions
are favored because of a direct dominance check with all of the solutions in
the population and the score assignment. Moreover, diversity is encouraged as
parents from less-crowded areas are preferred because fitness values are modified
by the niche count. For the full discussion on the system and first experimental
results, see [168].

This research was followed by hundreds of papers on different aspects of
multiobjective optimization. For a detailed survey, see [74].

In section 7.4 we indicated that an *elitist* rule (where the best solution(s) in
the population are certain to survive to the next generation) can by applied in
evolutionary algorithms. The vast majority of recent research in evolutionary
multiobjective optimization has incorporated some form of elitist selection.

In single–objective optimization, the best individual is usually easy to identify once it's discovered. However, as we saw earlier in this chapter, in Pareto-
based multiobjective optimization, we deal with a set of best (i.e., nondominated) solutions rather than the best (single) solution. The straightforward
implementation of elitism in a purely multiobjective environment would force
at least all of the nondominated solutions in a population into the next generation. But elitism in multiobjective optimization should be handled with caution
because after several generations, particularly in the face of a large number
of objectives, most of the individuals in the population will be nondominated.
That means they all qualify as elite solutions. Applying elitist selection to all
of them would not aid in maintaining diversity, and this is especially true if a
diversity-reducing variation operator (e.g., crossover) is used. It's important to
control elitism in this case.

In the mid-1990s, the second wave of papers started to appear introducing
elitist evolutionary algorithms for multiobjective problems where researchers
handled the issue of controlling elitism in different ways. For example, Rudolph

[399] proposed an elaborate mechanism for identifying the most promising individuals for the next generation: nondominated solutions are found in the offspring population and compared one by one with all the individuals in the parent population (these two populations are examined separately). If an offspring solution dominates some (at least one) solutions from the parent population then (a) the offspring is placed in a set P_1 and (b) dominated parents are removed from the population. If an offspring solution does not dominate any parent solution and also is not dominated by any parent solution, it is placed in a set P_2. The intuition behind P_1 and P_2 is that these are initial sets of promising individuals for the next generation; individuals in P_1 are more promising than those in P_2, however. Finally, the next generation is created by picking individuals in some preference order. A similar approach was investigated in Deb et al. [99], with the difference that the populations of parents and offspring were combined before nondominated solutions were identified. Osyczka and Kundu [341] experimented with an approach where two populations are maintained: a standard population and an elite population, where all nondominated solutions found so far are kept. A special fitness assignment in this algorithm emphasizes solutions that are closer to the Pareto front and also maintains diversity among nondominated solutions. Similarly, Zitzler and Thiele [509] proposed an algorithm that also maintains an external population of nondominated solutions that have been found so far. However, this elite set also participates in variation to direct the regular population towards promising areas of the search space.

There have been many good surveys of evolutionary algorithms applied to multiobjective optimization. Fonseca and Fleming [169] provided an overview of techniques that combine many criteria into one evaluation function and return a single value as well as those that are based on Pareto optimality and return a set of values. Zitzler et al. [507] provided an empirical comparison of eight evolutionary algorithms for multiobjective optimization on six test problems; Veldhuizen [479] compared four algorithms on seven test problems, whereas Knowles and Corne [262] compared 13 algorithms on six test problems. These studies indicated that elitism can be very valuable in implementing a quality evolutionary algorithm for multiobjective optimization. Coello [74] gave a comprehensive survey of evolutionary multiobjective optimization methods. Recent books [75, 98, 340] and the above survey articles provide excellent overviews on classical and evolutionary approaches to multiobjective optimization.

15.2.2 Evolutionary algorithm for multiobjective NLP

The most-studied multiobjective optimization problem is probably the one proposed by Schaffer in his early study [410]. It's a single-variable test problem with two objectives, f_1 and f_2, formulated as follows:

minimize $f_i(x)$, $i = 1, 2$,
subject to $-A \le x \le A$,

where

$$f_1(x) = x^2, \text{ and} \tag{15.1}$$
$$f_2(x) = (x-2)^2.$$

Note that $\mathbf{x} = (x_1)$, as the problem has a single variable only, and to simplify the notation, we drop the subscript, i.e., instead of x_1 we write just x.

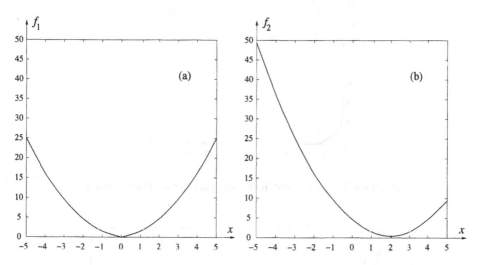

Fig. 15.7. Decision variable and two objectives: (a) f_1 and (b) f_2 for $A = 5$

Figure 15.7 displays the graphs for both objectives, f_1 and f_2, as functions of the decision variable x. It's clear that this problem has Pareto-optimal solutions for $x \in [0, 2]$: from -5 to 0 both functions decrease, between 0 and 2 the function f_1 increases while f_2 decreases, and between 2 and 5 both functions increase.

It is possible to plot the values of f_2 with respect to f_1:

$$f_2 = (\sqrt{f_1} - 2)^2, \tag{15.2}$$

by making a straightforward replacement of x in the formulas (15.1). Figure 15.8 displays the relationship between f_1 and f_2 as well as the Pareto-optimal front in the objective space.

Note that in this example the decision space is the one-dimensional range $[-5, 5]$. Any value from the range $[0, 2]$ is a Pareto-optimal solution. On the other hand, the objective space is two dimensional: the Pareto-optimal front is defined by formula (15.2), where $f_1 \in [0, 4]$. This is the corresponding range for $x \in [0, 2]$.

The algorithm we use here to illustrate some points discussed earlier in the chapter is Fonseca and Fleming's MOGA [168]. A binary encoding was adopted, together with two-point crossover and proportional selection. For this example, we used a crossover rate of 0.8, a mutation rate of $1/24$ (24 is the length of the data structure), and the maximum number of generations was 80.

The plot of the results obtained by MOGA compared with respect to the true Pareto front of the problem is shown in figure 15.9.

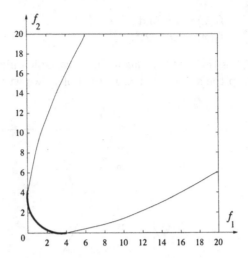

Fig. 15.8. Pareto-optimal front (bold line) in the objective space

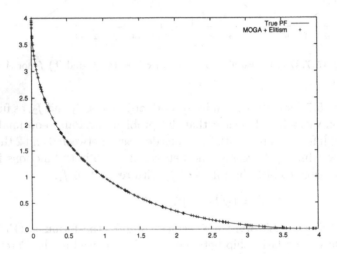

Fig. 15.9. Plot that shows the Pareto front generated by MOGA (shown with crosses) with respect to the true Pareto front of the problem (shown as a continuous line) for Schaffer's problem.

Two additional examples are given below. The first one was also proposed by Schaffer [411]. It also has only one variable, but in this case the Pareto front is disconnected.

$$\text{minimize } f_i(x), \; i = 1, 2,$$
$$\text{subject to } -5 \leq x \leq 10,$$

where

$$f_1(x) = \begin{cases} -x, & \text{if } x \le 1 \\ -2+x, & \text{if } 1 < x \le 3 \\ 4-x, & \text{if } 3 < x \le 4 \\ -4+x, & \text{if } 4 < x, \text{ and} \end{cases}$$
$$f_2(x) = (x-5)^2.$$

The second additional problem was proposed by Valenzuela-Rendón [475]. It has two variables. The true Pareto front is connected and is convex. It's formulated as follows:

minimize $f_i(x_1, x_2)$, $i = 1, 2$,
subject to $-3 \le x_i \le 3$, $i = 1, 2$,

where

$f_1(x_1, x_2) = x_1 + x_2 + 1$, and
$f_2(x_1, x_2) = x_1^2 + 2x_2 - 1.$

In both cases, the same parameters for MOGA were used as for the first problem. The plots of the results obtained by MOGA as compared with respect to the true Pareto fronts of the Shaffer's second problem and the Valenzuela-Rendón problem are shown in figures 15.10 and 15.11, respectively.

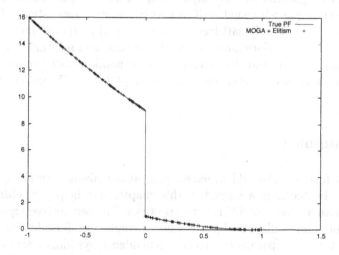

Fig. 15.10. Plot that shows the Pareto front generated by MOGA (shown with crosses) with respect to the true Pareto front of Schaffer's second problem (shown as a continuous line). Note that the vertical line is not part of the true Pareto front, and is included only for illustrative purposes.

These three test problems, as well as many others, are also discussed in Coello et al. [75]. The results displayed in figures 15.9, 15.10, and 15.11 illustrate that MOGA performed "quite well" on all three test cases. However, when

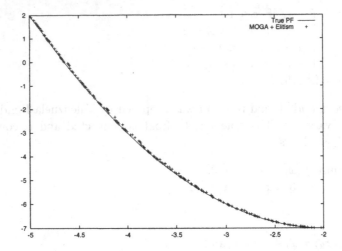

Fig. 15.11. Plot that shows the Pareto front generated by MOGA (shown with crosses) with respect to the true Pareto front of the Valenzuela-Rendón problem (shown as a continuous line).

evaluating an algorithm for multiobjective optimization problems, or when comparing algorithms, such visual verifications or general observations of the type "quite well" are not fully satisfactory. More formal methods should be introduced that involve performance metrics, the nearness of solutions found on the Pareto-optimal front, and the diversity among nondominated solutions. For a full discussion on several such metrics, see chapter 4 of [75] and section 8.2 of [98].

15.3 Summary

When you face a real-world problem, you almost always are facing multiple concerns. The techniques offered in this chapter can help you address those concerns from at least two different perspectives. One perspective aggregates the contributions from different objectives into a single overall evaluation function. The other treats the parameters more independently, yielding a set of solutions that are all potentially optimal in the sense of being nondominated. In the end, you'll almost always need to pick one solution to implement, but the way you discover this solution can be very different depending on which of the methods you choose, and the solution itself may be quite different too. Neither the single aggregate approach nor the Pareto-optimal approach is intrinsically better than the other. In the parlance of the chapter, each is "nondominated."

When choosing the Pareto approach on constrained problems, the problem is not just to find a Pareto-optimal front of solutions, but these Pareto-optimal solutions should also be *feasible*, i.e., they should satisfy all the specified constraints. As we discussed in chapter 9, there are many constraint-handling techniques that can be applied in the presence of constraints. These techniques can be modified for multiobjective optimization problems. Some of these methods are straightforward — reject infeasible solutions, apply penalty functions, evaluate the "nearness" to the boundary of a feasible region while comparing two infeasible solutions — but others can be extended by including additional features, such as using niche count to compare to infeasible solutions. In a more elaborate example, Ray et al. [381] used three different rankings of solutions based on the concept of dominance. These three rankings took into account the m values of the objective function (only), the $l+q$ values of constraint violations (where there was l inequality constraints and q equality constraints), and the $m + l + q$ values of the objective function *and* constraint violations. Then the algorithm selected solutions from these three sets to fill the population for the next generation by following a particular set of preference rules. Handling constraints in a multiobjective environment is an important and current research area.

There are many other open interesting issues that pertain to multiobjective optimization problems. These include the design of performance matrices and test problems, visualization methods for displaying nondominant solutions, and comparisons of various algorithms for multiobjective optimization problems. There are also interesting research issues connected with programmatic details such as controlling elitism, controlling diversity in the decision space versus the objective space, and constraint handling. The best current references that provide the widest surveys of evolutionary multiobjective methods are [75, 98]. You can also find significant contributions in special sessions of recent conferences [508, 373, 278, 374, 62, 171].

Whenever you face multiple objective problems, you have a lot to think about regardless of whether you choose the single aggregate approach or the Pareto-optimal approach. Be sure you give sufficient attention to the relative importance of parameters and their degrees of criticality in the single aggregate approach, and be careful to ensure diversity in your population if you seek to find the entire collection of solutions along the Pareto front before focusing in on one or more candidates. Perhaps most importantly, be certain that you can justify the solutions that you choose. Most people outside mathematics, economics, and engineering have never heard of Pareto, but they often know a good solution when they see one. Hopefully, you will too. Being able to explain a solution to a customer can be the difference between a successful implementation and a merely "interesting result."

XVI. Do You Like Simple Solutions?

We all have a tendency to complicate things. In particular, being mainly scientists and engineers, we work with mathematical equations. We're tempted to try to describe things in mathematical terms even when we don't have to. You'll often find that one of the main stumbling blocks to problem solving is making the solution overly complex, loaded with various components. Do yourself a favor and *think* before you write down an equation. Let's illustrate this point with a few examples.

In front of you is a glass of water and a glass of juice. Their volumes are identical. You take a teaspoon, put it in the glass of juice, take a full spoon, and put it in the water (figure XVI.1). After briefly stirring up the mixture, you put a full teaspoon of the mixture back into the glass of juice. The question is which amount is greater: the amount of juice in the water or the amount of water in the juice?

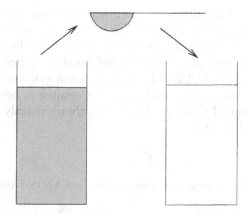

Fig. XVI.1. A glass of juice and a glass of water. A full teaspoon travels from juice to water.

You can set out to solve this problem by calculating the required volumes of liquid. You might assume that the volume of the water or juice in a glass is V and assume that the volume of a teaspoon is x. You'd then calculate the ratio between the water and juice after a teaspoon of juice is added to the water. You'd next calculate the amount of water coming back to the glass with the juice, and so forth. This method has a few disadvantages. First, it requires some calculations. Second, it's based on the assumption of a perfect mixture (i.e., the percentage of juice in the water in the glass is the same as in the teaspoon on its way back), which needn't be the case. We could take a teaspoon of the mixture from the surface, where the amount of juice is usually lower than in the rest of the glass.

We can solve the problem without thinking about ratios of water and juice. Think about the glass of water. We put in a teaspoon of juice. Then we remove some juice and some water. How much liquid remains in the glass? Exactly the same amount as we started with. No more, no less. The same thing goes for the glass of juice. We removed a teaspoon of juice and replaced it with a teaspoon of a juice–water mixture. There's just as much liquid in the glass of juice after we return the teaspoon.

So, without any ratios,

$$W + j_1 - j_2 - w_1 = W, \text{ and } J - j_1 + j_2 + w_1 = J,$$

where W and J are the volumes of water and juice in their glasses, respectively (and they are both equal to each other), j_1 is the amount of juice in a teaspoon during the first move (figure XVI.1), and j_2 is the amount of juice we returned to the first glass along with w_1 units of water. Think about it. The quantity $j_1 - j_2$ is the amount of juice in the water and w_1 is the amount of water in the juice. By either equation, $j_1 - j_2$ has to be the same as w_1, therefore the amount of water in the juice is exactly the same as the amount of juice in the water!

We can even think about the solution without using any equations at all. We just have to make the realization that on the way back from the mixture to the juice, the amount of water in the teaspoon has to be exactly the same as the amount of juice we are leaving behind because we just transported one teaspoon of juice (figure XVI.2). Moreover, you can make as many exchanges between two glasses as you like. As long as the volumes in both glasses remain the same, the amount of water in the juice is always exactly the same as the amount of juice in the water.

Fig. XVI.2. A teaspoon on its way back with a little bit of juice

The next example is also quite educational. There's a tennis tournament with 937 players. The player who wins a game advances further, while the loser leaves the competition. How many games are required to complete the tournament?

Again, we can calculate the answer with relative ease. Such a tournament can be visualized as a binary tree where each node represents a game. There's one game at the root (the final game), two games on the next level (semi-finals), four games one level down (quarter-finals), and so forth (figure XVI.3). If a player can have a free draw only in the first round, then 512 players must advance to the second round. Thus the first round must have $937 - 512 = 425$ games, which involve 850 players. The winners in these games (425 players) plus players who get a lucky draw ($937 - 850 = 87$) constitute the group of 512 players who advance to the second round. As $512 = 2^9$, the remaining calculations are easy. The total number of games in the tournament is

$$425 + 256 + 128 + 64 + 32 + 16 + 8 + 4 + 2 + 1 = 936.$$

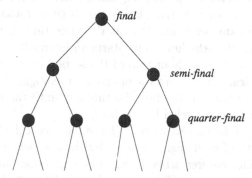

Fig. XVI.3. Tennis tournament

In much less time, however, we could have obtained the same result without any calculations. All we have to do is note that every game of the tournament eliminates a single player. During the tournament of n players we have to eliminate $n - 1$ of them to get a winner, therefore we need $n - 1$ games. So, for 937 players, we need 936 games. Game, set, and match!

The third example concerns a clock and its hands. The minute hand of an analog clock covers the hour hand at noon. At what time will both hands overlap again?

This is easy to visualize. As the minute hand advances around the clock, the hour hand moves much slower. At one o'clock the minute hand is at 12, and the hour hand is at 1. Five minutes later the minute hand almost covers the hour hand as it advances to 1, but the hour hand also moves forward a bit during these five minutes.

Suppose that the speed of the hour hand is v cm/h. Then the speed of the minute hand is $12v$ cm/h and the distance covered during one rotation (i.e., in one hour) is $12v$ cm. After one hour, i.e., at 1 p.m., the (quicker) minute hand is five minutes behind the (slower) hour hand (or the distance v cm). After the time t the minute hand catches up to the hour hand:

$$t \cdot 12v = v + t \cdot v,$$

as the distance covered by the minute hand (the product of time t and speed $12v$) is by v cm greater than the distance covered by the hour hand (again, product of time t and speed v). The above equation gives $t = 1/11$, so the two indicators overlap every $1\frac{1}{11}$ hours.

Again, we could arrive at this result much faster if we recall that clock hands overlap 11 times in every 12 hours. Since they overlap at regular intervals, this happens every $1\frac{1}{11}$ hours.

This observation also allows us to solve the following upgrade. The second, minute, and hour hands of an analog clock overlap at noon. At what time will all of the hands overlap again?

Here's a classic example of a problem that can be solved in either a long way or a short way. There are two cities, A and B, that are 400 km apart. Simultaneously, a train leaves city A going toward city B with a constant speed of 40 km per hour, and another train leaves city B going toward city A with a constant speed of 60 km per hour. Also, at the same time, a busy bee, which was resting at the front of the first train, starts an interesting trip. It flies with a constant speed of 75 km per hour toward the second train, and as soon as it touches the second train, it reverses the direction of its flight. The bee does this every time it meets a train. If the bee flies this way until the two trains meet, what is the total distance traveled by the bee?

At first glance, it looks like we're in for some nice calculations! We can determine the length of each segment flown by the bee, where a segment corresponds to a distance covered while flying in one direction, and then treat an infinite summation of segments of decreasing length. But somehow we know we can do better than this!

The two trains will meet after four hours since both of them cover 40 + 60 = 100 km every hour of the total distance of 400 km. In turn, the bee flies continuously for four hours, and at a constant speed of 75 km per hour, the total distance traveled by the bee is 300 km. Q.E.D.

Okay, here's another one for you. There's a regular chess board with two opposite corners removed (see figure XVI.4). There are 31 dominoes. Each rectangular domino can cover two squares of the chess board. Your challenge is to cover the chess board with these dominoes by placing them horizontally or vertically.

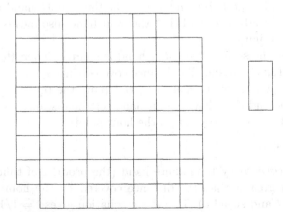

Fig. XVI.4. A chess board with two missing corners and a single domino

You can spend hours trying to arrange the dominoes on the board, but a simple observation can terminate this process very quickly. Note that if we color the squares of the board black and white in the usual manner, the board will have 30 squares of one color, and 32 squares of another. The squares at opposite corners must be of the same color (see figure XVI.5).

Since each domino always covers one white and one black square, regardless of its placement on the board, it's impossible to cover this chess board with these dominoes!

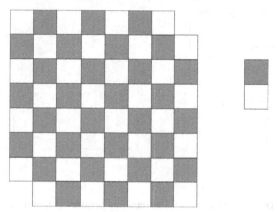

Fig. XVI.5. A chess board with two missing corners and a single domino. Black and white colors are added.

There is a three-dimensional instance of the above problem. The task is to fill a box of dimensions $6 \times 6 \times 6$ with bricks. Each brick has dimensions $1 \times 2 \times 4$. Can you do it?

The final puzzle is the most amazing. (Try it out on a friend!) There is a square divided into four identical smaller squares (figure XVI.6a).

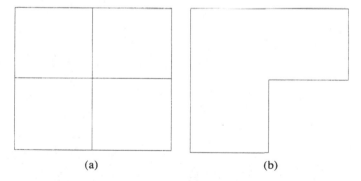

(a)   (b)

Fig. XVI.6. (a) A square divided into four identical squares. (b) One of the four smaller squares is removed.

One of the smaller squares is removed. Your task is to divide the remaining figure (figure XVI.6b) into four identical parts. You're allowed to use continuous lines of any shape. It takes some time, but sooner or later most people find the solution (see figure XVI.7a).

But this is only a first part of the whole puzzle which sets the mind of the problem solver into a particular way of thinking. The second (and really the main) part of the puzzle is to divide the remaining square (figure XVI.7b) into five identical pieces. Despite the fact that this part of the puzzle is *extremely* simple (see figure XVI.7c), it takes a significant amount of effort for some people to find the solution.[1]

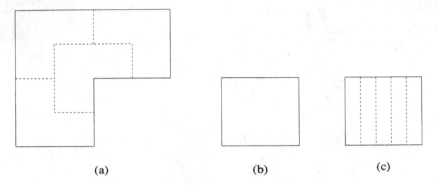

(a)   (b)   (c)

Fig. XVI.7. The required division (a), the remaining square (b), and the required division for the second part of the puzzle (c)

The same phenomenon can be observed on a larger scale. Past experiences needn't always point in the right direction! It's important to retain some flexibility, to have the ability to solve new problems in new ways! This is the true mark of intelligence.

---

[1]One of the math professors we gave this problem to actually "proved" that it's impossible to make such a division!

16. Hybrid Systems

'The time has come,' the Walrus said,
'To talk of many things:
Of shoes—and ships—and sealing wax—
Of cabbages—and kings—
And why the sea is boiling hot—
And whether pigs have wings.'

Lewis Carroll, *Through the Looking-Glass*

We know that no single algorithm can be the best approach to solve every problem. We have to incorporate knowledge about the problem at hand into our algorithm in some useful manner; otherwise, it may not be any better than a random search. One way to approach this issue is to hybridize an evolutionary algorithm with more standard procedures, such as hill-climbing or greedy methods. Individual solutions can be improved using local techniques and then placed back in competition with other members of the population. The initial population can be seeded with solutions that are found with classic methods. There are many ways to form hybrid approaches that marry evolutionary algorithms with other procedures.

In fact, there are so many ways to hybridize that there's a common tendency to overload hybrid systems with too many components. Some modern hybrid systems contain fuzzy-neural-evolutionary components, together with some other problem-specific heuristics. A typical joke at many of the computational intelligence conferences concerns when we will see the first paper titled "a self-adaptive greedy neural-fuzzy-evolutionary-annealing-approach for improved memory rules in tabu search over a tree that defines a recurrent rule-set for generating expert systems in data mining." You can invent any number of other possibilities. The trick is knowing how best to find synergies between different approaches and to keep the overall approach as parsimonious as possible.

Evolutionary algorithms are extremely flexible and can be extended by incorporating diverse concepts and alternative approaches. For example, we discussed the possibility of incorporating local search into an evolutionary algorithm in the case of the TSP (chapter 8). We also discussed various methods for handling constraints, which included Lamarckian learning and the Baldwin effect (chapter 7). We even discussed some of the ways you can incorporate control parameters into the individuals in an evolutionary algorithm so that they can tune themselves to the problem at hand (chapter 10). Evolutionary algorithms can also be enhanced by including memory into individuals or entire populations so as to better handle the case of time-varying environments (chapter 11). But these few examples are just the beginning.

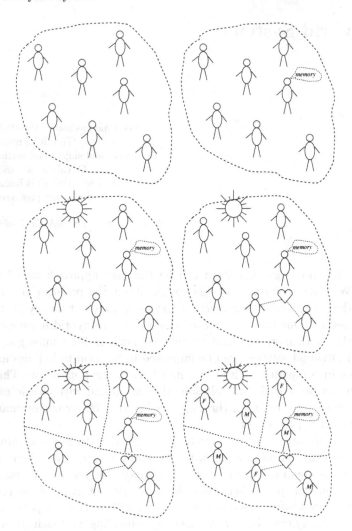

Fig. 16.1. Gradual enhancements of an evolutionary algorithm. Starting from a "plain" evolutionary algorithm, we successively incorporate memory, temperature, mating and attractiveness, subpopulations, and gender.

Let's consider a series of six figures (see figure 16.1). We start in the upper-leftmost figure with a population of individuals as you would find in any evolutionary algorithm. Then in the upper-rightmost figure we extend each individual by giving them a memory structure (whether explicit or implicit). This is a step toward a tabu type of search. Then, in the middle-left figure, we can introduce the concept of a temperature, where the population gradually cools down, as in simulated annealing. We might consider relating the size of a variation to apply to an individual to the temperature of their surrounding environment, with higher temperatures yielding larger steps.

Now we're ready for further enhancements. We can introduce a concept of *attractiveness* that affects the manner in which individuals mate and generate offspring solutions (middle-right figure). We can even talk about seduction [394]! And we can consider different ways for carefully choosing which individuals to mate (e.g., "your brains and my beauty" [222]), how many individuals should take part in producing an offspring (e.g., "orgies" discussed in [117]), or how many offspring they should produce [132]. Then we can split the population into several subpopulations (lower-left figure) and execute the evolutionary algorithm in a distributed mode. Finally, we can assign a gender (M = males, F = females in the lower-right figure), together with family relationships and even rules to prevent members from within the same family from mating [129]. We could even consider the possibility of having individuals in the population that are awake or asleep [128]!

As you'd expect, many of these extensions give rise to other possibilities. We can split the population in several subpopulations, which yields the so-called *island, migration, or coarse-grain* models [300]. We could run different evolutionary algorithms on each subpopulation, where variation and selection takes place in each subpopulation independently of the others. From time to time, one or more individuals might migrate from one subpopulation to another. In this way, we can share information between subpopulations. This poses many interesting issues to resolve:

- How many subpopulations should we have, and how big should they be?

- Should all of the subpopulations be the same size?

- Should the same parameters for the evolutionary algorithm be used to control each subpopulation, or should we somehow tailor each evolutionary algorithm to each subpopulation?

- What topology should we use to connect the subpopulations? Can an individual from any subpopulation migrate directly to any other?

- What are the mechanisms for migration:

  - How often will individuals migrate? Should this happen periodically or instead when it is trigged by some event, such as a decrease in population diversity?

  - Which individuals should migrate? The best? The worst? A typical individual? A random individual?

  - What are the rules for immigration? If a new individual **x** arrives in subpopulation S_k from subpopulation S_j, what are the consequences? Does a copy of **x** remain in S_j? Does it displace some existing member of S_k?

We can also think about organizing a population by assigning individuals to particular geographic locations (this is a so-called *diffusion, neighborhood, or fine-grain* model, see figure 16.2).

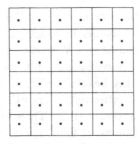

Fig. 16.2. An interconnecting grid topology in a classical diffusion model

This type of model maintains overlapping neighborhoods in a single population. Every individual has a local neighborhood that's relatively small, restricted, and defined by the topology of the neighborhood. Selection, variation, and replacement are restricted to this neighborhood. The overlap of neighborhoods contributes to the diffusion of information throughout the population.

Both of these approaches were motivated by the observation that, from a population genetics perspective, variation and selection don't usually occur in a global population sense. Instead, variation and selection are usually restricted to occur within *demes*: local subsets of interbreeding individuals. When using structured population models, the genetic makeup of individuals can spread throughout the population in a manner that's similar to the diffusion process [195].

One of the most important choices that designers of evolutionary algorithms face is the size of the population. It can be critical in many applications. If the population is too small, the evolutionary algorithm may converge too quickly. If it's too large, the algorithm may waste computational resources. The population size influences two important factors in evolutionary search: population diversity and selective pressure. Note, however, that both the diffusion and island models assume a population of fixed size. In the diffusion model, the population size is constant. In the island model, typically, each of N subpopulations has a fixed number $\mu = M/N$ individuals. Of course, each subpopulation could have a different population size μ_i, where $\sum_{i=1}^{N} \mu_i = M$. Migration may increase and decrease the size of subpopulations, thereby keeping M constant, but the values μ_i are kept constant in most implementations.

A new model was recently proposed [274] called a *patchwork* model, which combines some properties of the island and diffusion models. Each individual is modeled as a mobile agent that lives in a virtual ecological niche and interacts with their environment through their sensors and motors. Its decisions are based on local information that it gathers with its sensors, and its actions can only affect its local environment. Other properties are assigned to individuals, such as a maximum life span, the ability to breed, mortality that is based on its fitness, and the ability to have preferences in decision making. The patchwork model incorporates a spatial structure, where agents move and interact in a two-dimensional grid. Each cell of the grid (called a *patch*) has local spatial properties, such as the maximum number of individuals it can accommodate (figure 16.3).

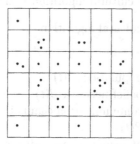

Fig. 16.3. The patchwork model

The mobile agents are extended by a set of adaptive behavioral rules. These rules determine the individuals' actions, for example, their desire to reproduce or migrate. Each individual makes a bid for its desired action. That action may or may not take place. Concurrent events are handled using a two-step process. First, agents make decisions and schedule actions, and second, the simulation executes the scheduled actions and resolves temporal conflicts. For instance, when agents within the same grid cell decide to mate, their intention is first scheduled, but not immediately executed. Here, a bid for reproduction might be unsuccessful because of the lack of a partner, or because there's no room for an offspring. Even if an offspring is produced, it might not be sufficiently competitive to be included in the population. Similarly, a bid for migration might be unsuccessful due to the bids from other individuals and/or overcrowding. Note that the patchwork model allows the entire system of individuals to adapt to changes by altering the appropriate local population sizes, decision rules, migration protocols, etc., while solving some problem. At different stages, different rules might be appropriate.

The idea to model the evolution of autonomous agents in an ecology has been introduced in a field of evolutionary computation called *artificial life*. Actually, some of the very earliest efforts in simulating evolution involved just this approach [32, 77, 14]. One common artificial life program is called SWARM [323], which consists of an Object C toolbox for modeling an artificial ecology. In ecology, so called individual-based models have gained considerable attention in recent years [92, 248], but very few approaches truly represent individuals as autonomous agents and mimic evolutionary processes [275, 106].

Another interesting development in evolutionary computation is the emergence of so-called *cultural algorithms* [388]. These procedures support two modes of inheritance: one takes place at the microevolutionary level in terms of behavioral or genetic traits, while the other acts at the macroevolutionary level in terms of the so-called *beliefs* of the population. The two modes interact via a communication channel that allows individuals to alter the population's belief structure, and also biases the individuals' behavior in light of the population's beliefs. Thus, cultural algorithms have three basic components: a belief structure, a population structure, and a communication channel [390].

The basic structure of a cultural algorithm is shown in figure 16.4.

```
procedure cultural algorithm
begin
    t ← 0
    initialize population P(t)
    initialize belief network B(t)
    initialize communication channel
    evaluate P(t)
    while (not termination-condition) do
    begin
        communicate (P(t), B(t))
        adjust B(t)
        communicate (B(t), P(t))
        modulate fitness P(t)
        t ← t + 1
        select P(t) from P(t − 1)
        evolve P(t)
        evaluate P(t)
    end
end
```

Fig. 16.4. The structure of a cultural algorithm

Cultural algorithms appear appropriate for adjusting the "evolution of evolution," where individuals in the population learn during their lifetime. The concept is a generalization of the idea that evolution underlies both phyletic learning within species and social learning within human societies [14].

Another technique that has been offered recently is *ant colony optimization* (ACO) [103, 104], which is a multiagent system where low-level interactions between artificial (i.e., simulated) ants result in a complex behavior of the larger ant colony. These algorithms were inspired by colonies of real ants [103] that deposit a chemical substance (*pheromone*) on the ground. This substance influences the behavior of individual ants. The greater the amount of pheromone is on a particular path, the larger the probability that an ant will select that path. Artificial ants in ACO algorithms behave similarly.

We can illustrate the basic concepts of the ACO algorithm introduced by Dorigo et al. [105] using the TSP as an example. In the TSP we minimize

$$COST(i_1, \ldots, i_n) = \sum_{j=1}^{n-1} d(C_{i_j}, C_{i_{j+1}}) + d(C_{i_n}, C_{i_1}), \tag{16.1}$$

where $d(C_x, C_y)$ is the distance between cities x and y.

Let $b_i(t)$ $(i = 1, ..., n)$ be the number of ants in city i at time t and let $m = \sum_{i=1}^{n} b_i(t)$ be the total number of ants. Let $\tau_{ij}(t+n)$ be the intensity of the pheromone trail connecting $(i\ j)$ at time $t+n$, given by

$$\tau_{ij}(t+n) = (1-\rho)\tau_{ij}(t) + \Delta\tau_{ij}(t, t+n), \qquad (16.2)$$

where $0 < \rho \le 1$ is a coefficient that represents pheromone evaporation. $\Delta\tau_{ij}(t, t+n) = \sum_{k=1}^{m} \Delta\tau_{ij}^k(t, t+n)$, where $\Delta\tau_{ij}^k(t, t+n)$ is the quantity per unit of length of trail substance (pheromone in real ants) laid on the connection $(i\ j)$ by the k-th ant at time $t+n$ and is given by the following formula:

$$\Delta\tau_{ij}^k(t+n) = \begin{cases} \frac{Q}{L_k} & \text{if } k\text{-th ant uses edge } (i\ j) \text{ in its tour} \\ 0 & \text{otherwise,} \end{cases}$$

where Q is a constant and L_k is the tour length found by the k-th ant. For each edge, the intensity of trail at time 0, $\tau_{ij}(0)$, is set to a very small value.

While building a tour, the probability that ant k in city i visits city j is

$$P_{ij}^k(t) = \begin{cases} [\tau_{ij}t]^\alpha [\eta_{ij}]^\beta / \sum_{h \in allowed_k(t)} [\tau_{ih}t]^\alpha [\eta_{ih}]^\beta, & \text{if } j \in allowed_k(t) \\ \\ 0, & \text{otherwise,} \end{cases} \qquad (16.3)$$

where $allowed_k(t)$ is the set of cities not visited by ant k at time t, and η_{ij} represents a local heuristic. For the TSP, $\eta_{ij} = 1/d(C_i, C_j)$ and is called the "visibility."

The parameters α and β control the relative importance of the pheromone trail versus the visibility. The transition probability is thus a trade-off between visibility, which says that closer cities should be chosen with a higher probability, and trail intensity, which says that if the connection $(i\ j)$ enjoys a lot of traffic then it's profitable to follow it.

Each ant has an associated data structure called a *tabu list* in order to avoid the case where ants visit a city more than once. This list $tabu_k(t)$ maintains a set of visited cities up to time t by the k-th ant. Therefore, the set $allowed_k(t)$ can be defined as follows: $allowed_k(t) = \{j | j \notin tabu_k(t)\}$. When a tour is completed, the $tabu_k(t)$ list $(k = 1, \ldots, m)$ is emptied and every ant is free to again choose an alternative tour for the next cycle.

By using the above definitions, we can provide a general outline of the ACO algorithm (figure 16.5). Such ant algorithms have been applied to many types of problems, ranging from the TSP to routing in telecommunications networks [103].

There are other related algorithms that seem to model "swarming" phenomena and use this for optimization. The *particle swarm* method [259] operates on a population of individuals where variation is applied, but without selection. Each individual has a position and velocity that is updated according to the relationship between the individual's parameters and the best location of the individual in the population found so far. The search is biased toward better regions of space, with the result being a sort of "flocking" toward the best solutions. Another somewhat similar procedure is *differential evolution* [364]. In its

```
procedure ant system
initialize
for t = 1 to number of cycles do
begin
    for k = 1 to m do
    begin
        repeat
            select city j to be visited next with
                probability P_ij^k given by eq. (16.3)
        until ant k has completed a tour
        calculate the length L_k of the tour generated by ant k
    end
    save the best solution found so far
    update the trail levels τ_ij on all paths according to eq. (16.2)
end
```

Fig. 16.5. The procedure ant system

most basic formulation, variations are applied by taking the vector difference between two randomly selected solutions and then scaling that difference before adding it to some other randomly selected individual. Over time, the population moves toward better regions of the search space.

In general, the area of agent-based systems is emerging as a new paradigm in computing, and its rapid development can have important and fundamental implications. The key general idea is that agents with differing capabilities may interact in a variety of ways to solve a problem. An agent can be viewed as a potential solution to a problem, as in evolutionary algorithms, or just as a decision-making unit, as in ant systems. Each agent's behavior is based on rules that determine its decisions. The interaction of the agents may result in *emergent behavior* of the whole system, which might be impossible to predict due to complexities of all interactions. Even a simple set of rules for each agent may result in a complex adaptive system.

Müller [331] wrote:

> The term *autonomous agent* appears to be a magic word of the 1990s. The concept of autonomous software programs that are capable of flexible action in complex and changing multiagent environments, has refocused the way artificial intelligence as a whole defines itself, and is about to find its way into industrial software engineering practice. Agent technology is used to model complex, dynamic, and open applications, e.g., in production planning, traffic control, workflow management, and increasingly, on the Internet.

Owing to this diversity, there's no general agreement on what an agent is, or what key features an agent should have. A variety of different architectures have been proposed. We can group them into three categories [331]:

- *Reactive agents* are built on the basis of a behavior-based paradigm. The decisions made by such agents are usually made at run time and are based on a limited amount of information (often just sensory input). Some research in artificial life follows this direction [277].

- *Deliberative agents* represent traditional artificial intelligence approaches. Each agent has (1) an internal (symbolic) representation of the environment and (2) logical inference mechanisms for decision making. Classical planning systems constitute examples of this category.

- *Interacting agents* are capable of coordinating their activities with some other agents through communication and negotiation. Such agents usually maintain some knowledge about each other and might be able to reason about the other agents. Distributed artificial intelligence research follows this direction.

Evolutionary algorithms can be applied to the design of agent-based systems, and can even be used to optimize them by applying variation and selection to the behavior of the agents in a simulation. In some ways, evolutionary algorithms represent one of the most general agent-based procedures.

We've already seen another possibility, in which evolutionary algorithms are used in coevolutionary settings and solutions compete and/or cooperate for survival. Just to refresh your memory, there are many opportunities for applying coevolution.

For example, we might consider a scenario with multiple populations and a separate evolutionary process running on each of them. These evolutionary processes might interact with each other and the evaluation function for any population may depend on the other individuals present in the population or on the state of the evolution processes in the other population(s).

Even within one population, there are many ways to use competing solutions. In the iterated prisoner's dilemma (chapter 11), we saw that solutions in a single population could act as strategies, where the fitness of each strategy was a function of the other strategies in the population. The process can be generalized to two or more populations, as in the examples with the sorting networks by Hillis, or the blood pressure controllers by Sebald and Schlenzig. Paredis [350] used a similar idea in the context of constraint satisfaction problems. His approached used a coevolutionary model where a population of potential solutions coevolved with a population of constraints. Fitter solutions satisfied more constraints, whereas fitter constraints were violated by more solutions. Individuals from the population of solutions were considered from the whole search space, and there's no distinction between feasible and infeasible individuals.

coevolutionary systems can be considered as useful tools for modeling some business applications. For example, a coevolutionary system has been used to model strategies of two competing bus and rail companies that compete for passengers on the same routes [333]. One company's profits depend on the current strategy (capacities and prices) of the other company. The study investigated the inter-relationship between various strategies over time. This approach can be generalized further into a scenario where there are m competing companies. An agent represents each company, and it follows the current best strategy of the evolutionary process in the m-th population. It would be interesting to observe the pattern of cooperation and competition between various companies. Is it possible to observe a cooperation (clustering) of weaker companies to compete better with larger ones? Under what circumstances (e.g., company resources) would one company eliminate others from the competition? There are many possibilities and many exciting experiments to make!

Note, that such an approach redefines the concept of an agent. Each agent can be seen as an evolutionary process, but interactions between agents is modeled by coevolution. Note that an agent represents a company's strategy and may contain many decision variables (e.g., pricing, investments). As the values of these variables evolve, the agent can be considered as being reactive, as its behavior depends on the current state of the environment defined by strategies of other agents. At the same time, such evolutionary agents can be considered deliberative, as they may be extended by a logical inference mechanism for decision making (these mechanisms might be coded in additional chromosomes of individuals and undergo evolution as well). Finally, these agents fall also into an interacting category, as they may communicate and even negotiate some activities.

coevolutionary systems can also be important for approaching large-scale problems [360], where a problem is decomposed into smaller subproblems, as we saw in cooperative coevolution. A separate evolutionary process is applied to each of the subproblems, but these evolutionary processes are connected. As before, the evaluation of individuals in one population depends also on developments in other populations. [1]

16.1 Summary

Evolutionary algorithms offer an especially flexible approach to problem solving. Not only can you hybridize them with more traditional algorithms, such as a greedy approach (e.g., in initializing a population), you can also extend them to include many of the facets we observe in nature. Multiple populations, memory, individual learning, social learning, and many more ideas can be incorporated. There really are no bounds on what you can do, except that, of course, you're limited by your own imagination.

---

[1] Many of the topics within evolutionary computation that are now receiving interest were actually first offered decades ago. coevolution is no exception. Barricelli [33], Reed et al. [384], Fogel and Burgin [165], and others all used coevolution in the 1960s to evolve strategies in games and simulated combat.

With this flexibility comes a price. Remember the no free lunch theorem [499]? No algorithm is best across all problems. No matter what extension you make, or how much better your algorithm seems to model some natural setting, there will always be problems for which it is particularly good, and problems for which it is particularly awful. Unfortunately, there's very little theory available to help make the appropriate decisions about which extensions to incorporate, and when. Each of the extensions we've described is an exciting possibility. Each also obligates you to use it wisely, otherwise you will head down the slippery slope toward the first "agent-based multipopulation coevolutionary ant colony system based on methods for particle swarm and differential evolution." Once disconnected from the problem at hand, hybrid systems have a certain appeal that makes them attractive for their own sake. Stay focused on solving problems rather than building elaborate but useless simulations.

17. Summary

Is it progress if a cannibal
uses knife and fork?

Stanisław Lec, *Unkempt Thoughts*

We've covered a great deal of material in this text, and as we approach the end, let's take the opportunity to review what we've discussed. The primary question to ask when facing a problem is "what's my objective?" If you don't have your goal well focused and plainly in mind, your hope of attaining the goal is minimal, and even if you do attain it, you might not even know that you did! Keep your eye on the objective whenever you're solving a problem. It's all too easy to simply march off in the quest for a solution and forget what you were trying to solve in the first place. It's also easy to be diverted or side-tracked into addressing some other concern that may or may not be of comparable importance.

There's a wonderful line from *Magnum, P.I.*, a popular television show from the 1980s about a private investigator in Hawaii that starred Tom Selleck. At one point, Magnum had to pick a lock to get into the main residence on the property where he stayed. (Magnum was relegated to the guest house.) The property was patrolled by two Doberman pincers. As Magnum started to pick the lock, he heard the dogs coming. "Work the lock. Don't look at the dogs. Work the lock. Don't look at the dogs," he said to himself. Of course, then he looked at the dogs. Give yourself the best chance to solve the problem by staying focused on the problem, and have a specific goal in mind.

Having specified the goal, you now have to address the manner in which you can model the problem. It's quite unlikely that you'll be able to treat the "problem" directly. Most problems aren't mathematical or computational in nature. We have to use our expertise to reformulate or model the problem into a mathematical or computational framework. It's at this point that we often face difficult issues that concern what to include in the model and what to leave out. If we rely on traditional methods of problem solving — linear programming, linear regression, and so forth — we might be tempted to leave out quite a lot. We might assume a squared error cost function, even when we don't really have that cost function in mind. We might approximate nonlinear constraints with linear constraints, just so we can generate an answer. These short cuts aren't always inappropriate. Sometimes you can use them to your advantage, but you have to be able to defend your decisions.

Once you've formulated a model, you have to decide on the technique to compute or search for the solution. If the space of possible solutions is small, there's no point in searching for it. Just enumerate everything and you can be completely sure that you've found the very best solution. Most real-world problems aren't so convenient. They pose large search spaces, with no analytic means for determining an optimal solution, and what's more, the problem can change over time, present noisy conditions that aren't stationary or Gaussian, and may even vary as a function of other competing problem solvers, people who want you to fail. The real world isn't always kind. You need to consider how each of these possibilities will affect the solutions that you offer. Are they robust, or will they work only in limited domains and conditions? Can you adapt on-the-fly as new situations present themselves? Can you anticipate what your opponents will do next? Having completed the material in this book, you can assuredly answer in the affirmative to all of these questions.

Problem solving is an art. As with all artistic endeavors, it requires practice.

- "Excuse me, how do I get to Carnegie Hall?"
- "Practice, practice, practice."

One way to become really good at problem solving is to continually challenge yourself with puzzles, as we've done here. They don't have to be extremely complicated or devious, but they do have to have an instructive lesson. Now that you've had a chance to read through the entire book, go back to each puzzle that precedes each chapter and review the problem it poses. Can you make new connections between the material in the subsequent chapter and that puzzle? Can you identify other salient points about problem solving from these puzzles? We recommend that you always find some new book of puzzles and brain teasers to keep your problem-solving abilities honed. If the lessons that each problem poses aren't obvious, see if you can create them. Better yet, create your own puzzles and challenge your friends; have them challenge you!

Most mathematicians who we know are good at proving things. The prime reason for this is that they practice proving things. They gain experience in how to prove things. They have an intuition for what things are provable, and what things are even worth proving in the first place. There's no algorithm for general problem solving. One was tried in the early days of computer science. Its performance was easily exceeded by the hype that surrounded it. Perhaps someday we'll have computers that can solve problems for us automatically. We'll just describe our problems in words to computers and let them do the work, but we seem to be a long way from building Commander Data from the *Star Trek – Next Generation* television series. Until then, our problems are our own to solve.

Here are ten heuristics for problem solving that we'd like you to keep in mind as you face real-world problems.

1. *Any problem worth solving is worth thinking about.* Don't rush to give an answer. Think about it. Think about different ways to manipulate the

information that you've got at hand. We offered a puzzle earlier called *What are the Prices in 7–11?* To solve this puzzle, you had to shift your thinking from the rational numbers (dollars and cents) to the integers. The problem required factoring a large number (711000000) and then splitting the possibilities into a few cases. These cases were organized by the largest prime factor which was 79. The approach required an exhaustive search, but many branches of the search tree were cut relatively early. Often times, it's that first formulation of the problem that will either lead you to the solution, or lots of frustration. Spend the time you need to think about how to best conceive and develop a useful approach.

2. *Concentrate on the essentials and don't worry about the noise.* It's often the case that we seem to be facing a monstrous problem with so many details that we don't know where to start. This is where having the goal clearly in mind is critical! When you know what you want to achieve, you can have a much better chance of filtering out the "noise" in a problem and focusing in on the things that really make a difference. Remember the problem *What is the Shortest Way?* There, you had to find the shortest path between two cities that were separated by a river. Much was made of the bridge that had to go across the river, but in the end, the best way to solve the problem is to forget about the bridge and the river entirely! The river was simply noise.

3. *Sometimes finding a solution can be really easy. Don't make it harder on yourself than you have to.* Some problems will make you feel a bit foolish, even after you've come up with the right answer, because someone else can show you how to solve it in a faster, easier way. We're so trained as engineers and scientists to think in terms of symbols and algorithms that we can easily forget about common sense! The problems that we offered in the section titled *Do You Like Simple Solutions?* were all like this (e.g., a glass of water and a glass of juice, the number of games in a tennis tournament, the overlapping hands of a clock, the bee traveling between two moving trains). Did you try to "calculate" the right answer to these problems? Give them to your friends and see how they do. If they start calculating, let them. Resist the temptation to stop them. Once they've finished, then help them out with some common sense. Be kind, though.

4. *Beware of obvious solutions. They might be wrong.* Sometimes the answers are so clear that they just *have to be right*, until someone shows you that things weren't quite as clear as you thought. We gave you some examples of counterintuitive puzzles in the section titled *How Good is Your Intuition?* Calculating the average speed of a car covering some distance seems pretty straightforward, but we saw that it might not be as obvious as first imagined. Oftentimes, these "simple" problems are incorporated in some larger system that's responsible for treating another problem. When

the larger system fails, we're tempted to look everywhere but exactly where we should!

5. *Don't be misled by previous experience.* Experience is usually an asset, but you have to use it carefully. Don't be led into solving all of your problems with an "old familiar" technique. Don't allow yourself to go about solving problems using the same methods every time. Your most recent experience can be misleading when applied to the next problem you face. Remember the case where you had to divide a figure into four identical parts and then had to divide a square into five identical parts? Be aware of your biases and prejudices, recognize them, and don't let them get the better of you.

6. *Start solving. Don't say "I don't know how."* Most people don't plan to fail, they just fail to plan. And when you have no plan at all, the world zips right by you and you can't solve anything. Even a bad plan is usually better than no plan. This principle was illustrated in two puzzles in chapter 1, where we needed to prove a property of a polyhedron, and where we had to determine the number of times Mrs. Smith shook hands. Both problems were easy once we got started, but they didn't seem to offer many footholds for taking that first step. The key is to invent one! Try something. If it doesn't work, that's no shame. Try something else. Keep trying. Persistence is everything.

   Evolutionary algorithms provide a penultimate example of this principle. Even when starting with completely random solutions to seemingly difficult problems, the process of random variation and selection can quickly lead to useful solutions. But you have to *start* with something! You can't start with an empty population.

   Another essential aspect is to "play" with the problem. Get an understanding of it. Perhaps you might guess at a solution: Even if you're wrong it might lead to some interesting discovery. For example, we discussed a puzzle that involved four numbers where the product of the first two was equal to twice the sum of the other two, and vice versa. By playing with the problem, we found that it's simply impossible for all the numbers in question to be large (*What Are the Numbers?*). After that, it was an elementary exercise to prove that the smallest of the four numbers couldn't be larger than four. This reduced the problem to four relatively easy subcases.

   Just keep this saying in mind: "If you aren't moving forward, you're moving backward." Get going!

7. *Don't limit yourself to the search space that's "defined" by the problem. Expand your horizon.* In our haste to get going, we can overlook things or miss opportunities to short cut difficult aspects of a problem. The problem might be very clearly stated and we think that we have an unambiguous

understanding of it. But then we get stuck. This is a good time to introduce auxiliary variables, ponder what it would be like if you were able to reach some other point along the path to the solution; could you then solve it? If so, maybe you can build the bridge you need to get to that intermediate point.

Remember the case with the ball on the billiard table (chapter 2). If you have to constrain your thinking to the bounds of the table, the billiard ball not only bounces all over the table, but all over your brain as well! But if you expand the horizon, or even invert the perspective — have the table move as the ball proceeds in a straight line — then the answer becomes clear. "Thinking outside the box" is what it's called in management, or at least outside the plane. Remember the case where you had to build four triangles out of six matches (again, chapter 2)? Even the early puzzle about a triangle and an interior point can be solved easily if you expand your horizon a bit, as you'll see shortly.

8. *Constraints can be helpful.* Constraints seem to be pesky annoyances: if only we could get rid of them! Actually, constraints don't always have to be interpreted as additional elements that just increase the complexity of a problem. After all, they imply a feasible subset of the search space, so we can concentrate our search right there. The nice puzzle where we had to find a six-digit number with some interesting properties (i.e., constraints) illustrated the point (*Can You Tune to the Problem?*). In chapter 9 which preceded that puzzle, we discussed some techniques (e.g., problem-specific operators and searching the boundary between feasible and infeasible search space) that take advantage of the problem constraints. If you face a heavily constrained problem, think positively. What have you got to lose? Negative thinking rarely solves any problem.

9. *Don't be satisfied with finding a solution.* Finding a solution doesn't have to be the end of the process. Sometimes it's just a new beginning! Is the solution unique? Are there other possibilities? Remember the case of Mr. Brown and the problem of determining the color of a bear he saw? If you identified all of the possible solutions to the geographical requirements of that puzzle, you definitely had a better understanding of the puzzle than someone else who just figured out one solution and guessed the bear's color, even if they were correct.

10. *Be patient. Be persistent.* Every problem solver gets stumped. In fact, this happens more often than not. You have to have patience to understand the problem, think about the problem, formulate a model, try out a method for treating that model, and still find out that you're wrong. Remember the problem about the length of a piece of rope that held a monkey and a weight in equilibrium? It took real patience to get through that problem. If you find yourself ready to give up in frustration, go back to point 6 and try again. If you find that your track record in problem solving doesn't

seem very good, keep this in perspective: In major league baseball, you'd be a hall-of-fame player if you could be successful as a hitter just one in three times. If that doesn't cheer you up, check out this guy's track record:

(a) He lost his job in 1832.

(b) He was defeated for the state legislature in 1832.

(c) He failed in business in 1833.

(d) He was elected to the state legislature in 1834.

(e) His sweetheart died in 1835.

(f) He had a nervous breakdown in 1836.

(g) He was defeated for speaker in 1838.

(h) He was defeated for nomination to Congress in 1843.

(i) He was elected to Congress in 1846.

(j) He lost renomination in 1848.

(k) He was defeated for the United States Senate in 1854.

(l) He was defeated for nomination for Vice President in 1856.

(m) He was again defeated for United States Senate in 1858.

(n) He was elected President of the United States in 1860.

If Abraham Lincoln was nothing else, he was persistent! Don't give up!

If you keep those ten "commandments" in mind, your problem-solving prowess will become formidable, but don't forget about the danger of making easy generalizations. Sometimes we may notice a pattern that leads to a correct solution. For example, consider the following scenario. There are 2^n tennis players, but two of them are twins. The tournament rules are as usual: There are 2^{n-1} games in the first round. The winners advance to the second round which consists of 2^{n-2} games, and so forth. Assuming, that each player has a 50-50 chance of winning any game, what's the probability that twins will play against each other in this tournament?

Let's denote the probability of their meeting in the tournament by p. Certainly, p is a function of n:

- If $n = 1$ (i.e., there are two players in the tournament), then $p = 1$ as they will meet for sure.

- If $n = 2$ (i.e., there are four players in the tournament), then $p = 1/2$. Do you see why? It's because there's a $1/3$ chance that the twins will be paired together in the first round and they will meet for sure, and a $2/3$ chance that they will play other opponents in the first round. In the latter case, there's a $1/4$ chance that they will meet in the second round, as both of them must win their games and there's a $1/2$ chance for that for each of them. So

$p = 1/2$ because $1/3 \cdot 1 + 2/3 \cdot 1/4 = 1/2$.

- Thus, a reasonable assumption is that $p(n) = 1/2^{n-1}$.

Indeed, this is the case and it can be proved easily by induction.
– The chances that the twins are in the same or opposite halves of the tournament tree are $p_1 = (2^{n-1} - 1)/(2^n - 1)$ and $p_2 = (2^{n-1})/(2^n - 1)$, respectively.
– The chances of their meeting in the former case (the same half) are $m_1 = 1/2^{n-2}$ (induction hypothesis).
– The chances of their meeting in the latter case (opposite halves) are $m_2 = 1/2^{n-1} \cdot 1/2^{n-1}$ as they can only meet in the final match. Thus both of them must win all $n - 1$ of their games.
– Thus the probability of their meeting is

$$p = p_1 \cdot m_1 + p_2 \cdot m_2 = \frac{2^{n-1} - 1}{2^n - 1} \frac{1}{2^{n-2}} + \frac{2^{n-1}}{2^n - 1} \frac{1}{2^{n-1}} \cdot \frac{1}{2^{n-1}} = \frac{1}{2^{n-1}}.$$

The hypothesis that $p(n)$ might equal to $1/2^{n-1}$ was correct; however, we should be careful in formulating such hypotheses, as we illustrate in the following example.

Figure 17.1 illustrates five initial cases (for $n = 1, 2, 3, 4,$ and 5), where n points are placed on a circle and every point is connected by a chord to every other point. Into how many pieces is each circle divided? It seems that adding the 6th point should result in 32 pieces. In actuality, however, we only get 31 pieces (see figure 17.2) because the formula for the number of pieces is $(n^4 - 6n^3 + 23n^2 - 18n + 24)/24$.

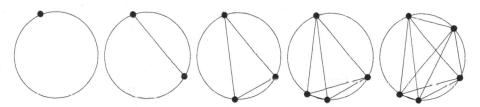

Fig. 17.1. Five initial cases for dividing a circle

Figure 17.1 suggests that the number of pieces grows as 2^{n-1} since

n	Number of pieces
1	1
2	2
3	4
4	8
5	16

Fig. 17.2. Dividing a circle for $n = 6$

You have to be careful not to leap to easy conclusions that just "have to" be right. As humans, we're great at seeing patterns. We're so good, in fact, that we can even invent them when they don't exist! Always do your best to rely on solid evidence rather than leaps of faith.

Now, let's return to the two puzzles given in the Introduction. The first problem was to prove that

$$AD + DB \le AC + CB$$

in a triangle ABC, where D is an arbitrary point in the interior of the triangle (figure 17.3).

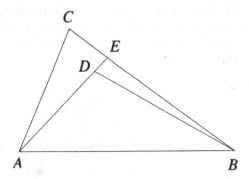

Fig. 17.3. Triangle ABC with an interior point D and extension of AD

Let's extend the segment AD towards the side of the triangle and denote the intersection point of the extension with BC by E. As the sum of two sides of any triangle is longer than the third side, we have

$$AD + DB \le AD + (DE + EB) = AE + EB \le (AC + CE) + EB = AC + CB,$$

which concludes this proof. You could alternatively construct a line segment that goes through D and is parallel to AB, and then pursue a very similar set of inequalities.

The second puzzle was about a well, three meters in diameter, into which we threw two sticks of length four and five meters, respectively, such that they landed in a special configuration. The problem was to find the distance h from the bottom of the well to the point where the two sticks intersect. Let's introduce the following notation. The ends of the shorter stick are A and B, and the ends of the longer stick are C and D (points A and D touch the bottom of the well). The intersection point is E. The height h of the intersection point is the length of the segment EF. Let's also denote the length of the segment CE by x (figure 17.4).

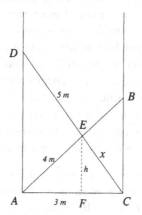

Fig. 17.4. Two sticks in a well

We know that $AB = 4$ and $CD = 5$. It's an easy exercise to calculate that $AD = 4$ and $BC = \sqrt{7}$ (Pythagorean theorem). Triangles CEF and CDA are similar, so

$$\frac{EF}{DA} = \frac{EC}{CD} \implies \frac{h}{4} = \frac{x}{5},$$

so $h = 4x/5$. Triangles EBC and EAD are also similar, so

$$\frac{EC}{BC} = \frac{DE}{DA} \implies \frac{x}{\sqrt{7}} = \frac{5-x}{4}.$$

The last equation results in $x = 5\sqrt{7}/(4 + \sqrt{7})$, so:

$$h = \frac{4x}{5} = \frac{4\sqrt{7}}{4 + \sqrt{7}}.$$

It's difficult to appreciate the solutions to the above two puzzles unless you have spent some time solving them. Indeed, many points discussed earlier in this section can be illustrated by these lines of reasoning.

We've learned about a great many algorithms for solving problems. Undoubtedly, we spent the most time considering the possibilities for applying evolutionary algorithms to our problems. That's because evolutionary algorithms have numerous advantages:

1. *Conceptually, evolutionary algorithms are very simple.* The basic flow of
   the procedure is straightforward. A population of contending solutions
   undergoes random variation and selection over a number of generations in
   light of some evaluation function. As compared with many calculus-based
   methods, evolutionary algorithms are easy to implement and understand.

2. *Evolutionary algorithms have a broad applicability.* You can apply evolutionary algorithms to just about every problem we've discussed in this
   text. Admittedly, there are many better ways to address some of these
   problems, and for those problems we suggest that you not use evolutionary algorithms. When you have a method that is perfect for your particular
   problem, use it! But very often you'll find that the real world poses problems that don't fit the classic algorithms and you'll be forced to either
   compromise on your assumptions about the problem in order to use those
   off-the-shelf procedures, or adopt a different approach. The evolutionary
   approach is pliable. You can adapt it to many different circumstances, and
   indeed it can even be used to adapt itself as it learns about a problem.

3. *Evolutionary algorithms offer the potential for creating useful hybrid methods.* Most classic algorithms come in a "take-it-or-leave-it" form. You either use them, or you don't. In contrast, evolutionary algorithms offer
   you the potential to incorporate domain-specific knowledge into the procedure. You can combine evolutionary methods with deterministic and/or
   local techniques. You can devise variation operators that utilize specific
   information about the problem, as we saw with the geometric crossover
   operator in chapter 9. You can even couple evolutionary computation with
   neural networks and fuzzy systems. In all of the applications that we've
   constructed to solve real-world problems, and that's over 30 years of combined experience, we've never used an evolutionary algorithm that didn't
   incorporate some other facet about the problem in some way.

4. *Evolutionary algorithms are highly parallel.* As computers become cheaper
   and faster, the prospects for distributed processing machines become more
   reasonable. Soon, most desktop computers will be multiprocessor machines. Evolutionary algorithms can often take advantage of this architecture because many designs allow for the evaluation and variation of
   solutions in a parallel mode. Furthermore, as we saw in chapter 16, there
   are evolutionary algorithms that work on subpopulations, so imagine each
   subpopulation on its own processor. The execution time for evolutionary
   algorithms can be accelerated greatly in this manner.

5. *Evolutionary algorithms are robust to dynamic changes.* We saw in chapter
   11 that we could take advantage of the knowledge that an evolutionary
   algorithm had gleaned from a problem to continue to tackle that problem
   after it has changed. Sometimes the problem might change so much that
   our last best set of solutions might not be worthwhile. But starting with
   those solutions probably won't cost much because the alternative is to

start from scratch anyway. The real-world poses problems that are always changing. It's never enough to just solve a static problem. While you are admiring your solution, someone else is already solving the next problem. Evolutionary algorithms offer a population-based reservoir of knowledge for you to rely on.

6. *Evolutionary algorithms offer the potential for self-adaptation.* The flexibility of evolutionary algorithms is manifest in the many control parameters at your disposal. Rather than face tuning these parameters by hand, we've seen ways to allow an evolutionary algorithm to evolve its own parameter values online as it learns from the problem. These self-adaptive methods are a form of reinforcement learning, where strategies for solving the problem that have worked well in the past are rewarded. As the conditions change, the evolutionary algorithm is able to update its search strategies and continue to explore new ways of solving the problem.

7. *Evolutionary algorithms can learn to solve problems for which there are no known solutions.* Perhaps the greatest advantage of evolutionary algorithms comes from the ability to address problems for which there are no human experts. Although human expertise should be used when it's available, it often proves less than adequate for automating problem-solving routines. Experts may not always agree, may not be self-consistent, may not be qualified, or may simply be wrong. If expertise is indeed available, it can be incorporated as hints to be placed in the population and competed with other ideas. In the end, natural selection can determine what works and what doesn't.

Evolutionary algorithms aren't a panacea, but they are often able to generate practical solutions to difficult problems in time for them to be useful. With a solid foundation of all of the problem-solving techniques that we've described in this book, and an appreciation for their advantages and limitations, you're armed and ready for action.

Let's conclude by describing a delightful problem that involves preparing three slices of hot buttered toast. (You didn't think you'd get away easy did you?!) The problem specifications are as follows:

- There's an old toaster with two hinged doors on each side. It can take two pieces of toast at a time, and it only toasts one side of a piece at a time.

- The times for the required activities are:
  - It takes 30 seconds to toast one piece of bread (that's one side).
  - It takes 3 seconds to put a piece in the toaster.
  - It takes 3 seconds to take a piece out from the toaster.
  - It takes 3 seconds to reverse a piece without removing it from the toaster.
  - It takes 12 seconds to butter a side of a toast.

- In addition, the activities of inserting a slice, reversing the sides of a slice, removing a slice, or buttering a slice, require both hands, so they can't be performed at the same time.

- Each piece is buttered only on one side; moreover, the butter can only be applied to a side after that side has already been toasted.

All these specifications are, as in real-world problems, very important. According to the above requirements it's possible to (1) put a piece of bread that is toasted and buttered on one side back into the toaster, or (2) to toast a side of a piece for, say, 15 seconds, take it out for a while, and then to continue toasting it for the remaining 15 seconds at a later time. We have to assume that when we start, the three pieces of bread to toast are out of the toaster, and we have to complete the toasting with the three pieces once again out of the toaster.

The problem is, as you've guessed by now, to develop a schedule so that the three pieces of bread are toasted and buttered in the shortest possible time! A straightforward solution will yield a total time of 120 seconds, but you can do much better! Ah, but how will you find this schedule? Will you use a greedy algorithm? Exhaustive search? An evolutionary algorithm? Some hybrid?

We were able to find a solution, just using our commonsense (!), that gives a total time of 114 seconds. Curiously, the famous writer Martin Gardner claimed on page 17 of [178]:

"... the total time can be reduced to 111 seconds ...".

We've tried to find a solution with a total time of 111 seconds, but we've failed. We feel that Gardner's claim is like the famous sentence by Pierre de Fermat, who (around 1637) wrote in the margin of *Arithmetica*:

It is impossible for a cube to be written as a sum of two cubes or a fourth power to be written as a sum of two fourth powers, or, in general, for any number which is a power greater than the second to be written as a sum of two like powers.

In other words, the equation $x^n + y^n = z^n$ has no whole number solution for n greater than 2. This note was followed by the famous sentence:

I have a truly marvelous demonstration of this proposition which this margin is too narrow to contain.

indexFermat's Last Theorem As this proposition (known as Fermat's Great Theorem or Fermat's Last Theorem) was proved recently — after more than 350 years! — by Andrew Wiles [498, 465], we too hope to discover a solution to the toasting problem that takes only 111 seconds. We also hope it won't take a few centuries![1]

---

[1]This is just a joke: In further printings of [178], 111 seconds were changed to 114, as the original edition took into account only the time required to finish the toasting, and three seconds must be added for removing the slice from the toaster. Also, Walter Kosters from Leiden University in the Netherlands sent us an elegant proof that it's impossible to construct a solution that's shorter than 114 seconds. The space of this footnote, however, is too small to contain this proof ...

Appendix A: Probability and Statistics

> I shall never believe that
> God plays dice with the world.
>
> Albert Einstein

The concepts of probability and statistics allow us to treat many real-world settings that involve apparently random effects. Much of the early work in these disciplines arose out of a need to understand different games of chance. Indeed, few people pay more attention to probability and statistics than do the operators of the gambling casinos in Las Vegas. Nevertheless, the framework of probability and statistics extends to almost every endeavor that concerns problem solving and it is important for you to have a basic understanding of the underlying fundamental concepts.

A.1 Basic concepts of probability

Probability is a branch of mathematics that treats the uncertainty that is inherent to the real world. Intrinsic to probability is the concept of an *experiment*. A *sample space*, S, defines the set of all possible outcomes of the experiment. Probabilities are then assigned to sets that contain outcomes. These sets are called *events*. For example, consider rolling a six-sided die. There are six possible outcomes that define the sample space of the experiment: $S = \{1, 2, 3, 4, 5, 6\}$. We can then define events such as $A = \{\text{The die roll is odd}\}$ or $B = \{\text{The die roll is less than four}\}$. Formally, probabilities are only assigned to events, not to outcomes. To describe the probability of rolling a 1, we must declare a set $A = \{1\}$, and then address the probability of A, which is denoted by $P(A)$ or $\Pr(A)$. Note that the sample space S is itself an event: the event that contains every possible outcome. Similarly, the empty set $\emptyset$ is also an event: the event that contains no outcome of the experiment.

It is useful to define probabilities primarily by three axioms:

1. $P(A) \geq 0$.

2. $P(S) = 1$.

3. (a) If $A_1, A_2, \ldots, A_n$ is a finite collection of mutually exclusive events, then the probability $P(A_1 \cup A_2 \cup \ldots \cup A_n) = \sum_{i=1}^{n} P(A_i)$.

(b) If $A_1, A_2, \ldots$ is an infinite collection of mutually exclusive events, then the probability $P(A_1 \cup A_2 \cup \ldots) = \sum_{i=1}^{\infty} P(A_i)$.

The basic axioms provide the basis for addressing likelihood in an intuitive manner. The probability that we will get some result (i.e., *any* result) from the experiment is 1.0. This is the maximum probability and corresponds to the event that contains all possible outcomes. The probability of any other event A that is not S must be greater than or equal to zero; thus, probabilities always range in $[0, 1]$. Finally, if two or more events are disjoint, the probability of the union of those events is the sum of their individual probabilities, regardless of whether there is a finite or infinite number of such disjoint events.

These axioms lead directly to a fundamental consequence. Let A' be defined as the complement of A; that is, the set that contains all of the outcomes in S that are not in A. Then by definition A and A' are disjoint and their union is S. Thus, by the second and third axioms: $P(A) + P(A') = P(S) = 1$. Therefore, the probability of the complement of A is $1 - P(A)$.

We often want to address the probability that an event A occurs given that some other event B has already occurred. This is called a *conditional probability*, and is written $P(A|B)$, which is read as "the probability of A given B." The situation can be visualized in figure A.1 using what is called a *Venn diagram*. The quantity $P(A|B)$ can't be derived from the axioms directly, so it's defined as:

$$P(A|B) = \frac{P(A \cap B)}{P(B)}, \text{ for } P(B) > 0.$$

This is a reasonable conceptualization in that for the event $A|B$ to occur, B must occur, which happens with probability $P(B)$, and the outcome must be in the intersection of A and B, which happens with probability $P(A \cap B)$.

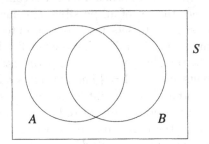

Fig. A.1. A Venn diagram that indicates the sample space S and two events A and B that have common outcomes.

Two events A and B are defined to be *independent* if and only if

$$P(A \cap B) = P(A)P(B).$$

This presents an initial paradox because it's natural to consider two things to be independent if they don't share any common elements. That concept has nothing to do with independence in terms of probabilities. Independence in the mathematical sense means that the fact of event B occurring provides no information to alter the probability of event A occurring. To see this, substitute the product $P(A)P(B)$ in place of $P(A \cap B)$ in the definition of $P(A|B)$. The probability of B cancels in the numerator and denominator, so the result is that when A and B are independent, the conditional probability $P(A|B) = P(A)$. When A and B are disjoint, the probability of the intersection of A and B is zero. Therefore, when A and B are disjoint, A and B are not independent except in the trivial case where $P(A) = 0$.

A.2 Random variables

With the underlying framework of the probability of events from a sample space it is possible to define functions that map from outcomes in the sample space to the real numbers. Such functions are called *random variables* (see figure A.2). Random variables are often denoted by italicized capital letters such as X or Y, with the value that they assume for a particular outcome $s \in S$ being described by the corresponding italicized lowercase letter (e.g., x or y). Formally then, a random variable is a function: $X : S \to \mathbb{R}$, where $X(s) = x$.

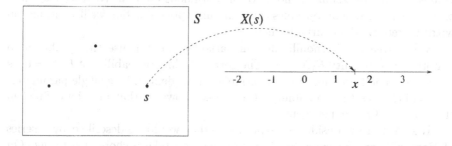

Fig. A.2. A random variable X is a function that maps outcomes from S to the real numbers.

A.2.1 Discrete random variables

We can ask questions about random variables in terms of their likelihood of generating particular values based on the outcome of an experiment. For example, suppose our experiment is defined by the toss of a coin. There are two outcomes: heads or tails. The sample space $S = \{\text{heads, tails}\}$. Let's assume the coin is a fair coin and therefore the probability of getting a head is the same as a tail. Since these are disjoint and partition S, each has probability 0.5. Formally, we need to define the events $H = \{\text{heads}\}$ and $T = \{\text{tails}\}$ and then $P(H) = P(T) = 0.5$. Now let's define a random variable X that maps the outcome "heads" to the real value 1.0 and the outcome "tails" to the real value 0.0. We could ask: What is the probability that X takes on the value 1.0? This is equivalent to asking: What is the probability of the event A that contains all of the outcomes in S such that $X(s \in A) = 1$? Here, there is only one outcome in S for which X maps that outcome to the value 1: heads. Thus, the probability that X will take on the value 1 is $P(H)$ which equals 0.5.

Formally, the probability that a random variable X takes on a value x is written as $P(X(s) = x)$, but it's often more convenient to write $P(X = x)$. We might even simplify the notation further and just write $p(x)$. (Note the use of a lowercase p in this shorthand form, which is common.) It's easy to forget when asking, say, what is the value of $p(1)$, that this implies an entire framework of an experiment, with outcomes and events, axioms of probability, and a random variable. Just remember that the random variable X is a function, not a number, so any time you are asking about some probability in relationship with a random variable, you are really talking about result of a function based on some outcome in an experimental sample space.

In the above example, the random variable X could take on two values: 0 or 1. This is a common form of random variable because we are often interested in classifying the outcome of experiments as one of two possibilities (e.g., we passed or failed the final exam, we hit or missed the target, or we won or lost the game). A random variable that maps every outcome in S to 0 or 1 is called a *Bernoulli* random variable. This is an example of a *discrete* random variable: the values that it can assume do not cover a continuous interval. We'll see examples of *continuous* random variables later in this appendix, but we'll continue on with discrete random variables here.

When treating Bernoulli random variables, we can use the symbol p to denote the probability $P(X = 1)$. The corresponding probability of $P(X = 0)$ is then $1 - p$. The Bernoulli random variable is thus defined by a single parameter, p, that indicates the probability of a *success*, however that may be defined in the context of the experiment.

It's natural to consider an experiment that would be described by a series of Bernoulli random variables. For example, you take n shots at a target. On each shot, you have some probability p of hitting the target (success). If we assume that each shot is independent of the others (i.e., the result of any shot does not affect the probability of success on any future shot), this experiment is described by a *binomial* random variable, which can take on integer values from 0 to n. Thus, for a binomial random variable, we might ask about the probability $P(X = 5)$ when we take 10 shots at a target with a probability of hitting the target (success) of 0.4 and the assumption that each shot is independent of the others. To determine this probability, we have to ask about all of the outcomes in the experiment that have 5 successes. The experiment is a series of 10 shots at a target. The outcomes of the experiment are things like SSSFSFFSFF, where S denotes a success and F signifies a failure. We need to determine the probability of the event that contains all of the outcomes with exactly 5 successes.

Based on the assumption that each shot is independent, we know that the probability of, say, SSSFSFFSFF is the product of $p(S)$, $p(S)$, $p(S)$, $p(F)$, $p(S)$, $p(F)$, $p(F)$, $p(S)$, $p(F)$, and $p(F)$. And since $p(S)$ is a constant p for each shot because we assumed that it had to be, this product simplifies to $p^5(1-p)^5$. But this is just one way of getting 5 successes in 10 shots. There are other sequences that also offer 5 successes: SSSSSFFFFF, SSSSFFFFFS, and so forth. Since we don't care about the order of the successes but rather simply the number of successes, this is a *combination* not a *permutation*. The number of combinations of 5 successes in 10 shots is $\frac{10!}{(5!)(5!)} = 252$. Thus, the probability of obtaining 5 successes in 10 shots given a probability of success $p = 0.4$ is

$$P(X = 5) = 252(0.4)^5(0.6)^5 = 0.200658.$$

To review, the binomial random variable describes the situation when the number of trials, n, is fixed before you start the experiment and you have a constant probability of success, p, on each trial. Furthermore, each trial is independent of the others, and we want to know the probability of obtaining some number of successes, typically denoted by k, in the n trials. If the number of trials isn't fixed beforehand, or the probability of success is not constant, or if the trials aren't independent, then you don't have a binomial random variable.

Random variables are used to model processes from the real world. As with all models, the fidelity of a random variable to some particular condition depends on how well the assumptions hold. A Bernoulli random variable with $p = 0.5$ is a good model for a typical flip of a coin. A binomial random variable is a good model for a case where we might want to know the likelihood of seeing 3 people show adverse reactions to a certain vaccine that is administered to 1000 people, where we assume that the chance of showing a reaction is some probability p that is constant for each person. Is it also a good model for asking about the likelihood that a student will pass, say, two or three of their four tests in a semester? We could set $n = 4$ and then we would need to calculate $P(X = 2 \text{ or } 3)$. The trouble with this is that we don't know the probability that the student will pass a test and, what's more, the likelihood of passing each test isn't constant. Each test might not be of equal difficulty and the student will also be learning along the way. The degree of correspondence between the real-world setting and the binomial random variable may or may not be sufficient. This is a matter that requires subjective judgment. The more experience you have with modeling processes, the better your judgment in these regards will be.

Suppose you wanted to model the number of meteors you might see in an hour. At once, you know that this is not a binomial-type variable because there is no preset number of trials. The outcome of the experiment could be any integer number from zero to, in theory, infinity. When facing such a circumstance, one possible random variable to choose is called a *Poisson*. It models a case where occurrences transpire with a certain rate, called λ and the number of occurrences $x \in 0, 1, \ldots, \infty$. The Poisson random variable assigns the likelihood of observing x occurrences of some process under a rate parameter λ as

$$P(X = x) = \frac{e^{-\lambda}\lambda^x}{x!}.$$

As with the Bernoulli and binomial random variables, the Poisson is discrete because it only takes on non-negative integer values.

At this point it's useful to introduce an additional property of random variables. From the axioms of probability, the probability that a random variable X will take on any particular value must be at least 0.0 and no more than 1.0. Thus, the probability that a Poisson random variable takes on a value from zero to infinity has to be 1.0. In other words, $P(X = 0) + P(X = 1) + \ldots = 1.0$. Let's verify that this is true.

By the Maclaurin infinite series expansion of e^λ,

$$e^\lambda = 1 + \lambda + \frac{\lambda^2}{2!} + \frac{\lambda^3}{3!} + \ldots = \sum_{x=0}^{\infty} \frac{\lambda^x}{x!}.$$

Therefore,

$$\sum_{x=0}^{\infty} \frac{e^{-\lambda}\lambda^x}{x!} = e^{-\lambda}e^\lambda = 1.$$

There are many other discrete random variables that you should know but they are beyond the scope of this appendix. Please see [101, 100] for a more complete treatment.

A.2.2 Continuous random variables

In contrast with discrete random variables, continuous random variables can take on any number in a continuous interval, even to the extent of being able to take on any real value. Immediately, we must switch from thinking about separate discrete masses of probability to a concept of *density*. If a random variable is continuous, then the probability that it will take on any particular value exactly is zero. We can only really speak about positive probabilities assigned to a range of values. Thus the probability that the continuous random variable X assumes a value between $[a, b]$ is defined as

$$P(a \le X \le b) = \int_a^b f(x)dx.$$

Here, $f(x)$ is the *probability density function* of the random variable X. The corresponding concept in discrete random variables is a *probability mass function*, where probability density is concentrated into measurable mass at single points. Since the probability of obtaining some outcome from the sample space S is 1.0, the probability that a continuous random variable will take on a value between $-\infty$ and ∞ must be 1.0. Also, we require that a probability density function always be non-negative.

Perhaps the simplest case of a continuous random variable is the *uniform* random variable with probability density function

$$f(x) = 1/(b - a), \qquad a \le x \le b.$$

To be explicit, this is sometimes defined with the additional notation that $f(x) = 0$ outside the interval $[a, b]$. If left unstated, it is generally presumed that a probability density function is zero anywhere it is not defined explicitly. In this case, it's easy to verify that $f(x)$ is a valid probability density function: (1) it is always non-negative, and (2) it integrates to 1.0 from $-\infty$ to ∞. The uniform random variable is the basis for many other random variables in that, through a series of manipulations, it is possible to transform a uniform random variable into other probability density functions that are useful in different contexts.

The most widely applicable continuous random variable is the Gaussian (also known as *normal*) random variable, which has a probability density function:

$$f(x) = \frac{1}{\sigma\sqrt{2\pi}}e^{-\frac{1}{2}(x-\mu)^2/\sigma^2}$$

for x in the reals. This distribution is called "normal" because it's what we normally observe when measuring a variety of natural phenomena: people's heights, the stem length of a species of flowers, the distance between eyes of a fruit fly. There is a theoretical justification for this observation in what is called the *central limit theorem*, which will be discussed later in this appendix. The normal random variable can also be used to approximate other random variables, even when they are discrete, as will also be presented later. Note that the normal random variable is defined by two parameters: μ and σ. These have physical interpretations: μ is the point of symmetry (also the *expected value*, see below) and σ is a measure of the degree of spread of the function (also the *standard deviation*, see below).

There are many other useful continuous random variables. The *exponential* random variable

$$f(x) = \lambda e^{-\lambda x}, \quad x > 0$$

is often used to model the lifetime of components and equipment (e.g., light bulbs). Note that the exponential random variable only takes on positive density for positive values of x, and that λ must also be positive. The *chi-square* random variable:

$$f(x) = \frac{1}{\sqrt{2\pi}}x^{-1/2}e^{-x/2}, \quad x > 0$$

describes what results when you take the square of a normal random variable with zero mean and a variance of one. This is sometimes useful in modeling errors that are assumed to be distributed according to a Gaussian distribution and we are interested in describing the squared error. The chi-square random variable actually can be used to describe the sum of a series of Gaussian random variables, and in that case the chi-square random variable is parameterized by the symbol ν which signifies the so-called *degrees of freedom* of the random variable (see [101] for more on the parameterized chi-square random variable). Both the exponential and chi-square random variables are special cases of what are called *Gamma* random variables (because they involve Gamma functions), but further description of these random variables is beyond the scope of this appendix.

One other important continuous random variable is the *Cauchy* random variable, which has a probability density function that is similar to the Gaussian, but has much fatter tails:

$$f(x) = \frac{a}{\pi(x^2 + a^2)}, \qquad a > 0, \qquad -\infty < x < \infty.$$

The Cauchy random variable with $a = 1$ is a special case of a so-called t-distribution, which will be used later in the appendix when discussing statistical hypothesis testing.

A.3 Descriptive statistics of random variables

It's often useful to characterize random variables in terms of their statistical properties. For example, what is the most likely value for a certain random variable? This quantity is termed the *mode*. In the case of a continuous random variable, the mode corresponds to the value that is associated with the maximum of the probability density function. It may be that a random variable has more than one mode, i.e., more than one value that occurs with maximum probability or that maximizes the probability density function. In fact, there may be an infinite number of such values, as occurs with a uniform random variable.

Another useful descriptive statistic is the *median*, which indicates the midpoint for which 50 percent of the probability mass or density lies below (and above) that value. Mathematically, the median is that value x such that $P(X \leq x) = P(X \geq x) = 0.5$. A unique median value may not exist for certain discrete random variables. For example, for a Bernoulli random variable with $p = 0.5$ there is no single value x that satisfies the above equality. Any number greater than zero and less than one suffices. One convention in such a case is to choose the midpoint of this range as the median, but this is not a steadfast rule.

By far, the most common descriptive statistic to summarize a random variable is its *mean* or synonymously its *expectation*. The mean is essentially the average of the values that the random variable assumes, weighted by the likelihood that it assumes each such value. For discrete random variables then, the expectation is defined as

$$E(X) = \sum_{x \in D} x P(X = x),$$

where D is the domain of X. Specifically, for a Bernoulli random variable, this is calculated as

$$E(X) = 0 \cdot P(X = 0) + 1 \cdot P(X = 1) = 0 \cdot (1 - p) + 1 \cdot p = p.$$

By similar calculations, it can be verified that the expected value of a binomial random variable is np and the expected value of a Poisson random variable is λ.

The continuous analog of the discrete expected value is the integral

$$E(X) = \int_{-\infty}^{\infty} x f(x) dx.$$

For a uniform random variable that takes on positive density over the range $[a, b]$,

$$E(X) = \int_{-\infty}^{\infty} x \frac{1}{b-a} dx = \int_{a}^{b} x \frac{1}{b-a} dx = \frac{1}{b-a} \frac{x^2}{2}\Big|_{a}^{b} = \frac{1}{b-a} \frac{b^2 - a^2}{2} = \frac{b+a}{2},$$

as could be intuited from visual inspection. The expectation of a Gaussian random variable is μ, the expectation of the exponential random variable is λ^{-1}, and the expectation of a chi-square random variable is ν, its degrees of freedom. The Cauchy random variable presents a pathological case where even though the probability density function is symmetric, the expectation does not exist because the integral

$$\int_{-\infty}^{\infty} x \left(\frac{a}{\pi(x^2 + a^2)} \right) dx$$

diverges for $a > 0$. Make a particular mental note that not every random variable has a well-defined expectation, and that you cannot simply "eyeball" a probability density function and determine the mean of the corresponding random variable.

The mode, median, and mean are all measures of *central tendency*, statistics that provide information on what to expect when sampling from a random variable (which, remember, means that an experiment is defined, an outcome occurs, and the random variable maps that outcome to a real value). Measures of central tendency, while useful, do not describe the uncertainty inherent in that expectation. In addition to this measure, it is often useful to describe the degree to which the probability mass or density is spread across the range of possible values. For example, both the uniform random variable from $[-a, a]$ and Gaussian random variable with $\mu = 0$ and $\sigma = a$ have the same expectation, but the likelihood of obtaining a value within a units of the mean (zero) is quite different in each case (i.e., it is 1.0 for the uniform random variable and is approximately 0.68 for the Gaussian).

One way to describe the variability of a random variable is in terms of its *variance*, which is defined for discrete random variables as

$$V(X) = \sum_{x \in D} (x - E(X))^2 P(X = x),$$

and for continuous random variables as

$$V(X) = \int_{-\infty}^{\infty} (x - E(X))^2 f(x) dx.$$

This is essentially the average squared distance from the mean to the value that the random variable assumes, weighted by the likelihood of that value (or a limiting case of the likelihood of an infinitesimally small interval dx in the continuous case). Instead of working with squared units of error, we can take the square root of the variance, which is called the *standard deviation*.

The variance of a Bernoulli random variable is

$$V(X) = (0 - p)^2(1 - p) + (1 - p)^2 p = p^2(1 - p) + (1 - p)^2 p = (1 - p)p(p + 1 - p) = (1 - p)p.$$

The value $(1 - p)$ is often referred to by the symbol q, thus, the variance of a Bernoulli random variable is pq. The variance of the other random variables discussed above are

> binomial: npq
> Poisson: λ
> uniform: $(b - a)^2/12$
> Gaussian: σ^2
> exponential: λ^{-2}
> chi-square: 2ν
> Cauchy: undefined

Because the Gaussian random variable is so fundamental for work in probability and statistics (as will be shown shortly), the typical symbols for the mean and variance of a general random variable are μ and σ^2, respectively.

Computationally, the variance is equal to

$$V(X) = E(X^2) - E(X)^2,$$

where the expectation of a function of X, say $g(X)$, is

$$E(g(X)) = \int_{-\infty}^{\infty} g(x)f(x)dx.$$

Thus,

$$E(X^2) = \int_{-\infty}^{\infty} x^2 f(x)dx.$$

Just as we can consider taking the expectation of X^2, we can also consider the expectations of X^3, X^4, and so forth. These are called the *moments* of a random variable: $E(X^n)$ is the n-th moment. Alternatively, we can consider the n-th *central moment*: $E((X - E(X))^n)$. There is a geometric interpretation of the third and fourth central moments (called *skewness* and *kurtosis*) where these indicate the degree to which a probability density function is skewed to the right or left of its mean, and how peaked or squashed it is, respectively. For further consideration of moments, their interpretation, and moment generating functions, see [100].

A.4 Limit theorems and inequalities

One of the most important theorems of probability is the *central limit theorem*. There are several versions of the theorem, but in the most basic case the theorem states

> *Theorem:* If $X_1, X_2, \ldots, X_n$ are independent random variables that are identically distributed and have finite means and variances, then the normalized sum $(S_n - n\mu)/(\sigma\sqrt{n})$, where $S_n = X_1 + X_2 + \ldots + X_n$, as $n \to \infty$ converges to a Gaussian random variable with zero mean and a variance of one.

In words, as long as you are adding up a sufficient number of independent and identically distributed random variables with finite means and variances, the sum will tend to be Gaussian.

This is a truly amazing and powerful result. If you take the sum of a sufficient number of, say, Poisson random variables each with a rate parameter λ, this sum will have a probability (mass) function that is closely approximated by a Gaussian with a mean of $n\lambda$ and a variance of $n\lambda$. The sum of a sufficient number of binomial random variables is approximately Gaussian with a mean of np and a variance of npq. Note that in both these cases, the values of the sum of Poisson or binomial random variables cannot be negative and can only take on integer values. Nevertheless, the correspondence to the Gaussian random variable can be very strong.

Likewise, the sum of exponential random variables, chi-square random variables, or even uniform random variables, in the limit, tends to a Gaussian distribution. The sum of Cauchy random variables does *not* approach a Gaussian distribution because the Cauchy random variable does not have finite means and variances. The proof of the theorem is outlined in [100].

A related theorem is termed the *law of large numbers*. If $X_1, \ldots, X_n$ are independent random variables each having finite means and variances, the sum $S_n = X_1 + \ldots + X_n$ satisfies

$$\lim_{n \to \infty} P\left(\left|\frac{S_n}{n} - E(X)\right| \geq \epsilon\right) = 0$$

for $\epsilon > 0$. To say this in words, think first of a random variable Y that is equal to the random variable S_n divided by n. Then, the probability that the magnitude of the difference between Y and its expected value will be greater than any positive value is zero in the limit. The more random variables you add up, the smaller the variance of the normalized sum of those random variables. There are two versions of the theorem, termed *weak* and *strong* that relate to the type of convergence (in probability, or with probability 1). For more details on modes of stochastic convergence see [345].

The central limit theorem offers a means for approximating the distribution of a sum of a sufficient number of random variables, even when very little is known about their individual properties. Other theorems are available to provide bounds on the probability that a random variable differs from its mean under very slight or even no assumptions about its distribution. One important result that does not impose any restriction beyond finite means and variances is the *Chebyshev inequality*:

$$P(|X - E(X)| \geq \epsilon) \leq \sigma^2/\epsilon^2$$

or with $\epsilon = k\sigma$

$$P(|X - E(X)| \geq k\sigma) \leq 1/k^2.$$

In words, the probability that the sample from a random variable will be at least k standard deviations away from its expectation is always less than or equal to $1/k^2$. If you know something about the distribution of the random variable in question, this bound may prove to be very conservative. For instance, with a Gaussian random variable, the probability of generating a value that is more than two standard deviations away from the mean is about 0.05. The Chebyshev inequality guarantees that this probability is less than 0.25. Nevertheless, when you know very little about the distribution of a random variable, the Chebyshev inequality can provide useful information.

A.5 Adding random variables

The central limit theorem and law of large numbers provide guidance on the limits of sums of random variables, but it is also important to know the distributions of sums of a smaller collection of random variables. Of primary importance is understanding that the random variable $Y = X_1 + X_2$ is not the sum of the probability density functions for X_1 and X_2 but rather the sum of the values that are obtained by sampling from X_1 and X_2. When X_1 and X_2 are independent, i.e., their joint density function $f(x_1, x_2) = f(x_1)f(x_2)$, the probability density function of Y, $g(y)$, is described by a convolution integral:

$$g(y) = \int_{-\infty}^{\infty} f_{X_1}(y - z) f_{X_2}(z) dz, \qquad -\infty < y < \infty.$$

For many of the canonical forms of random variables described above, however, it's helpful to know the following relationships:

- The sum of two independent Gaussian random variables with means μ_1 and μ_2 and variances σ_1^2 and σ_2^2 is also Gaussian with a mean of $\mu_1 + \mu_2$ and a variance of $\sigma_1^2 + \sigma_2^2$.

- The sum of two independent Poisson random variables with rate parameters λ_1 and λ_2 is a Poisson random variable with rate parameter $\lambda_1 + \lambda_2$.

- The sum of two independent exponential random variables with parameters λ_1 and λ_2 is an exponential random variable with parameter $\lambda_1 \lambda_2 / (\lambda_1 + \lambda_2)$.

- The sum of two independent chi-square random variables with degrees of freedom ν_1 and ν_2 is a chi-square random variable with degrees of freedom $\nu_1 + \nu_2$.

Don't be lulled into thinking that if you add two random variables of the same type that you always get another random variable of that type. For example, if you add two independent uniform random variables with positive density over the range [0,1], the result is a random variable with a triangular probability density function over [0,2].

A.6 Generating random numbers on a computer

You will need to generate random numbers with particular, specified distributions in order to implement many of the problem-solving methods described in this book. It's important to note that almost all random number generating procedures actually create series of *pseudorandom* numbers. They aren't random in the strict mathematical sense because the entire sequence of numbers is a deterministic function based on an initial seed value. Nevertheless, sequences of numbers can be created such that they appear very close to random.

Many programming development environments include random number generators as built-in functions. Most include a function that will return values in the range [0,1] with a uniform distribution. Some also include functions for returning values that are Gaussian distributed with a given mean and variance. You should know that not all random number generators have the same quality. The most basic measure of quality is the cycle length: the number of numbers that are generated before the generator cycles and repeats the same sequence again. The longer the cycle length, the better. If you aren't sure about the properties of the random number generator that you are using, you should consider using one that is well documented. Many random number generators are detailed in programming texts (e.g., [363]).

If you can generate uniform random numbers between [0,1] then you have the basis for generating many other random variables as well. For example, if you want to generate a Bernoulli random variable with parameter p, then call a uniform random number between [0,1] and determine if it is less than p. If so, return the value 1.0, otherwise return the value 0.0. Binomial random variables can be constructed by successively calling Bernoulli random variables.

Uniform random variables can also be used to create other important random variables, such as the Gaussian and exponential. You can generate two Gaussian random numbers with zero mean and variance one by taking two independent uniform random numbers, U_1 and U_2, and transforming them as

$$\sqrt{-2\log U_1}\sin(2\pi U_2) \text{ and } -\sqrt{-2\log U_1}\cos(2\pi U_2).$$

You can generate an exponential random number with parameter λ by taking the transform $\frac{1}{-\lambda}\log(U_1)$.

From a Gaussian random variable, you can generate a chi-square random variable. If you take the square of the Gaussian number, then you get the chi-square number (with one degree of freedom). Recall that the sum of independent chi-square random variables is another chi-square random variable with degrees of freedom equal to the sum of the degrees of freedom of the two independent chi-square random variables. Thus, you can create any chi-square random variable that you want by adding up a sufficient number of squared independent Gaussian random variables.

You can also generate a Cauchy random variable from two independent Gaussian random variables simply by taking their ratio. Alternatively, you can use the formula

$$\tan[\pi(U_1 - 0.5)],$$

which relies only on a uniformly distributed random variable.

One other random variable to consider here is the Poisson, which is a bit more difficult to generate. A recommended procedure [133] is to determine some arbitrary cutoff value for the maximum number that you can observe, N, and then determine the probability (possibly from a statistical table or by brute-force calculation from the probability mass function) of observing 0, 1, 2, ... , N, observations. Next, generate a uniform random number and determine the first number from 0, 1, 2, ... , N, such that the cumulative probability is greater than the number obtained from the uniform distribution. In the unlikely event that all N numbers are considered without obtaining a cumulative probability that is larger than the sampled uniform distribution, simply choose N.

A.7 Estimation

One of the fundamental activities of statistics is to gain insight into the parameters of a population through sampling and estimation.[2] For example, suppose that we want to determine the mean height of men in Sweden. There are too many men to locate and measure, and we could never be certain that we sampled everyone (this is always a problem when taking any census). Instead, suppose that we could assume that these heights were distributed according to a Gaussian distribution with some mean μ and variance σ^2, but these parameters were unknown. In notational shorthand, this is written as $N(\mu, \sigma)$, where N stands for "normal."[3] Our task is to estimate these parameters based on a sample of n people. If each person's height is then a sample from a random variable X that is distributed as $N(\mu, \sigma)$ we can estimate μ by taking the sample mean: $\bar{x} = (x_1 + \ldots + x_n)/n$.[4] The larger the sample size n the better our estimate will

---

[2]Professor Will Gersch of University of Hawaii at Manoa once said "the purpose of statistics is insight, not data."

[3]Or alternatively $N(\mu, \sigma^2)$. The question of whether the second parameter is the standard deviation or the variance should be made explicit.

[4]Remember that $\bar{x}$ is the realized sample value of the random variable $\bar{X}$.

likely be by the law of large numbers (we'll talk about measuring confidence in an estimate shortly).

In our case of sampling here, the random variable $\overline{X}$ is the sum of the random variables $X_1, \ldots, X_n$ divided by n. Therefore $\overline{X}$ has its own probability density function. In this case, it is Gaussian because the sum of Gaussians is also Gaussian, but it would be Gaussian by the central limit theorem if n were large and each X_i, $i = 1, \ldots, n$ had finite means and variances. Since $\overline{X}$ is a random variable, it's of interest to determine its expectation, if it exists. Here,

$$E(\overline{X}) = E\left(\frac{X_1 + \ldots + X_n}{n}\right) = \frac{1}{n}\left(E(X_1) + \ldots + E(X_n)\right) = \frac{1}{n}n\mu = \mu.$$

Thus, the expectation of the sample mean is the same as the mean of the population. Note that we didn't use the fact that we are assuming the population has a Gaussian distribution here. This property will hold regardless of the distribution of the population (assuming finite means). Whenever an estimator, like $\overline{X}$ here, has the property that its expectation is equal to the parameter it estimates then it is said to be *unbiased*.

With regard to estimating the variance of a population, the sample variance

$$s^2 = \frac{\sum_{i=1}^{n}(x_i - \overline{x})^2}{n - 1}$$

is an unbiased estimator of σ^2. Note that the denominator is $n - 1$ and not n, which yields a biased estimate of the variance.

An alternative method of estimating parameters is based on finding the values of the parameters that maximize the likelihood of the data. Given a set of observed values $x_1, \ldots, x_n$ and some assumed underlying random variable, if the samples are independent and identically distributed then the *likelihood function* is

$$L(\mathbf{x}, \theta) = f(x_1, \theta)f(x_2, \theta)\ldots f(x_n, \theta),$$

where θ is the parameter(s) that governs the probability density function $f(x)$ (e.g., λ in an exponential random variable). The idea is to find the value of θ that maximizes $L(\mathbf{x}, \theta)$. A typical procedure for many random variables that involve exponential functions (e.g., Gaussian and exponential random variables) is to take the natural logarithm of the likelihood function, then take the derivative with respect to θ, set the result equal to zero, and solve for θ. In the case of multiple parameters, the maximum likelihood estimate of each is found by taking the partial derivative of the likelihood function (or the loglikelihood function) with respect to that parameter (e.g., the parameters μ and σ in a Gaussian random variable).

The use of a single estimate of a parameter (whether it is generated using maximum likelihood principles or any other means) does not afford a measure of confidence in the accuracy of the estimate. To do this, we need to take into account the variance of the estimate. If the population is assumed to be distributed with a Gaussian probability density function, then the distribution of $\overline{X}$ is also Gaussian with a mean of $E(X)$ and a variance of $V(X)/n$. Therefore,

$$\frac{\overline{X} - E(X)}{\sigma/\sqrt{n}} \sim N(0,1),$$

where $E(X)$ is the mean of the population, σ is the standard deviation of the population, n is the sample size, "$\sim$" is read as "is distributed as," and $N(0,1)$ represents a *standard* Gaussian random variable. Consider the probability:

$$P(\overline{X} - 1.96\sigma/\sqrt{n} < E(X) < \overline{X} + 1.96\sigma/\sqrt{n}).$$

This is the likelihood that $\overline{X}$ will generate a value that is within 1.96 standard deviations of its mean. (Note that $\frac{\sigma}{\sqrt{n}}$ is the standard deviation of $\overline{X}$.) Since $\overline{X}$ is a Gaussian random variable, this probability is about 0.95. The interval

$$[\overline{X} - 1.96\sigma/\sqrt{n}, \overline{X} + 1.96\sigma/\sqrt{n}]$$

is termed a *95 percent confidence interval*. The confidence associated with the interval can be increased or decreased by increasing or decreasing the scaling factor 1.96. The way to interpret the interval is that, in the limit of repeating values of $\overline{X}$, 95 percent of the intervals generated would bound the value of $E(X)$. The way *not* to interpret the interval is to say that the probability that $E(X)$ lies within the bounds of the interval is 0.95. The subtle difference is that the latter statement implies that $E(X)$ is a random variable. It is not; it's a parameter of the population. The interval itself is what is random.

If the original population is not assumed to be Gaussian, if n is sufficiently large, we know from the central limit theorem that the distribution of $\overline{X}$ will be approximately Gaussian, and we can still use the same formula for a 95 percent confidence interval. The problem that you might have noticed, however, is that we need to know the value of σ (or equivalently $V(X)$), and this is another population parameter that is almost always unknown. Thus, we must estimate σ with the sample standard deviation s. This presents a potential difficulty, however, because s is itself a random variable, and the distribution of

$$\frac{\overline{X} - E(X)}{s/\sqrt{n}}$$

is not Gaussian. Fortunately, the distribution is known and is described by the so-called t-distribution, which has a parameter termed the *degrees of freedom* that is equal to $n - 1$. The t-distribution with one degree of freedom is the Cauchy distribution; with infinite degrees of freedom, it becomes equivalent to a Gaussian. Inbetween, the shape of the t-distribution gradually changes from one to the other. This means that to have an interval around a parameter at a specified confidence we must change the scaling coefficient. As n gets large, the scaling coefficient tends to 1.96 (the same as used with a Gaussian). For small n, however, the scaling coefficient can be quite large. For example, if $n = 2$ then the scaling coefficient for a 95 percent confidence limit on $E(X)$ is 12.706.

The more samples are taken, the shorter the confidence interval. If n is very large, you might still use the scaling value of 1.96 by assuming that the estimate of σ that is obtained by s is very good. For small n, however, you'll need to (1) assume that the population is normally distributed, and (2) use the t-distribution to calculate the appropriate scaling term. Like the Gaussian distribution, the t-distribution is tabulated in most books on statistics (see [101]).

If the data are not normally distributed and n is small, then you will violate the required assumptions if you use a Gaussian approximation or the t-distribution. The degree to which this violation is important depends on the problem at hand. This is an issue that requires judgment, and again this is really best obtained through experience.

It is also possible to generate confidence intervals for the population variance (or other parameters). See [101, 100] for further information.

A.8 Statistical hypothesis testing

Probability and statistics provide a framework for testing various beliefs about the real world. In particular, we are often concerned with testing hypotheses of the form $H_0 : \mu = \mu_0$ versus $H_a : \mu \neq \mu_0$, where μ_0 is some specified value and μ is the mean of a population or process. H_0 is described as the *null hypothesis* and H_a is known as the *alternative hypothesis*. The challenge is to make a judgment as to whether there is sufficient reason to reject H_0 in the face of available data. There are two types of errors that we could make. The first is known as a *Type I* error, which occurs when we reject H_0 when it is in fact true. This is also known as a *false alarm*. The second error is a *Type II* error, which occurs when we fail to reject H_0 when it is false. This is also known as a *miss*. In general, for any statistical test, there are nonzero probabilities of both Type I and Type II errors. In practice, we set the likelihood of a Type I error before we collect the data. Tradition has dictated that this probability (also known as the *level of significance*, and usually described by the symbol α) be set to 0.05. The proper choice of the level of significance depends on the application and the cost of a Type I error. Once the level of significance is set, the only ways to control the probability of a Type II error are typically by selecting the appropriate statistical test and by adjusting the sample size.

Statistical hypothesis testing generally begins with a belief that the null hypothesis is true. Under that belief, we determine the statistical distribution of a sampling statistic that estimates the parameter of interest in the null hypothesis. When treating the mean of a population or process, we use the sample mean $\bar{x}$ as an estimator of μ. We have already seen that $\overline{X}$ is an unbiased estimator of μ and when the original population has a Gaussian distribution, so does $\overline{X}$. Thus, if we assume a Gaussian distribution for a population, with parameters μ and σ, and we are interested in testing whether or not μ equals some particular value, we rely on the distribution of $\overline{X}$ to help make a determination. For example, we might want to know if there is sufficient evidence to reject the belief that the level of pollutants in a river is greater than, say, 100 parts per million (ppm), under the assumption that the ppm has an approximate Gaussian distribution. We start with a null hypothesis of $H_0 : \mu = 100$ versus the alternative hypothesis $H_a : \mu \neq 100$. If we know the value of σ then we also know that under the null hypothesis, the distribution of $\overline{X}$ will be N($100, \sigma/\sqrt{n}$). Let's say we take n samples from the river. We can ask: for what values of $\overline{x}$ will we be inclined to reject the null hypothesis? The answer is that we'd want to reject the null hypothesis if the resulting value of $\overline{x}$ was much larger than 100, or much smaller than 100. The only way to measure what *much* means in this case is in terms of the standard deviation of $\overline{x}$: $\sigma/\sqrt{n}$.

Since $\overline{X} \sim$ N($100, \sigma/\sqrt{n}$) under the null hypothesis, there is a 0.95 probability that $\overline{x}$ will fall between $100 \pm 1.96\sigma/\sqrt{n}$. Thus, we can adopt the decision rule that if $\overline{x}$ falls outside this range we will state that there is *statistically significant* evidence to reject H_0. When H_0 is true, this will occur with 0.05 probability. Alternatively, we can calculate the quantity $\frac{\overline{x}-\mu}{\sigma/\sqrt{n}}$ and then reject H_0 if this quantity is greater than 1.96 or less than -1.96. The region of values for which we reject H_0 is called the *critical region*. If our test statistic fails to come up in the critical region then we *fail to reject H_0*. Note that we can never *accept H_0*. For any sample size n, there are an infinitude of hypotheses for which we would fail to reject H_0 (e.g., here, if $\overline{x}$ were 100.4 with $\sigma = 10$ and $n = 50$ we would fail to reject $H_0 : \mu = 100$, but we would also fail to reject a null hypothesis that claimed $\mu = 100.1$). No amount of evidence can convince us that H_0 is true. If you ever read a published account that *accepts* the null hypothesis, you can safely surmise that the author does not understand statistical hypothesis testing, or they are being sloppy with shorthand.

When σ is unknown, just as with generating confidence limits, we can substitute the sample standard deviation s as an estimate of σ. If n is sufficiently large we can again use the Gaussian approximation for $\overline{X}$. If n is small, then we must assume that the population we are sampling is Gaussian and use the t-distribution with $n - 1$ degrees of freedom to model $\overline{X}$. Please see [101, 100] for more details on statistical hypothesis testing.

When we cannot make assumptions about the distribution of the underlying population, we must rely on *nonparametric* methods for testing hypotheses. A treatment of these methods is beyond the scope of this appendix, but see [433] for a thorough discussion of these techniques.

A.9 Linear regression

A common recurring problem in engineering is that of modeling the response of a dependent variable y based on a collection of m independent variables $x_1, \ldots, x_m$. As a first approximation we often work within the class of linear models. Without loss of generality, consider the model

$$y = a + bx.$$

Given a set of n observations y_i, x_i, $i = 1, \ldots, n$, the task is to find the best values of a and b. To specify the meaning of "best," we can choose a and b such that the value of

$$\sum_{i=1}^{n}[y_i - (a + bx_i)]^2 \tag{17.1}$$

is minimized. In words, we can choose a and b such that the sum of the squared differences between the model's predicted value of each y_i and the actual value of y_i is as small as possible. Taking the partial derivatives of (17.1) with respect to a and b and setting them equal to zero yields a system of two linear equations in two variables:

$$\sum_{i=1}^{n} y_i = an + b \sum_{i=1}^{n} x_i,$$
$$\sum_{i=1}^{n} x_i y_i = a \sum_{i=1}^{n} x_i + b \sum_{i=1}^{n} x_i^2.$$

The solution for a and b are

$$a = \frac{(\sum y)(\sum x^2) - (\sum x)(\sum xy)}{n \sum x^2 - (\sum x)^2}, \quad b = \frac{\sum(x - \bar{x})(y - \bar{y})}{\sum(x - \bar{x})^2}.$$

Note that the subscripts have been omitted for the sake of simplifying the presentation. Computationally, it is easy to first find b and then solve for a as

$$a = \bar{y} - b\bar{x}.$$

Similar procedures can be employed when multiple independent variables are included in a linear model.

The quality of a linear model of data can be measured in terms of the variation of the dependent variable that is explained by that model. The so-called *total variation* of the dependent variable y

$$\sum_{i=1}^{n}(y_i - \bar{y})^2,$$

can be divided into two parts,

$$\sum_{i=1}^{n}(y_i - \bar{y})^2 = \sum_{i=1}^{n}(y_i - \hat{y}_i)^2 + \sum_{i=1}^{n}(\hat{y}_i - \bar{y})^2,$$

where $\hat{y}_i$ is the estimate of y_i. The first part is called the *unexplained variation* and the second part is termed the *explained variation*. Taking the ratio of the explained variation to the total variation,

$$\frac{\sum_{i=1}^{n}(\hat{y}_i - \bar{y})^2}{\sum_{i=1}^{n}(y_i - \bar{y}_i)^2}$$

yields the percentage of variation of the data that is explained by the model. This quantity is described by the symbol R^2.

A related statistic, r, which is the signed square root of R^2, measures the degree of linear correlation between the dependent variable and the model's estimates. A value of r that is either ± 1 indicates an exactly linear correlation (either positively or negatively correlated). The value of r can be computed directed as

$$r = \frac{\sum(x - \bar{x})(y - \bar{y})}{\sqrt{\sum(x - \bar{x})^2}\sqrt{\sum(y - \bar{y})^2}},$$

where the summations are taken over all n samples and the subscripts on variables x and y have again been omitted.

It is often useful to posit a model of the form

$$y = a + bx + \epsilon,$$

where ϵ represents exogenous noise terms. When these are distributed as a Gaussian random variable with zero mean and a constant variance, it becomes possible to use statistical hypothesis tests to assay whether or not there is statistically significant eviden   o reject a belief that the coefficient on any independent variable is zero. ¬ ⌐ procedure is described in [101, 100], as are many other important details re⌐⌐⌐ling the design of linear and nonlinear models of data.

A.10 Summary

This appendix provides only a cursory overview of the important concepts of probability and statistics that are typically covered in a first course at the undergraduate level. Additional topics include: (1) the fundamental cumulative distribution function, which is the integral of the probability density (mass) function, (2) statistical hypothesis tests based on analysis of variance and likelihood ratios, (3) Bayesian estimation of parameters, (4) sampling in surveys, (5) statistical quality control, (6) factorial design, and (7) resampling statistics. Certainly, this list is incomplete. You're encouraged to learn more about the development, theory, and application of these and other topics in probability and statistics, particularly because they provide an invaluable foundation for the design and application of many of the stochastic algorithms that are described in the main body of the book.

Appendix B: Problems and Projects

> The relationship between the teaching and research
> is the same as between the confession and sin:
> If you have not sinned, then you have nothing to confess!
>
> Anonymous

The best way to learn about problem solving is hands on. You have to experiment with different ideas, apply them to problems, and assess the results. In this appendix, we provide some suggestions for you to consider. The first set is intended for those who are fairly new to problem solving with computers, but even if you're more experienced you should still think about the answers. The second set provides some possibilities for projects that you might try. We hope that if you are taking a formal course in heuristics at the university level, then this course will be organized in a project-oriented fashion. It will give you the opportunity to learn from experience, and also to share that experience with your fellow classmates, and even the professor too!

Writing your own software from scratch is always the best way to learn about an algorithm, but there are many books that offer source code for traditional optimization algorithms (e.g., [363]) and there are several sources for free software in evolutionary algorithms that are available on the web. We've placed some software that you can download on our own personal web pages. The current URLs are http://www.coe.uncc.edu/~zbyszek/ and http://www.natural-selection.com/people/dbf.html, for each of us. If they should change or become unavailable for any reason, you can always contact either of us by email, or through the publisher. We're easy to find on the World Wide Web.

If you've never tried any problem-solving techniques and this text is your introduction to the field then you might start with these basic problems:

1. Your first step in solving a problem is always to identify what it is that you want to achieve. What's more, you must identify what you want to avoid too! When properly framed, your purpose can be quantified to indicate the relative worth of all of the possible solutions to your problem. Let's put you in a few different situations. Your task is to invent suitable evaluation functions for each case:

   (a) Let's start off easy. You want to find the real value x that minimizes $f(x) = x^2$.

   (b) You want to find the string of 1s or 0s such that the sum of the entries in the string is maximized.

(c) Now things get a bit more involved. You have a pattern classification problem. A physician has rated features of mammograms that they've studied as well as the associated outcome of whether or not the patient in question has a malignancy. You have 100 samples, where each sample contains ten input features and one output classification.

(d) A farmer wants to know the best model for predicting the amount of wheat that they will harvest based on the number of sunny days, the precipitation, and the amount of fertilizer they use each month.

(e) You need to keep a cart with a pole that's hinged on the top stabilized by pushing on the cart. The cart can't exceed the limits of the track and the pole can't exceed a specified angle of deflection. You're able to ascertain the cart's and pole's position and velocity at all times.

(f) Here's a tough one. Invent a suitable evaluation function that describes your own personal purpose. (Hint: Think about the parameters of concern and how you would weight them in relative importance and aggregate the degree of achievement with respect to each parameter appropriately.)

2. Here are some basic things to consider just as a review of the basic concepts that are presented in the text:

   (a) What is a random variable?

   (b) Of the following algorithms, which are deterministic?
       i. linear programming
       ii. simulated annealing
       iii. tabu search
       iv. dynamic programming
       v. evolutionary algorithms

   (c) What makes simulated annealing a randomized algorithm?

   (d) What facets of evolutionary algorithms are randomized?

   (e) If you are using a stochastic algorithm to solve a problem, will you get the same result every time you try it? How about when you use a deterministic algorithm?

   (f) Can you think of times when it would be better not to generate the same solution to the same problem in repeated trials?

3. When modeling a problem, you have to consider all the "players" that affect you and that you affect. Think about each of the following circumstances and enumerate all the players involved:

   (a) You're scheduling the production of shoes in a factory.

   (b) You're managing a baseball team.

(c) You have to get to work in the morning.

(d) You're an author and want to write a best-selling book.

(e) You're a graduate student and want to become a university professor.

(f) You're a graduate student and want to avoid becoming a university professor.

(g) You're a fireman and have arrived at the scene of a burning building.

Now go back to each of these cases and define some reasonable purpose to achieve and see if you can express that purpose in quantitative terms. What do "solutions" look like? For example, when scheduling the production of shoes, one thing to consider is the profitability of the factory which is the difference between revenues and costs. What are the things that comprise the revenues from the factors? What are the costs?

B.1 Trying some practical problems

Once you've tried your hand at modeling some situations, you're ready for tackling some practical problems that you might find in the real world. The following list of tasks make good projects to help you explore the fundamentals of problem solving and specific algorithms for finding solutions.

1. *Algorithm scaling on the TSP.* How well do various algorithms scale on the traveling salesman problem? We've seen a great many approaches to the TSP in the book. Some were based on deterministic techniques, such as a greedy algorithm or 2-opt, whereas others, like simulated annealing or evolutionary algorithms, were stochastic. It's interesting to find out how different approaches scale with the number of cities in the problem. Try the following experiment. For a given algorithm, distribute n cities at random in a unit square. There are formulas in chapter 8 that you can use to estimate the expected best tour length. Pick a threshold, like ten percent above the expected best length, and try your algorithm on 100 different randomly sampled TSPs with n cities. How many trials are able to find a solution that meets the threshold? As you increase n what happens to the number of successes? Can you find a functional relationship between the frequency of success and the number of cities? If you are using stochastic methods, you might try to ascertain things like:

- How many tours do you have to evaluate to find one that meets the threshold on average?

- How many trials are successful at meeting the threshold when the number of evaluated tours is fixed?

- What is the relationship between the success rate and the population size in an evolutionary algorithm as a function of the number of generations?

2. *Pattern classification.* The Internet provides an opportunity to easily obtain data sets that other people have collected for testing different algorithms on a variety of problems. One problem concerns classifying cells for evidence of breast cancer. If you visit the web site:

http://www.ics.uci.edu/AI/ML/MLDBRepository.html,

you can download a set of data called, "breast-cancer-wisconsin." Your task is to devise a procedure that can classify correctly as many cases as possible. You might try using a neural network or a linear discriminant function. Remember to separate your data into training and testing sets. Use the test set for a final assessment of your method. If you know of other sites where similar data are stored, in any field of interest, try out your method for building a suitable model on those data. After you've tried a few cases, go back and add some random noise to the input data. What happens as you increase the variability of the noise? Does the performance improve or suffer?

3. *Constrained optimization.* Chapter 9 provided several functions that pose constraints along with a variety of approaches that are tailored for the conditions that each one presents. Can you determine how much benefit you can get from these specially designed operators? Suppose you were to try an evolutionary algorithm that used a binary encoding of a real-valued constrained optimization problem, and used, say, one-point crossover and bit mutation as variation operators, along with proportional or rank-based selection. How would this approach, which doesn't use any information about the problem, compare with the results offered in chapter 9? Remember, you'll need to execute many trials in each case in order to estimate the average performance. What if you changed the representation to continuous (floating-point) vectors and then also changed the variation operators to, say, arithmetic crossover and Gaussian mutation? How does this framework compare? What if you take out mutation altogether, or crossover? Remember not to attribute the success or failure of any method to the inclusion or exclusion of any particular facet. It's the integration of all of the facets that's important.

4. *Explore adaptive and self-adaptation methods.* We saw in chapter 10 that evolutionary algorithms can incorporate means for adjusting their own parameters as a function of the evolutionary search. Can you invent new methods for controlling evolutionary parameters online? Try the following. Suppose a parent generates an offspring that's an improvement. Might it be reasonable to continue to generate offspring in the same direction? How could you implement that variation operator in a continuous domain? Might it also be reasonable to place less emphasis on moving in the opposite direction? Can you think of some counter-examples to these heuristics?

We saw many approaches for self-adaptation where each individual solution incorporated its own strategy parameters for searching the solution space. Do you think there might be some benefit in allowing individuals to use the strategy parameters of other solutions in the population? Maybe the population contains useful information that's not found in any single solution. Can you invent a self-adaptive operator that would use the "knowledge" of the whole population to guide the search for improved solutions? Try to frame this within a continuous optimization context, or maybe within the TSP, but it might be more of a challenge when working with discrete data structures.

5. *Combine different types of variation operators in modeling.* If you are familiar with modeling systems then you know that one of the fundamental approaches is for you to posit a model form, given some data, and then use a gradient-based algorithm to optimize the parameters of that model. The drawback to this is two-fold: (1) you have to choose a model first, and (2) the use of a gradient algorithm often can trap you in local optima. With stochastic search algorithms, however, you can make variation operators that don't act on gradient information, so you can search for both the model structure and the parameters at the same time. Think about how you might set this up in the context of evolving neural networks or fuzzy systems. You'll need to vary not only the weights of the network or locations of the fuzzy membership functions, but also the number of nodes and functions, respectively. You might try executing your algorithm on some data (e.g., see the pattern classification example above) and compare your results with the standard technique of hand tuning followed by gradient optimization.

6. *Combine local search operators into an evolutionary search.* Instead of comparing two algorithms, such as a greedy algorithm and an evolutionary algorithm, see if you can discover ways to improve both of them by incorporating different facets from each. Try this out on the TSP or perhaps the SAT problem. Think about how you would design an evolutionary approach to the problem. How would you apply some classic technique? What are the advantages and disadvantages of each one? Are there things that the evolutionary approach is good at that the classic approach fails on, or vice versa? Can you see ways to incorporate other facets, such as temperature or memory from annealing or tabu search, that would help? Execute a series of trials on the problem you choose and quantify the improvement you can obtain by hybridizing methods as compared to using either one alone.

7. *How important is initialization?* Many of the algorithms we've considered require an initial starting point or an entire collection of points. Are there any useful rules for choosing these points? Say you're facing a TSP. Would it be better to start with a solution or perhaps a population of solutions that are created by some local optimization algorithm, like a greedy search or 2-opt? How about when facing an NLP? Should you first perform some local optimization and then start a stochastic search using annealing or evolutionary algorithms? Does this buy you short-term gains at the expense of long-term losses, meaning that your algorithm exhibits better performance earlier in the run than it would otherwise, but then it doesn't seem to find solutions that are as good later?

8. *Explore the robustness of different algorithms in time-varying and noisy environments.* We identified a number of ways to handle time-varying environments in chapter 11. Try to implement a niching or sharing technique on some NLPs. Start with unconstrained problems that have multiple local optima. Can you invent heuristics for setting the scaling parameters based on the function? How does the situation change if you are facing a constrained optimization problem? What if the feasible regions are disjoint, separated by infeasible regions?

Another project involves comparing the effectiveness of different approaches to time-varying and noisy environments. Are some algorithms more "brittle" than others? How does annealing compare to tabu search on the TSP when the cities are given noisy positions (e.g., a city is located at (50,50), but you modify that to be a pair of random values each with a mean of 50 at every iteration)? Do evolutionary algorithms offer any advantages over annealing or tabu search here? Can you think of ways to combine all three methods to give better results?

9. *Financial forecasting.* There's a lot of interest in making money, particularly in predicting the future values of stocks and commodities. Due to the Internet, you can now gain easy access to many company's complete stock-price history. Your task is to search for sufficient data from a variety of companies in different market sectors (e.g., banking, computer chip manufacturers, home improvement stores, Internet, automotive). Once you acquire the data and have examined it, try to develop models that will help you predict the future of the stock with sufficient accuracy for you to turn a profit. This involves not only defining the models that you will use (e.g., linear functions of previous values, neural networks, fuzzy systems) but also the payoff function for different outcomes. Furthermore, you need to think about how to validate your models. Be sure to test your models on brand new data to assay their reliability, and be sure that your sample size is adequate. And remember, this is just for "entertainment" purposes!

In addition to the above projects (and of course you can invent your own too), as you begin to learn more about evolutionary algorithms, there are a number of interesting experiments you can try. See if your intuition matches up with reality. Remember, you learn the most when your hypotheses are proved false!

10. *Comparing variation operators.* What do you think about the relative importance of different variation operators? It depends on the problem, and it also depends on the stage of the evolution. Different operators can change in their relative importance. Suppose you use the TSP as a testbed. Try out different evolutionary algorithms and vary the rates at which different variation operators are utilized. There are many options to try, as indicated in chapter 8. Then move to NLPs, say even using something as simple as linear systems of equations. Does the importance of single-parent variation operators appear to be more or less significant in this context than with the TSP? What about

at different stages of the evolution? Are two-parent operators seemingly more useful early in the evolution? How could you invent a method for allowing the evolutionary algorithm to set the rates of various operators in light of how well they appear to be performing?

11. *Headless chicken crossover.* One of the basic ideas underlying most two-parent operators is that it might be a good idea to take a piece of one solution and put it together with a piece of another solution. But what if you tried to crossover some parent from your population with a completely random solution? This protocol tests the hypothesis that crossover is really working on "building blocks" of good subsolutions, or instead is just offering a "macromutation." If you can crossover existing solutions with random solutions and you find that you get better answers (or at least as good) than you get when you crossover solutions within the same population, then you probably haven't used a representation that suggests useful building blocks. This might suggest tailoring some different variation operators for your task.

This operator was called "headless chicken" crossover by Terry Jones in 1995, who said the randomized crossover was much like a chicken running around with its head cut off. It's still a chicken, but it's missing an important part. Headless chicken crossover has been seen to outperform swapping branches of parse trees that encode programs in certain examples. Can you find others?

12. *Tagging individuals and memory.* In chapter 16 we discussed many different hybrid methods for extending evolutionary algorithms. One thing to try is to incorporate a tag (or set of tags) on each individual so that they can recognize other individuals in the population. Then you can also develop some rules for governing which solutions will recombine with others based on these tags, and of course which solutions will refuse to recombine with others too. The tags could be updated using self-adaptation. In this way, individuals could build up a memory of which solutions to "embrace" and which to avoid. Does this sort of approach offer computational advantages? Try it out on some NLPs and TSPs. What is the required computational overhead?

The above list is really just a set of suggestions. The field of evolutionary computation, as well as all the other problem-solving methods we've discussed in the text, still provide the possibility for raising many more questions than answers. Your projects can vary in complexity to fit your own programming background and the time available for completion. Some are quite simple, others might require a team approach.

Here are just a few more ideas for you:

- Compare the performance of several algorithms (e.g., hill-climbing, stochastic hill-climbing, simulated annealing, evolutionary algorithms) on several test functions.

- Compare different selection methods (i.e., proportional, ranking, tournament) in evolutionary algorithms on a variety of problems.

- Implement different constraint-handling techniques for constrained NLPs (decoders, repair algorithms, penalty functions, etc.).

- Test self-adaptive methods for adjusting variation operators of evolutionary algorithms on NLPs. See if you can export those same ideas to annealing or tabu search.

- Implement a diploid representation to an NLP, TSP, or SAT. See what different methods for determining the dominance relationship you can invent.

- Implement the ideas of Lamarckian evolution or the Baldwin effect into an existing evolutionary algorithm. Conduct trials with different problems and determine when these additions might be useful.

- Explore the possibilities for addressing constrained NLPs or perhaps traveling salesman problems with multiple salesmen using coevolution, where solutions compete directly against each other but don't have some extrinsic evaluation function. Can you think of ways to implement this approach?

- Take an application that you've worked on before in some other class or at your job and hybridize that application using annealing, tabu, or evolutionary algorithms. Test the results and see what sort of improvements you can make. If you find that the approach seems really interesting and you want to share it with others, write it up as a technical paper and submit it to a conference or a journal!

Above all else, it's important that your projects be challenging, interesting, and fun!

B.2 Reporting computational experiments with heuristic methods

Many techniques are evaluated by experimenting with several test cases. It's often quite difficult to generalize these experimental results to make some "global" claim about a particular technique. It's possible, however, to demonstrate the utility of a new method on several, carefully selected cases, by comparing the method against other well-established techniques. Following [31], contributions of a new heuristic method may include the following:

- It produces high-quality solutions faster than other approaches.

- It finds higher-quality solutions than other approaches.

- It's less sensitive to differences in problem characteristics, data quality, or tuning parameters than other approaches.

- It's easier to implement.

- It has a broader range of applicability.

Furthermore [31],

> Research reports about heuristics are valuable if they are *revealing* — offering insight into general heuristic design or the problem structure by establishing the reasons for an algorithm's performance and explaining its behavior, and *theoretical* — providing theoretical insights, such as bounds on solution quality.

Barr et al. [31] gives a useful overview on how to design and report on computational experiments with heuristic methods. For example, in preparing and reporting your experiments, it might be desirable if you follow five steps (listed in [31]):

- Define the goals of the experiment.

- Choose measures of performance and factors to explore.

- Design and execute the experiment.

- Analyze the data and draw conclusions.

- Report the experiment's results.

It's important to address all these issues. For example, the goal of experiments may vary [31]:

> Computational experiments with algorithms are usually undertaken (a) to compare the performance of different algorithms for the same class of problems, or (b) to characterize or describe an algorithm's performance in isolation. While these goals are somewhat interrelated, the investigator should identify what, specifically, is to be accomplished by the testing (e.g., what questions are to be answered, what hypothesis are to be tested).

Also, you might choose to measure performance in terms of the quality of the best solution found, the time to discover it, the time to reach an "acceptable" solution, or the robustness of the method, to list a few possibilities. In most cases it's essential to compare the new method with established techniques for a given class of problems. It's important to remember to analyze the key factors (like the influence of the problem size on the quality of the solution and the computational effort). The final report should also contain sufficient information to allow the reader to reproduce the results, at least if they have access to essentially the same equipment.

There are many libraries of standard test problems available on the World Wide Web. These should be used frequently. For a collection of operations research problems, see, for example, the OR-Library at http://mscmga.ms.ic.ac.uk.

References

1. Aarts, E. and J. Korst (1989). *Simulated Annealing and Boltzmann Machines.* John Wiley, Chichester, UK.

2. Aarts, E.H.L. and J.K. Lenstra, eds. (1995). *Local Search in Combinatorial Optimization,* John Wiley, Chichester, UK.

3. Adami, C. (1997). *Introduction to Artificial Life.* Telos Press, New York, NY.

4. Akl, S.G. (1989). *The Design and Analysis of Parallel Algorithms,* Prentice-Hall, Englewood Cliffs, NJ.

5. Angeline, P.J. (1994). An Alternative Interpretation of the Iterated Prisoner's Dilemma and the Evolution of Non-mutual Cooperation. *Artificial Life IV,* R.A. Brooks and P. Maes, eds., MIT Press, Cambridge, MA, pp.353-358.

6. Angeline, P.J. (1995). Adaptive and Self-Adaptive Evolutionary Computation. In *Computational Intelligence: A Dynamic System Perspective,* M. Palaniswami, Y. Attikiouzel, R.J. Marks, D. Fogel, and T. Fukuda, eds., IEEE Press, Piscataway, NJ, pp.152-161.

7. Angeline, P.J. (1996). The Effects of Noise on Self-Adaptive Evolutionary Optimization In [163], pp.433-439.

8. Angeline, P.J. (1997). Tracking Extrema in Dynamic Environments. In [10], pp.335-345.

9. Angeline, P.J. (1998). Evolving Predictors for Chaotic Time Series. In *Applications and Science of Computational Intelligence,* SPIE Vol. 3390, S.K Rogers, D.B. Fogel, J.C. Bezdek, and B. Bossachi, eds., SPIE, Bellingham, WA, pp.170-180.

10. Angeline, P.J., R.G. Reynolds, J.R. McDonnell, and R. Eberhart, eds. (1997). *Evolutionary Programming VI,* Lecture Notes in Computer Science, Vol.1213, Springer, Berlin.

11. Applegate, D., R.E. Bixby, V. Chvatal, and W. Cook (1995). Finding Cuts in the TSP: A Preliminary Report. Report 95-05, DIMACS, Rutgers University, NJ.

12. Arabas, J., Z. Michalewicz, and J. Mulawka (1994). GAVaPS — A Genetic Algorithm with Varying Population Size. In [365], pp.73-78.

13. Arthur, W.B. (1994). Inductive Reasoning and Bounded Rationality. *Amer. Econ. Assn. Papers Proc.,* Vol.84, pp.406-411.

14. Atmar, J.W. (1976). Speculation on the Evolution of Intelligence and Its Possible Realization in Machine Form. PhD Dissertation, New Mexico State University, Las Cruces, NM.

15. Axelrod, R. (1987). Evolution of Strategies in the Iterated Prisoner's Dilemma. In *Genetic Algorithms and Simulated Annealing.* L. Davis, ed., Pitman, London, pp. 32-41.

16. Bäck, T. (1992). The Interaction of Mutation Rate, Selection, and Self-Adaptation within a Genetic Algorithm. In [299], pp.85-94.

17. Bäck, T. (1992). Self-Adaption in Genetic Algorithms. *Toward a Practice of Autonomous Systems: Proceedings of the 1st European Conference on Artificial Life*, F.J. Varela and P. Bourgine, eds., MIT Press, Cambridge, MA, pp.263-271.

18. Bäck, T. (1993). Evolutionary Programming and Evolution Strategies: Similarities and Differences. In [153], pp.11-22.

19. Bäck, T. (1993). Optimal Mutation Rates in Genetic Search. In [172], pp.2-8.

20. Bäck, T. (1994). Selective Pressure in Evolutionary Algorithms: A Characterization of Selection Mechanisms. In [365], pp.57-62.

21. Bäck, T. (1995). *Evolutionary Algorithms in Theory and Practice.* Oxford University Press, New York, NY.

22. Bäck, T. (1997). Mutation Parameters. In [26], page E1.2.1:7.

23. Bäck, T., ed. (1997). *Proceedings of the Seventh International Conference on Genetic Algorithms.* Morgan Kaufmann, San Mateo, CA.

24. Bäck, T. (1998). On the Behavior of Evolutionary Algorithms in Dynamic Environments. In [369], pp.446-451.

25. Bäck, T., A.E. Eiben, and M.E. Vink (1998). A Superior Evolutionary Algorithm for 3-SAT. In [359], pp.125-136.

26. Bäck, T., D.B. Fogel, and Z. Michalewicz, eds. (1997). *Handbook of Evolutionary Computation.* Oxford University Press, New York, and Institute of Physics, London, UK.

27. Bäck, T. and M. Schütz (1996). Intelligent Mutation Rate Control in Canonical Genetic Algorithms. Technical Report, Center for Applied Systems Analysis, Dortmund, Germany.

28. Bagley, J.D. (1967). The Behavior of Adaptive Systems which Employ Genetic and Correlation Algorithms. PhD Dissertation, University of Michigan, Ann Arbor, MI. *Dissertation Abstracts International*, 28(12), 5106B. (University Microfilms No. 68-7556).

29. Bak, P. (1996). *How Nature Works*, Oxford University Press, London, UK.

30. Banzhaf, W., P. Nordin, R.E. Keller, and F.D. Francone (1997). *Genetic Programming — An Introduction.* Morgan Kaufmann, San Mateo, CA.

31. Barr, R.S., B.L. Golden, J.P. Kelly, M.G.C. Resende, and W.R. Stewart (1995). Designing and Reporting on Computational Experiments with Heuristic Methods. In *Proceedings of the International Conference on Metaheuristics for Optimization*, Kluwer Publishing, Norwell, MA, pp.1-17.

32. Barricelli, N.A. (1954). Esempi Numerici de Processi di Evoluzione. *Methodos*, pp.45-68.

33. Barricelli, N.A. (1963). Numerical Testing of Evolution Theories. Part II: Preliminary Tests of Performance, Symbiogenesis and Terrestrial Life. *Acta Biotheoretica*, Vol.16, No.3-4, pp.99-126.

34. Batcher, K.E. (1964). A new internal sorting method. Goodyear Aerospace Report GER-11759.

35. Bazaraa, M.S., M.D. Sherali, and C.M. Shetty (1993). *Nonlinear Programming: Theory and Algorithms.* 2nd edition, John Wiley, New York, NY.

36. Bean, J.C. and A.B. Hadj-Alouane (1992). A Dual Genetic Algorithm for Bounded Integer Programs. Department of Industrial and Operations Engineering, The University of Michigan, TR 92-53.

37. Beasley, D., D.R. Bull, and R.R. Martin (1993). An Overview of Genetic Algorithms: Part 2, Research Topics. *University Computing*, Vol.15, No.4, pp.170-181.

38. Belegundu, A.D., D.V. Murthy, R.R. Salagame, and E.W. Constans (1994). Multi-objective Optimization of Laminated Ceramic Composites Using Genetic Algorithms. *Proceedings of the 5th AIAA/NASA/USAF/ISSMO Symposium on Multidisciplinary Analysis and Optimization*, Panama City, Florida, pp.1015-1022.

39. Belew, R.K. and L.B. Booker, eds. (1991). *Proceedings of the Fourth International Conference on Genetic Algorithms.* Morgan Kaufmann Publishers, San Mateo, CA.

40. Bellman, R. (1957). *Dynamic Programming.* Princeton University Press, Princeton, NJ.

41. Bertsekas, D.P. (1987). *Dynamic Programming. Deterministic and Stochastic Models.* Prentice-Hall, Englewood Cliffs, NJ.

42. Bezdek, J.C. (1981). *Pattern Recognition with Fuzzy Objective Function Algorithms.* Plenum Press, NY.

43. Bezdek, J.C., J. Keller, R. Krishnapuran, and N. Pal (1999). *Fuzzy Models and Algorithms for Pattern Recognition and Image Processing.* Kluwer Academic Publishers, Norwell, MA.

44. Bezdek, J.C. and S.K. Pal, eds. (1992). *Fuzzy Models for Pattern Recognition: Methods that Search for Structures in Data.* IEEE Press, Piscataway, NJ.

45. Bilchev, G. and I. Parmee (1995). Ant Colony Search vs. Genetic Algorithms. Technical Report, Plymouth Engineering Design Centre, University of Plymouth, UK.

46. Beyer, H.-G. (1995). Toward a Theory of Evolution Strategies: On the Benefits of Sex — the (μ, λ) Theory. *Evolutionary Computation*, Vol.3, No.1, pp.81-111.

47. Blair, A.D. (1999). Co-evolutionary Learning — Lessons for Human Education? In *Proceedings of the Fourth Conference of the Australasian Cognitive Science Society*, Newcastle, Australia.

48. Blair, A.D. and J.B. Pollack (1997). What Makes a Good Co-evolutionary Learning Environment? *Australian Journal of Intelligent Information Processing Systems*, Vol.4, pp.166-175.

49. Bland, R.G. and D.F. Shallcross (1989). Large Traveling Salesman Problems Arising from Experiments in X-Ray Crystallography: A Preliminary Report on Computation. *Operations Research Letters*, Vol.8, pp.125-128.

50. Bledsoe, W.W. (1961). The Use of Biological Concepts in the Analytical Study of Systems. Technical Report, Panoramic Research, Inc., Palo Alto, CA, November.

51. Bonomi, E. and J.-L. Lutton (1984). The n-city Traveling Salesman Problem: Statistical Mechanics and the Metropolis Algorithm. *SIAM Review*, Vol.26, pp.551-568.

52. Bowen, J. and G. Dozier (1995). Solving Constraint Satisfaction Problems Using a Genetic/Systematic Search Hybrid that Realizes When to Quit. In [130], pp.122-129.

53. Box, G.E.P. (1957). Evolutionary Operation: Method for Increasing Industrial Productivity. *Applied Statistics*, Vol.6, No.2, pp.81-101.

54. Box, G.E.P. and P.V. Voule (1955). The Exploration and Exploitation of Response Surfaces: An Example of the Link between the Fitted Surface and the Basic Mechanism of the System. *Biometric*, Vol.11, pp.287-323.

55. Branke, J. (1999). Enhanced Evolutionary Algorithms for Changing Optimization Problems. In [370], pp.1877-1884.

56. Braun, H. (1990). On Solving Traveling Salesman Problems by Genetic Algorithms. In [420], pp.129-133.

57. Bremermann, H.J. (1958). The Evolution of Intelligence. The Nervous System as a Model of Its Environment. Technical Report No.1, Contract No.477(17), Dept. Mathematics, University of Washington, Seattle, July.

58. Bremermann, H.J., M. Rogson, and S. Salaff (1966). Global properties of evolution processes. In *Natural Automata and Useful Simulations*, H.H. Pattee, E.A. Edlsack, L. Fein, and A.B. Callahan, eds., Spartan Books, Washington DC, pp.3-41.

59. Brindle, A. (1981). Genetic Algorithms for Function Optimization. PhD Dissertation, University of Alberta, Edmonton, Canada.

60. Budinich, M. (1996). A Self-Organizing Neural Network for the Traveling Salesman Problem that is Competitive with Simulated Annealing. *Neural Computing*, Vol.8, pp.416-424.

61. Burke, L.I. (1994). Neural Methods for the Traveling Salesman Problem: Insights from Operations Research. *Neural Networks*, Vol.7, No.4, pp.681-690.

62. Cantu-Paz, E. et al. (2003). *Proceedings of the Genetic and Evolutionary Conference 2003*, Lecture Notes in Computer Science, Vol.2723 and Vol. 2724, Springer, Berlin.

63. Cartwright, H.M. and G.F. Mott (1991). Looking Around: Using Clues from the Data Space to Guide Genetic Algorithm Searches. In [39], pp.108-114.

64. Chakraborty, U., K. Deb, and M. Chakraborty (1996). Analysis of Selection Algorithms: A Markov Chain Approach. *Evolutionary Computation*, Vol.4, No.2, pp.132-167.

65. Chellapilla, K. (1998). Combining Mutation Operators in Evolutionary Programming. *IEEE Transactions on Evolutionary Computation*, Vol.2, No.3, pp.91-96.

66. Chellapilla, K. and D.B. Fogel (1997). Exploring Self-Adaptive Methods to Improve the Efficiency of Generating Approximate Solutions to Traveling Salesman Problems Using Evolutionary Programming. In [10], pp.361-371.

67. Chellapilla, K. and D.B. Fogel (1999). Evolution, neural networks, games, and intelligence. In *Proceedings of the IEEE*, Vol.87, No.9, pp.1471-1496.

68. Chellapilla, K. and D.B. Fogel (1999). Evolving neural networks to play checkers without expert knowledge. *IEEE Transactions on Neural Networks*, Vol.10, No.6, pp.1382-1391.

69. Chellapilla, K. and D.B. Fogel (2001). Evolving an expert checkers playing program without using human expertise. *IEEE Transactions on Evolutionary Computation*, Vol.5, No.4, pp. 422-428.

70. Chellapilla, K., D.B. Fogel, and S.S. Rao (1997). Gaining Insight into Evolutionary Programming Through Landscape Visualization: An Investigation into IIR Filtering. In [10], pp. 407-417.

71. Chen, S.J. and C.L. Hwang (1993). Fuzzy Multiple Attribute Decision-Making: Methods and Applications. *Lecture Notes in Economics and Mathematical Systems*, Vol.375, Springer, Heidelberg.

72. Christofides, N. (1976). Worst-Case Analysis of a New Heuristic for the Traveling Salesman Problem. Report 388, Graduate School of Industrial Administration, Carnegie Mellon University, Pittsburgh, PA.

73. Clarke, G. and J.W. Wright (1964). Scheduling of Vehicles from a Central Depot to a Number of Delivery Points. *Operations Research*, Vol.12, pp.568-581.

74. Coello Coello, C.A. (1999). A Comprehensive Survey of Evolutionary-Based Multiobjective Optimization Techniques. *Knowledge and Information Systems. An International Journal*, Vol.1, No.3, pp.269-308.

75. Coello Coello, C.A., Van Veldhuizen, D.A., and Lamont, G.B. (2002). *Evolutionary Algorithms for Solving Multi-Objective Problems.* Kluwer Academic, New York, NY.

76. Conrad, M. (1969). Computer Experiments on the Evolution of Coadaptation in a Primitive Ecosystem. PhD dissertation, Standford University, Stanford, CA.

77. Conrad, M. and H.H. Pattee (1970). Evolution Experiments with an Artificial Ecosystem. *Journal of Theoretical Biology,* Vol.28, pp.393-409.

78. Cook, W.J., W.H. Cunningham, W.R. Pulleyblank, and A. Schrijver (1998). *Combinatorial Optimization.* John Wiley, Chichester, UK.

79. Costa, L. and Oliveira, P., An Adaptive Sharing Elitist Evolution Strategy for Multi-objective Optimization. Evolutionary Computation, Vol.11, No.4, pp.417-438.

80. Coveyou. R.R. and J.G. Sullivan (1961). Permutation (Algorithm 71). *Communications of the ACM,* Vol.4, No.11, p.497.

81. Craighurst, R. and W. Martin (1995). Enhancing GA Performance through Crossover Prohibitions Based on Ancestry. In [130], pp.130-135.

82. Darwen, P.J. and X. Yao (2001). Why More Choices Cause Less Cooperation in Iterated Prisoner's Dilemma. In *Proceedings of the 2001 Congress on Evolutionary Computation,* IEEE, Piscataway, NJ, pp.987-994.

83. Dasgupta, D. and Z. Michalewicz, Z., eds. (1997). *Evolutionary Algorithms in Engineering Applications.* Springer, New York, NY.

84. Davidor, Y., H.-P. Schwefel, and R. Männer, eds. (1994). *Proceedings of the 3rd Conference on Parallel Problem Solving from Nature.* Lecture Notes in Computer Science, Vol.866, Springer, Berlin.

85. Davis, L. (1985). Job Shop Scheduling with Genetic Algorithms. In [199], pp.136-140.

86. Davis, L. (1995). Applying Adaptive Algorithms to Epistatic Domains. *Proceedings of the International Joint Conference on Artificial Intelligence,* pp.162-164.

87. Davis, L., ed. (1987). *Genetic Algorithms and Simulated Annealing.* Morgan Kaufmann Publishers, San Mateo, CA.

88. Davis, L. (1989). Adapting Operator Probabilities in Genetic Algorithms. In [199], pp. 61-69.

89. Davis, L. (1991). Bit-Climbing, Representational Bias, and Test Suite Design. In [39], pp.18-23.

90. Davis, L. (1998). Private communication.

91. Davis, L., ed. (1991). *Handbook of Genetic Algorithms.* Van Nostrand Reinhold, NY.

92. DeAngelis, D.L. and L. Godbout (1991). An Individual-Based Approach to Predicting Density-Dependent Dynamics in Smallmouth Bass Populations. *Ecological Modelling,* Vol.57, pp.91-115.

93. De Jong, K.A. (1975). An Analysis of the Behavior of a Class of Genetic Adaptive Systems. Doctoral Dissertation, University of Michigan, Ann Arbor, MI. *Dissertation Abstract International,* 36(10), 5140B. (University Microfilms No 76-9381).

94. De Jong, K.A. (2002). *Evolutiuonary Computation.* MIT Press, Cambridge, MA.

95. De Jong K.A. and W.M. Spears (1989). Using Genetic Algorithms to Solve NP-Complete Problems. In [409], pp.124-132.

96. de la Maza, M. and B. Tidor (1993). An Analysis of Selection Procedures with Particular Attention Payed to Boltzmann Selection. In [172], pp.124-131.

97. Deb, K. (2000). An Efficient Constraint Handling Method for Genetic Algorithms. *Computer Methods in Applied Mechanics and Engineering*, Vol.186, pp.311-338.

98. Deb, K. (2001). *Multi-Objective Optimization using Evolutionary Algorithms.* John Wiley, New York, NY.

99. Deb, K., S. Agrawal, A. Pratap, and T. Meyarivan (2001). A Fast Elitist Non-dominated Sorting Genetic Algorithm: NSGA-II. In [414], pp.849-858.

100. DeGroot, M.H. (1975). *Probability and Statistics.* Addison-Wesley, Reading, MA.

101. Devore, J.L. (1995). *Probability and Statistics for Engineering and the Sciences.* 4th edition, Duxbury Press, Belmont, CA.

102. Dhar, V. and N. Ranganathan (1990). Integer Programming vs. Expert Systems: An Experimental Comparison. *Communications of ACM*, Vol.33, No.3, pp.323-336.

103. Dorigo M. and G. Di Caro (1999). The Ant Colony Optimization Meta-Heuristic. *New Ideas in Optimization*, D. Corne, M. Dorigo and F. Glover, eds., McGraw-Hill, New York, NY. (Also available as: Tech. Rep. IRIDIA/99-1, Université Libre de Bruxelles, Belgium.)

104. Dorigo M. and L.M. Gambardella (1997). Ant Colony System: A Cooperative Learning Approach to the Traveling Salesman Problem. *IEEE Transactions on Evolutionary Computation*, Vol.1, No.1, pp.53-66.

105. Dorigo M., V. Maniezzo, and A. Colorni (1991). Positive Feedback as a Search Strategy. Tech. Rep. No.91-016, Politecnico di Milano, Italy.

106. Downing, K. (1997). EUZONE: Simulating the Evolution of Aquatic Ecosystems. *Artificial Life*, Vol.3, pp.307-333.

107. Doyle, A.C. (1905). *The Adventure of the Priory School.* In *The Complete Sherlock Holmes*, Vol.II, Barns & Noble Classics, New York, NY, 2003.

108. Dozier, G., J. Bowen, and D. Bahler (1994). Solving Small and Large Constraint Satisfaction Problems using a Heuristic-based Microgenetic Algorithm. In [365], pp.306-311.

109. Dozier, G., J. Bowen, and D. Bahler (1994). Solving Randomly Generated Constraint Satisfaction Problems using a Micro-evolutionary Hybrid that Evolves a Population of Hill-Climbers. In [366], pp.614-619.

110. Efron, B. (1982). *The Jackknife, the Bootstrap, and other Resampling Plans*, SIAM, Philadelphia, PA.

111. Eiben, A.E. (1999). Multi-Parent Recombination. In *Handbook of Evolutionary Computation*, 2nd edition, T. Bäck, D.B. Fogel, and Z. Michalewicz, eds., Institute of Physics, Philadelphia, PA, pp.289-290.

112. Eiben, A.E. and T. Bäck (1997). Empirical Investigation of Multiparent Recombination Operators in Evolution Strategies. *Evolutionary Computation*, Vol.5, No.3, pp.347-365.

113. Eiben, A.E., T. Bäck, M. Schoenauer, and H.-P. Schwefel, eds. (1998). *Proceedings of the 5th Parallel Problem Solving from Nature Conference.* Lecture Notes in Computer Science, Vol.1498, Springer, Berlin.

114. Eiben, A.E., R. Hinterding, and Z. Michalewicz (1999). Parameter Control in Evolutionary Algorithms. *IEEE Transactions on Evolutionary Computation*, Vol.3, No.2, pp.124-141.

115. Eiben, A.E., C.H.M. van Kemenade, and J.N. Kok (1995). Orgy in the Computer: Multi-Parent Reproduction in Genetic Algorithms. In *Third European Conference on Artificial Life*, Springer, Berlin, pp.934-945.

116. Eiben, A.E., P.-E. Raue, and Zs. Ruttkay (1994). Solving Constraint Satisfaction Problems Using Genetic Algorithms. In [365], pp.542-547.

117. Eiben, A.E., P.-E. Raue, and Zs. Ruttkay (1994). Genetic Algorithms with Multi-Parent Recombination. In [84], pp.78-87.

118. Eiben, A.E. and Zs. Ruttkay (1996). Self-Adaptivity for Constraint Satisfaction: Learning Penalty Functions. In [367], pp.258-261.

119. Eiben, A.E. and C.A. Schippers (1996). Multi-Parent's Niche: N-ary Crossovers on NK-Landscapes. In [480], pp.319-328.

120. Eiben, A.E. and J.E. Smith (2003). *Introduction to Evolutionary Computing*, Springer, Berlin.

121. Eiben, A.E., I.G. Sprinkhuizen-Kuyper, and B.A. Thijssen (1998). Competing Crossovers in an Adaptive GA Framework. In [369], pp.787-792.

122. Eiben, A.E. and J.K. van der Hauw (1997). Adaptive Penalties for Evolutionary Graph-Coloring. *Artificial Evolution '97*, J.-K. Hao, E. Lutton, E. Ronald, M. Schoenauer, and D. Snyers, eds., Lecture Notes in Computer Science, Vol.1363, Springer, Berlin, pp. 95-106.

123. Eiben, A.E. and J.K. van der Hauw (1997). Solving 3-SAT with Adaptive Genetic Algorithms. In [368], pp.81-86.

124. Eiben, A.E., J.K. van der Hauw, and J.I. van Hemert (1998). Graph Coloring with Adaptive Evolutionary Algorithms. *Journal of Heuristics*, Vol.4, pp.25-46.

125. English, T.M. (1996). Evaluation of Evolutionary and Genetic Optimizers: No Free Lunch. In [163], pp.163-169.

126. Eriksson, R. (1996). Applying Cooperative Coevolution to Inventory Control Parameter Optimization. Master's thesis, University of Sk'ovde, Sweden.

127. Eriksson, R. and B. Olsson (1997). Cooperative Coevolution in Inventory Control Optimization. In *Proceedings of Third Int. Conf. Artif. Neural Networks and Genetic Algorithms*, G.D. Smith, N. C. Steele, and R. F. Albrecht, eds., Springer, Berlin.

128. Escazut, C. and Ph. Collard (1997). Genetic Algorithms at the Edge of a Dream. In *Artificial Evolution '97*, J.-K. Hao, E. Lutton, E. Ronald, M. Schoenauer, and D. Snyers, eds., Lecture Notes in Computer Science, Vol.1363, Springer, Berlin, pp.69-80.

129. Eshelman, L. (1990). The CHC Adaptive Search Algorithm: How to Have Safe Search when Engaging in Nontraditional Genetic Recombination. In [380], pp.265-283.

130. Eshelman, L.J., ed. (1995). *Proceedings of the Sixth International Conference on Genetic Algorithms*, Morgan Kaufmann, San Mateo, CA.

131. Eshelman, L.J. and J.D. Schaffer (1993). Crossover's Niche. In [172], pp.9-14.

132. Esquivel S., A. Leiva, and R. Gallard (1997). Multiple Crossover per Couple in Genetic Algorithms. In [368], pp.103-106.

133. Evans, M., N. Hastings, and B. Peacock (1993). *Statistical Distributions*. 2nd edition, John Wiley, New York, NY.

134. Falkenauer, E. (1994). A New Representation and Operators for GAs Applied to Grouping Problems. *Evolutionary Computation*, Vol.2, No.2, pp.123-144.

135. Fiesler, E. and R. Beale, eds. (1997). *Handbook of Neural Computation*, IOP Press, Philadelphia, PA.

136. Floreano, D. and S. Nolfi (1997). God Save the Red Queen! Competition in Co-evolutionary Robotics. In *Genetic Programming 1997*, J.R. Koza, K. Deb, M. Dorigo, D.B. Fogel, M. Garzon, H. Iba, and R.L. Riolo, eds., Morgan Kaufmann, San Mateo, CA, pp.398-406.

137. Floudas, C.A. and Pardalos, P.M. (1992). *Recent Advances in Global Optimization*. Princeton Series in Computer Science, Princeton University Press, Princeton, NJ.

138. Floyd, R.W. and D.E. Knuth (1967) Improved Constructions for the Bose-Nelson Sorting Problem. *Notices American Mathematical Society*, Vol.14, p.283.

139. Fodor, J.C., J.-L. Marichal, and M. Roubens (1995). Characterization of the Ordered Weighted Averaging Operators. *IEEE Transactions on Fuzzy Systems*, Vol.3, pp.236-240.

140. Fodor, J.C. and M. Roubens (1994). *Fuzzy Preference Modelling and Multicriteria Decision Support*. Kluwer, Dordrecht.

141. Fogarty, T. (1989). Varying the Probability of Mutation in the Genetic Algorithm. In [409], pp.104-109.

142. Fogel, D.B. (1988). An Evolutionary Approach to the Traveling Salesman Problem. *Biological Cybernetics*, Vol.60, pp.139-144.

143. Fogel, D.B. (1990). System Identification through Simulated Evolution. MSc. Thesis, UC San Diego, CA.

144. Fogel, D.B. (1991). An Information Criterion for Optimal Neural Network Selection. *IEEE Transactions on Neural Networks*, Vol.2, No.5, pp.490-497.

145. Fogel, D.B. (1992). Evolving Artificial Intelligence. PhD Dissertation, University of California at San Diego, La Jolla, CA.

146. Fogel, D.B. (1993). Evolving Behaviors in the Iterated Prisoner's Dilemma. *Evolutionary Computation*, Vol.1, No.1, pp.77-97.

147. Fogel, D.B. (1993). Applying Evolutionary Programming to Selected Traveling Salesman Problems. *Cybernetics and Systems*, Vol.24, No.1, pp.27-36.

148. Fogel, D.B. (1995). *Evolutionary Computation: Toward a New Philosophy of Machine Intelligence*. IEEE Press, Piscataway, NJ.

149. Fogel, D.B., ed. (1998). *Evolutionary Computation: The Fossil Record*, IEEE Press, Piscataway, NJ.

150. Fogel, D.B. (2002). *Blondie 24: Playing At The Edge of AI*. Morgan Kaufmann, San Francisco, CA.

151. Fogel, D.B. and J.W. Atmar (1990). Comparing Genetic Operators with Gaussian Mutations in Simulated Evolutionary Processes Using Linear Systems. *Biological Cybernetics*, Vol.63, No.2, pp.111-114.

152. Fogel, D.B. and W. Atmar, eds. (1992). *Proceedings of the 1st Annual Conference on Evolutionary Programming*, Evolutionary Programming Society, La Jolla, CA.

153. Fogel, D.B. and W. Atmar, eds. (1993). *Proceedings of the 2nd Annual Conference on Evolutionary Programming*. Evolutionary Programming Society, La Jolla, CA.

154. Fogel, D.B., K. Chellapilla, and P.J. Angeline (1999). Inductive Reasoning and Bounded Rationality Reconsidered. *IEEE Transactions on Evolutionary Computation*, Vol.3, No.2, pp.142-146.

155. Fogel, D.B., L.J. Fogel, and J.W. Atmar (1991). Meta-Evolutionary Programming In *Proceeding of the 25th Asilomar Conference on Signals, Systems and Computers*, R.R. Chen, ed., Maple Press, San Jose, CA, pp.540-545.

156. Fogel, D.B. and A. Ghozeil (1997). A Note on Representations and Variation Operators. *IEEE Transactions on Evolutionary Computation*, Vol.1, No.2, pp.159-161.

157. Fogel, D.B. and A. Ghozeil (1996). Using Fitness Distributions to Design More Efficient Evolutionary Computations. In [367], pp.11-19.

158. Fogel, D.B. and T.J. Hays (2003). New results in evolving strategies in chess. In *Applications and Science of Neural Networks, Fuzzy Systems, and Evolutionary Computation VI*, Vol.5200, B. Bosacchi, D.B. Fogel, and J.C. Bezdek (chairs), SPIE, Bellingham, WA, pp.56-63.

159. Fogel, D.B., E.C. Wasson, and E.M. Boughton (1995). Evolving Neural Networks for Detecting Breast Cancer. *Cancer Letters*, Vol.96, pp.49-53.

160. Fogel, D.B., E.C. Wasson, E.M. Boughton, and V.W. Porto (1998). Evolving Artificial Neural Networks for Screening Features from Mammograms. *Artificial Intelligence in Medicine*, Vol.14, pp.317-326.

161. Fogel, L.J. (1964). On the Organization of Intellect. PhD Dissertation, University of California at Los Angeles, Los Angeles, CA.

162. Fogel, L.J. (1995). The Valuated State Space Approach and Evolutionary Computation for Problem Solving. In *Computational Intelligence: A Dynamic Systems Perspective*, M. Palaniswami, Y. Attikiouzel, R.J. Marks, D. Fogel, and T. Fukuda, eds., IEEE Press, NY, pp. 129-136.

163. Fogel, L.J., P.J. Angeline, and T. Bäck, eds. (1996). *Proceedings of the 5th Annual Conference on Evolutionary Programming*, MIT Press, Cambridge, MA.

164. Fogel, L.J., P.J. Angeline, and D.B. Fogel (1995). An Evolutionary Programming Approach to Self-Adaptation on Finite State Machines. In [304], pp.355-365.

165. Fogel, L.J. and G.H. Burgin (1969). Competitive Goal-Seeking through Evolutionary Programming. Final Report, Contract AF 19(628)-5927, AF Cambridge Research Lab.

166. Fogel, L.J., D.B. Fogel, and W. Atmar (1993). Evolutionary Programming for ASAT Battle Management. In *Proceedings of the 27th Asilomar Conference on Signals, Systems, and Computers*, A. Singh, ed., IEEE Computer Society Press, Los Alamitos, CA, pp.617-621.

167. Fogel, L.J., A.J. Owens, and M.J. Walsh (1966). *Artificial Intelligence Through Simulated Evolution*. John Wiley, New York, NY.

168. Fonseca, C.M. and P.J. Fleming (1993). Genetic Algorithm for Multiobjective Optimization: Formulation, Discussion, and Generalization. In [172], pp.416-423.

169. Fonseca, C.M. and P.J. Fleming (1995). An Overview of Evolutionary Algorithms in Multiobjective Optimization. *Evolutionary Computation*, Vol.3, No.1, 1995, pp.165-180.

170. Fonseca, C.M. and P.J. Fleming (1997). Multiobjective Optimization. In *Handbook of Evolutionary Computation*, T. Bäck, D.B. Fogel, and Z. Michalewicz, eds., Oxford/IOP, NY, pp. C4.5:1-9.

171. Fonseca, C.M., P.J. Fleming, E. Zitzler, K. Deb, and L. Thiele, eds. (2003). *Evolutionary Multi-Criterion Optimization. Second International Conference EMO 2003*. Lecture Notes in Computer Science, Vol.2632, Springer, Berlin.

172. Forrest, S., ed. (1993). *Proceedings of the Fifth International Conference on Genetic Algorithms*. Morgan Kaufmann, San Mateo, CA.

173. Fourman, M. (1985). Compaction of Symbolic Layout using Genetic Algorithms. In [199], pp.141-153.

174. Fox, B.R. and M.B. McMahon (1990). Genetic Operators for Sequencing Problems. In [380], pp.284-300.

175. Fox, M.S. (1987). *Constraint-Directed Search: A Case Study of Job-Shop Scheduling.* Morgan Kaufmann, San Mateo, CA.

176. Freville, A. and G. Plateau (1993). Heuristics and Reduction Methods for Multiple Constraint 0-1 Linear Programming Problems. *European Journal of Operational Research,* Vol.24, pp.206-215.

177. Fuller, R. (1996). OWA Operators in Decision Making. In *Exploring the Limits of Support Systems*, C. Carlsson, ed., TUCS General Publications, No.3, Turku Centre for Computer Science, Abo, pp.85-104.

178. Gardner, M. (1978). *Aha! Insight.* Scientific American, Inc./W.H. Freeman, New York, NY.

179. Garey, M. and D. Johnson (1979). *Computers and Intractability.* W.H. Freeman, San Francisco, CA.

180. Gee, A.H. and R.W. Prager (1995). Limitations of Neural Networks for Solving Traveling Salesman Problems. *IEEE Transactions on Neural Networks*, Vol.6, No.1, pp.280-282.

181. Gehlhaar, D.K., G.M. Verkhivker, P.A. Rejto, C.J. Sherman, D.B. Fogel, L.J. Fogel, and S.T. Freer (1995). Molecular Recognition of the Inhibitor AG-1343 by HIV-1 Protease: Conformationally Flexible Docking by Evolutionary Programming. *Chemistry and Biology*, Vol.2, No.5, pp.317-324.

182. Gent, I.P. and T. Walsh (1992). The Enigma of SAT Hill-climbing Procedures. Technical Report 605, Department of Computer Science, University of Edinburgh.

183. Ghozeil, A. and D.B. Fogel (1996). A Preliminary Investigation into Directed Mutations in Evolutionary Algorithms. In [480], pp.329-335.

184. Glover, F. (1995). Tabu Search Fundamentals and Uses. Graduate School of Business, University of Colorado, CO.

185. Glover, F. and G. Kochenberger (1995). Critical Event Tabu Search for Multidimensional Knapsack Problems. In *Proceedings of the International Conference on Metaheuristics for Optimization,* Kluwer Publishing, Norwell, MA, pp.113-133.

186. Glover, F. and M. Laguna (1997). *Tabu Search.* Kluwer, London.

187. Goldberg, D.E. (1989). *Genetic Algorithms in Search, Optimization and Machine Learning.* Addison-Wesley, Reading, MA.

188. Goldberg, D.E., K. Deb, and J.H. Clark (1992). Accounting for Noise in the Sizing of Populations. In [490], pp.127-140.

189. Goldberg, D.E., K. Deb, and J.H. Clark (1992). Genetic Algorithms, Noise, and the Sizing of Populations. *Complex Systems*, Vol.6, pp.333-362.

190. Goldberg, D.E., K. Deb, and B. Korb (1991). Don't Worry, be Messy. In [39], pp.24-30.

191. Goldberg, D.E., K. Deb, and D. Thierens (1992). Toward a Better Understanding of Mixing in Genetic Algorithms. [39], pp.190-195.

192. Goldberg, D.E. and R. Lingle (1985). Alleles, Loci, and the TSP. In [199], pp.154-159.

193. Goldberg, D.E. and R.E. Smith (1987). Nonstationary Function Optimization Using Genetic Algorithms with Dominance and Diploidy. In [200], pp.59-68.

194. Goldberg, D.E. et al. (1999). *Proceedings of the Genetic and Evolutionary Conference 1999*, Morgan Kaufmann, San Mateo, CA.

195. Gorges-Schleuter, M. (1991). ASPARAGOS: An Asynchronous Parallel Genetic Optimization Strategy. In [39], pp.422-427.

196. Greene, F. (1994). A Method for Utilizing Diploid and Dominance in Genetic Search. In [365], pp. 439-444.

197. Greene, F. (1997). Performance of Diploid Dominance with Genetically Synthesized Signal Processing Networks. In [23], pp.615-622.

198. Grefenstette, J.J. (1984). GENESIS: A System for Using Genetic Search Procedures. *Proceedings of the 1984 Conference on Intelligent Systems and Machines*, Rochester, MI, pp.161-165.

199. Grefenstette, J.J., ed. (1985). *Proceedings of the First International Conference on Genetic Algorithms.* Lawrence Erlbaum Associates, Hillsdale, NJ.

200. Grefenstette, J.J., ed. (1987). *Proceedings of the 2nd International Conference on Genetic Algorithms.* Lawrence Erlbaum Associates, Hillsdale, NJ.

201. Grefenstette, J.J. (1986). Optimization of Control Parameters for Genetic Algorithms. *IEEE Transactions on Systems, Man, and Cybernetics*, Vol.16, No.1, pp.122-128.

202. Grefenstette, J.J. (1987). Incorporating Problem Specific Knowledge into Genetic Algorithms. In [87], pp.42-60.

203. Grefenstette, J.J., R. Gopal, B. Rosmaita, and D. Van Gucht (1985). Genetic Algorithms for the TSP. In [199], pp.160-168.

204. Gregory, J. (1995). Nonlinear Programming FAQ. Usenet sci.answers, 1995. Available at ftp://rtfm.mit.edu/pub/usenet/sci.answers/nonlinear-programming-faq.

205. Hadj-Alouane, A.B. and J.C. Bean (1992). A Genetic Algorithm for the Multiple-Choice Integer Program. Department of Industrial and Operations Engineering, University of Michigan, TR 92-50.

206. Hajela, P. and C.-Y.Lin (1992). Genetic Search Strategies in Multicriterion Optimal Design. it Structural Optimization, Vol.4, pp.99-107.

207. Harik, G., E. Cantu-Paz, D.E. Goldberg, and B.L. Miller (1997). The Gambler's Ruin Problem, Genetic Algorithms, and the Sizing of Populations. In [368], pp.7-12.

208. Harrald, P.G. and D.B. Fogel (1996). Evolving Continuous Behaviors in the Iterated Prisoner's Dilemma. *BioSystems.* Vol.37, pp.135-145.

209. Hart, W.E. and R.K. Belew (1991). Optimizing an Arbitrary Function is Hard for Genetic Algorithms. In [39], pp.190-195.

210. Haykin, S. (1994). *Neural Networks: A Comprehensive Foundation*, IEEE Press, Piscataway, NJ.

211. Held, M. and Karp, R.M. (1970). The Traveling Salesman Problem and Minimum Spanning Trees. *Operations Research*, Vol.18, pp.1138-1162.

212. Held, M. and Karp, R.M. (1971). The Traveling Salesman Problem and Minimum Spanning Trees: Part II. *Mathematical Programming*, Vol.1, pp.6-25.

213. Herdy, M. (1990). Application of the Evolution Strategy to Discrete Optimization Problems. In [420], pp.188-192.

214. Herrera, F., E. Herrera-Viedma, and J.L. Verdegay (1996). Direct Approach Processes in Group Decision Making using Linguistic OWA Operators. *Fuzzy Sets and Systems*, Vol.79, pp.175-190.

215. Hesser, J. and R. Männer (1991). Towards an Optimal Mutation Probability for Genetic Algorithms. In [420], pp.23-32.

216. Hilliard, M.R., G.E. Liepins, M. Palmer, and G. Rangarajen (1989). The Computer as a Partner in Algorithmic Design: Automated Discovery of Parameters for a Multiobjective Scheduling Heuristic. In *Impacts of Recent Computer Advances on Operations Research*, R. Sharda, B.L. Golden, E. Wasil, O. Balci, and W. Stewart, eds. North-Holland, New York, NY, 1989.

217. Hillier, F.S. and G.J. Lieberman (1980). *Introduction to Operations Research*. 3rd edition, Holden-Day, Inc., Oakland, CA.

218. Hillis, W.D. (1992). Co-evolving Parasites Improve Simulated Evolution as an Optimization Procedure. In *Artificial Life II*, C.G. Langton, C. Taylor, J.D. Farmer, and S. Rasmussen, eds., Addison-Wesley, Reading, MA, pp.313-324.

219. Himmelblau, D. (1992). *Applied Nonlinear Programming*. McGraw-Hill, NY.

220. Hinterding, R. (1995). Gaussian Mutation and Self-Adaption in Numeric Genetic Algorithms. In [366], pp.384-389.

221. Hinterding, R. (1997). Self-Adaptation Using Multi-Chromosomes. In [368], pp.87-91.

222. Hinterding, R. and Z. Michalewicz (1998). Your Brains and My Beauty: Parent Matching for Constrained Optimisation. In [369], pp.810-815.

223. Hinterding, R., Z. Michalewicz, and A.E. Eiben (1997). Adaptation in Evolutionary Computation: A Survey. In [368], pp.65-69.

224. Hinterding, R., Z. Michalewicz, and T.C. Peachey (1996). Self-Adaptive Genetic Algorithm for Numeric Functions. In [480], pp.420-429.

225. Hinton, G.E. and T. Sejnowski (1983). Optimal Perceptual Inference. In *Proceedings of the IEEE Computer Society Conference on Computer Vision and Pattern Recognition*, IEEE, Washington D.C., pp.448-453.

226. Hochbaum, D.S. (1997). *Approximation Algorithms for NP-hard Problems*, PWS Publishing, Boston, MA.

227. Hock, W. and Schittkowski, K. (1981). *Test Examples for Nonlinear Programming Codes*. Lecture Notes in Economics and Mathematical Systems, Vol.187, Springer, Berlin.

228. Holland, J.H. (1975). *Adaptation in Natural and Artificial Systems*. University of Michigan Press, Ann Arbor, MI.

229. Hollstein, R.B. (1971). Artificial Genetic Adaption in Computer Control Systems. PhD Thesis, Department of Computer and Communication Sciences, University of Michigan, Ann Arbor, MI.

230. Homaifar, A. and S. Guan (1991). A New Approach on the Traveling Salesman Problem by Genetic Algorithms. Technical Report, North Carolina A & T State University.

231. Homaifar, A., S. H.-Y. Lai, and X. Qi (1994). Constrained Optimization via Genetic Algorithms. *Simulation*, Vol.62, pp.242-254.

232. Hopfield, J.J. (1982). Neural Networks and Physical Systems with Emergent Collective Computational Abilities. *Proceedings of the National Academy of Sciences*, Vol.79, pp.2554-2558.

233. Hopfield, J.J. and D.W. Tank (1985). Neural Computation of Decisions in Optimization Problems. *Biological Cybernetics*, Vol.52, pp.141-152.

234. Horn, J. and N. Nafpliotis (1993). Multiobjective Optimization Using the Niched Pareto Genetic Algorithm. Department of Computer Science, University of Illinois at Urbana-Champaign, Urbana, IL, Technical Report IlliGAL 93005.

235. Hornik, K., M. Stinchcombe, and H. White (1990). Universal Approximation of an Unknown Mapping and Its Derivatives Using Multilayer Feedforward Networks. *Neural Networks*, Vol.3, pp.551-560.

236. Jacob, W., M. Gorges-Schleuter, and C. Blume (1992). Application of Genetic Algorithms to Task Planning and Learning. In [299], pp.291-300.

237. Janikow, C. and Z. Michalewicz (1991). An Experimental Comparison of Binary and Floating Point Representations in Genetic Algorithms. In [39], pp.151-157.

238. Jog, P., J.Y. Suh, and D.V. Gucht (1989). The Effects of Population Size, Heuristic Crossover, and Local Improvement on a Genetic Algorithm for the Traveling Salesman Problem. In [409], pp.110-115.

239. Johnson, D.S. (1990). Local Optimization and the Traveling Salesman Problem. In *Proceedings of the 17th Colloquium on Automata, Languages, and Programming*, M.S. Paterson, ed., Lecture Notes in Computer Science, Vol.443, Springer, Berlin, pp.446-461.

240. Johnson, D.S. (1995). Private communication.

241. Johnson, D.S. (1995). The Traveling Salesman Problem: A Case Study in Local Search. Presented during the Metaheuristics International Conference, Breckenridge, Colorado, July 22-26, 1995.

242. Johnson, D.S. (1995). The Traveling Salesman Problem: A Case Study. In [2], pp.215-310.

243. Johnson, D.S., L.A. Bentley, L.A. McGeoch, and E.E. Rothberg (1999). Near-Optimal Solutions to Very Large Traveling Salesman Problems. In preparation.

244. Johnson, D.S., L.A. McGeoch, and E.E. Rothberg (1996). Asymptotic Experimental Analysis for the Held-Karp Traveling Salesman Bound. In *Proceedings of the Seventh Annual ACM-SIAM Symposium on Discrete Algorithms*, ACM, New York, and SIAM, Philadelphia, PA, pp.341-350.

245. Johnson, L.W. and R.D. Riess (1982). *Numerical Analysis*, 2nd edition, John Wiley, New York, NY.

246. Johnson, R.W., M.E. Melich, Z. Michalewicz, and M. Schmidt (2004). Coevolutionary "TEMPO" Game. Technical Report, University of North Carolina at Charlotte.

247. Joines, J. and C. Houck (1994). On the Use of Non-stationary Penalty Functions to Solve Nonlinear Constrained Optimization Problems with GAs. In [365], pp.579-584.

248. Judson, O.P. (1994). The Rise of the Individual-Based Model in Ecology. *Trends in Ecology and Evolution*, Vol.9, pp.9-14.

249. Julstrom, B.A. (1995). What Have You Done for me Lately? Adapting Operator Probabilities in a Steady-State Genetic Algorithm. In [130], pp.81-87.

250. Kampfer, R.R. and M. Conrad (1983). Computational Modeling of Evolutionary Learning Processes in the Brain. *Bulletin of Mathematical Biology*, Vol.45, No.6, pp.931-968.

251. Karp, R.M. (1977). Probabilistic Analysis of Partitioning Algorithm for the Traveling Salesman Problem in the Plane. *Mathematics of Operations Research*, Vol.2, No.3, pp.209-224.

252. Kauffman, S.A. (1994). *The Origins of Order: Self-Organization and Selection in Evolution*. Oxford University Press, London, UK.

253. Kaufman, H. (1967). An Experimental Investigation of Process Identification by Competitive Evolution. *IEEE Transactions on Systems, Science and Cybernetics*, Vol.SSC-3, No.1, pp.11-16.

254. Keane, A.J. (1996). A Brief Comparison of Some Evolutionary Optimization Methods. In *Modern Heuristic Search Methods,* V. Rayward-Smith, I. Osman, C. Reeves and G. D. Smith, eds., John Wiley, New York, NY, pp.255-272.

255. Keeney, R. and H. Raiffa (1976). *Decisions with Multiple Objectives: Preferences and Value Tradeoffs.* John Wiley, New York, NY.

256. Kelly, J.P., B.L. Golden, and A.A. Assad (1993). Large Scale Controlled Rounding Using Tabu Search with Strategic Oscillation. *Annals of Operations Research,* Vol.41, pp.69-84.

257. Kelly, J. and M. Laguna (1996). Article posted in *Genetic Algorithms Digest,* Vol.10, No.16.

258. Kendall, G. and G. Whitwell (2001). An Evolutionary Approach for the Tuning of a Chess Evaluation Function Using Population Dynamics. In *Proceedings of the 2001 Congress on Evolutionary Computation,* IEEE Press, Piscataway, NY, pp.995-1002.

259. Kennedy, J. and R. Eberhart (1995). Particle Swarm Optimization. In *Proceedings IEEE International Conference on Neural Networks,* IEEE Press, Piscataway, NJ, pp.1942-1948.

260. Kita, H., Y. Yabumoto, N. Mori, and Y. Nishikawa (1996). Multi-Objective Optimization by Means of the Thermodynamical Genetic Algorithm. In [480], pp.504-512.

261. Klir, G.J. and B. Yuan (1995). *Fuzzy Sets and Fuzzy Logic: Theory and Applications.* Prentice-Hall, Upper Saddle River, NJ.

262. Knowles, J.D. and D.W. Corne (2000). Approximating the Non-dominated Front using the Pareto archived evolution strategy. *Evolutionary Computation,* Vol.8, No.2, pp.149-172.

263. Knox, J. (1994). Tabu Search Performance on the Symmetric Traveling Salesman Problem. *Computer Operations Research,* Vol.21, No.8, pp.867-876.

264. Knuth, D.E. (1973). *Sorting and Searching,* Vol.3. *The Art of Computer Programming.* Addison-Wesley, New York, NY.

265. Konhauser, J.D.E., D. Velleman, and S. Wagon (1996). *Which Way Did the Bicycle Go?* Mathematical Association of America, Washington CD.

266. Korte, B. (1988). Applications of Combinatorial Optimization. Talk at the 13th International Mathematical Programming Symposium, Tokyo.

267. Koza, J.R. (1992). *Genetic Programming.* MIT Press, Cambridge, MA.

268. Koza, J.R. (1994). *Genetic Programming – 2.* MIT Press, Cambridge, MA.

269. Koza, J.R., D.E. Goldberg, D.B. Fogel, and R.L. Riolo, eds. (1996). *Proceedings of the 1st Annual Conference on Genetic Programming.* MIT Press, Cambridge, MA.

270. Koza, J.R., K. Deb, M. Dorigo, D.B. Fogel, M. Garzon, H. Iba, and R.L. Riolo, eds. (1997). *Proceedings of the 2nd Annual Conference on Genetic Programming.* MIT Press, Cambridge, MA.

271. Koza, J.R., W. Banzhaf, K. Chellapilla, K. Deb, M. Dorigo, D.B. Fogel, M. Garzon, D.E. Goldberg, H. Iba, and R.L. Riolo, eds. (1998). *Proceedings of the 3rd Annual Conference on Genetic Programming.* MIT Press, Cambridge, MA.

272. Kozieł, S. and Z. Michalewicz (1998). A Decoder-Based Evolutionary Algorithm for Constrained Parameter Optimization Problems. In [113], pp.231-240.

273. Kozieł, S. and Z. Michalewicz (1999). Evolutionary Algorithms, Homomorphous Mappings, and Constrained Parameter Optimization. *Evolutionary Computation,* Vol.7, No.1, pp.19-44.

274. Krink, T., B.H. Mayoh, and Z. Michalewicz (1999). A PATCHWORK Model for Evolutionary Algorithms with Structured and Variable Size Populations. In *Proceedings of the Genetic and Evolutionary Computation Conference*, Orlando, Florida, pp.1321-1328.

275. Krink, T. and F. Vollrath (1997). Analysing Spider Web-Building Behaviour with Rule-based Simulations and Genetic Algorithms. *Journal of Theoretical Biology*, Vol.185, pp.321-331.

276. Kursawe, F. (1991). A Variant of Evolution Strategies for Vector Optimization. In [420], pp.193-197.

277. Langton, C.G., ed. (1989). *Artificial Life*. Addison-Wesley, Reading, MA.

278. Langdon, W.B. et al. (2002). *Proceedings of the Genetic and Evolutionary Conference 2002*, Morgan Kaufmann, San Mateo, CA.

279. Lawler, E.L., J.K. Lenstra, A.H.G. Rinnooy Kan, and D.B. Shmoys (1985). *The Traveling Salesman Problem*. John Wiley, Chichester, UK.

280. Le Riche, R., C. Knopf-Lenoir, and R.T. Haftka (1995). *A Segregated Genetic Algorithm for Constrained Structural Optimization*. In [130], pp.558-565.

281. Lewis, J., E. Hart, and G. Ritchie (1998). A Comparison of Dominance Mechanism and Simple Mutation on Non-stationary Problems. In [113], pp.139-148.

282. Lidd, M.L. (1991). Traveling Salesman Problem Domain Application of a Fundamentally New Approach to Utilizing Genetic Algorithms. Technical Report, MITRE Corporation.

283. Licpins, G.E., M.R. Hilliard, J. Richardson, and M. Palmer (1990). Genetic Algorithms Application to Set Covering and Traveling Salesman Problems. In *Operations Research and Artificial Intelligence: The Integration of Problem-solving Strategies*, D.E. Brown, and C.C. White, eds. Kluwer Academic, Norwell, MA, pp.29-57.

284. Lin, S. and B.W. Kernighan (1973). An Effective Heuristic Algorithm for the Traveling-Salesman Problem. *Operations Research*, Vol.21, pp.498-516.

285. Lis, J. (1996). Parallel Genetic Algorithm with Dynamic Control Parameter. In [367], pp.324-329.

286. Lis, J. and M. Lis (1996). Self-adapting Parallel Genetic Algorithm with the Dynamic Mutation Probability, Crossover Rate and Population Size. In *Proceedings of the 1st Polish National Conference on Evolutionary Computation*, J. Arabas, ed., Politechnika Warszawska, pp.324-329.

287. Litke, J.D. (1984). An Improved Solution to the Traveling Salesman Problem with Thousands of Nodes. *Communications of the ACM*, Vol.27, No.12, pp.1227-1236.

288. Liu, Y. and X. Yao (1999). Ensemble Learning via Negative Correlation. *Neural Networks*, Vol.12, No.10, pp.1399-1404.

289. Liu, Y. and X. Yao (1999). Simultaneous Training of Negatively Correlated Neural Networks in an Ensemble. *IEEE Transactions on Systems, Man, and Cybernetics, Part B: Cybernetics*, Vol.29, No.6, pp.716-725.

290. Liu, Y, X. Yao, and T. Higuchi (2000). Evolutionary Ensembles with Negative Correlation Learning. *IEEE Transactions on Evolutionary Computation*, Vol.4, No.4, pp.380-387.

291. Lord, W. (1955). *A Night to Remember*. Holt, Rinehart, & Winston, Austin, TX.

292. Louis, S.J. and Z. Xu (1996). Genetic Algorithms for Open Shop Scheduling and Rescheduling. In *Proceedings of the ISCA Eleventh International Conference on Computers and Their Applications*, M.E. Cohen and D.L. Hudson, eds., pp.99-102.

293. Luhandjula, M.K. (1989). Fuzzy Optimization: An Appraisal. *Fuzzy Sets and Systems*, Vol.30, pp.257-282.

294. Maa, C. and M. Shanblatt (1992). A Two-Phase Optimization Neural Network. *IEEE Transactions on Neural Networks*, Vol.3, No.6, pp.1003-1009.

295. Macready, W.G. and D.H. Wolpert (1998). Bandit Problems and the Exploration/ Exploitation Tradeoff. *IEEE Transactions on Evolutionary Computation*, Vol.2, No.1, pp.2-22.

296. Madsen, K. and S. Zertchaninov (1998). A New Branch-and-Bound Method for Global Optimization. Technical Report IMM-REP-1998-05, Department of Mathematical Modelling, Technical University of Denmark.

297. Maekawa, K., N. Mori, H. Tamaki, H. Kita, and Y. Nishikawa (1996). A Genetic Solution for the Traveling Salesman Problem by Means of a Thermodynamical Selection Rule. In [367], pp.529-534.

298. Mahfoud, S.W. (1997). Boltzmann Selection. In [26], pp.C2.5:1-4.

299. Männer, R. and B. Manderick, eds. (1992). *Proceedings of the 2nd Conference on Parallel Problem Solving from Nature 2*. North-Holland, Amsterdam, The Netherlands.

300. Martin, W.N., J. Lienig, and J.P. Cohoon (1997). Island (Migration) Models: Evolutionary Algorithms Based on Punctuated Equilibria. In [26], Sect. C6.3.

301. Mathias, K. and D. Whitley (1993). Remapping Hyperspace During Genetic Search: Canonical Delta Folding. In [490], pp.167-186.

302. Mathias, K. and D. Whitley (1992). Genetic Operators, the Fitness Landscape and the Traveling Salesman Problem. In [299], pp.219-228.

303. McCulloch, W.S. and W. Pitts (1943). A Logical Calculus of the Ideas Immanent in Nervous Activity. *Bulletin of Mathematical Biophysics*, Vol.5, pp.115-133.

304. McDonnell, J.R., R.G. Reynolds, and D.B. Fogel, eds. (1995). *Proceedings of the 4th Annual Conference on Evolutionary Programming*, MIT Press, Cambridge, MA.

305. Merelo, J.-J., P. Adamidis, H.-G. Beyer, J.-L. Fernandez-Villacañas, and H.-P. Schwefel, eds. (2002). *Proceedings of the 7th Parallel Problem Solving from Nature Conference*. Lecture Notes in Computer Science, Vol.2439, Springer, Berlin.

306. Merz, P. and B. Freisleben (1997). Genetic Local Search for the TSP: New Results. In [368], pp.159-164.

307. Michalewicz, Z. (1996). *Genetic Algorithms + Data Structures = Evolution Programs*. 3rd edition, Springer, Berlin.

308. Michalewicz, Z. (1995). Genetic Algorithms, Numerical Optimization and Constraints. In [130], pp.151-158.

309. Michalewicz, Z. (1993). A Hierarchy of Evolution Programs: An Experimental Study. *Evolutionary Computation*, Vol.1, No.1, pp.51-76.

310. Michalewicz, Z. (1994). Evolutionary Computation Techniques for Nonlinear Programming Problems. *International Transactions in Operational Research*, Vol.1, No.2, pp.223-240.

311. Michalewicz, Z. (1995). Heuristic Methods for Evolutionary Computation Techniques. *Journal of Heuristics*, Vol.1, No.2, pp.177-206.

312. Michalewicz, Z. and N. Attia (1994). Evolutionary Optimization of Constrained Problems. In [424], pp.98-108.

313. Michalewicz, Z., D. Dasgupta, R.G. Le Riche, and M. Schoenauer (1996). Evolutionary Algorithms for Constrained Engineering Problems. *Computers & Industrial Engineering Journal*, Vol.30, No.4, pp.851-870.

314. Michalewicz, Z., K. Deb, M. Schmidt, and T. Stidsen (1999). Towards Understanding Constrained-Handling Methods in Evolutionary Algorithms. In [370], pp.583-590.

315. Michalewicz, Z., K. Deb, M. Schmidt, and T. Stidsen (2000). Test Case Generator for Constrained Parameter Optimization Techniques. *IEEE Transactions on Evolutionary Computation*, Vol.4, No.3, pp.197-215.

316. Michalewicz, Z. and C. Janikow, C. (1996). GENOCOP: A Genetic Algorithm for Numerical Optimization Problems with Linear Constraints. *Communications of the ACM*, December, p.118.

317. Michalewicz, Z. and G. Nazhiyath (1995). GENOCOP III: A Coevolutionary Algorithm for Numerical Optimization Problems with Nonlinear Constraints. In [366], pp.647-651.

318. Michalewicz, Z., G. Nazhiyath, and M. Michalewicz (1996). A Note on Usefulness of Geometrical Crossover for Numerical Optimization Problems. In [163], pp.305-312.

319. Michalewicz, Z. and M. Schoenauer (1996). Evolutionary Algorithms for Constrained Parameter Optimization Problems. *Evolutionary Computation*, Vol.4, No.1, pp.1-32.

320. Michalewicz, Z., G.A. Vignaux, and M. Hobbs (1991). A Non-standard Genetic Algorithm for the Nonlinear Transportation Problem. *ORSA Journal on Computing*, Vol.3, No.4, pp.307-316.

321. Miettinen, K. (1999). *Nonlinear Multiobjective Optimization*. Kluwer, Boston, MA.

322. Miller, G.A. (1956). The Magic Number Seven, Plus or Minus Two: Some Limits on our Capacity for Processing Information. *Psychological Review*, Vol.63, pp.81-97.

323. Minar, N., R. Burkhart, C. Langton, and M. Askenazi (1996). The Swarm Simulation System: A Toolkit for Building Multi-Agent Simulations. Working Paper, The Santa Fe Institute, rep. no.: 96-06-042.

324. Minsky, M.L. (1961). Steps toward articial intelligence. In *Proceedings of the IEEE*, Vol.49, pp. 8-30.

325. Mizumoto, M. (1998). Defuzzification Methods. In [403], pp.B6.1:1-7.

326. Mori, N., S. Imanishi, H. Kita, and Y. Nishikawa (1997). Adaptation to Changing Environments by Means of the Memory Based Thermodynamical Genetic Algorithm. In [23], pp.299-306.

327. Morris, P. (1993). The Breakout Method for Escaping from Local Minima. In *Proceedings of the 11th National Conference on Artificial Intelligence, AAAI-93*, AAAI Press/The MIT Press, pp.40-45.

328. Mühlenbein, H. (1992). How Genetic Algorithms Really Work: I. Mutation and Hill-climbing. In [299], pp.15-25.

329. Mühlenbein, H. (1991). Evolution in Time and Space — The Parallel Genetic Algorithm. In [380], pp.316-337.

330. Mühlenbein, H., M. Gorges-Schleuter, and O. Krämer (1988). Evolution Algorithms in Combinatorial Optimization. *Parallel Computing*, Vol.7, pp.65-85.

331. Müller, J.P. (1998). Architectures and Applications of Intelligent Agents: A Survey. *The Knowledge Engineering Review*, Vol.13, No.4, pp.353-380.

332. Myung, H., J.-H. Kim, and D.B. Fogel (1995). Preliminary Investigation Into a Two-stage Method of Evolutionary Optimization on Constrained Problems. In [304], pp.449-463.

333. Nadhamuni, P.V.R. (1995). Application of Co-evolutionary Genetic Algorithm to a Game. Master's Thesis, Department of Computer Science, University of North Carolina, Charlotte, NC.

334. Nagata, Y. and S. Kobayashi (1997). Edge Assembly Crossover: A High-Power Genetic Algorithm for the Traveling Salesman Problem. In [23], pp.450-457.

335. Ng, K.P. and K.C. Wong (1995). A New Diploid Scheme and Dominance Change Mechanism for Non-stationary Function Optimization. In [130], pp.159-166.

336. Nix, A.E. and M.D. Vose (1992). Modelling Genetic Algorithms with Markov Chains. *Annals of Mathematics and Artificial Intelligence*, Vol.5, pp.79-88.

337. Nurnber, H.-T. and H.-G. Beyer (1997). The Dynamics of Evolution Strategies in the Optimization of Traveling Salesman Problem. In [10], pp.349-359.

338. Oliver, I.M., D.J. Smith, and J.R.C. Holland (1987). A Study of Permutation Crossover Operators on the Traveling Salesman Problem. In [200], pp.224-230.

339. Orvosh, D. and L. Davis (1993). Shall We Repair? Genetic Algorithms, Combinatorial Optimization, and Feasibility Constraints. In [172], p.650.

340. Osyczka, A. (2002). *Evolutionary Algorithms for Single and Multicriteria Design Optimization*. Physica-Verlag, Heidelberg, Germany.

341. Osyczka, A. and S. Kundu (1995). A New Method to Solve Generalized Multicriteria Optimization Problems using the Simple Genetic Algorithm. *Structural Optimization*, Vol.10, No.2, pp.94-99.

342. Padberg, M. and G. Rinaldi (1986). Optimization of a 532-City Symmetric Traveling Salesman Problem. Technical Report IASI-CNR, Italy.

343. Palmer, C.C. and A. Kershenbaum (1994). Representing Trees in Genetic Algorithms. In [365], pp.379-384.

344. Papadimitriou, Ch.H. and K. Steiglitz (1982). *Combinatorial Optimization*. Prentice-Hall, Englewood Cliffs, HJ.

345. Papoulis, A. (1991). *Probability, Random Variables, and Stochastic Processes*. 3rd edition, McGraw-Hill, New York, NY.

346. *Parabola*, University of New South Wales, Vol.14, pp.30-31, 1978, problem PARAB 348.

347. Pardalos, P. (1994). On the Passage from Local to Global in Optimization. In *Mathematical Programming*, J.R. Birge and K.G. Murty, eds., University of Michigan, 1994.

348. Paredis, J. (1992). Exploiting Constraints as Background Knowledge for Genetic Algorithms: A Case-Study for Scheduling. In [299], pp.229-238.

349. Paredis, J. (1993). Genetic State-Space Search for Constrained Optimization Problems. In *Proceedings of the 13th International Joint Conference on Artificial Intelligence*, Morgan Kaufmann, San Mateo, CA.

350. Paredis, J. (1994). Co-Evolutionary Constraint Satisfaction. In [84], pp.46-55.

351. Paredis, J. (1995). The Symbiotic Evolution of Solutions and Their Representations. In [130], pp.359-365.

352. Paredis, J. (1999). Coevolutionary Algorithms. In *Handbook of Evolutionary Computation*, 2nd edition, T. Bäck, D.B. Fogel, and Z. Michalewicz, eds., Institute of Physics, pp.224-238.

353. Parmee, I. and G. Purchase (1994). The Development of Directed Genetic Search Technique for Heavily Constrained Design Spaces. In *Proceedings of the Conference on Adaptive Computing in Engineering Design and Control*, University of Plymouth, UK, pp. 97-102.

354. Pearl, J. (1984). *Heuristics.* Addison-Wesley, Reading, MA.

355. Pollack, J.B. and A.D. Blair (1998). Co-evolution in the Successful Learning of Backgammon Strategy. *Machine Learning,* Vol.32, pp.225-240.

356. Pollack, J.B., A.D. Blair, and M. Land (1997). Coevolution of a Backgammon Player. In *Proceedings of the Fifth International Conference on Artificial Life,* MIT Press, Cambridge, MA, pp.92-98.

357. Polya, G. (1945). *How to Solve It: A New Aspect of Mathematical Method.* Princeton University Press, Princeton, NJ.

358. Poggio, T. and F. Girosi (1990). Networks for Approximation and Learning. *Proceedings IEEE,* Vol.78, pp.1481-1497.

359. Porto, V.W., N. Saravanan, D. Waagen, and A.E. Eiben, eds. (1998). *Evolutionary Programming VII,* Lecture Notes in Computer Science, Vol.1447, Springer, Berlin.

360. Potter, M. and K.A. De Jong (1994). A Cooperative Coevolutionary Approach to Function Optimization. In [84], pp.249-257.

361. Potter, M.A. and K.A. De Jong (2000). Cooperative Coevolution: An Architecture for Evolving Coadapted Subcomponents. *Evolutionary Computation,* Vol.8, No.1, pp.1-29.

362. Powell, D. and M.M. Skolnick (1993). Using Genetic Algorithms in Engineering Design Optimization with Non-linear Constraints. In [172], pp.424-430.

363. Press, W.H., B.P. Flannery, S.A. Teukolsky, and W.T. Vettering (1989). *Numerical Recipes in Pascal: The Art of Scientific Computing.* Cambridge University Press, Cambridge, MA.

364. Price, K.V. (1997). Differential Evolution vs. the Functions of the 2nd ICEO. In [368], pp.153-157.

365. *Proceedings of the 1994 IEEE Conference on Evolutionary Computation.* IEEE Press, Piscataway, NJ, 1994.

366. *Proceedings of the 1995 IEEE Conference on Evolutionary Computation.* IEEE Press, Piscataway, NJ, 1995.

367. *Proceedings of the 1996 IEEE Conference on Evolutionary Computation.* IEEE Press, Piscataway, NJ, 1996.

368. *Proceedings of the 1997 IEEE Conference on Evolutionary Computation.* IEEE Press, Piscataway, NJ, 1997.

369. *Proceedings of the 1998 IEEE Conference on Evolutionary Computation.* IEEE Press, Piscataway, NJ, 1998.

370. *Proceedings of the 1999 Congress on Evolutionary Computation.* IEEE Press, Piscataway, NJ, 1999.

371. *Proceedings of the 2000 Congress on Evolutionary Computation.* IEEE Press, Piscataway, NJ, 1999.

372. *Proceedings of the 2001 Congress on Evolutionary Computation.* IEEE Press, Piscataway, NJ, 1999.

373. *Proceedings of the 2002 Congress on Evolutionary Computation.* IEEE Press, Piscataway, NJ, 1999.

374. *Proceedings of the 2003 Congress on Evolutionary Computation.* IEEE Press, Piscataway, NJ, 1999.

375. Radcliffe, N.J. (1991). Equivalence Class Analysis of Genetic Algorithms. *Complex Systems,* Vol.5, No.2, pp.183-205.

376. Ramsey, C.L. and J.J. Grefenstette (1993). Case-Based Initialization of Genetic Algorithms. In [172], pp.84-91.

377. Ranjithan, S., J.W. Eheart, and J.C. Liebman (1992). Incorporating Fixed-cost Component of Pumping into Stochastic Groundwater Management: A Genetic Algorithm-based Optimization Approach. *Eos Transactions AGU*, Spring meeting supplement, Vol.73, No.14, p.125.

378. Ratschek, H. and J. Rokne (1995). Global Unconstrained Optimization. In *Handbook of Global Optimization*, R. Horst and P.M. Pardalos, eds., Kluwer Publishing, Norwell, MA, pp.779-796.

379. Raup, D.M. (1991). *Extinction: Bad Genes or Bad Luck?* W.W. Norton and Company, New York, NY.

380. Rawlins, G.J.E., ed. (1991). *Foundations of Genetic Algorithms*, Morgan Kaufmann, San Mateo, CA.

381. Ray, T., K. Tai, and K.C. Seow (2001). An Evolutionary Algorithm for Multiobjective Optimization. *Engineering Optimization*, Vol.33, No.3, pp.399-424.

382. Ray, T.S. (1992). An Approach to the Synthesis of Life. In *Artificial Life II*, C.G. Langton, C. Taylor, J.D. Farmer, and S. Rasmussen, eds., Addison-Wesley, Reading, MA, pp.371-408.

383. Rechenberg, I. (1973). *Evolutionsstrategie: Optimierung technischer Systeme nach Prinzipien der biologischen Evolution*. Frommann-Holzboog Verlag, Stuttgart.

384. Reed, J., R. Toombs, and N.A. Barricelli (1967). Simulation of Biological Evolution and Machine Learning. I. Selection of Self-Reproducing Numeric Patterns by Data Processing Machines, Effects of Hereditary Control, Mutation Type and Crossing. *Journal of Theoretical Biology*, Vol.17, pp.319-342.

385. Reinelt, G. (1991). TSPLIB - A Traveling Salesman Problem Library. *ORSA Journal on Computing*, Vol.3, No.4, pp.376-384.

386. Resende, M.G.C. and J.P. de Sousa, eds. (2003). *Metaheuristics: Computer Decision-Making*. Kluwer Academic Publishers, New York, NY.

387. Reynolds, C. (1994). Competition, Coevolution and the Game of Tag. In *Proceedings of Artificial Life IV*, R. Brooks and P. Maes, eds., MIT Press, Cambridge, MA, pp.56-69.

388. Reynolds, R.G. (1994). An Introduction to Cultural Algorithms. In [424], pp.131-139.

389. Reynolds, R.G., Z. Michalewicz, and M. Cavaretta (1995). Using Cultural Algorithms for Constraint Handling in GENOCOP. In [304], pp.298-305.

390. Reynolds, R.G. and W. Sverdlik (1993). Solving Problems in Hierarchically Structured Systems Using Cultural Algorithms. In [153], pp.144-153.

391. Richardson, J.T., M.R. Palmer, G. Liepins, and M. Hilliard (1989). Some Guidelines for Genetic Algorithms with Penalty Functions. In [409], pp.191-197.

392. Riesbeck, C.K. and R.C. Schank (1989). *Inside Case-Based Reasoning*. Lawrence Erlbaum Associates, Hillsdale, NJ.

393. Romero, C. (1991). *Handbook of Critical Issues in Goal Programming*. Pergamon Press, Oxford, UK.

394. Ronald, E. (1995). When Selection Meets Seduction. In [130], pp.167-173.

395. Rosenberg, R.S. (1967). Simulation of Genetic Populations with Biochemical Properties. PhD Dissertation, University of Michigan, Ann Arbor, MI. *Dissertation Abstracts International*, 28(7), 2732B. (University Microfilms No. 67-17, 836).

396. Rosenblatt, F. (1958). The Perceptron: A Probabilistic Model for Information Storage and Organization in the Brain. *Psychological Review*, Vol.65, pp.386-408.

397. Rosin, C.D. and R.K. Belew (1995). Methods for Competitive Co-evolution: Finding Opponents Worth Beating. In [130], pp.373-380.

398. Rosin, C.D. and R.K. Belew (1997). New Methods for Competitive Evolution. *Evolutionary Computation*, Vol.5, pp.1-29.

399. Rudolph, G. (1996). Convergence of Evolutionary Algorithms in Genaral Search Space. In [367], pp.50-54.

400. Rudolph, G. (1998). On a Multi-Objective Evolutionary Algorithm and Its Convergence to the Pareto Set. In [369], pp.511-516.

401. Rudolph, G. (2001). Evolutionary Search under Partially Ordered Fitness Sets. In *Proceedings of the International Symposium on Information Science Innovations in Engineering of Natural and Artificial Intelligent Systems (ISI 2001)*, pp.818-822.

402. Rudolph, G. and J. Sprave (1995). A Cellular Genetic Algorithm with Self-Adjusting Acceptance Threshold. In *Proceedings of the First IEE/IEEE International Conference on Genetic Algorithms in Engineering Systems: Innovations and Applications*, IEE, London, pp.365-372.

403. Rupsini, E.H., P.P. Bonissone, and W. Pedrycz, eds. (1998). *Handbook of Fuzzy Computation*, IOP, Philadelphia, PA.

404. Ryan, C. (1996). The Degree of Oneness. In *Proceedings of the First Online Workshop on Soft Computing*, Furuhashi, T., Morikawa, K., Miyata, Y. and Tkeuchi, I., eds., pp.43-48.

405. Sakawa, M. (1993). *Fuzzy Sets and Interactive Multiobjective Optimization*. Plenum Press, New York, NY.

406. San Pedro, J. and F. Burstein (2003). A Framework for Case-based Fuzzy Multicriteria Decision Support for Tropical Cyclone Forecasting. In *Proceedings of the 36th Hawaii International Conference on Systems Sciences*, IEEE, New York, NY.

407. Saravanan, N. and D.B. Fogel (1994). Learning Strategy Parameters in Evolutionary Programming: An Empirical Study. In [424], pp.269-280.

408. Saravanan, N., D.B. Fogel, and K.M. Nelson (1995). A Comparison of Methods for Self-Adaptation in Evolutionary Algorithms. *BioSystems*, Vol.36, pp.157-166.

409. Schaffer, J.D., ed. (1989). *Proceedings of the 3rd International Conference on Genetic Algorithms*. Morgan Kaufmann, San Mateo, CA.

410. Schaffer, J.D. (1984). Some Experiments in Machine Learning Using Vector Evaluated Genetic Algorithms. PhD Dissertation, Vanderbilt University, Nashville, TN.

411. Schaffer, J.D. (1985). Multiple Objective Optimization with Vector Evaluated Genetic Algorithms. In [199], pp.93-100.

412. Schaffer, J.D., R. Caruana, L. Eshelman, and R. Das (1989). A Study of Control Parameters Affecting Online Performance of Genetic Algorithms for Function Optimization. In [409], pp.51-60.

413. Schaffer, J.D. and A. Morishima (1987). An Adaptive Crossover Distribution Mechanism for Genetic Algorithms. In [200], pp.36-40.

414. Schoenauer, M., K. Deb, G. Rudolph, X. Yao, E. Lutton, J.-J. Merelo, and H.-P. Schwefel, eds. (2000). *Proceedings of the 6th Parallel Problem Solving from Nature Conference*. Lecture Notes in Computer Science, Vol.1917, Springer, Berlin.

415. Schoenauer, M. and Z. Michalewicz (1996). Evolutionary Computation at the Edge of Feasibility. In [480], pp.245-254.

416. Schoenauer, M. and S. Xanthakis (1993). Constrained GA Optimization. In [172], pp.573-580.

417. Schraudolph, N. and R. Belew (1992). Dynamic Parameter Encoding for Genetic Algorithms. *Machine Learning*, Vol.9, No.1, pp.9-21.

418. Schwefel, H.-P. (1981). *Numerical Optimization for Computer Models*, John Wiley, Chichester, UK.

419. Schwefel, H.-P. (1995). *Evolution and Optimum Seeking*, John Wiley, New York, NY.

420. Schwefel, H.-P. and R. Männer, eds. (1991). *Proceedings of the 1st Conference on Parallel Problem Solving from Nature*. Lecture Notes in Computer Science, Vol.496, Springer, Berlin.

421. Sebald, A.V. and J. Schlenzig (1994). Minimax Design of Neural-net Controllers for Uncertain Plants. *IEEE Transactions on Neural Networks*, Vol.5, No.1, pp.73-82.

422. Sebald A.V. and K. Chellapilla (1998). On Making Problems Evolutionarily Friendly. Part 1: Evolving the Most Convenient Representations. In [359], pp.271-280.

423. Sebald, A.V. and D.B. Fogel (1992). Design of Fault Tolerant Neural Networks for Pattern Classification. In [153], pp.90-99.

424. Sebald, A.V. and Fogel, L.J., eds. (1994). *Proceedings of the Third Annual Conference on Evolutionary Programming*, World Scientific, River Edge, NJ.

425. Sedgewick, R. (1988). *Algorithms*. 2nd edition, Addison-Wesley, Reading, MA.

426. Sedgewick, R. (1977). Permutation Generation Methods. *Computing Surveys*, Vol.9, No.2, pp.137-164.

427. Selman, B., H.A. Kautz, and B. Cohen (1993). Local Search Strategies for Satisfiability Testing. Presented at the 2nd DIMACS Challenge on Cliques, Coloring, and Satisfiability, Rutgers University, NJ.

428. Selman, B., H.J. Levesque, and D.G. Mitchell (1992). A New Method for Solving Hard Satisfiability Problems. In *Proceedings AAAI-92*, San Jose, CA, pp.440-446.

429. Selman, B. and H.A. Kautz (1993). Domain-Independent Extensions to GSAT: Solving Large Structured Satisfiability Problems. In *Proceedings IJCAI-93*, Chambery, France.

430. Seniw, D. (1991). A Genetic Algorithm for the Traveling Salesman Problem. MSc Thesis, University of North Carolina at Charlotte, NC.

431. Shaefer, C.G. (1987). The ARGOT Strategy: Adaptive Representation Genetic Optimizer Technique. In [200], pp.50-55.

432. Shapiro, J.F. (1979). *Mathematical Programming*. John Wiley, Chichester, UK.

433. Siegel, S. (1956). *Non-parametric Statistics*. McGraw-Hill, New York, NY.

434. Šmierzchalski, R. and Z. Michalewicz (2000). Modeling of Ship Trajectory in Collision Situations by an Evolutionary Algorithm. *IEEE Transactions on Evolutionary Computation*, Vol.4, No.3, pp.227-241.

435. Smith, A. and D. Tate (1993). Genetic Optimization Using A Penalty Function. In [172], pp.499-503.

436. Smith, J.E. (1997). Self Adaptation in Evolutionary Algorithms. PhD thesis, University of the West of England, Bristol, UK.

437. Smith, J.E. and T.C. Fogarty (1996). Self-Adaptation of Mutation Rates in a Steady-State Genetic Algorithm. In [367], pp.318-323.

438. Smith, J.E. and T.C. Fogarty (1996). Adaptively Parameterised Evolutionary Systems: Self Adaptive Recombination and Mutation in a Genetic Algorithm. In [480], pp.441-450.

439. Smith, J.E. and T.C. Fogarty (1997). Operator and Parameter Adaptation in Genetic Algorithms. Soft Computing, Vol.1, No.2, pp.81-87.

440. Smith, R. (1993). Adaptively Resizing Populations: An Algorithm and Analysis. In [172], page 653.

441. Smith, R. (1997). Population Size. In [26], pp.E1.1:1-5.

442. Soule, T. and J.A. Foster (1997). Code Size and Depth Flows in Genetic Programming. In [270], pp.313-320.

443. Spears, W.M. (1995). Adapting Crossover in Evolutionary Algorithms. In [304], pp.367-384.

444. Spears, W.M. (1996). Simulated Annealing for Hard Satisfiability Problems. In Cliques, Coloring, and Satisfiability: Second DIMACS Implementation Challenge, D.S. Johnson and M.A. Trick, eds., DIMACS Series in Discrete Mathematics and Theoretical Computer Science, Vol.26, American Mathematical Society, pp.533-558.

445. Spears, W.M. (1994). Simple Subpopulation Schemes. In [424], pp.296-307.

446. Spears, W.M. and K.A. De Jong (1991). On the Virtues of Parametrized Uniform Crossover. In [39], pp.230-236.

447. Spector, L. et al. (2001). Proceedings of the Genetic and Evolutionary Conference 2001, Morgan Kaufmann, San Mateo, CA.

448. Srinivas, N. and K. Deb (1994). Multiobjective Optimization Using Nondominated Sorting in Genetic Algorithms. Evolutionary Computation, Vol.2, No.3, 1994, pp.221-248.

449. Srinivas, M. and L.M. Patnaik (1994). Adaptive Probabilities of Crossover and Mutation in Genetic Algorithms. IEEE Transactions on Systems, Man, and Cybernetics, Vol.24, No.4, pp.17-26.

450. Standish, R., M.A. Bedau, and H. Abbass, eds. (2003). Artificial Life VII: Proceedings of the Eighth International Conference on Artificial Life. MIT Press, Cambridge, MA.

451. Stanley, K.O. and R. Miikkulainen (2002). Continual Coevolution through Complexification. In Proceedings 2002 Genetic and Evolutionary Computation Conference, Morgan Kaufmann, San Francisco, CA, pp.113-120.

452. Starkweather, T., S. McDaniel, K. Mathias, C. Whitley, and D. Whitley (1991). A Comparison of Genetic Sequencing Operators. In [39], pp.69-76.

453. Steele, J.M. (1986). Probabilistic Algorithm for the Directed Traveling Salesman Problem. Mathematics of Operations Research, Vol.11, No.2, pp.343-350.

454. Stein, D. (1977). Scheduling Dial a Ride Transportation Systems: An Asymptotic Approach. PhD Dissertation, Harvard University, MA.

455. Steinhaus, H. (1964). One Hundered Problems in Elementary Mathematics. Dover Publications, New York, NY.

456. Stewart, I. (1999) A Puzzle for Pirates. Scientific American, May 1999, pp.98-99.

548   References

457. Suh, J.-Y. and Van D. Gucht (1987). Incorporating Heuristic Information into Genetic Search. In [200], pp.100-107.

458. Surry, P.D., N.J. Radcliffe, and I.D. Boyd (1995). A Multi-Objective Approach to Constrained Optimization of Gas Supply Networks. In *Proceedings of the AISB-95 Workshop on Evolutionary Computing*, T. Fogarty, ed., Lecture Notes in Computer Science, Vol.993, Springer, Berlin, pp. 166-180.

459. Suzuki, J. (1993). A Markov Chain Analysis on a Genetic Algorithm. In [172], pp.146-153.

460. Syswerda, G. (1989). Uniform Crossover in Genetic Algorithms. In [409], pp.2-9.

461. Syswerda, G. (1991). Schedule Optimization Using Genetic Algorithms. In [91], pp.332-349.

462. Syswerda, G. and J. Palmucci (1991). The Application of Genetic Algorithms to Resource Scheduling. In [39], pp.502-508.

463. Tao, G. and Z. Michalewicz (1998). Evolutionary Algorithms for the TSP. In [113], pp.803-812.

464. Tate, D.M. and E.A. Smith (1993). Expected Allele Coverage and the Role of Mutation in Genetic Algorithms. In [172], pp.31-37.

465. Taylor, R. and A. Wiles (1995). Ring-Theoretic Properties of Certain Hecke Algebras. *Annals of Mathematics*, Vol.142, pp.553-573.

466. Thierens, D. (1996). Dimensional Analysis of Allele-Wise Mixing Revisited. In [480], pp.255-265.

467. Thierens, D. and D.E. Goldberg (1993). Mixing in Genetic Algorithms. In [172], pp.38-45.

468. Trojanowski, K., Z. Michalewicz, and J. Xiao (1997). Adding Memory to the Evolutionary Planner/Navigator. In [368], pp.483-487.

469. Trojanowski, K. and Z. Michalewicz (1999). Searching for Optima in Non-stationary Environments. In [370], pp.1845-1852.

470. Tuson, A. and P. Ross (1996). Cost Based Operator Rate Adaptation: An Investigation. In [480], pp.461-469.

471. Ulder, N.L.J., E.H.L. Aarts, H.-J. Bandelt, P.J.M. van Laarhoven, and E. Pesch (1990). Genetic Local Search Algorithms for the Traveling Salesman Problem. In [420], pp.109-116.

472. Ursem, R.K., T. Krink, M.T. Jensen, and Z. Michalewicz (2000). Analysis and Modeling of Control Tasks in Dynamic Systems. *IEEE Transactions on Evolutionary Computation*, Vol.6, No.4, pp.378-389.

473. Valenzuela, C.L. and A.J. Jones (1994). Evolutionary Divide and Conquer (I): A Novel Genetic Approach to the TSP. *Evolutionary Computation*, Vol.1, No.4, pp.313-333.

474. Valenzuela, C.L. and L.P. Williams (1997). Improving Simple Heuristic Algorithms for the Traveling Salesman Problem Using a Genetic Algorithm. In [23], pp.458-464.

475. Valenzuela-Rendón, M. and E. Uresti Charre (1997). A Non-Generational Genetic Algorithm for Multiobjective Optimization. In [23], pp.658-665.

476. Van Laarhoven, P.J.M. and E.H.L. Aarts (1987). *Simulated Annealing: Theory and Applications*. D. Reidel, Dordrecht, The Netherlands.

477. Vavak, F., T.C. Fogarty, and K. Jukes (1996). A Genetic Algorithm with Variable Range of Local Search for Tracking Changing Environments. In [480], pp.376-385.

478. Vavak, F., K. Jukes, and T.C. Fogarty (1997). Learning the Local Search Range for Genetic Optimisation in Nonstationary Environments. In [368], pp.355-360.

479. Veldhuizen, D.V. (1999). Multiobjective Evolutionary Algorithms: Classifications, Analysis, and New Innovations. PhD Dissertation, Air Force Institute of Technology, Dayton, OH.

480. Voigt, H.-M., W. Ebeling, I. Rechenberg, and H.-P. Schwefel, eds. (1996). *Proceedings of the 4th Conference on Parallel Problem Solving from Nature.* Lecture Notes in Computer Science, Vol.1141, Springer, Berlin.

481. Von Neumann, J. and O. Morgenstern (1944). *Theory of Games and Economic Behavior.* Princeton University Press, Princeton, NJ.

482. Vose, M.D. (1992). Modeling Simple Genetic Algorithms. In [490], pp.63-74.

483. Voss, S. (1993). Tabu Search: Applications and Prospects. Technical Report, Technische Hochschule, Darmstadt, Germany.

484. Waagen, D., P. Diercks, and J. McDonnell (1992). The Stochastic Direction Set Algorithm: A Hybrid Technique for Finding Function Extrema. In [152], pp.35-42.

485. Watson, J., C. Ross, V. Eisele, J. Denton, J. Bins, C. Guerra, D. Whitley, and A. Howe (1998). The Traveling Salesrep Problem, Edge Assembly Crossover, and 2-opt. In [113], pp.823-832.

486. White, T. and F. Oppacher (1994). Adaptive Crossover Using Automata. In [84], pp.229-238.

487. Whitley, D. (1988). GENITOR: A Different Genetic Algorithm. *Proceedings of the Rocky Mountain Conference on Artificial Intelligence*, Denver, CO.

488. Whitley, D. (1989). The GENITOR Algorithm and Selection Pressure: Why Rank-Based Allocation of Reproductive Trials is Best. In [409], pp.116-121.

489. Whitley, D. (1990). GENITOR II: A Distributed Genetic Algorithm. *Journal of Experimental and Theoretical Artificial Intelligence*, Vol.2, pp.189-214.

490. Whitley, L.D., ed. (1992). *Foundations of Genetic Algorithms 2*, Morgan Kaufmann, San Mateo, CA.

491. Whitley, D., V.S. Gordon, and K. Mathias (1996). Lamarckian Evolution, the Baldwin Effect and Function Optimization. In [84], pp.6-15.

492. Whitley, D., K. Mathias, and P. Fitzhorn (1991). Delta Coding: An Iterative Strategy for Genetic Algorithms. In [39], pp.77-84.

493. Whitley, D., K. Mathias, S. Rana, and J. Dzubera (1995). Building Better Test Functions. In [130], pp.239-246.

494. Whitley, D., T. Starkweather, and D'A Fuquay (1989). Scheduling Problems and Traveling Salesman: The Genetic Edge Recombination Operator. In [409], pp.133-140.

495. Whitley, D., T. Starkweather, and D. Shaner (1990). Traveling Salesman and Sequence Scheduling: Quality Solutions Using Genetic Edge Recombination. In [91], pp.350-372.

496. Whitley, D. et al. (2000). *Proceedings of the Genetic and Evolutionary Conference 2000*, Morgan Kaufmann, San Mateo, CA.

497. Wienke, P.B., C. Lucasius, and G. Kateman (1992). Multicriteria Target Optimization of Analytical Procedures using a Genetic Algorithm. *Analytical Chimica Acta*, Vol.265, No.2, pp.211-225.

498. Wiles, A. (1995). Modular Elliptic Curves and Fermat's Last Theorem. *Annals of Mathematics*, Vol.142, pp.443-551.

499. Wolpert, D.H. and W.G. Macready (1997). No Free Lunch Theorems for Optimization. *IEEE Transactions on Evolutionary Computation*, Vol.1, No.1, pp.67-82.

500. Wu, A., R.K. Lindsay, and R.L. Riolo (1997). Empirical Observation on the Roles of Crossover and Mutation. In [23], pp.362-369.

501. Xiao, J., Z. Michalewicz, L. Zhang, and K. Trojanowski (1997). Adaptive Evolutionary Planner/Navigator for Mobile Robots. *IEEE Transactions on Evolutionary Computation*, Vol.1, No.1, pp.18-28.

502. Yager, R.R. (1988). On Ordered Weighted Averaging Aggregation Operators in Multicriteria Decision Making. *IEEE Transactions on Systems, Man, and Cybernetics*, Vol.18, pp.183-190.

503. Yager, R.R. and D.P. Filev (1994). *Essentials of Fuzzy Modeling and Control*, John Wiley, New York, NY.

504. Yao, X., G. Lin, and Y. Liu (1997). An Analysis of Evolutionary Algorithms Based on Neighbourhood and Step Sizes. In [10], pp.297-307.

505. Zadeh, L.A. (1965). Fuzzy Sets. *Information and Control*, Vol.8, pp.338-352.

506. Zadeh, L.A. (1973). Outline of a New Approach to the Analysis of Complex Systems and Decision Processes. *IEEE Transactions on Systems, Man, and Cybernetics*, Vol.SMC-3:1, pp.28-44.

507. Zitzler, E., K. Deb, and L. Thiele (2000). Comparison of Multiobjective Evolutionary Algorithms: Empirical Results. *Evolutionary Computation*, Vol.8, No.2, pp.173-195.

508. Zitzler, E., K. Deb, L. Thiele, C.A. Coello Coello, and D. Corne, eds. (2001). *Evolutionary Multi-Criterion Optimization. First International Conference EMO 2001*. Lecture Notes in Computer Science, Vol. 1993, Springer, Berlin.

509. Zitzler, E. and L. Thiele (1998). Multiobjective Optimization using Evolutionary Algorithms — A Comparative Case Study. In [113], pp.292-301.

Index